Advances in Intelligent Systems and Computing

Volume 811

Series editor

Janusz Kacprzyk, Systems Research Institute, Polish Academy of Sciences,
Warsaw, Poland
e-mail: kacprzyk@ibspan.waw.pl

The series "Advances in Intelligent Systems and Computing" contains publications on theory, applications, and design methods of Intelligent Systems and Intelligent Computing. Virtually all disciplines such as engineering, natural sciences, computer and information science, ICT, economics, business, e-commerce, environment, healthcare, life science are covered. The list of topics spans all the areas of modern intelligent systems and computing such as: computational intelligence, soft computing including neural networks, fuzzy systems, evolutionary computing and the fusion of these paradigms, social intelligence, ambient intelligence, computational neuroscience, artificial life, virtual worlds and society, cognitive science and systems, Perception and Vision, DNA and immune based systems, self-organizing and adaptive systems, e-Learning and teaching, human-centered and human-centric computing, recommender systems, intelligent control, robotics and mechatronics including human-machine teaming, knowledge-based paradigms, learning paradigms, machine ethics, intelligent data analysis, knowledge management, intelligent agents, intelligent decision making and support, intelligent network security, trust management, interactive entertainment, Web intelligence and multimedia.

The publications within "Advances in Intelligent Systems and Computing" are primarily proceedings of important conferences, symposia and congresses. They cover significant recent developments in the field, both of a foundational and applicable character. An important characteristic feature of the series is the short publication time and world-wide distribution. This permits a rapid and broad dissemination of research results.

More information about this series at http://www.springer.com/series/11156

Mohuya Chakraborty · Satyajit Chakrabarti
Valentina Emilia Balas · J. K. Mandal
Editors

Proceedings of International Ethical Hacking Conference 2018

eHaCON 2018, Kolkata, India

Springer

Editors
Mohuya Chakraborty
Department of Information Technology
Institute of Engineering and Management
Kolkata, India

Satyajit Chakrabarti
Institute of Engineering and Management
Kolkata, India

Valentina Emilia Balas
Department of Automation and Applied Informatics
Aurel Vlaicu University of Arad
Arad, Romania

J. K. Mandal
Department of Computer Science and Engineering
University of Kalyani
Kalyani, West Bengal, India

ISSN 2194-5357 ISSN 2194-5365 (electronic)
Advances in Intelligent Systems and Computing
ISBN 978-981-13-1543-5 ISBN 978-981-13-1544-2 (eBook)
https://doi.org/10.1007/978-981-13-1544-2

Library of Congress Control Number: 2018948600

This Springer imprint is published by the registered company Springer Nature Singapore Pte Ltd.
The registered company address is: 152 Beach Road, #21-01/04 Gateway East, Singapore 189721, Singapore

Preface

The rapid growth of computer networks and the Internet has changed the prospect of network security in all spheres of life. An easy accessibility condition has caused the computer networks to be vulnerable against numerous and potentially devastating threats from hackers. Globalization does not solely depend on the world's economies but to a large extent relies on the world's computerization era. By exploiting the worldwide integration, the computer networks in various applications are dealing with anonymous antagonists that may attack from any side on the Internet. Today's endeavor for security administration must extend its degree to screen the malignant actions on the Internet.

International Ethical Hacking Conference (eHaCON) 2018 gave an open platform where people were able to discuss the implications of new technologies for a secured society. The most substantial new findings about computer network attacks and defenses, commercial security solutions, and pragmatic real-world security experiences were presented in the two-day informative research paper presentations and invited talks.

This book is split into nine parts. Part I discusses the keynote talk on "Hadamard Modulo Prime Matrices and Their Application in Cryptography." Part II and Part III present "Ethical Hacking and Cloud Computing" and "Cryptography" respectively. Part IV highlights "Modeling and Simulation" whereas Part V elaborates "Network Security." Part VI and Part VII present "Artificial Intelligence." Part VIII presents "Internet of Things." Part IX concludes the book with "Data Analytics."

Part I: Keynote Talk

The present paper surveys some recent results on Hadamard modulo prime matrices by revealing their connections with coding theory, combinatorics, and elementary number theory and discusses its attractive application in the modern cryptography.

Part II: Ethical Hacking and Cloud Computing

In this part, there are six papers dealing with social engineering attack, vulnerability scanner, AI chatbot, efficient task management in multi-cloud environment, dynamic load balancing in cloud computing environment, and cryptanalysis.

Part III: Cryptography
This part consists of six papers that deal with various cryptographic and steganographic techniques.

Part IV: Modeling and Simulation
There are four papers in this part. This part deals with modeling techniques based on fear detection using brain—computer interface, proton-exchange membrane fuel cell, cognitive radio sensor node, and cost-effective LED driver.

Part V: Network Security
This part has three papers and highlights various concepts on network trust models as well as convolution coder-based encryption algorithm for cognitive radio sensor node.

Part VI and Part VII: Artificial Intelligence
There are ten papers in these two parts that present various application areas of artificial intelligence.

Part VIII: Internet of Things
This part has four papers that deal with IoT-based healthcare monitoring, irrigation monitoring, medical body sensor network, and brain—computer-interface-oriented multi-target-based cursor movement.

Part IX: Data Analytics
There are six papers in this part of the book. The papers are based on data analytics, predictive modeling, and deep neural network.

The editors believe that this book is unique and significant in that it provides various research papers on end-to-end perspective on security issues in today's world as well as gives ideas of various application areas of artificial intelligence, deep learning, and data analytics which can be of great assistance to a large group of scientists, engineers, and computer network community with regard to the fast-growing era of ethical hacking and network security.

Kolkata, India — Mohuya Chakraborty
Kolkata, India — Satyajit Chakrabarti
Arad, Romania — Valentina Emilia Balas
Kalyani, India — J. K. Mandal

Message from Chief Patron

Satyajit Chakrabarti
President
Institute of Engineering and Management
Kolkata, India
April 2018

A very warm welcome to eHaCON 2018 International Ethical Hacking Conference, which is the first of its series. eHaCON 2018 is an annual event of the Department of Information Technology at the Institute of Engineering and Management, Kolkata, India. The main objective of this flagship event is to provide a platform to leading researchers from academia and practitioners from industry in India and abroad to share their innovative ideas, experiences, and cutting-edge research in the areas of cyber security, ethical hacking and network security.

eHaCON 2018 has been made possible with the generous support of our sponsors: Springer, Indian School of Ethical Hacking, ACM Chapter, IEEE IEM ComSoc Student Branch, IEEE IEM CIS Student Branch, State Bank of India, HDFC Bank. I thank all the sponsors, supporters, and the members of the Department of Information Technology for the grand success of this event.

Message from Patrons

Satyajit Chakrabarti
Director
Institute of Engineering and Management
Kolkata, India
April 2018

Amlan Kusum Nayak
Principal
Institute of Engineering and Management
Kolkata, India
April 2018

In this challenging world, technically stagnant has no survival. Also, moving forward directionless cannot help. Every effort should be taken with deep thought and understanding and also should be taken under the guidance of the masters.

I am glad that the Department of Information Technology at the Institute of Engineering and Management has organized eHaCON 2018 International Ethical

Hacking Conference. I believe that in this competitive world only innovation can stand at the top. Hence, every step has to be in the direction of excellence and research while walking with the pace of the world.

Department of Information Technology in this respect has tried to contribute to the society by creating the awareness of cyber security through this two-day conference comprising of workshops, paper presentations, coding competition, and live hacking demonstration. I appreciate the members of the organizing committee for their conscious and consistent efforts for the technical and overall success of the conference.

Message from Conference Chair

Mohuya Chakraborty
Head, Department of Information Technology
Institute of Engineering and Management
Kolkata, India
April 2018

It was my great pleasure to extend an affable welcome to all the attendees of eHaCON 2018—International Ethical Hacking Conference organized by the Department of Information Technology at the Institute of Engineering and Management (IEM), Kolkata, held at Gurukul Campus of IEM on April 6–7, 2018. The aim of eHaCON 2018 was to give an open platform where people were able to discuss the implication of new technologies for a secured society. The conference was a balanced mix consisting of technical paper presentations, live demonstrations, workshops, and online coding competitions on hacking. The goal was to kick-start efforts to fully automate cyber defense. The most substantial new findings about computer network attacks and defenses, commercial security solutions, and pragmatic real-world security experiences were presented in a two-day informative workshop, research paper presentations, and invited talks. Research papers were submitted from eight different countries around the world.

I express my sincerest thanks to the keynote speakers, Yuri Borrisov and Debdeep Mukhopadhyay, for delivering speeches on various cutting-edge topics of applications of cryptography and hardware monitoring of malware, respectively. I am thankful to Sanjeeb Sengupta and Arijit Bhattacharyya for delivering outstanding speeches on various aspects of security solutions of the industry.

I am immensely grateful to Lopa Mandal for performing the outstanding job for conducting the technical programs. With the help of an excellent committee of international and national experts, very rigorous principles were followed for selecting only the very best technical papers out of a large number of submissions in order to maintain the high quality of the conference.

I would also like to thank Avijit Bose for outstanding contribution in managing the workshop. Participants were there from various schools, colleges, and industry. Hope they were immensely benefited from the two-day workshop on ethical hacking.

My heartiest regards are due to Tapan Kumar Hazra, Arup Kumar Chattopadhyay, and Swagatam Basu for creating the online coding portal and conducting various pre-conference workshops for coding competition "De-Cipher" for the benefit of the participants.

I would like to thank Maumita Chakraborty for her dedicated contribution in the preparation of conference proceedings. My sincere thanks are due to Aditya Ray, Partha Bhattacharya, Subhabrata Sengupta, Animesh Kairi, and Subindu Saha, for arranging event participation from various organizations.

My sincere thanks are due to Satyajit Chakrabarti, Director of Institute of Engineering and Management, for co-sponsoring this conference as well as providing both financial and infrastructural support. I gratefully acknowledge the support of Springer, ACM Chapter, IEEE ComSoc Student Branch Chapter of IEM, IEEE Computational Intelligence Student Branch Chapter of IEM, Computer Society of India, Indian School of Ethical Hacking, Hack Cieux, ITOrizon, State Bank of India, HDFC Bank in sponsoring this event, without which the conference could not have been organized on this scale.

I am grateful to all the members of the advisory and technical committees comprising of professors and researchers from various parts of the world like Bulgaria, Romania, Pakistan, California, Portugal, UK, Switzerland, Japan, and India for providing their excellent service. I am also thankful to all the local organizing committee comprising of Moutushi Singh, Debalina Ghosh, Koyel Mukherjee, PulakBaral (publicity team), Pralay Kar, Sumon Mondal (print team), Sourav Mukherjee (Web site management), Ankit Anand, Sohan Datta and Sourav Ghosh (Web site development and maintenance), Kajari Sur, Dhriti Barua, Shreyashi Datta, Amit Kumar Mandal, Partha Sarathi Paul, Rabi Narayan Behera, Satyasaran Changdar, and Paramita Mitra (hospitality team) for their hard work and effort to make this conference a grand success.

Last but not least, thanks to all the participants and authors. I hope that they appreciated the conference, and I anticipate that they liked our culturally lively city of joy—Kolkata as well!

Message from Organizing Chairs

Lopa Mandal
Institute of Engineering and Management
Kolkata, India
April 2018

Tapan Kumar Hazra
Institute of Engineering and Management
Kolkata, India
April 2018

On behalf of the organizing committee of eHaCON 2018 International Ethical Hacking Conference, it was our pleasure to welcome the attendees to the Institute of Engineering and Management, Kolkata, India.

The conference consisted of eight technical sessions with 42 contributed papers, four keynote addresses, two-day workshop, and two-day online coding competition on ethical hacking. eHaCON 2018 Program Committee, comprising of 30 distinguished members, worked hard to organize the technical program. Following rigorous review process, out of about 100 submissions, only 42 full papers were accepted for the presentation in the technical sessions.

Behind every successful event, there lies the hard work, commitment, and dedication of many personalities. Firstly, we wish to thank the entire program committee for the excellent job it did in organizing the technical sessions. Special thanks are due to all the reviewers for their obligation in reviewing the papers within a very short time.

We are indebted to Avijit Bose for managing the two-day workshop on ethical hacking where participants from various schools, colleges, and industries were benefited. We wish to convey thanks to Arup Kumar Chattopadhyay and Swagatam Basu for creating, managing, and conducting the online coding competition "De-Cipher" where more than 80 teams comprising of three members per team participated. Pre-conference workshops conducted by IEEE ComSoc Student Branch Chapter IEM proved to be very successful.

A special thank goes to the conference chair Mohuya Chakraborty for giving us immense support and encouragement throughout this period. Once again, we hope that all the delegates from India and abroad found the program beneficial and enjoyed the historic city of joy—Kolkata.

We sincerely thank Satyajit Chakrabarti, Director of Institute of Engineering and Management, for his constant support throughout the event.

Programme Committee

Chief Patron

Dr. Satyajit Chakrabarti, President, Institute of Engineering and Management, India

Patron

Dr. Satyajit Chakrabarti, Director, Institute of Engineering and Management, India
Dr. Amlan Kusum Nayak, Principal, Institute of Engineering and Management, India

Conference Chair

Dr. Mohuya Chakraborty, Institute of Engineering and Management, Kolkata, India

Convener

Dr. Lopa Mandal, Institute of Engineering and Management, Kolkata, India
Mr. Tapan Kumar Hazra, Institute of Engineering and Management, Kolkata, India

Co-Convener

Ms. Maumita Chakraborty, Institute of Engineering and Management, Kolkata, India
Mr. Arup Chattopadhyay, Institute of Engineering and Management, Kolkata, India

Advisory Committee

Dr. Ioan Dzitac, Agora University of Oradea, Romania
Dr. Angappa Gunasekaran, California State University, Bakersfield, California
Dr. Valentina Emilia Balas, Aurel Vlaicu University of Arad, Romania
Dr. Joao Manuel RS Tavares, University of Porto, Portugal
Dr. Antonio Pescape, University of Napoli Federico II, Italy, and University of Bradford, UK
Dr. Yuri Borissov, Bulgarian Academy of Science, Bulgaria
Dr. Gautam Sanyal, National Institute of Technology, Durgapur, India
Dr. Atta Ur Rehman, Air University, Islamabad, Pakistan
Dr. Pramatha Nath Basu, Former Professor, Jadavpur University, Kolkata, India

Dr. Samar Bhattacharya, Former Professor, Jadavpur University, Kolkata, India
Dr. Subir Kumar Sarkar, Jadavpur University, Kolkata, India
Dr. Nabanita Das, Indian Statistical Institute, Kolkata, India
Dr. Debotosh Guha, Calcutta University, Kolkata, India
Dr. Nandini Mukherjee, Jadavpur University, Kolkata, India
Dr. Jyoti Prakash Singh, National Institute of Technology, Patna, India
Dr. Jyotsna Kumar Mandal, Kalyani University, Kalyani, India
Dr. Ujjwal Maulik, Jadavpur University, Kolkata, India
Dr. Rajat Kumar Pal, Calcutta University, India
Dr. Arindam Biswas, Indian Institute of Engineering Science and Technology, Shibpur, India
Dr. Sankhayan Choudhury, Calcutta University, India
Dr. Nabendu Chaki, Calcutta University, India
Dr. Amlan Chakraborty, Calcutta University, India
Dr. Sujata Dash, North Orissa University, Orissa, India
Mr. Basudev Gangopadhyay, ITOrizon, India
Mr. Kushal Banerjee, Tata Consultancy Services, India
Dr. Karan Singh, Jawaharlal Nehru University, Delhi, India
Dr. Partha Pratim Roy, Indian Institute of Technology, Roorkee, India
Dr. Pranab Kumar Banerjee, Former Professor, Jadavpur University, Kolkata, India
Dr. Amitava Mukherjee, Former Senior Manager, IBM, Kolkata, India
Ms. Gopa Goswami, Institute of Engineering and Management, Kolkata, India
Ms. Banani Chakrabarti, Institute of Engineering and Management, Kolkata, India

Invited Speakers

Dr. Yuri Borissov, Institute of Mathematics and Informatics, Bulgarian Academy of Science, Bulgaria
Dr. Rajeeva Karandikar, Chennai Mathematical Institute, India
Dr. Debdeep Mukhopadhyay, Indian Institute of Technology, Kharagpur, India
Mr. Sandeep Sengupta, Indian School of Ethical Hacking, India
Mr. Arijit Bhattacharyya, Virtual Infocom, India

Technical Programme Committee Chair

Dr. Rajeeva Karandikar, Chennai Mathematical Institute, Chennai, India
Dr. Yuri Borissov, Institute of Mathematics and Informatics, Bulgarian Academy of Science, Bulgaria
Dr. Antonio Pescape, University of Napoli Federico II, Italy, and University of Bradford, UK
Dr. Raphael M. Reischuk, Zühlke Engineering AG, Bern, Switzerland
Dr. Debdeep Mukhopadhyay, Indian Institute of Technology, Kharagpur, India

Editorial Board

Dr. Valentina E. Balas, Professor, "Aure Vlaicu" University of Arad, Romania
Dr. Jyotsna Kumar Mandal, Professor, Kalyani University, Kalyani, India

Dr. Mohuya Chakraborty, Institute of Engineering and Management, Kolkata, India
Dr. Satyajit Chakrabarti, Institute of Engineering and Management, Kolkata, India

Guest Editors

Dr. Yuri Borissov, Institute of Mathematics and Informatics, Bulgarian Academy of Science, Bulgaria
Dr. Rajeeva Karandikar, Chennai Mathematical Institute, India
Dr. Ioan Dzitac, Agora University, Romania
Dr. Raphael M. Reischuk, Zühlke Engineering AG, Bern, Switzerland
Dr. Joao Manuel R. S. Tavares, University of Porto, Portugal
Dr. Atta Ur Rehman Khan, University, Islamabad, Pakistan
Dr. Rana Barua, Indian Statistical Institute, Kolkata, India
Mr. Sandeep Sengupta, Indian School of Ethical Hacking, Kolkata, India

Technical Programme Committee

Dr. Valentina E. Balas, "Aure Vlaicu" University of Arad, Romania
Dr. Rana Barua, Statistics and Mathematics Unit, Indian Statistical Institute, Kolkata, India
Dr. Gautam Sanyal, National Institute of Technology, Durgapur, India
Dr. Jyotsna Kumar Mandal, Kalyani University, Kalyani, India
Dr. Matangini Chattopadhyay, Jadavpur University, Kolkata, India
Dr. Shingo Yamaguchi, Graduate School of Science and Engineering, Yamaguchi University, Japan
Dr. IoanDzitac, Agora University, Romania
Dr. Nandini Mukherjee, Jadavpur University, Kolkata, India
Dr. Joao Manuel R. S. Tavares, University of Porto, Portugal
Dr. Nabanita Das, Indian Statistical Institute, Kolkata, India
Dr. Debasish Jana, TEOCO Software Pvt Ltd., Kolkata, India
Dr. Mohuya Chakraborty, Institute of Engineering and Management, Kolkata, India
Dr. Rajat Subhra Chakraborty, Indian Institute of Technology, Kharagpur, India
Dr. Satyajit Chakrabarti, Institute of Engineering and Management, Kolkata, India
Dr. Atta Ur Rehman, Air University, Islamabad, Pakistan
Mr. Kushal Banerjee, Tata Consultancy Services, India
Dr. Parama Bhaumik, Jadavpur University, Kolkata, India
Dr. Santanu Phadikar, Maulana Abul Kalam Azad University of Technology, Kolkata, India
Mr. Basudev Gangopadhyay, ITOrizio, India
Dr. Lopa Mandal, Institute of Engineering and Management, Kolkata, India
Dr. Amitava Nag, Central Institute of Technology, Kokrajhar, India
Mr. Tapan Kumar Hazra, Institute of Engineering and Management, Kolkata, India
Dr. Karan Singh, Jawaharlal Nehru University, Delhi, India
Mr. Sandeep Sengupta, Indian School of Ethical Hacking, Kolkata, India
Ms. Maumita Chakraborty, Institute of Engineering and Management, Kolkata, India

Dr. Kaushik Majumder, Maulana Abul Kalam Azad University of Technology, Kolkata, India
Ms. Moutushi Biswas Singh, Institute of Engineering and Management, Kolkata, India
Mr. Kirit Sankar Gupta, Indian School of Ethical Hacking
Mr. Rabi Narayan Behera, Institute of Engineering and Management, Kolkata, India
Mr. Satyasaran Changdar, Institute of Engineering and Management, Kolkata, India

Organizing Committee Chair

Dr. Lopa Mandal, Institute of Engineering and Management, Kolkata, India
Mr. Tapan Kumar Hazra, Institute of Engineering and Management, Kolkata, India

Organizing Committee Co-Chair

Ms. Maumita Chakraborty, Institute of Engineering and Management, Kolkata, India
Mr. Shubhabrata Sengupta, Institute of Engineering and Management, Kolkata, India
Mr. Avijit Bose, Institute of Engineering and Management, Kolkata, India
Mr. Arup Kumar Chattopadhyay, Institute of Engineering and Management, Kolkata, India

Organizing Committee

Ms. Dhriti Barua, Institute of Engineering and Management, Kolkata, India
Ms. Debalina Ghosh, Institute of Engineering and Management, Kolkata, India
Mr. Aditya Ray, Institute of Engineering and Management, Kolkata, India
Mr. Partha Bhattacharyya, Institute of Engineering and Management, Kolkata, India
Mr. Sourav Mukherjee, Institute of Engineering and Management, Kolkata, India
Mr. Animesh Kairi, Institute of Engineering and Management, Kolkata, India
Dr. Amit Kumar Mandal, Institute of Engineering and Management, Kolkata, India
Mr. Partha Sarathi Paul, Institute of Engineering and Management, Kolkata, India
Ms. Koyel Mukherjee, Institute of Engineering and Management, Kolkata, India
Ms. Paramita Mitra, Institute of Engineering and Management, Kolkata, India
Ms. Kajari Sur, Institute of Engineering and Management, Kolkata, India
Ms. Shreyashi Dutta, Institute of Engineering and Management, Kolkata, India
Mr. Swagatam Basu, Institute of Engineering and Management, Kolkata, India
Mr. Subindu Saha, Institute of Engineering and Management, Kolkata, India

Contents

About the Editors

Mohuya Chakraborty presently holds the post of Professor and Head of the Department of Information Technology at the Institute of Engineering and Management (IEM), Kolkata. She also holds the post of Head of Human Resource Development Centre, IEM. She has done her B.Tech. and M. Tech. from the Institute of Radio Physics and Electronics, Calcutta University, in the year 1994 and 2000, respectively, and her Ph.D. (engg.) in the field of mobile computing from Jadavpur University in 2007. She is the recipient of prestigious Paresh Lal Dhar Bhowmik Award. She is the member of editorial board of several International journals. She has published three patents and over 80 research papers in reputed international journals and conferences. She has handled many research projects funded by the DST, AICTE, CSIR, and NRDC and has published a number of papers in high-impact journals. Her research areas include network security, cognitive radio, brain–computer interface, and parallel computing. She is Member of IEEE Communication Society, IEEE Computer Society, and IEEE Computational Intelligence Society as well as Faculty Adviser of IEEE Communication Society and IEEE Computational Intelligence Student Branch Chapters of IEM, Kolkata Section.

Satyajit Chakrabarti is Pro-Vice Chancellor, University of Engineering and Management, Kolkata and Jaipur Campus, India, and Director of Institute of Engineering and Management, IEM. As the Director of one of the most reputed organizations in Engineering and Management in Eastern India, he launched a PGDM Programme to run AICTE Approved Management courses, Toppers Academy to train students for certificate courses, and software development in the field of ERP solutions. He was Project Manager in TELUS, Vancouver, Canada, from February 2006 to September 2009, where he was intensively involved in planning, execution, monitoring, communicating with stakeholders, negotiating with vendors and cross-functional teams, and motivating members. He managed a team of 50 employees and projects with a combined budget of $3 million.

Valentina Emilia Balas is currently Associate Professor at the Department of Automatics and Applied Software at the Faculty of Engineering, "Aurel Vlaicu" University of Arad, Romania. She holds a Ph.D. in applied electronics and telecommunications from the Polytechnic University of Timisoara. She is the author of more than 160 research papers in refereed journals and international conferences. Her main research interests are in intelligent systems, fuzzy control, soft computing, smart sensors, information fusion, modeling, and simulation. She is Editor-in-Chief of the *International Journal of Advanced Intelligence Paradigms (IJAIP)*, an editorial board member of several national and international journals, and an expert evaluator for various national and international projects. She is Member of EUSFLAT, ACM and the IEEE, TC-Fuzzy Systems (IEEE CIS), TC-Emergent Technologies (IEEE CIS), TC-Soft Computing (IEEE SMCS), and IFAC's TC 3.2-Computational Intelligence in Control.

J. K. Mandal received his M.Sc. in physics from Jadavpur University, Kolkata, West Bengal, in 1986 and his M.Tech. in computer science from the University of Calcutta. He was awarded his Ph.D. in computer science and engineering by the Jadavpur University in 2000. Presently, he is working as Professor of computer science and engineering and Former Dean, Faculty of Engineering, Technology and Management at Kalyani University, Kalyani, Nadia, West Bengal, for two consecutive terms. He started his career as Lecturer at NERIST, Arunachal Pradesh, in September 1988. With 28 years of teaching and research experience, his major research areas include coding theory, data and network security, remote sensing and GIS-based applications, data compression, error correction, and visual cryptography. He has been Member of the Computer Society of India since 1992, CRSI since 2009, ACM since 2012, IEEE since 2013 and Fellow Member of IETE since 2012, as well as Honorary Chairman of the CSI Kolkata Chapter. He has chaired more than 30 sessions in various international conferences and delivered more than 50 expert/invited lectures in the past five years. He has acted as program chair for several international conferences and edited more than 15 volumes of proceedings from Springer Series, ScienceDirect, etc. He is a reviewer of various international journals and conferences. He has over 360 articles and six published books to his credit.

Part I
Keynote Talk

Hadamard Modulo Prime Matrices and Their Application in Cryptography: A Survey of Some Recent Works

Yuri L. Borissov

Abstract The notion of Hadamard modulo prime (HMP) matrix inherits in basics that of classical real Hadamard matrix. Namely, by definition, HMP modulo odd prime p matrix $\mathbf{H}$ of size n, is a $n \times n$ non-singular over $\mathbb{Z}_p$ matrix of ± 1's satisfying the equality: $\mathbf{H}\mathbf{H}^T = n(modp)\,\mathbf{I}$ where $\mathbf{I}$ is the identity matrix of same size. The HMP matrices have an attractive application in the modern cryptography due to the fact of their efficient employment in constructing of some all-or-nothing transform schemes. The present paper surveys some recent results on this kind of matrices by revealing their connections with coding theory, combinatorics, and elementary number theory.

1 Introduction

The HMP matrices can be considered in the broader context of modular Hadamard matrices introduced by Marrero and Butson [1] in 1973. Notice as well that the concept of modular Hadamard matrices has recently resurfaced in the engineering literature during the course of investigation of jacket transforms [2].

In this paper, the focus of attention is on the prime modular matrices motivated by their important application in cryptography: the so-called all-or-nothing transform (AONT).

Usually, an AONT scheme is a public (non-confidential, keyless) preprocessing step when encrypting data with some block cipher encryption. Its essence consists of providing a certain amount of additional security over and above the block cipher encryption since to determine any one of the message blocks embedded by that

Y. L. Borissov (✉)
Institute of Mathematics and Informatics, Bulgarian Academy of Sciences,
8 G. Bonchev Street, 1113 Sofia, Bulgaria
e-mail: youri@math.bas.bg

M. Chakraborty et al. (eds.), *Proceedings of International Ethical Hacking Conference 2018*, Advances in Intelligent Systems and Computing 811,
https://doi.org/10.1007/978-981-13-1544-2_1

transform into a single large block, the potential adversary has to break (somehow) all corresponding blocks of the cryptogram [3].

In [4] it is shown (among other things) how to construct an efficient AONT scheme of linear type by exploiting in appropriate way conventional real Hadamard matrix. Later on, the authors of [5] have proposed an extension of that construction employing instead of conventional matrix such a matrix of HMP type which enables the size not restricted to 2 or multiples of 4.

Recently, some newly obtained classification and (non-)existence results on matrices of the latter kind have been presented in [6] and [7]. On the other hand, the mathematical concept of AONT scheme has evolved as well (see, the newest articles [8, 9] devoted to that topic).

The outline of the present survey is as follows. In the next section, the necessary definitions and preliminary facts are recalled. In Sect. 3, some general results on HMP matrices, and in the subsequent section some results on HMP matrices whose size is relatively small with respect to their modulo, are exposed. In Sect. 5, the results concerning HMP matrices derived by the finite projective planes are exhibited. In Sect. 6, after a brief reminder of the basic concept and construction of AONT scheme presented in [4], it is indicated how matrices of the considered kind can be employed in such a scheme. Finally, some conclusions and directions for future research are drawn.

2 Preliminaries

Definition 1 ([5, 7]) A HMP modulo odd prime p matrix $\mathbf{H}$ of size n is a $n \times n$ non-singular over $\mathbb{Z}_p$ matrix of ± 1's such that

$$\mathbf{H}\mathbf{H}^T = n(mod\ p)\,\mathbf{I}, \tag{1}$$

where $\mathbf{I}$ is the identity matrix of size n.

As usual, $\mathbf{H}^T$ denotes the transpose matrix of a given matrix $\mathbf{H}$. Also, further on $HMP(n, p)$ stands for the set of HMP modulo p matrices of size n.

It is necessary to set out two simple but essential remarks.

Remark 1 Although some authors do not impose invertibility on the (modular) matrices considered [6], I prefer to do because of the aforesaid application of corresponding linear transforms. A necessary and sufficient condition for that is the matrix size n not to be a multiple of the chosen modulo p. So, further on it is always assumed that $p \neq n$.

Remark 2 Apparently, each conventional Hadamard matrix is a HMP modulo arbitrary prime $p > 2$ matrix, provided p does not divide the matrix size.

Example 1 The simplest non-trivial HMP matrix is obtained for $n = 7$, $p = 3$, e.g.,

$$\begin{pmatrix} 1 & 1 & 1 & 1 & 1 & 1 & 1 \\ 1 & - & 1 & 1 & 1 & 1 & - \\ 1 & 1 & - & 1 & 1 & 1 & - \\ 1 & 1 & 1 & - & 1 & 1 & - \\ 1 & 1 & 1 & 1 & - & 1 & - \\ 1 & 1 & 1 & 1 & 1 & - & - \\ 1 & - & - & - & - & - & 1 \end{pmatrix},$$

where—has been written instead of -1.

It is easy to see that by permuting the rows/columns or multiplying a row/column with -1 of a HMP matrix one gets again HMP matrix. This motivates the following definition relevant in the context of Hadamard matrices (see, e.g., [10, Ch. 14]).

Definition 2 The matrix $\mathbf{A}$ of ±1s is called equivalent to the matrix $\mathbf{B}$ if the former is obtained from the latter by the following transformations:

- permuting the set of rows/columns of $\mathbf{B}$;
- multiplying each row/column from a certain subset of rows/columns in $\mathbf{B}$ with -1.

Remark 3 W.l.o.g. when performing these equivalence transformations one can apply at the beginning all permutations and then all transformations of the second kind (for details consult [7]).

3 Some Constructions of HMP Matrices

The results exposed in this section are from [5, 6].

First, a construction of HMP matrices which extends Example 1 is described.

Construction 1 *Let* $\mathbf{E}_n$ *where* $n = pk + 4, k \geq 0$, *be a square matrix of size* n *consisting of* ±1*'s with the following description: its first row and column consist entirely of* 1*'s; its last row and column consist of* -1*'s with exception of the corner entries, and all other entries besides those on the main diagonal are equal to* 1. *Notice that* $\mathbf{E}_4$ *is the Sylvester-type Hadamard matrix of size* 4.

The proof that $\mathbf{E}_n$ is a HMP modulo prime p matrix is straightforward [5]. By multiplying the last row and column with -1 and then swapping the first and last column, one deduces that the matrix $\mathbf{E}_n$ is equivalent to the "diagonal" matrix $\mathbf{D}_n = \mathbf{J} - 2\mathbf{I}$ where $\mathbf{J}$ and $\mathbf{I}$ are the all-ones matrix and the identity matrix, respectively, both of size n.

Analogously to the case of conventional Hadamard matrices, the Kronecker product of two HMP modulo the same prime matrices of sizes n and m is a HMP matrix of size nm. This property allows starting from the matrix $\mathbf{G}_1 = \mathbf{E}_q, q = p + 4$, to construct an infinite sequence of odd size HMP matrices defined recursively by: $\mathbf{G}_t = \mathbf{G}_1 \otimes \mathbf{G}_{t-1}, t \geq 2$. Clearly, for the size of $\mathbf{G}_t$ it holds: $q^t \ (mod\ p) = 4^t \ (mod\ p)$, and of course, the set $\{4^t \ (mod\ p) | \ t \geq 1\}$ is a subset of the set QR_p of quadratic residues modulo p. Some sufficient conditions for the prime p such that these two sets coincide (i.e., the order of 4 in the group $\mathbb{Z}_p^*$ to be equal to $|QR_p| = (p-1)/2$), are presented in [5] as follows:

Proposition 1 *Let p' be a prime number.*

- *if $p = 2p' + 1$ is also a prime number then $ord_p(4) = (p-1)/2$;*
- *if $p = 4p' + 1$ is also a prime number then $ord_p(4) = (p-1)/2$.*

The proof of Proposition 1 is based on facts which can be found, e.g., in [11, p. 123, p. 197].

Remark 4 An odd prime p is called a Sophie Germain prime if $2p + 1$ is also a prime. The first few S. Germain primes are: 3, 5, 11, 23, 29, 41, 53, 83, 113, 131, ... If both p and $4p + 1$ are primes, p is called sometimes a Stern prime. The first few such primes are: 3, 7, 13, 37, ...

The next necessary condition for the existence of HMP matrix of odd size is well-known.

Proposition 2 *[5] If the size n of HMP modulo p matrix is odd, then $n(mod\ p) \in QR_p$.*

Consider the case $p = 3$. Then the above proposition implies that for odd n the set $HMP(n, 3)$ can be non-empty only if $n \ (mod\ 6) = 1$. In fact, Construction 1 provides for any such n the matrix $\mathbf{E}_n \in HMP(n, 3)$ when k is odd. The even size case is split into two subcases: for $n \ (mod\ 6) = 4$ the same construction provides HMP matrix whenever k is even; while for $n \ (mod\ 6) = 2, n > 2$, a matrix of this kind can be constructed by the Kronecker product of Hadamard matrix of size 2 and $\mathbf{E}_{n/2}$. Results of similar type about 5−modular Hadamard matrices are presented in [6], where it is shown that such matrices do exist if and only if the size n satisfies constraints: $n \ (mod\ 10) \neq 3, 7$ or $n \neq 6, 11$.

Remark 5 Proposition 2 can be generalized for the m−modular Hadamard matrices of odd size n whenever n and m are co-primes (see, [6, Lemma 2.2] or earlier [1, Th 2.2]).

4 HMP Matrices of Small Size with Respect to Their Modulo

For basic definitions and facts from coding theory, the reader is referred to [12]. The classification and (non-)existence results about HMP matrices of the type considered in this section and obtained in [7] are based on the following two lesser-known facts:

(Hereinafter, $dist(\mathbf{x}, \mathbf{y})$ denotes the (Hamming) distance between the two vectors $\mathbf{x}$ and $\mathbf{y}$ of ± 1s while $wt(\mathbf{x}) \triangleq dist(\mathbf{x}, \mathbf{1})$, where $\mathbf{1}$ is the all-ones vector, is called weight of $\mathbf{x}$.)

- **observation for parity**: For the (real) inner product of any two length n vectors $\mathbf{x}$ and $\mathbf{y}$ of ± 1s, it holds: $(\mathbf{x}, \mathbf{y}) = n - 2dist(\mathbf{x}, \mathbf{y})$, so $(\mathbf{x}, \mathbf{y}) \equiv n \ (mod\ 2)$;
- **intersection lemma**: For any two vectors $\mathbf{x}$ and $\mathbf{y}$ of ± 1s with same length, it holds: $dist(\mathbf{x}, \mathbf{y}) = wt(\mathbf{x}) + wt(\mathbf{y}) - 2wt(\mathbf{x} * \mathbf{y})$, where $\mathbf{x} * \mathbf{y}$ is the vector having -1s only where both $\mathbf{x}$ and $\mathbf{y}$ do.

The first proposition to pay attention is the following.

Proposition 3 *Let* $\mathbf{H} \in HMP(n, p)$*, where* $n \le p + 1$*. Then* $\mathbf{H}$ *is a conventional Hadamard matrix.*

Corollary 1 *If* $p \equiv 1 (mod\ 4)$*, then the set* $HMP(p + 1, p)$ *is the empty one.*

Proof When $p \equiv 1 \ (mod\ 4)$ the existence of conventional Hadamard matrix of size $n = p + 1$ contradicts the well-known fact that n must be $1,\ 2$, or $n \equiv 0 (mod\ 4)$ (see, e.g., [13, Sect. 2.2]). □

Example 2 In particular, the above corollary implies that there does not exist $HMP(6, 5)$ matrix. The interested reader is referred to [6] for another proof of this particular case.

The next proposition considers HMP matrices of even sizes less than twice the modulo.

Proposition 4 *Let* $\mathbf{H} \in HMP(n, p)$*, where* n *is an even number such that* $n < 2p$*. Then* $\mathbf{H}$ *is a conventional Hadamard matrix.*

The proof of Proposition 4 given in [7] is based on the observation for parity. Correspondingly, it holds:

Corollary 2 *If* $2 < n < 2p$ *and* $n \equiv 2 (mod\ 4)$*, then* $HMP(n, p)$ *is the empty set.*

Next, an assertion with respect to odd size HMP matrices extending a bit the region where that size varies is given by the following.

Proposition 5 *Let* $\mathbf{H} \in HMP(n, p)$*, where* n *is an odd number such that* $n < 3p$*, and let* $\omega = (n - p)/2$*. Then the matrix* $\mathbf{H}$ *is equivalent to a matrix* $\mathbf{M}$ *having the following properties:*

(i) the first row of $\mathbf{M}$ *is the all-ones vector* $\mathbf{1}$ *(i.e.,* $\mathbf{M}$ *is a normalized matrix);*

(ii) all remaining rows are of weight ω*;*

(iii) for arbitrary two distinct rows $\mathbf{r}'$ *and* $\mathbf{r}''$ *of* $\mathbf{M}$*, it holds:* $dist(\mathbf{r}', \mathbf{r}'') = \omega$*.*

In addition, $n - p \equiv 0 \ (mod\ 4)$ *and* $wt(\mathbf{r}' * \mathbf{r}'') = \omega/2$*.*

The proof of Proposition 5 given in [7] makes use of both the observation for parity and the intersection lemma. An immediate consequence is the following.

Corollary 3 *The set $HMP(p+2l, p)$, where $l \equiv 1(mod\ 2)$ and $1 \leq l < p$, is the empty one for arbitrary prime p.*

In particular,

Corollary 4 *If $p \equiv 1(mod\ 4)$ then $HMP(2p+1, p) = \emptyset$; If $p \equiv 3(mod\ 4)$, then $HMP(2p-1, p) = \emptyset$.*

Example 3 The set $HMP(11, 5)$ is the empty one (see, also [6] about that case).

Remark 6 The first claim of Corollary 4 cannot be inferred by Proposition 2 because 1 is always quadratic residue, while the second could be as well derived by that proposition since -1 is a quadratic non-residue modulo $p \equiv 3\ (mod\ 4)$.

Remark 7 Properties (iii)–(ii) from Proposition 5 mean that the binary code behind the rows (excepting the first one) of the matrix $\mathbf{M}$ is an equidistant constant weight code. Note, as well, that a theorem on the equivalence of a conventional Hadamard matrix of any admissible size and a certain constant weight binary code was proved in [14].

Proposition 3 shows the non-existence of matrices in $HMP(n, p)$ apart from the conventional ones when $n \leq p+1$. And, putting $l = 1$ in Corollary 3, it is concluded that $HMP(p+2, p)$ is empty for each p. Further, the case of even size $p+3$ (except the trivial $p = 3$) is managed by Proposition 4. So, the simplest case when a HMP matrix distinct from conventional one may exist is that of size $p+4$. Observe that the matrix $\mathbf{D}_{p+4}$, equivalent to the matrix $\mathbf{E}_{p+4}$ given by Construction 1, is an instance of $p+4$ size matrix. Finally, the following theorem completely characterizes all HMP matrices of this size.

Theorem 1 ([7]) *Let $n = p+4$ where p is an odd prime. Then*
(i) Every $\mathbf{H} \in HMP(n, p)$ is equivalent to the matrix $\mathbf{D}_n$;
(ii) The cardinality of $HMP(n, p)$ equals to $2^{2n-1}\ n!$

For the proof of this theorem, based on Proposition 5 and Remark 3, the interested reader is referred to [7].

Remark 8 A careful analysis of the proofs of Propositions 4, 5, and Theorem 1 shows that their assertions remain valid if instead of prime p it is put an arbitrary odd modulo m, while Proposition 3 is true for any invertible modular matrix.

5 HMP Matrices Derived by Finite Projective Planes

For basic definitions and facts about the finite projective planes, the reader is referred to [13, Sect. 1.2] or [15, Sect. 13.3.2]. Herein, for his/her convenience, recall the following:

Theorem 2 *Let $(\mathcal{P}, \mathcal{L})$ be a finite projective plane with $\mathcal{P}$ and $\mathcal{L}$ being the sets of its points and lines, respectively. Then there exists a constant s, called order of the plane, such that:*

- *Every line contains exactly $s + 1$ points;*
- *Every point lies on exactly $s + 1$ lines;*
- $|\mathcal{P}| = |\mathcal{L}| = v = s^2 + s + 1$.

Let $P_1, P_2, \ldots, P_v$; $l_1, l_2, \ldots, l_v$ be lists of the points and lines (arranged in some way) in the finite projective plane $(\mathcal{P}, \mathcal{L})$ of order s.

Definition 3 A binary $v \times v$ matrix $\mathcal{I} = (b_{km})$ with entry $b_{km} = 1$ if and only if the point $P_m \in l_k$ is called incidence matrix of the finite projective plane $(\mathcal{P}, \mathcal{L})$.

The matrix obtained from $\mathcal{I}$ by replacing 1 with -1 and 0 with 1 will be referred as $(\mathbf{-1}, \mathbf{1})-$image of the matrix $\mathcal{I}$.

Proposition 6 *The* $(\mathbf{-1}, \mathbf{1})$*-image of incidence matrix of a finite projective plane of order $s > 3$ is a HMP matrix modulo any prime factor of $s^2 - 3s + 1$.*

The proof follows by Theorem 2 and the definition of finite projective plane which together imply that the rows of incidence matrix constitute an equidistant constant weight code with parameters: length v, weight $s + 1$, and distance $2s$.

It is necessary to remind some background from elementary number theory in order to set out further results on HMP matrices considered in this section.
A particular case of the well-known law of quadratic reciprocity (see, e.g., Sect. 6 in [16]) is the following fact.

Lemma 1 *The number* 5 *is a quadratic residue modulo odd prime p if and only if $p \equiv \pm 1 \pmod{10}$.*

Recall also that the famous Dirichlet's theorem on primes in arithmetic progressions (see, e.g., [17, p. 16, Th. 15]) states that a progression $a + dt,\ t = 0, 1, \ldots$ with two positive co-prime integers a and d contains infinitely many primes.

Lemma 2 ([7]) *Let $T(x) = x^2 - 3x + 1$.*

(i) If p is a prime factor of $T(s)$ for some integer $s > 3$, then p is either equal to 5 *or $p \equiv \pm 1 \pmod{10}$;*

(ii) For any prime p either equal to 5 *or $p \equiv \pm 1 \pmod{10}$, there exist infinite many primes q's such that p divides $T(q)$.*

The proof of claim (i) is based on some elementary number-theoretic considerations and Lemma 1 while that of (ii) relies, in addition, on the Dirichlet prime number theorem.

The main result of this section is the following theorem.

Theorem 3 ([7]) *For $p = 5$ or any prime p of the form $p \equiv \pm 1 \pmod{10}$, there exist infinite class of HMP modulo p matrices each one of them being the* $(\mathbf{-1}, \mathbf{1})-$*image of incidence matrix of some finite projective plane of prime order.*

Table 1 HMP matrices derived by finite projective planes of prime orders ≤31

Order	5	7	11	13	17	19	23	29	31
Size	31	57	133	183	307	381	553	871	993
Modulo	11	29	89	131	239	5, 61	461	5, 151	11, 79

The proof is carried out taking into consideration Proposition 6, Lemma 2, and the existence of finite projective plane of order arbitrary prime power (see, e.g., [15, Sect. 13.3.2] for a construction).

For the prime numbers in the interval [5, 31]considered as orders of finite projective planes, Table 1 presents the corresponding sizes of HMP matrices with all possible modulus.

6 Application of HMP Matrices in Some AONT Schemes

Hereinafter, it is given a brief reminder of the description of all-or-nothing transform (AONT) scheme presented in [4].

Let X be a finite set, called alphabet. Let n be a positive integer, and suppose that $\phi : X^n \to X^n$, i.e., ϕ maps an input $n-$tuple, say $\mathbf{x} = (x_1, \ldots, x_n)$ to an output $n-$tuple, say $\mathbf{y} = (y_1, \ldots, y_n)$, where $x_i, y_i \in X$ for $1 \leq i \leq n$. Informally, the mapping ϕ is an *all-or-nothing transform* provided that the following properties are satisfied:

- ϕ is a bijection;
- If the values of any $n - 1$ of the output variables $y_1, \ldots, y_n$ are fixed, then the value of each one input variable x_i, $(1 \leq i \leq n)$ is completely undetermined.

The mapping ϕ is referred as to a $(n, v)-$AONT, where $v = |X|$.

In [4], D.R. Stinson has given an easy method of constructing unconditionally secure linear AONT by the following theorem.

Theorem 4 ([4], 2001) *Suppose that q is a prime power, and $\mathbf{M}$ is an invertible square matrix of size n with entries from the field $\mathbb{F}_q$, such that no entry of $\mathbf{M}$ is equal to 0. Then the mapping $\phi : \mathbb{F}_q^n \to \mathbb{F}_q^n$ defined by $\phi(\mathbf{x}) = \mathbf{x}\mathbf{M}^{-1}$ is a linear $(n, q)-$AONT.*

As an illustration of his method, Stinson has presented example of a linear $(n, p)-$AONT, for $n \equiv 0$ *mod* 4 and p odd prime, where in place of the matrix $\mathbf{M}$ is taken a conventional Hadamard matrix of size n with entries reduced to modulo p.
The contribution of [5] to the topic of interest can be expressed by the following:

Claim 1 ([5]) *The existence of HMP matrices (and corresponding constructions present so far) with sizes $\not\equiv 0$ mod 4 affords the scope of the aforesaid AONTs to be extended, e.g., for odd sizes. Also, note that such an AONT scheme is highly efficient* *requiring only additions, subtractions and (eventually) multiplication by constant modulo prime and even can provide opportunity to apply fast transform if such an algorithm is available.*

7 Conclusion

As it is pointed out in [6], dealing with several exceptional cases of relatively small size and presenting infinite constructions of HMP matrices initialized in the surveyed works might be non-trivial in principal. An example in this direction is the infinite class of odd size HMP matrices derivable from finite projective planes and presented in [7].

The HMP matrices inherit the useful properties of the classical Hadamard matrices. However, an advantage in applications might be the existence among them of such species whose sizes are not restricted to multiples of 4. For instance, the employment of HMP matrix instead of conventional Hadamard in implementation of AONT scheme can extend essentially the scope of that cryptographic application while keeping its efficiency.

Acknowledgements The author is grateful to Prof. Moon Ho Lee for his helpful discussions on this topic and hospitality of the Department of Electrical and Electronics Engineering of Chonbuk National University, Republic of Korea, where most of this research was done during the years 2010–2012.

References

1. Marrero, O., Butson, A.T.: Modular Hadamard matrices and related designs. J. Comb. Theory A **15**, 257–269 (1973)
2. Lee, M.H.: A new reverse jacket transform and its fast algorithm. IEEE Trans. Circuits Syst. II **47**(6), 39–47 (2000)
3. Rivest, R.L.: All-or-nothing encryption and the package transform. In: Biham, E. (ed.) Fast Software Encryption. Lecture Notes Computer Science, vol. 1267, pp. 210–218 (1997)
4. Stinson, D.R.: Something about all or nothing (transforms). Des. Codes Cryptogr. **22**, 133–138 (2001)
5. Lee, M.H., Borissov, Y.L., Dodunekov, S.M.: Class of jacket matrices over finite characteristic fields. Electron. Lett. **46**(13), 916–918 (2010)
6. Lee, M.H., Szollosi, F.: Hadamard matrices modulo 5. J. Comb. Des. 171–178 (2013)
7. Borissov, Y.L.: Some new results on Hadamard modulo prime matrices. Probl. Inf. Transm. **52**(2), 134–141 (2016)
8. D'Arco, P., Nasr Esfahani, N., Stinson, D.R.: All or nothing at all. Electron. J. Comb. **23**(4), paper # P4.10, 24 pp (2016)
9. Nasr Esfahani, N., Goldberg, I., Stinson, D.R.: Some results on the existence of t-all-or-nothing transforms over arbitrary alphabets. IACR Cryptol. ePrint Archive **177** (2017)
10. Hall, M.: Combinatorial Theory. Blaisdell Publishing Company (1967)
11. Cusick, T.W., Ding, C., Revall, A.: Stream Ciphers and Number Theory. Elsevier, Amsterdam, The Netherlands (2004)

12. MacWilliams, F.J., Sloane, N.J.A.: The Theory of Error-correcting Codes. North-Holland Publishing Company (1977)
13. Tonchev, V.D.: Combinatorial Configurations: Designs, Codes, Graphs. Longman Scientific & Technical (1988)
14. Zinoviev, V.A.: On the equivalence of certain constant weight codes and combinatorial designs. J. Stat. Plan. Inference **56**(2), 289–294 (1996)
15. van Tilborg, H.C.A.: Fundamentals of Cryptology, a Professional Reference and Interactive Tutorial. Kluwer Academic Publishers, Boston, Dordrecht, London (2000)
16. Vinogradov, I.M.: Elements of Number Theory (translated from the fifth revised edition by Saul Kravetz), 227 pp. Dover Publications Inc., Mineola, N.Y. (1954)
17. Hardy, G.H., Wright, E.M.: An Introduction to the Theory of Numbers, 6th edn. Clarendon Press, Oxford, England (2008)

Part II
Session 1A: Ethical Hacking and Cloud Computing

Social Engineering Attack Detection and Data Protection Model (SEADDPM)

Arindam Dan and Sumit Gupta

Abstract Our modern life has been influenced in myriad ways by the Internet and digital technologies along with its every ameliorating advancement. But its omnipresence has consequently turned out be a boon for cyber attackers and intruders. The colossal impact of the Internet and the widespread growth of E-business have cleared the way for cyber fraudulence whereby attackers tend to target various public agencies especially the employees of Call Centre. The intruders have been using various techniques and tools of Social Engineering for the purpose of security breach and data leakage. This paper proposes a Social Engineering Attack Detection and Data Protection Model which can be used by the employees of any agency to not only detect the social engineering attacks but also to protect their files containing sensitive data and information from an attacker. Hence, this model will be very helpful and effective in resisting the attacker from manipulating himself or herself, in ensuring data protection and in safeguarding the security of employees.

Keywords Social engineering · Attack detection · Data protection · Encryption Decryption

1 Introduction

In cyber-security, the use of deception to control or influence an individual into divulging confidential information that may be used for fraudulent purposes is known as social engineering. In simple words, the 'art' of influencing people to divulge sensitive information is known as social engineering and the process of doing so is known as social engineering attack [1]. As per [2], social engineering is the act

A. Dan (✉) · S. Gupta (✉)
University Institute of Technology, The University of
Burdwan, Golapbag (North), Burdwan 713104, West Bengal, India
e-mail: danarindam1233@gmail.com

S. Gupta
e-mail: sgupta@uit.buruniv.ac.in

M. Chakraborty et al. (eds.), *Proceedings of International Ethical Hacking Conference 2018*, Advances in Intelligent Systems and Computing 811,
https://doi.org/10.1007/978-981-13-1544-2_2

of manipulating a person to take an action that may or may not be in the target's best interest. In order to be successful, a large amount of technical knowledge is not necessarily required in social engineering. Instead, social engineering explores vices such as lies, impersonation, tricks, bribes, blackmail, threats and preys on common emotions and aspects of human psychology such as courtesy, curiosity, greed, gullibility, thoughtlessness, shyness, and apathy. In other words, social engineering can be described as the art of gathering information that others should not disclose and share under normal conditions by taking advantage of and utilizing methods of influencing and convincing [3].

According to a survey [4], there were many social engineering attacks performed in 2010 in the USA in which nearly 15 companies have been targeted and a total of 135 conversations or phone calls have been made during social engineering attacks.

Most people think that they have a low chance of being deceived. Being aware of this shared conviction, the attacker presents his desire so cleverly that he/she arouses no suspicion and exploits the victim's trust [5–8]. Thus, it is very essential that we must design better models and devise efficient systems to deal with the attacks of social engineering.

Different researches have been carried out to deal with social engineering but till now an optimal model that could claim to solve the problem with utmost perfection and accuracy is not discovered. We have few models that help in detection of attacks on call centre employees by using a decision tree structure, questionnaire-based tests, neural network approach, etc., but each one of them has certain limitations and drawbacks. We have attempted to provide a novel and better model that could serve two purposes—detection of social engineering attacks and protection of sensitive data.

This paper is organized as follows—Sect. 2 discusses the previous related works on social engineering attacks by different researchers. In Sect. 3, we have presented our proposed model that helps in not only detecting social engineering attacks, but also protects data. Section 4 includes the implementation and result of our proposed model. In Sect. 5, we have highlighted few of the future scope of improvements of our work followed by the conclusion and the references.

2 Previous Related Works

A good amount of work has been done in the area of social engineering attack detection. This section discusses a few of the most popular works related to this domain.

In paper [9], the authors had proposed a social engineering attack detection model (SEADM) in which they had used a decision tree structure where a process had been broken down into multiple manageable components. This model had been proposed to make a user aware about the social engineering threat and provides guidelines to aid in the decision-making process.

Paper [10] presents the Social Engineering Attack Detection Model version 2 (SEADMv2) that had been proposed to achieve detection of social engineering attacks by using a series of states. Here a user is required to provide answers to questions in the form of yes or no. Based upon the reply, the model predicts the occurrence of an attack. SEADMv2 can be used to detect textual, verbal, unidirectional, bidirectional or indirect social engineering attacks.

In paper [11], the authors had implemented SEADMv2 as an Android application called social engineering prevention training tool (SEPTT). The authors had tested this application on 20 subjects to determine the probability of a user getting victimized by a social engineering attack. It was claimed that the use of their Android application reduced the number of subjects falling prey to the malicious, bidirectional and indirect communication social engineering attacks.

In paper [12], the authors had utilized the notion of neural networks to detect social engineering attacks in call centres. They have selected few attributes of the caller or a phone call to predict whether any social engineering attack had occurred or not.

Paper [13] focuses on detecting social engineering attacks committed over phone lines. The architecture proposed in this paper is the Social Engineering Defense Architecture (SEDA). The primary purpose of SEDA is to make recipients of phone calls within an organization aware of unauthorized and malicious callers by analysing conversations over phone in real-time fashion. Through this paper, the authors tend to produce real-time signatures that could aid in forensic investigation.

3 Our Proposed Work

The overall working of our proposed Social Engineering Attack Detection and Data Protection Model (SEADDPM) is depicted in Fig. 1. Throughout this paper, the term individual is defined as the person receiving the phone call (caller) and the term requester is defined as the person who is making the call and requesting for information (caller). Our model has been designed to work on two layers—attack detection layer and data protection layer.

Attack Detection Layer

The most crucial step in this model deals with making an individual aware and alert of an imminent danger. The individual has to be conscious enough in understanding the emotional state and must consummately evaluate through the questions posed in the conversation about the intention of the requester. The decision tree used in our model contains a set of questions that would aid the individual in detecting whether the requester is trying to deceive or lure him/her into revealing sensitive information. The questions are arranged in a way so as to help an individual in making a correct guess and predict whether the requester is trying to dupe him/her or is trying to perform on him/her any social engineering attack.

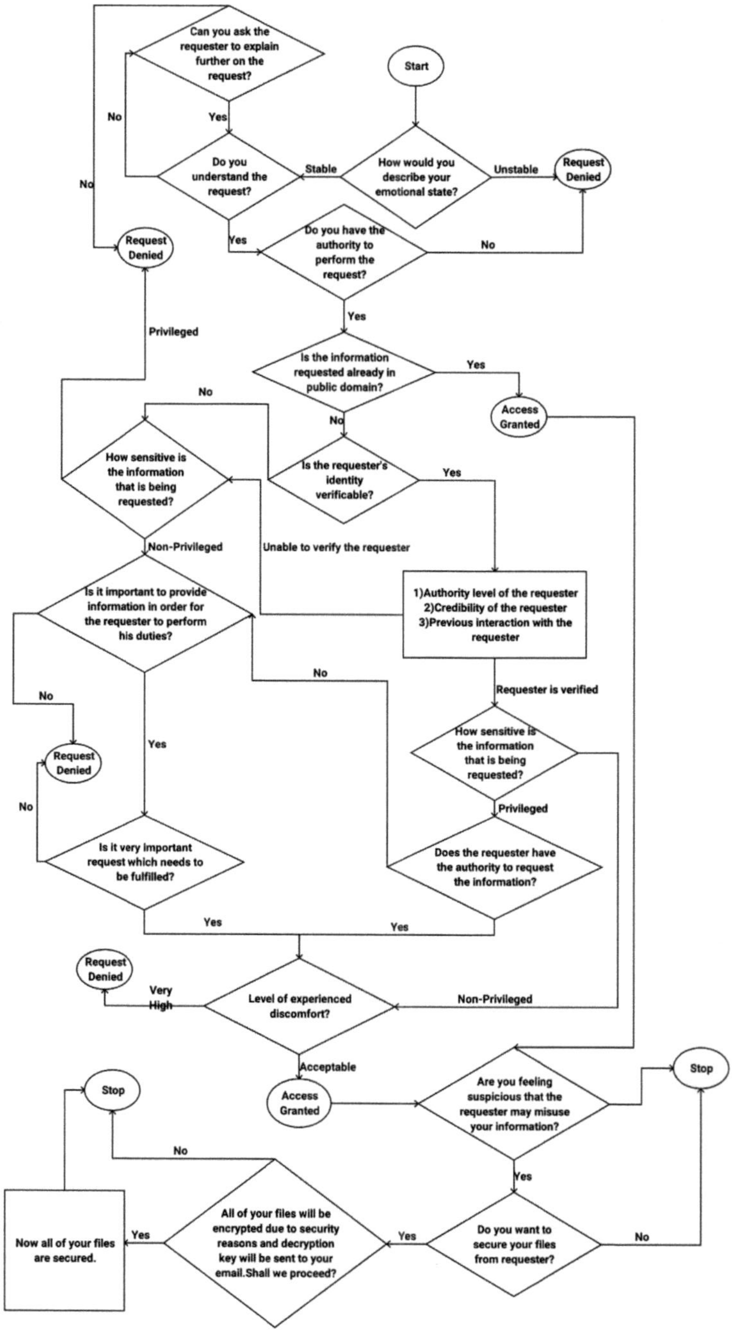

Fig. 1 Social Engineering Attack Detection and Data Protection Model (SEADDPM)

Data Protection Layer

The second layer of our model deals with protection of sensitive data from the attacker. Here if the individual has a sense of doubt on the requester or feels something fishy, then after providing information to the requester, the individual can transfer his/her files comprising sensitive information to the cloud platform in an encrypted manner. This would prevent the attacker in accessing the original data even after getting information from the individual. We have used the process of encryption and decryption to provide data protection and only an authorized individual would be able to decrypt the information. Figure 2 depicts the encryption and the decryption process.

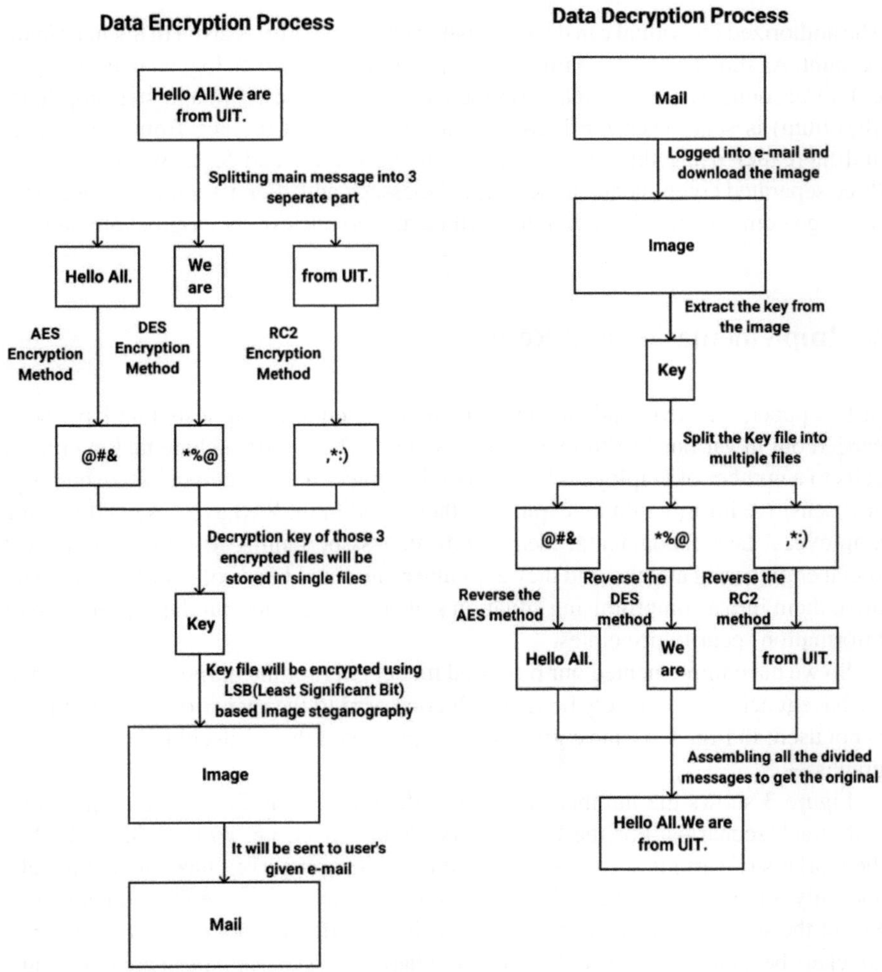

Fig. 2 Data encryption and decryption process for ensuring data protection

(A) *How does the Encryption process work?*

We have used a hybridized cryptographic mechanism as shown in Fig. 2 to encrypt data. In this mechanism, the original message is divided into three separate message components and stored in three separate files. Now these three files will be encrypted individually using three different encryption methods, viz. Advanced Encryption Standard (AES), Data Encryption Standard (DES) and Rivest Cipher or Ron's Code (RC2) encryption methods to increase the level of security. Next, the keys of these three encrypted files will be stored together in one file. Now it is necessary to encrypt that key file too. We have used least significant bit (LSB)-based image steganography algorithm for improving the security level. And finally the image will be sent to the individual's e-mail id.

(B) *How does the Decryption process work?*

The authorized individual can decrypt his/her encrypted file received in his/her e-mail account. As shown in Fig. 2, initially, the individual needs to be logged in into his/her e-mail account where the image (embedded via LSB-based image steganography algorithm) is sent. Then the individual needs to extract the key from that image, and thereafter using the reverse process of AES, DES and RC2, we will get the three separated components of the original message and after assembling those three message components, the individual will be able to retrieve the original message.

4 Implementation and Result

In this paper, we have made an analysis of social engineering tests that have been carried out in various institutions in our location. These tests include making phone calls to a number of employees by the social engineer in an attempt to seize employees' sensitive information by exploiting their good faith. We have observed that the employees' lack fundamental idea and awareness pertaining to protection against social engineering attacks, and they are vulnerable at the hands of cyberattacker who drive them into a compromising situation and cleverly carries out the exploitation of information security principles.

So we have implemented our proposed model only for the purpose of testing, and we have received a relatively better result compared to the scenario when our model is not used. In Fig. 3, we have graphically represented the results obtained using bar graphs.

Figure 3 shows the number of errors with and without the model for harmless and attack scenarios. For the first case (without model, i.e. without SEADDPM), the workers of various agencies or institution as per our survey faced lots of attacks and only a few of them were able to keep their information safe from the attacker. But in the second case (with model, i.e. with SEADDPM), not only the number of attackers became less but also the number of harmless scenarios was increased. Thus, it can be inferred that our proposed model SEADDPM offers better security and data protection.

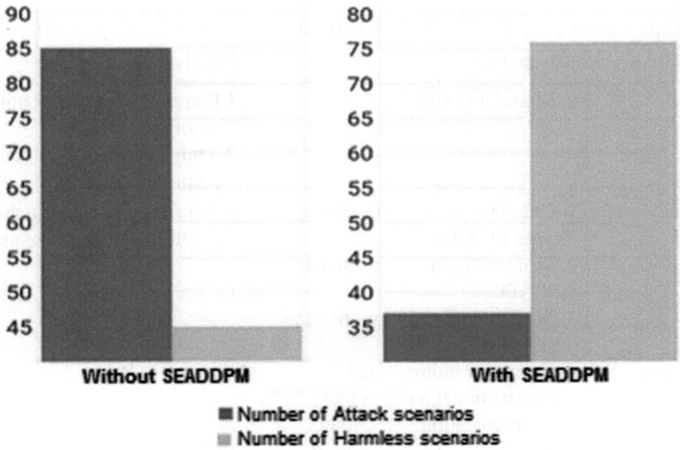

Fig. 3 Bar graphs showing comparison of number of errors with and without SEADDPM for harmless and attack scenarios

A comparative study based upon the advantages and disadvantages offered by different social engineering models has been given in Table 1 to comprehend how our proposed model SEADDPM is better than other existing approaches. It can be easily observed that unlike other models, our proposed model incorporates the extra feature of data protection to protect the integrity of data and provide the user with a sense of safety and security. Moreover, our model discusses more state transitions in order to portray the consummate picture as to what decisions the user should take for a variety of scenarios.

In our model, we have proposed a more robust and secure encryption technique whereby we have used three encryption techniques, viz. AES, DES and RC2. The original data (message) will be split into three parts, and each of these subparts will go through the three mentioned encryption processes to generate three encryption keys. As we know that each of these techniques has a few pros and cons in terms of block size, number of rounds required in the encryption process, speed of execution and key generation. But, it is also worth mentioning that each one of them offers a different level of security. Individually, AES technique outperforms the other two encryption mechanisms because it has a block size of 128 bits unlike 64 bits as in case of DES and RC2 and so using any attack such as brute force will not be of any help for the attacker. Further, the decryption keys of these three encrypted files will be stored in a file which will be protected using the LSB-based image steganography method. Though our approach seems to consume more effort and time because of the usage of multiple encryption algorithms but with all the labour, it results in creating a difficult and challenging task for the social engineer to break the code and attack the system. We can conclude by saying that on the performance front, and our model tends to not just detect but also protect from social engineering attack in such a way

Table 1 Comparison of our proposed model SEADDPM with other existing security models

Model used	Advantages	Disadvantages
SEADM	1. Modular design	1.Requires user to determine own emotional state 2. Only caters for bidirectional communication
SEADMv2	1. Colour codes to differentiate types of states 2. More state transitions than the SEADM 3. More modular design than the SEADM 4. Caters for bidirectional, unidirectional and indirect communication	1. No states to examine the emotional state of the user
Social engineering detection using neural networks	1. Accurate at detecting attacks	1. Never been tested in a real-world scenario 2. Tedious for the user to enter values into the input nodes
SEDA	1. No user interaction required 2. Prevents same social engineer targeting different employees	1. Social engineer could trick the system by using voice recordings of the person they are imitating 2. Only works for verbal social engineering attacks
Our proposed model (SEADDPM)	1. Modular design 2. More state transitions than the SEADM and SEADMv2 3. Incorporates and implements the data protection process	1. Requires the user to determine own emotional state 2. No AI is used to activate data protection process

that it becomes a daunting and Herculean task for the social engineer or cyberattacker to break into our system and get access to the sensitive information by deceiving an individual.

5 Future Work

Through this paper, we have attempted to design a novel model named Social Engineering Attack Detection and Data Protection Model (SEADDPM) that aims at detecting social engineering attacks along with ensuring protection of data from cyberattackers. But sometimes it is more important to teach the intruder a lesson so that he/she does not happen to torment or trouble the individual in future. Our model can play a vital role in offering data protection and attack detection features but it does not provide the complete solution to stop this type of fraudulent activity in future. So, as a future scope of improvement, we are working on enhancing our proposed model so that it becomes capable of not only defending an individual against social engineering attacks but also of punishing and penalizing the attacker who had carried out any malicious activity.

6 Conclusion

Social engineering has turned out be a real threat to the society, and it is high time that we escalate the ongoing research activities in this domain to tackle the imminent challenges and make our future secure. Apart from detecting the attacks, we are also offering different mechanisms by which sensitive information can be protected from leakage or exploitation. Our proposed model SEADDPM is just a small step in dealing with the hazards of social engineering. We have miles to cover in order to safeguard sensitive information in this age of digitalization where every passing day bestows upon us a new security-related task to deal with.

References

1. Mouton, F., Leenen, L., Venter, H.S.: Social engineering attack examples, templates and scenarios. In: Computers and Security (2016)
2. Hadnagy, C.: Social Engineering: The Art of Human Hacking, 1st edn. Wiley Publishing Inc., Indianapolis, Indiana (2011)
3. Mataracioglu, T., Ozkan, S., Hackney, R.: Towards a security lifecycle model against social engineering attacks: SLM-SEA. In: Proceedings of the Nineteenth Americas Conference on Information Systems, Chicago, Illinois, pp. 1–10 (2013)
4. Hadnagy, C.J., Aharoni, M., O'Gorman, J.: Social engineering capture the flag results. In: The Defcon 18 Social Engineering CTF Report. http://www.social-engineer.org/resources/sectf/Social-Engineer_CTF_Report.pdf. Accessed 02 Aug 2010
5. Bican, C.: Social Engineering Attacks. TUBITAK UEKAE—National Gate of Information Security. http://www.bilgiguvenligi.gov.tr/teknik-yazilar-kategorisi/sosyal-muhendislik-saldirilari.html?Itemid=6. Accessed May 2008
6. Huber, M., Kowalski, S., Nohlberg, M., Tjoa, S.: Towards automating social engineering using social networking sites. In: International Conference on Computational Science and Engineering, vol. 3, pp. 117–124. IEEE, Vancouver, BC, Canada (2009)
7. Nohlberg, M.: Why humans are the weakest link. In: Gupta, M., Sharman, R. (eds.) Social and Human Elements in Information Security: Emerging Trends and Countermeasures, pp. 15. IGI Global, Hershey, PA (2008)
8. Tarakcioglu, T.: Analysis of social engineering attacks in Turkey. J. Natl. Res. Inst. Electron. Cryptol. (UEKAE) **2**(4), 88–95 (2010)
9. Bezuidenhout, M., Mouton, F., Venter, H.S.: Social engineering attack detection model: SEADM. In: Information Security for South Africa (ISSA), pp. 1–8. IEEE, Johannesburg, South Africa (2010)
10. Mouton, F., Leenen, L., Venter, H.S.: Social engineering attack detection model: SEADMv2. In: International Conference on Cyberworlds, pp. 216–223. IEEE, Visby, Sweden (2015)

11. Mouton, F., Teixeira, M., Meyer, T.: Benchmarking a mobile implementation of the social engineering prevention training tool. In: Information Security for South Africa (ISSA), pp. 1–9. IEEE, Johannesburg, South Africa (2017)
12. Sandouka, H., Cullen, A.J., Mann, I.: Social engineering detection using neural networks. In: International Conference on CyberWorlds, pp. 273–278. IEEE, Bradford, UK (2009)
13. Hoeschele, M., Rogers, M.: Detecting Social Engineering. In: Pollitt, M., Shenoi, S. (eds.) Advances in Digital Forensics. Digital Forensics 2005. IFIP—The International Federation for Information Processing, vol. 194, pp. 67–77. Springer, Boston, MA (2006)

OnlineKALI: Online Vulnerability Scanner

Parthajit Dholey and Anup Kumar Shaw

Abstract OnlineKALI is a Web framework for vulnerability assessment which allows you to quickly do a security audit of your own websites and network infrastructures from a remote location without having to set up external pen-testing operating system and with very high-speed network capability and processing power. It allows you to scan, enumerate the security loopholes, and vulnerability with full customization of the open-source tools. It uses a chroot or Docker environment to launch an attack without affecting the main system. It uses the features of Django, PostgreSQL, Jinja2, and python to be secure as far as possible. This paper is basically to take a maximum of open-source tools of Kali Linux and put into the cloud so that all can work without any hardware or network issues.

Keywords Vulnerability scanner · Open-source tools · Pen-testing framework Web application vulnerabilities · Security testing

1 Introduction

Penetration testing is a series of activities undertaken to identify and exploit security vulnerabilities. It helps to confirm the effectiveness or ineffectiveness of the security measures that have been implemented [1]. Today day by day, the threats are increasing in the digital world. Everyone is coming online to increase their business to the whole world, but many of them lack basic security which can lead to loss of money and business. So lots of manual human effort and infrastructure and network setup are required to scan the servers [2]. The attacks and breaches harm an organization reputation. To eradicate threats, a lot of open-source tools are available to analyze the servers if they lack some security hole which can be harmful [3]. We are trying

P. Dholey (✉) · A. K. Shaw
HackCieux, Cyber Space Security: Consultancy, Training, and Solutions, Kolkata, India
e-mail: p.dholey@hackcieux.com; parthajit1994@gmail.com

A. K. Shaw
e-mail: anup@hackcieux.com

M. Chakraborty et al. (eds.), *Proceedings of International Ethical Hacking Conference 2018*, Advances in Intelligent Systems and Computing 811,
https://doi.org/10.1007/978-981-13-1544-2_3

here to move such open-source tools to the clouds and provide users a usable Web framework which has high network speed and CPU performance that can let the users scan their servers and find a known vulnerability that exists [4]. We try to provide as much as customization for the tools and real-time output to the users. We are building a platform that helps users with no or intermediate knowledge in security to help find the vulnerabilities in their server [5].

2 Background

A research paper stated few questions and concerns about (a) network vulnerability scanning, (b) security vulnerabilities, (c) Is System Security a Concern? (d) application security [6]. So, we are trying to come up with one solution for fighting against cyberattacks. A description of "Main stages of a pen testing" [7], i.e., Step 1. Planning and preparation, Step 2. Reconnaissance, Step 3. Discovery, Step 4. Analyzing Information and Risks, Step 5. Active Intrusion Attempts, Step 6. Final Analysis, Step 7. Report Preparation. So our work focuses on mainly from steps 2–5 and step 7.

3 Related Work

Pen-testing your application server or network has become an important area, and several studies are being developed to improve security in data systems and networks. A live exercise was presented to help students with a live case study to understand the importance and complexity of security. E-voting is getting popular day by day and expects to assess and enhance critical thinking in the field of security [2]. A discussion was made with all the different network security tools that are available and developed in the market that are widely used to test a system network hardening [4]. They have developed one of the most used networks scanning and banner grabbing tools called as Nmap which is widely used and most advanced useful network sniffing tool called Wireshark. A person writes and shows what are the current rate and effectiveness of the current security and vulnerability scanning tools that are available in the market and how much of the prices are we trying to reduce [8]. It compares one of the best to most widely used tools for doing vulnerability assessment and penetration testing. Finding vulnerability has been the target of many of the tools and system available. But they start having a problem with configuration issues and network issues, and a very limited scanning and vulnerability assessment access are there until you have installed on the local system [9]. An implementation was widely used Kali Linux OS on Raspberry Pi [10]. Here we are focusing to bring all tools installed in Kali Linux to online instead of installing it locally or Raspberry Pi. This framework will provide users all features of open-source pen-testing tools (command line tools only) with highly improved performance of the integrated tools.

4 Technical Description

4.1 Overview

OS: Ubuntu 16.04, Nginx version: Nginx/1.10.3 (Ubuntu), Django 1.8.7, PostgreSQL Version: (PostgreSQL) 9.5.11, Pgbouncer: pgbouncer version 1.7 (Debian), Chroot Jail: Ubuntu Artful High-Level Design.

We have developed a framework which is responsible for handling the tools and the request that the user is making to make them available in a browser and sending a request to scan the target and synchronously getting output. The Nginx web server is accepting connection request and sending it over to Django web server, where the highly secured front end is responsible for validating user request [11]. Depending upon validation, it passes the request to a chroot jailed environment, where it has very limited environment access and it executes the scan requests and the commands that the user gives and sends the results back to users synchronously [12]. The diagrammatic representation of the framework can be seen in Fig. 1, where the arrow represents the flow of data. If ever an RCE is done or the environment is compromised, we have built a jailed environment where the basic functionality is removed. It cannot revert to main Django server because of the environment access commands, such as chmod, ls, cd, whoamiand muchmore, will not be available [13]. The configuration we have tested are Memory: 1 GB, CPU = 1 vCPU, SSD Disk = 25 GB, Network Speed (Download: 1053.43 Mbit/s, Upload: 1118.61 Mbit/s) Memory: 1 GB, CPU = 1 vCPU, SSD Disk = 25 GB.

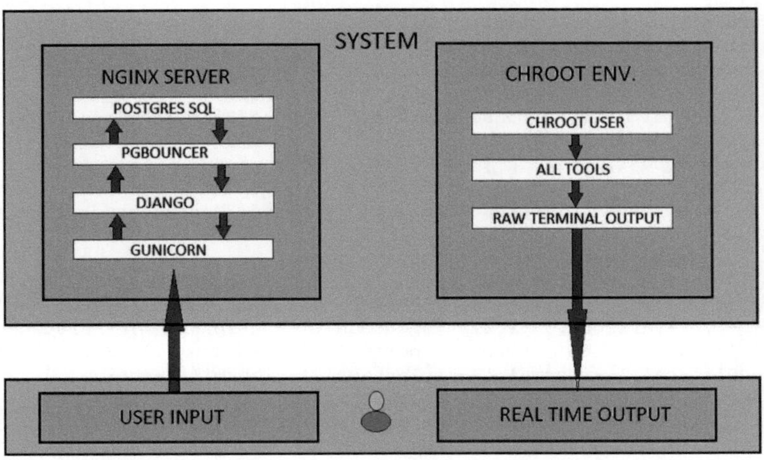

Fig. 1 Overview of the design

4.2 Technology Used

Django is a high-level web development framework built in python that encourages rapid and clean development, pragmatic design keeping security in mind. Built by experienced developers, it takes care of much of the hassle of Web development, so one can focus on writing our application without needing to reinvent the wheel for the development of its security protocols [11]. Its free and open source do same for PostgreSQL and pgbouncer and Nginx [14].

4.3 Algorithm Overview

The framework provides the power to customize the scans as far as possible. Every scan request coming from the user is handled by Django, and it creates a shell file of the scan command and starts the shell file to run in a jailed environment and feed with the output to the users at the same time so that they can see the raw results. As soon as the request is sent by the user, the web server checks if the user has a valid session and also check if any other scan is running. The program code is

```
def invokenmap(request):
 if 'username' in request.COOKIES:
  try:
          value = request.COOKIES['username']
          ses_email = request.session['member_id']
  except:
          return HttpResponseRedirect('/ohmygod/?message=Not Allowed')
 else:
  return HttpResponseRedirect('/')
 if authenticate(value ,ses_email) == True:
  if captcha_validation( request.POST.get('g-recaptcha-response')) ==
False:
          return HttpResponseRedirect('/ohmygod/?message=Cannot Validate
Captcha&tags=hidden')
  address=request.POST['ip']
  if address == "":
          return HttpResponseRedirect('/ohmygod/?message=address blank')
  try:
          data=request.POST['flags']
          not_acceptable_strings = ['&&','all',';', '-p-','|','*','-iL',
'-iR','-e','>','>>','--exclude','--excludefile','-sL','--system-dns','-
sI','-p-','--version','--script-updatedb','--script-all','--osscan-
guess','--max-retries','--source-port','-g','--badsum','-oN','-oX','-
oS','-oG','-oA','--resume','--stylesheet','--webxml','--no-
stylesheet','--datadir','--interactive','!sh','127.0.0.1']
          if any(x in address for x in not_acceptable_strings):
                    return HttpResponseRedi-
rect('/ohmygod/?message=Invalid Strings at the Flag parameter')
          if any(x in data for x in not_acceptable_strings):
                    return HttpResponseRedi-
rect('/ohmygod/?message=Invalid Strings at the Flag parameter')
  except Exception:
          data = ""
  value = request.COOKIES['username']
```

```
  if one_at_a_time (value) == False:
          muluser = value + ".txt"
          open(muluser, 'w').close()
          scanner_input = 'nmap '+data+' '+address+'
          make_file ( value , scanner_input)
          myprocess = subprocess.Popen(['schroot -c artful -u django --
directory=/home/django/'+value+'/ -- "./magic.sh"'],shell=True,
stdout=subprocess.PIPE, bufsize=1)
          scan_log (get_ip(request) , address , value , scanner_input)
          t = threading.Thread(target=process_output,
args=(myprocess,muluser,value))
          t.daemon = True
          t.setName(value)
          t.start()
          dbupdate(value , True , myprocess.pid)
          return HttpResponseRedirect("/result")
  else:
        ip = get_ip(request)
        objects = UserDatabase.objects.get(mail=value )
          return ren-
der(request,'ltscan.html',{'name':objects.first_name,'user_ip':ip})
 else:
          return HttpResponseRedirect('/')
```

Another thread starts running and synchronously reads the shell output and starts sending to the user. Restarting process or blocking the user to run multiple scans at the same time, as this can lead to the exhaustive use of resources. The output is formatted to support the HTML front end so the user gets the output synchronously and hassle-free. The below code is responsible for taking the execution output in real time and sending it to the user by sending the output to a file.

```
def process_output(myprocess,muluser,value): #output-consuming thread
    nextline = None
    buf = ''
    while True:
        out = myprocess.stdout.read(1)
        if out == '' and myprocess.poll() != None: break
        if out != '':
            buf += out
            if out == '\n':
                nextline = buf
                buf = ''
        if not nextline: continue
        line = nextline
        nextline = None
        with open(muluser,"a") as test_file:
                    line = line.encode("utf-8")
                line = line + ' <br>'
                    test_file.write(line)
          test_file.close()
myprocess.stdout.close()
```

The tools that we have integrated with the framework has been modified to make it compatible such as not to ask any further input, run independently, remove vulnerable functionality inside the tools source code, and tried to make it as much optimized as possible. We can see the below graph Fig. 2 which shows the performance and

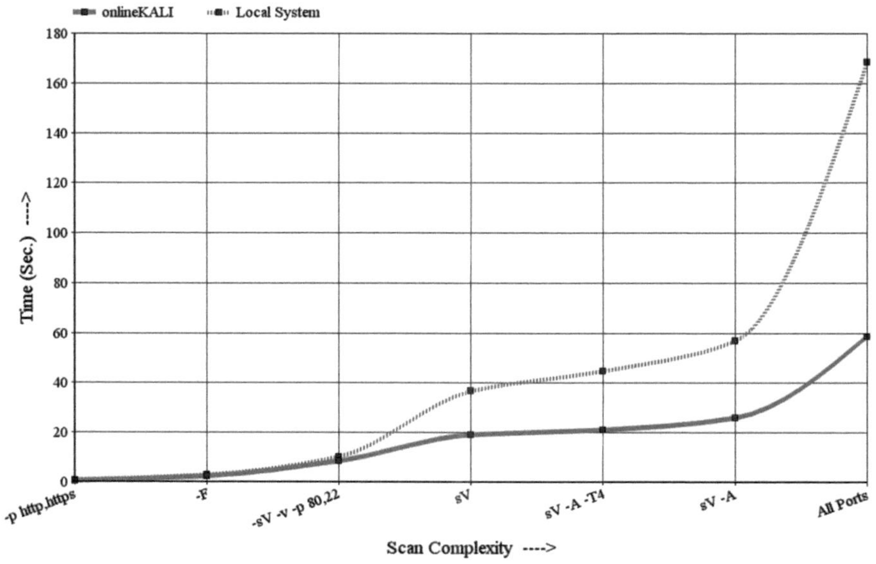

Fig. 2 Comparison between locally running Nmap against web framework

timings of Nmap scans running from a local machine compared with the framework. We drastically see a performance improvement when running by the framework.

We are also working on building the customized layout of the raw output and also a personalized report for a particular IP with all the necessary vulnerability assessment done with a single click.

5 Experimental Results and Findings

5.1 Performance Graph

Performance of the framework is quite impressive. See the below performance graph on specific machine configuration. Performance level will increase with machine configuration. Currently, we are using the lowest configuration of server available for our development purpose, as soon as we move into our production we hope to have a high-end CPU and RAM with load balancer which may work ever faster and reliable than the current results. We are using servers to test from DigitalOcean [15]

Scan: nmap -A -p- -v scanme.nmap.org

We will show three graphs to explain this section.

As in Figs. 1, 2, 3, we can see the CPU, memory, and disk usage when we are running 10 simultaneous Nmap full port services scan at the same time. The curve in Fig. 3 describes us about the load that the CPU and memory are having when very

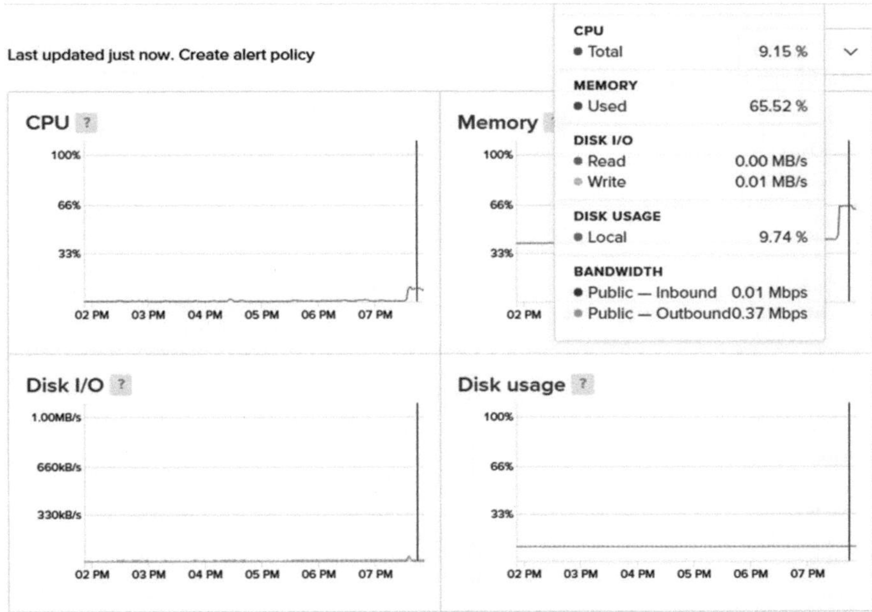

Fig. 3 Performance load graph for 10 simultaneous Nmap scan while scanning on scanme.nmap.org

high complexity scan is running 10 times at same time. Figures 4 and 5 show the load on the system when high complexity scan is running 30 and maximum scans at same time. The scan we choose is of very high complexity that uses the maximum functionality of the Nmap framework, which shows how well our front end is capable of handling load of the request that will be provided.

5.2 Security Level

Security level of this web framework has been kept at very high priority, so we chose the Django framework to develop it. The attacks are launched with a jailed chroot environment, which has all the environment connection ability and user movability and commands are disabled so that if any user by loophole or command injection gets any access, he is jailed in a non-movable environment [12]. The front is based on Django framework which is known as the most secure web development framework currently in the market [11]. We also use the encryption or hashing of some input and output flags and working on currently to put our own custom encryption algorithm to all the messages flowing throughout the system. We have also made some changes to the legacy tools to remove some flags which can impact on the performance and

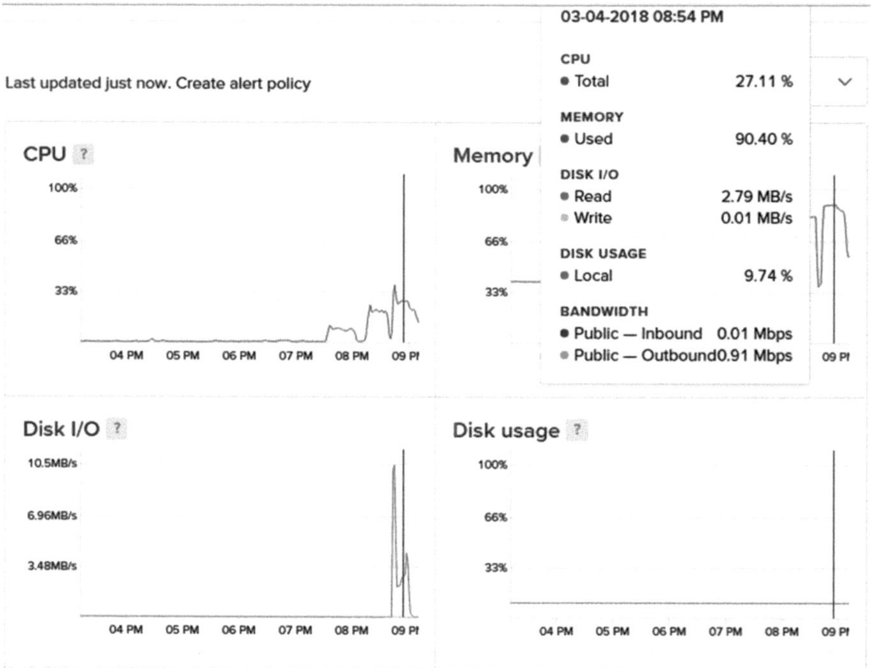

Fig. 4 Performance load graph for 30 simultaneous Nmap scan while scanning on scanme.nmap.org

system performance. This implementation improves the performance of the scanning process of all the tools implemented.

5.3 *User Experience*

This framework will help cybersecurity professionals (beginner and intermediate) to do a security testing and vulnerability assessment on the go without any hardware configuring issues or network hassle. Figure 6 shows the UI that the framework provides to the user. We recommend the users to have a proper knowledge of what they are doing, and they have the permissions of the things doing through our framework. We do not encourage data loss or hacking of any computer or IOT devices. Continuous research is going on for the advancement of this framework so that it can be used as a single platform for cybersecurity professionals. Also, this framework includes flag field for all the tools with as much as customization that we can provide to the user for input for the doing customized scan. We are also continuously working on to put new tools in our system and compatibility with the framework [16].

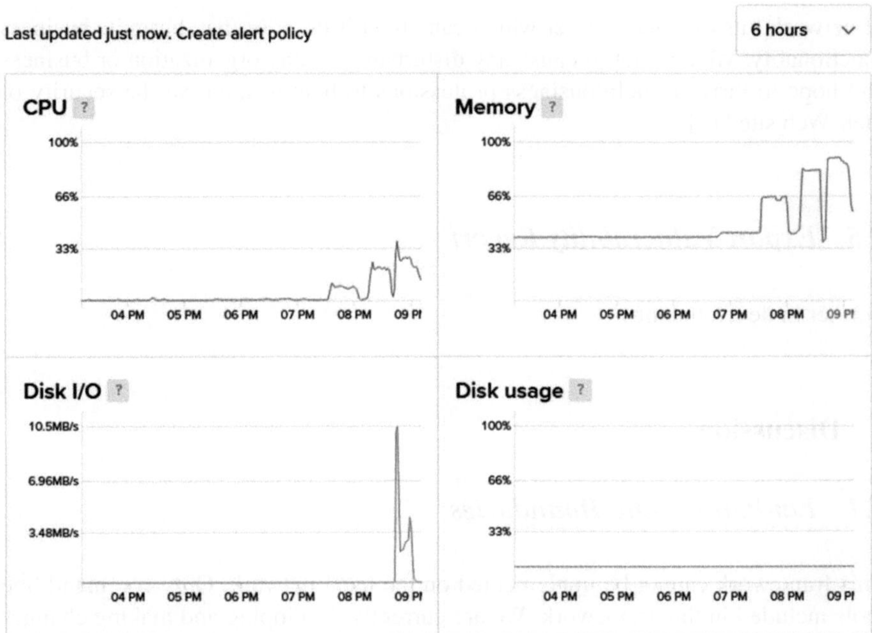

Fig. 5 Performance load graph for 10–35 simultaneous Nmap scan while scanning on scanme.nmap.org

Fig. 6 Nmap scan page from the framework (flags filed for user input)

5.4 *Time Consumption*

Time consumption for a particular scan depends on the flags of scan and complexity level. As simple scan would take less time, while full port scan will take maximum time [17]. We also have some time limit for some scans as it can through too much of network hits on a web server which can affect it users or may harm its business functionality. We aim not to cause any disturbance to any organization or business and hope to thrive to help business professions to help to improve the security of their Web site [13].

5.5 Export Vulnerability Report

Further in development.

6 Discussion

6.1 Limitations and Boundaries

This framework cannot be implemented on the local network. Only command line tools included in this framework. We are currently developing and making changes to the tools so that the tools can run faster and remove the flags from its basic functionality

And user interaction code so that it can perform smoothly with the framework.

6.2 Future Scope

Automation: For future development and research, this framework can be used as an automate vulnerability and penetration testing.
Agent: This framework can come up with an agent so that internal network pen test can be performed (still under research process).

7 Conclusion

The automated penetration test plays an important role in the security professional's toolkit. As part of a comprehensive security program, these tools can quickly evaluate the security of systems, networks, and applications against a wide variety of threats. But security pros should view them as a supplement, rather than a replacement, for traditional manual testing techniques.

So, this framework is all about bringing all open-source security tools into a single online framework. So only IP/Domain name required in IP field and already basic flags options are there in form of the checkbox to perform a simple quick scan.

As stated before this framework will help cybersecurity professionals (beginner and intermediate) to do a security testing and vulnerability assessment on the go without any hardware configuring issues or network hassle.

Acknowledgements This research is supported by HackCieux. We are thankful to our colleagues who provided expertise that greatly assisted the research, although they may not agree with all of the interpretations provided in this paper.

References

1. Bacudio, A.G., Yuan, X., Bill Chu, B.-T., Jones, M.: An overview of penetration testing. Int. J. Netw. Secur. Appl. (IJNSA) **3**(6) (2011). http://airccse.org/journal/nsa/1111nsa02.pdf
2. Bishop, M., Frincke, D.A.: Achieving Learning Objectives Through E-Voting Case Studies. http://ieeexplore.ieee.org/document/4085594/
3. Web Application Vulnerability Scanner Evaluation Project (Vulnerability Scanner Evaluation Project) (2012). http://code.google.com/p/wavsep/
4. Gordon Lyon. Top 125 Network Security Tools [EB] (2011). http://sectools.org/
5. Chroot Concept. https://en.wikipedia.org/wiki/Chroot
6. Ethical hacking and network defense: choose your best network vulnerability scanning tool. In: 2017 31st International Conference on Advanced Information Networking and Applications Workshops (WAINA). http://ieeexplore.ieee.org/document/7929663/
7. de Jimenez, R.E.L.: Pentesting on web applications. In: 2016 IEEE 36th Central American and Panama Convention (CONCAPAN XXXVI). http://ieeexplore.ieee.org/document/7942364/
8. Shay-Chen The Web Application Vulnerability Scanner Evaluation Project [EB] (2012). http://www.sectoolmarket.com/
9. W3af. http://w3af.sourceforge.net/
10. Yevdokymenko, M., Mohamed, E., Arinze, P.O.: Ethical hacking and penetration testing using Rasberry PI. In: 2017 4th International Scientific-Practical Conference Problems of Infocommunications, Science and Technology (PIC S&T). http://ieeexplore.ieee.org/document/8246375/
11. https://www.digitalocean.com/
12. PgBouncer. https://pgbouncer.github.io/
13. DDOS. https://www.arbornetworks.com/research/what-is-ddos
14. Django Server hosted in DigitalOcean. https://www.digitalocean.com/
15. Nginx Web Server Security and Hardening Guide. https://geekflare.com/nginx-webserver-security-hardening-guide/
16. Kali Linux Operating System (Specially designed for Pentesting). https://www.kali.org/
17. Web Application Vulnerability Scanner Evaluation Project (Vulnerability Scanner Evaluation Project) (2012). http://code.google.com/p/wavsep/
18. Offensive Security. https://www.offensive-security.com/
19. Open Source tools list which is inbuilt in Kali Linux. https://tools.kali.org/tools-listing
20. Ngnix. https://nginx.org/en/
21. Django. https://www.djangoproject.com/
22. PostgresSQL. https://www.postgresql.org/

Toward an AI Chatbot-Driven Advanced Digital Locker

Arindam Dan, Sumit Gupta, Shubham Rakshit and Soumadip Banerjee

Abstract The ongoing digital era is witnessing and endorsing online transactions and information exchange at the speed of light. But the increasing number of hackers and social engineers has made the digital environment susceptible and vulnerable to intrusion and attack. Also, because of the dearth of advanced security models, maintaining security and protecting integrity of sensitive information is at stake. What the world needs now is a robust and reliable security model to establish information security and secure all digital transactions. Through this project work, we are introducing an artificially intelligent chatbot that will provide a user with the menu for choosing an appropriate encryption method (out of AES, DES, RC2, and hybrid methods) for securing his/her crucial information. The proposed idea of a firewall-based advanced digital locker for authenticating user's digital signature before allowing access to the encrypted files provides the authorized user with a sense of more robustness, reliability, and a higher level of security.

Keywords Encryption · Decryption · Chatbot · Digital signature · Image steganography

A. Dan (✉) · S. Gupta (✉) · S. Rakshit · S. Banerjee
University Institute of Technology, The University of Burdwan, Golapbag (North), Burdwan 713104, West Bengal, India
e-mail: danarindam1233@gmail.com

S. Gupta
e-mail: sgupta@uit.buruniv.ac.in

S. Rakshit
e-mail: shubhamr238@gmail.com

S. Banerjee
e-mail: soumadipban000@gmail.com

M. Chakraborty et al. (eds.), *Proceedings of International Ethical Hacking Conference 2018*, Advances in Intelligent Systems and Computing 811,
https://doi.org/10.1007/978-981-13-1544-2_4

1 Introduction

The world is full of fascinating problems waiting to be solved. Security issue is one such problem that has been researched and analyzed since a long time, but due to the advent of latest technologies, there still remains a lot of horizons untraveled and unexplored. As we know, uploading of any file or document containing sensitive information to any server (like cloud storage) is a vulnerable process as chances of the file getting hacked or its content being altered comes into picture. Further, storing any sensitive information in any disk has several issues and risks involved. Thus, there arises a need for a system that could save the uploaded document in an encrypted form. If the original file is encrypted, then the hacker would not be able to decipher the contents even after getting access to the file. The presence of digital locker will provide a storage area where the encrypted file will be stored. The crucial aspect of the digital locker is that it can only be accessed via the digital signature of the authorized user. So, the chances of accessing such a locker will be a difficult task for any intruder or hacker.

The most important requirement in information security is the presence of a right person who would guide us through the entire process of file encryption. In real life, getting this person is a very difficult and challenging task. Even if we get hold of a person who will help us in understanding the intricate encryption techniques and procedures, the person will charge us a handsome amount. Moreover, the person would come to know about all the procedures used and the secret keys generated during the encryption process, thus posing a chance of blackmailing or threat in near future. Our AI chatbot proposed in this paper will be the best alternative to deal with this situation. The chatbot will act as a virtual assistant and will work as per the order and command of the user. Thus, the user will no longer have to depend on any physical entity for help.

This paper is organized as follows: Sect. 2 discusses the previous related works on various security models by different researchers. In Sect. 3, we have presented our proposed model that helps in protecting data. Section 4 discusses the implementation and results obtained. Section 5 highlights the future scope of improvements in our work. In Sect. 6, we have finally concluded our paper followed by references in the end.

2 Previous Related Work

Many researchers have proposed a variety of security models to protect and safeguard information by using the cryptographic techniques such as encryption and digital signature to name a few. This section discusses a few of the most popular works related to this domain.

The authors in the paper [1] have designed a Web services-based security model by integrating the watermark embedding and watermark detection technology components with the Web services. The proposed system architecture is based on Web services via SOAP service requester, digital certificates, XML encryption, and digital signatures to ensure secure exchange of online information between service providers while carrying out multimedia services, online services, and e-work applications.

Researchers in [2] have presented a patented security device by using a processor, an operating system (OS) software program loaded onto the processor, a type-II virtual machine monitor that will run on top of the host OS and create a user-definable number of sensitive, nonsensitive, encryption, and router virtual machines. This device is claimed to make the virtual computing environment secured and robust.

In paper [3], the authors have proposed an authentication algorithm based on the visual secret sharing scheme of the visual cryptography domain. They have offered a fool-proof lock-key mechanism in which every lock-key pair has a unique image associated with it. The lock can be opened by its paired key only, and the key cannot be duplicated. The lock has a memory and behaves like a safe door. It can be used to transmit and receive signals like the key. Here, the lock and the key can alter the pixel distribution of the secret image when an unauthorized access is triggered so that security can be established, and unauthorized access can be prevented.

Paper [4] provides a comparative study and evaluation of symmetric (AES, DES, Blowfish) as well as asymmetric (RSA) cryptographic algorithms by taking different types of files such as binary, text, and image files. Different evaluation parameters such as encryption time, decryption time, and throughput are considered by the author for performing the comparison, and AES algorithm was found to yield better performance.

The authors in the paper [5] have presented a review of digital lockers specifically based on the cloud platform. They have explained the features, objectives, and working of the digital locker which was released by the Department of Electronics and Information Technology (DeitY), Govt. of India [6]. Digital lockers have been created to provide a secure and dedicated personal electronic space for storing the documents on the cloud and to do away with the paperwork-based document storage for exploring the possibilities of the Digital India Campaign.

3 Our Proposed Work

Through this project work, we are introducing an AI chatbot-driven advanced digital locker for providing a user-friendly environment to a user for protecting one's sensitive information. In this work, we have designed a chatbot named Augusta which will help its user to secure data as per user's requirement. Here, we have primarily used

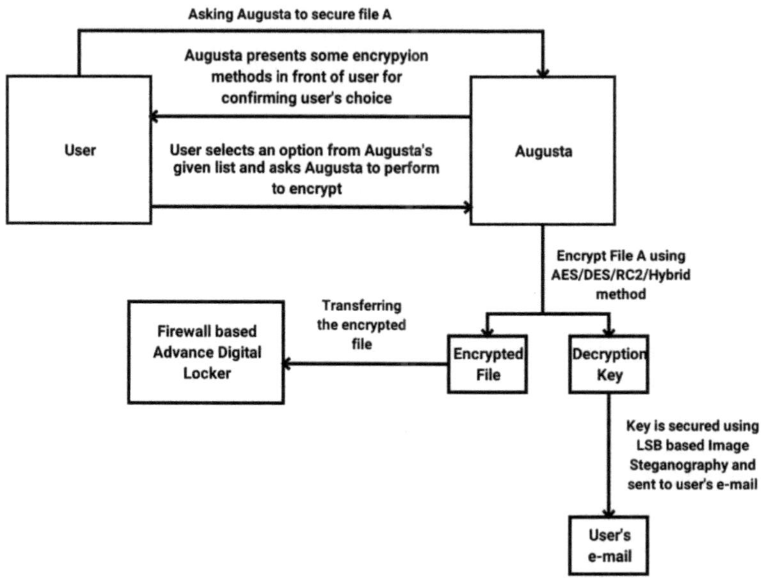

Fig. 1 Process of file encryption and transfer to advanced digital locker

four types of encryption methods, viz. Advanced Encryption Standard (AES), Data Encryption Standard (DES), Rivest Cipher Encryption (RC2), and hybrid method (four hybrid methods by combining all the three named encryption methods).

As seen in Fig. 1, users will have the choice to encrypt their files as per their need, and all the encrypted files will be transferred to the advanced digital locker. This locker will be protected by a firewall. The decryption key of the corresponding encrypted files will be secured by least significant bit (LSB)-based image steganography method. Further, the resultant image version of that key will be sent to the user's email account.

We have also attempted to add an extra feature to our firewall-based advanced digital locker for checking the user's digital signature while collecting the encrypted files and for applying the decryption method consequently (see Fig. 2).

The novelty of our work lies in how an artificially intelligent chatbot is providing a user with the menu for choosing an appropriate encryption method (out of AES, DES, RC2, and hybrid methods) for securing his/her crucial information. The firewall-based advanced digital locker for authenticating user's digital signature before allowing access to the encrypted files provides the authorized user with a sense of more robustness, reliability, and a higher level of security.

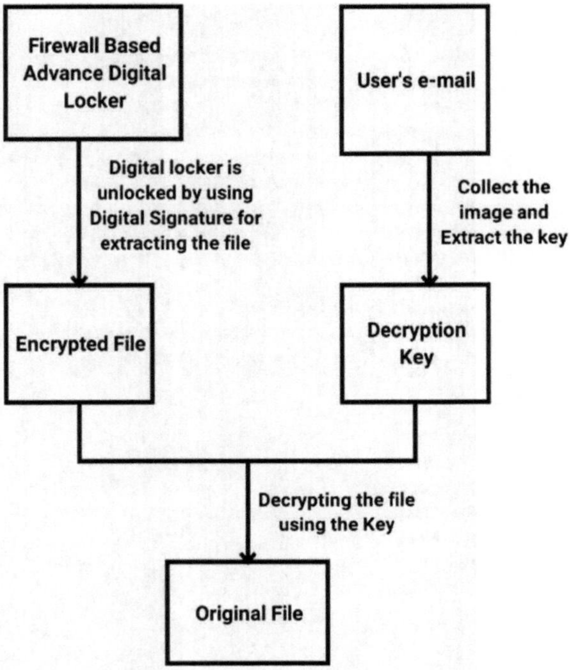

Fig. 2 Process of original file retrieval

4 Implementation and Results

To implement our proposed system, we have used Python 3.5 and SQLite. The algorithms such as AES [7, 8], DES [8, 9], RC2 [10], hybrid encryption, LSB-based image steganography [11, 12] are used in our project work. The operating system on which our project is built is Windows 10.

We have designed our AI-based chatbot named Augusta (see Fig. 3) which is capable of providing a user with a simple, understandable, and friendly environment for sharing and putting across his/her requirements. The chatbot provides the user with a menu of choices wherein a user can select which encryption algorithm the user wants to use for encrypting his/her file. Firstly, the user has to select the path from where the original file (where sensitive information is stored) is to be retrieved. After the path has been specified by the user, the chatbot asks the user to choose which encryption algorithm the user wants to use for encrypting the original file. Based on user's entry, the original file will be encrypted, and the user will be informed about successful completion of the encryption process. When the user chats with Augusta and opts for encrypting a file, he/she is prompted by Augusta to select the path where the original file is stored (see Fig. 4).

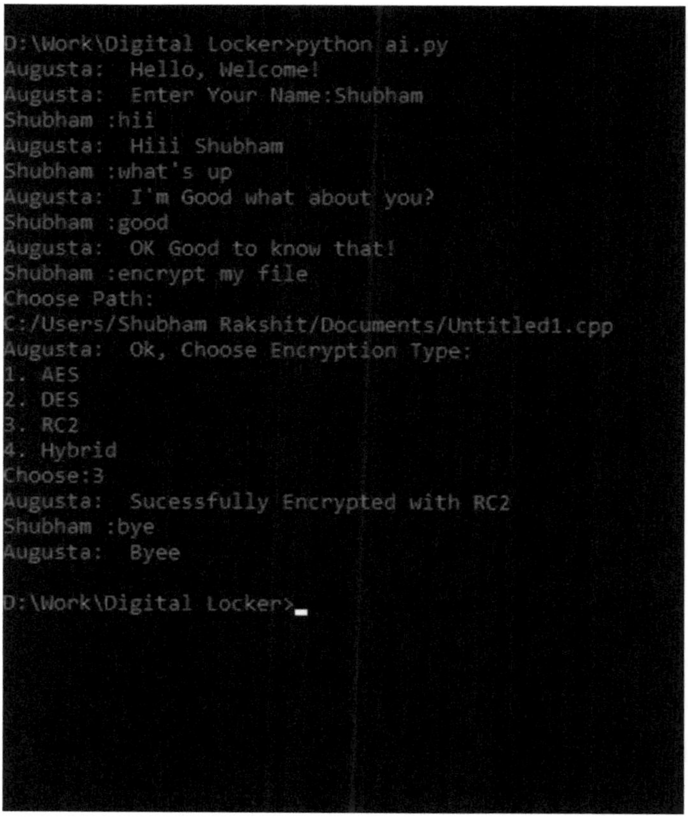

Fig. 3 Screenshot of AI chatbot Augusta

The creation of digital locker is shown in Fig. 5, and the encrypted files are stored in .dat format in the digital locker after the user chooses the encryption process, and the chatbot performs the encryption process successfully (see Fig. 6).

A comparative study based on the advantages and disadvantages offered by different security models has been given in Table 1 to comprehend how our proposed model is better than other existing approaches.

In Table 2, we have shown the performances of different encryption techniques, viz. AES, DES, and RC2 and their hybrid counterparts, viz. AES-DES, AES-RC2, DES-RC2, and AES-DES-RC2 on the basis of different factors such as key length, round(s), block size, speed, and security. On analyzing, it has been observed that as the hybrid models work in levels, they tend to offer more security than basic encryption techniques. The hybrid AES-DES-RC2 method offers the highest level of security at the cost of slow speed because three different levels L1, L2, and L3 of encryption is utilized in implementing this approach.

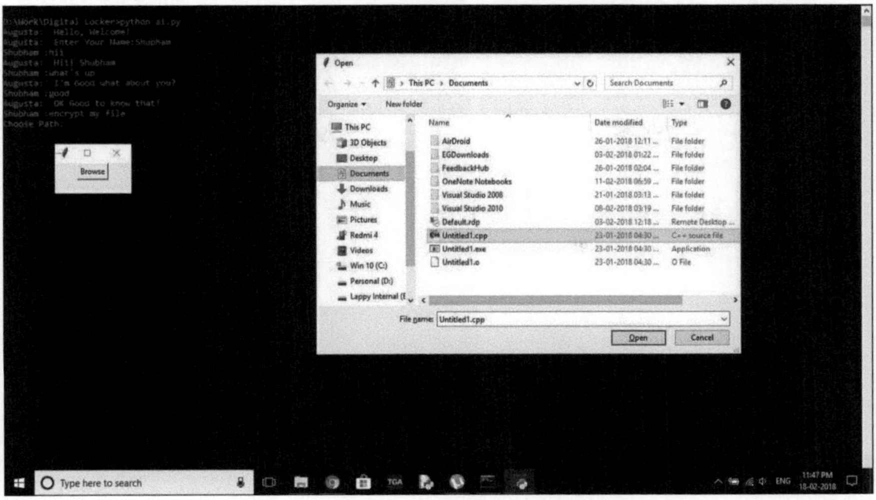

Fig. 4 Screenshot showing the selection of path where original file is stored

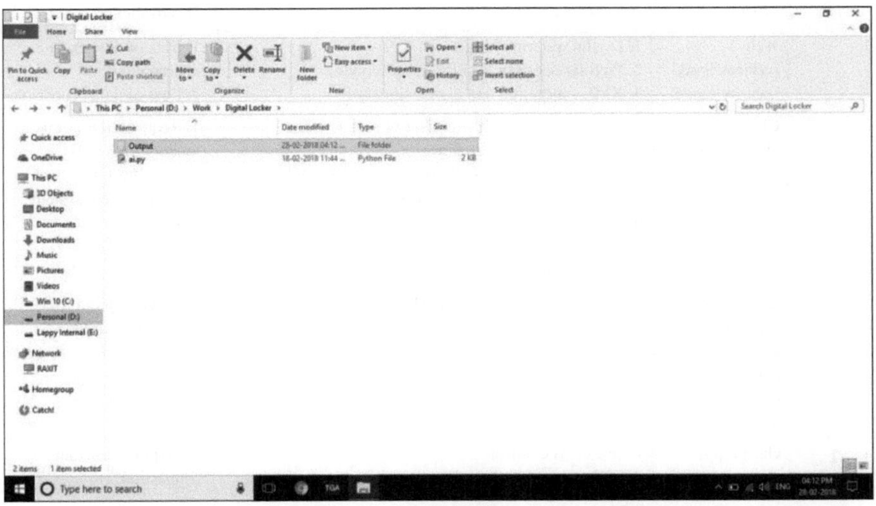

Fig. 5 Screenshot showing the folder of the digital locker

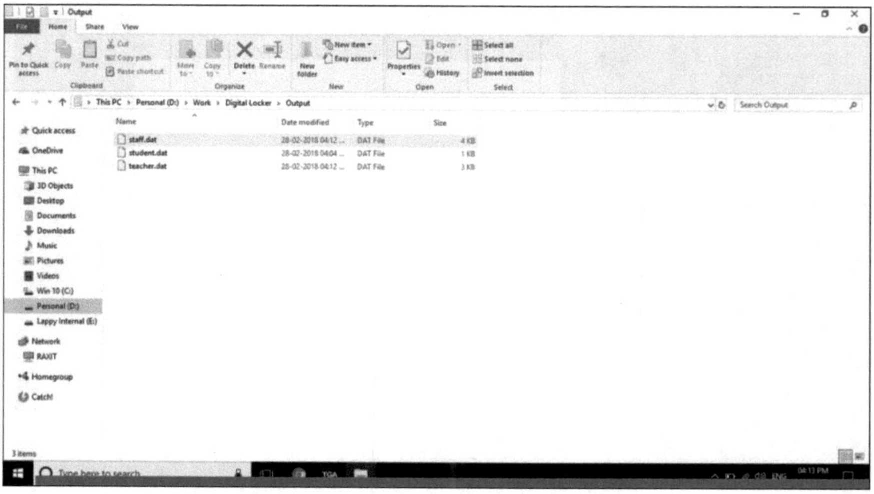

Fig. 6 Screenshot of encrypted files stored in digital locker

Table 1 Comparison of our proposed model with other existing security models

Sl. no.	Models	Advantages	Disadvantages
1	Web services-based security model	1. Digital watermarking 2. Web services via SOAP service requester 3. XML encryption and digital signatures	1. Risk of tampering and interception 2. Loss of data is possible
2	Cloud-based digital locker	1. Use of user-definable number of sensitive and nonsensitive virtual machines 2. Each encryption virtual machine is connected to one of the said user-definable number of sensitive virtual machines 3. Each encryption virtual machine includes at least one encryption algorithm	1. Requires server-client mode 2. If the host or server fails, the whole system fails
3	Device for and method of secure computing using virtual machines	1. Can be accessed from anywhere	1. No encryption technique is used 2. Chances of hacking is high
4	Our proposed model	1. Use of AI chatBot 2. Different encryption techniques including hybrid are used 3. Data is migrated to cloud storage-based digital locker 4. Decryption key is sent to user's email using LSB-based image steganography	1. Requires Internet connection

5 Future Work

Through this project work, we have aimed at creating an AI chatbot-driven advanced digital locker which would create a user-friendly environment to facilitate a user in encrypting any document by choosing any encryption algorithm (out of AES,

Table 2 Performance analysis of different encryption techniques

Factors	Basic			Hybrid				
	AES	DES	RC2	AES-DES		AES-RC2	DES-RC2	AES-DES-RC2
Key length	128, 192, or 256 bits	56 bits	8–1024 bits, in steps of 8 bits; default 64 bits	**L1**	128, 192, or 256 bits	128, 192, or 256 bits	56 bits	128, 192, or 256 bits
				L2	56 bits	8–1024 bits, in steps of 8 bits; default 64 bits	8–1024 bits, in steps of 8 bits; default 64 bits	56 bits
				L3	N/A	N/A	N/A	8–1024 bits, in steps of 8 bits; default 64 bits
Round(s)	10–128 bit key, 12–192 bit key, 14–256 bit key	16	16	**L1**	10–128 bit key, 12–192 bit key, 14–256 bit key	10–128 bit key, 12–192 bit key, 14–256 bit key	16	10–128 bit key, 12–192 bit key, 14–256 bit key
				L2	16	16	16	16
				L3	N/A	N/A	N/A	16
Block size	128 bits	64 bits	64 bits	**L1**	128 bits	128 bits	64 bits	128 bits
				L2	64 bits	64 bits	64 bits	64 bits
				L3	N/A	N/A	N/A	64 bits
Speed	Fast	Medium	Slow	Faster than AES-RC2, DES-RC2, and AES-DES-RC2		Faster than DES-RC2 and AES-DES-RC2	Faster than AES-DES-RC2	Very Slow
Security	High	Medium	Medium	Higher than AES-RC2 and DES-RC2 but lower than AES-DES-RC2		Higher than DES-RC2 but lower than AES-DES and AES-DES-RC2	Higher than basic encryption methods but lower than hybrid methods	Highest

DES, RC2 or hybrid encryption algorithms) as per user's choice or requirement. Till now, we have completed making the chatbot, and this chatbot is capable of encrypting a file based on user's choice. But our objective in future is to develop the firewall-based advanced digital locker for storing the encrypted file. We are also aiming at sending the secured decryption key in user's email via LSB-based image steganography method. In our next endeavor, we will incorporate the digital signature-based authentication mechanism so that only the authorized user could be able to extract the encrypted file from the locker and use the decryption key (received in his/her email) to finally decrypt the encrypted file and get back the original file.

6 Conclusion

This proposed system offers an advanced user-friendly application which offers an efficient, reliable, and secured platform for file access, storage, and retrieval with a high level of data protection. The use of various encryption algorithms added up with the presence of a chatbot (which acts as a virtual assistant) enhances the acceptability and popularity of our work among the masses.

References

1. Zhang, J.: A web services-based security model for digital watermarking. In: International Conference on Multimedia Technology (ICMT), pp. 4805–4808. IEEE, Hangzhou, China (2011)
2. Meushaw, R.V., Schneider, M.S., Simard, D.N., Wagner, G.M.: Device for and method of secure computing using virtual machines. In: United States Patent, Patent number-US6922774B2, Filing date: May 14, 2001, Issue date: Jul. 26, 2005. Application number: 09/854,818, United States (2005)
3. Tunga, H., Mukherjee, S.: Design and implementation of a novel authentication algorithm for fool-proof lock-key system based on visual secret sharing scheme. Int. J. Comput. Sci. Iss. (IJCSI) **9**(3), 182–186 (2012)
4. Panda, M.: Performance analysis of encryption algorithms for security. In: International Conference on Signal Processing, Communication, Power and Embedded System (SCOPES), pp. 278–284. IEEE, Paralakhemundi, India (2016)
5. Vaijawade, V., Khelkar, N., Thakare, D.: Review on "cloud based digital locker". In: International Journal of Research in Science & Engineering (IJRISE), Special Issue: Techno-Xtreme 16, pp. 682–686 (2016)
6. National eGovernance Division, Ministry of Electronics & Information Technology (MeitY), Government of India. http://digitallocker.gov.in. Accessed 28 Feb 2018
7. Daemen, J., Rijmen, V.: Rijndael: the advanced encryption standard. Dr. Dobb's J. **26**(3), 137–139 (2001)
8. Singh, G., Supriya: A study of encryption algorithms (RSA, DES, 3DES and AES) for information security. Int. J. Comput. Appl. **67**(19), 33–38 (2013)
9. Nadeem, A., Javed, M.Y.: A performance comparison of data encryption algorithms. In: First International Conference on Information and Communication Technologies (ICICT), pp. 84–89. IEEE, Karachi, Pakistan (2006)
10. Knudsen, L.R., Rijmen, V., Rivest, R.L., Robshaw, M.J.B.: On the design and security of RC2. In: Vaudenay, S. (ed.) Fast Software Encryption (FSE), LNCS, pp. 206–221. Springer, Berlin, Heidelberg (1998)
11. Thangadurai, K., Devi, G.S.: An analysis of LSB based image steganography techniques. In: International Conference on Computer Communication and Informatics (ICCCI). IEEE, Coimbatore, India (2014)
12. Singh, A., Singh, H.: An improved LSB based image steganography technique for RGB images. In: International Conference on Electrical, Computer and Communication Technologies (ICECCT). IEEE, Coimbatore, India (2015)

A Hybrid Task Scheduling Algorithm for Efficient Task Management in Multi-cloud Environment

Asmita Roy, Sadip Midya, Debojyoti Hazra, Koushik Majumder and Santanu Phadikar

Abstract Cloud computing is an emerging area in the field of computation where various IT infrastructures are leveraged to users based on their requirement. With ever-growing number of cloud users, the number of task requests that needs to be handled in one-time instance is huge. At the same time, to deliver a good QoS, CSPs need to achieve best performance in a cost-efficient manner with minimal completion time along with reduced delay and latency. Thus, an efficient task scheduling algorithm in a multi-cloud environment needs to deploy a hybrid approach considering multiple factors for allocating tasks among several available clouds. In this work, a well-defined efficient hybrid task scheduling algorithm is developed where tasks are scheduled in a multi-cloud environment. The task scheduler employs a priority-based algorithm that determines the priority level of every task based on the computation cost, time required and power consumed to execute the task. The simulation result shows that our approach attains better efficiency in comparison with other existing approaches.

Keywords Cloud computing · Task scheduling · Cost · Deadline · Power
Multi-cloud environment

1 Introduction

With the growth of information and communication technology (ICT), the vision of computing being the 5th utility is perceived more strongly [1–3]. Several computing paradigms like grid computing, cluster computing and more recently cloud computing aim to deliver this service. Cloud computing provides a vast virtualized environment that delivers the following services—(a) software as a service (SaaS), (b) platform as a service (PaaS) and (c) infrastructure as a service (IaaS) [4]. Services

A. Roy · S. Midya · D. Hazra · K. Majumder (✉) · S. Phadikar
Maulana Abul Kalam Azad University of Technology, BF 142 Sector 1,
Salt Lake City, Kolkata 700064, India
e-mail: koushikwbutcse@gmail.com

M. Chakraborty et al. (eds.), *Proceedings of International Ethical Hacking Conference 2018*, Advances in Intelligent Systems and Computing 811,
https://doi.org/10.1007/978-981-13-1544-2_5

are requested by cloud user based on their demand [5]. Cloud resources are then allocated to user based on the task requested. This requires a task scheduling algorithm that arranges tasks in a way such that the allocations of resources are done efficiently [6–8].

In [9, 10], priority-based task scheduling mechanism is proposed. Authors arrange the tasks based on its execution cost. In [11–13], a priority-scheduling algorithm is proposed that works with cost and deadline of a task to define priority. After calculating priority, tasks are allocated to some VM according to min-min heuristic approach. These approaches reduce the deadline miss of a task. Task deadline is an important parameter as frequent missing of task deadline causes users' dissatisfaction, thus affecting users' QoE. In [14], an improved version of cost-based algorithm is proposed. This scheme calculates priority of a task according to cost. Then, these prioritized tasks are allocated to some VM according to that VM's instruction handling capacity. Here, deadline of a task is not considered and as a result tasks' deadline miss is not optimal. In [15], Chunling Cheng et al. proposed a power-saving strategy. In this strategy, they are keeping processors of a CSP in three modes like sleep, alive and idle. In sleep and idle mode, power consumption by a processor is low. In [16, 17], author Lee, Y. et al. have proposed a power management scheme to reduce power consumption by VMs for executing the tasks. All the above works are either concentrated on reducing cost of task execution or aimed to reduce power consumed while executing the task.

In the real-world scenario, there are multiple clouds available to provide various kinds of services like web service, file operations, PDF tools, online gaming, storing user data and image operations to their customers. According to Alsughayyir and Erlebach [17], for various kinds of services, a CSP sets different charges. With increasing number of cloud users and their various application demands, it is not possible for a single cloud to provide all types of services. At the same time, if a single cloud provides service in one area, it results in performance bottleneck and the system becomes susceptible to single-point failure. This requires the necessity to develop a multi-cloud architecture. In the multi-cloud architecture, a task scheduler must consider multiple parameters including cost of task execution, various network parameters and power requirement before taking scheduling decision. Thus, a hybrid scheduling policy is required to be developed that considers all of the above-mentioned parameters. In this work, a priority-based hybrid task scheduling policy is developed that calculates task priority taking into consideration its execution cost, network congestion and power required to execute the task.

2 Proposed Multi-cloud System Architecture

In this work, a multi-cloud architecture is proposed in Fig. 1. It consists of chief scheduler which takes scheduling decision based on generated task priority and various geographically distributed cloud connected to the chief scheduler.

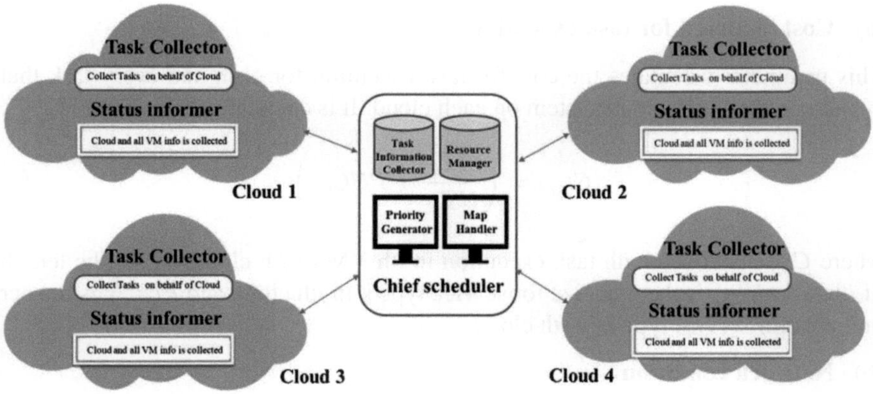

Fig. 1 Proposed multi-cloud architecture

2.1 Components of the Proposed Multi-cloud System

Chief Scheduler: The chief scheduler consists of four components, namely—

(a) Task information assortment—This is responsible for collecting all information about the requested task from each cloud.
(b) Resource manager—This module is responsible for collecting resource-related information from the various interconnected clouds.
(c) Priority generator—This module is responsible for generating the priority list of all requested tasks based on the priority function.
(d) Map handler—Based on the priority list generated, the tasks are mapped to each resources.

Cloud Server. Every geographically distributed cloud consists of the following modules:

(a) Task collector—Information about every task request appearing are stored in task collector queue.
(b) Status informer—This stores the detailed information about the status of cloud like its available memory, the services it provides, cost of each service.

3 Proposed Priority-Based Task Scheduling Approach

Every cloud user is registered to its home cloud. Task requests are sent to each home cloud. At every time interval *t,* the chief scheduler collects the requested task from each home cloud and resources are assigned based on the generated priority which considers three parameters—(a) the cost of executing the task in a VM, (b) the network condition and (c) the power required to execute the task.

(a) **Cost incurred for task execution**

This parameter calculates the cost for task execution for each and every task that appears in the multi-cloud system on each cloud. It is calculated using Eq. (1)

$$Ct_{i,j,l} = \left(\frac{TL_i}{US_{j,k}} * PUC_{j,k}\right) \tag{1}$$

where $Ct_{i,l,j}$ is cost for ith task execution in lth VM of jth cloud, TL_i is the length of ith task, $US_{j,k}$ is the unit size for service type k in jth cloud and $PUC_{j,k}$ is the per unit cost for service type k in jth cloud.

(b) **Network condition**

The network condition between the cloud user and every cloud helps in determining the time required for completion of the requested task. An efficient scheduler always aims to successfully execute the task. This is because to deliver a good QoS; it must be ensured the requested task completes before reaching its deadline. The completion time for the requested task of the proposed system model is calculated using Eq. (2)

$$FT_{i,j,l} = T1_{j,l} + T2_{i,j,l} + Comm(CU)_{i,j} + \alpha * Comm(CC)_{i,j,z} + Comm(CCM)_{i,j} \tag{2}$$

where $FT_{i,j,l}$ is the finish time of ith task's successful execution on lth VM of jth cloud. $T1_{j,l}$ is the expected completion time of current load of lth VM of jth cloud. $T2_{i,j,l}$ is the expected completion time of ith task that is going to be executed in lth VM of jth cloud. $Comm(CU)_{i,j}$ is the communication time between user of ith task and its home cloud j. $Comm(CC)_{i,j,z}$ is the communication time to exchange data of ith task between jth home cloud and zth destination cloud where it will be executed. A flag parameter α is used to determine if the application is executed in home cloud or not. When a task is going to be executed in home cloud then the value of α is set to 0. Otherwise, if the task is going to be executed in any other destination cloud except the home cloud then the value of α is set to 1. $Comm(CCM)_{i,j}$ is the communication time to exchange metadata of ith task by jth cloud with the chief scheduler.

The time required by a VM to complete its queued task is calculated using Eqs. (3) and (4) that calculates the expected completion time of ith task.

$$T1_{j,l} = \frac{VM\_Load_{j,l}}{VM\_MIPS_{j,l}} \tag{3}$$

where $VM\_Load_{j,l}$ is the current load of lth VM of jth cloud. $VM\_MIPS_{j,l}$ is the MIPS value of lth VM of jth cloud.

$$T2_{i,j,l} = \frac{Task\_Length_i}{VM\_MIPS_{j,l}} \tag{4}$$

where $Task\_Length_i$ denotes the length of that task.

In the multi-cloud architecture, three communications are required. The first communication is time required to communicate with user and its home cloud. Communication time period between user and home cloud is calculated as per Eq. (5).

$$Comm(CU)_{i,j} = \frac{Length_i}{UB_{CU_j}} + \frac{DownloadableTaskLength_i}{DB_{CU_j}} + \frac{Dist_{CU_{i,J}}}{Propagation\ Speed} \tag{5}$$

$Length_i$ is the length of ith task; UB_{CU_j} is the upload bandwidth between user and jth cloud. Similarly, $DownloadableTaskLength_i$ is the length of the result of ith task. DB_{CU_j} is the download bandwidth between user and jth cloud. $Dist_{CU_{i,J}}$ is the geographical distance between user of ith task and jth cloud. *PropagationSpeed* is the speed of light.

The second communication is time required to send and receive data from home cloud to any other destination cloud. Communication between home cloud and other destination cloud is calculated in Eq. (6).

$$Comm(CC)_{i,j,z} = \frac{Length_i}{UB_{CC_{j,z}}} + \frac{DownloadableTaskLength_i}{DB_{CC_{j,z}}} + \frac{Dist_{CC_{j,z}}}{Propagation\ Speed} \tag{6}$$

$UB_{CC_{j,z}}$ is the upload bandwidth between jth and zth cloud. $DB_{CC_{j,z}}$ is the download bandwidth between jth and zth cloud. $Dist_{CC_{j,z}}$ is the geographical distance between jth and zth cloud.

The last communication occurs between the chief scheduler and home cloud. Equation (7) calculates the communication time required for metadata exchange between home cloud and chief scheduler.

$$Comm(CCM)_{i,j} = \frac{MetaLength_i}{UB_{CCM_j}} + \frac{MapLength_i}{DB_{CCM_j}} + \frac{Dist_{CCM_j}}{Propagation\ Speed} \tag{7}$$

$MetaLength_i$ is the length of metadata for ith task. UB_{CCM_j} is the upload bandwidth between jth cloud and chief scheduler. $MapLength_i$ is the length of the mapped tuple for ith task. DB_{CCM_j} is the download bandwidth between jth cloud and chief scheduler. $Dist_{CCM_j}$ is the geographical distance between jth cloud and chief scheduler.

(c) **Power required to execute the task**

Total power consumption is also very crucial in determining the priority of task, and it is calculated using Eq. (8).

$$Power_{i,j,l} = P\_Ex_{i,j,l} + P\_COMM\,(CU)_{i,j} + \alpha * P\_COMM\,(CC)_{i,j,z} + P\_COMM\,(CCM)_{i,j} \tag{8}$$

where $Power_{i,j,l}$ is the total power consumption by lth VM of jth cloud for successful execution of ith task. $P\_EX_{i,j,l}$ is the execution power consumption by lth VM of jth cloud for execution of ith task. $P\_COMM\,(CU)_{i,j}$ is the power consumption between user of ith task and jth cloud to send task data in cloud and receives the executed result. $P\_COMM\,(CC)_{j,z}$ is the communication power required between jth cloud and zth destination cloud to send task data to zth destination cloud and receives executed result for ith task and the value of α is same as in Eq. (2). When a task is going to be executed in home cloud then it values 0, otherwise 1. $P\_COMM\,(CCM)_{i,j}$ is the power consumed by jth cloud to communicate with the chief scheduler to send ith task's metadata to it and receives mapped information from there for that task.

The power required to execute a task on a certain VM is calculated using Eq. (9).

$$P\_Ex_{i,j,l} = P\_EXEC_{j,l} * \frac{TL_i}{VM\_MIPS_{j,l}} \tag{9}$$

where $P\_EXEC_{j,l}$ is the power consumption by the lth VM during execution in unit time.

Different communication power is also required for uploading and downloading the data. Equation (10) denotes the power consumed by a user to communicate with the home cloud.

$$P\_COMM\,(CU)_{i,j} = \frac{Length_i}{UB_{CU_j}} * P\_RC_{CU_{i,j}} + \frac{DownloadableTaskLength_i}{DB_{CU_j}} * P\_TS_{CU_{i,J}} \tag{10}$$

where $P\_RC_{CU_{i,j}}$ is the power consumption to upload task in jth cloud by user of ith task in unit time. $P\_TS_{CU_{i,j}}$ is the power consumption to download the executed task result from jth cloud by user of ith task in unit time.

Power consumed for communication between home cloud and any other destination cloud is calculated using Eq. (11).

$$P\_COMM\,(CC)_{i,j,z} = \frac{Length_i}{UB_{CC_{j,z}}} * P\_TS_{CC_{j,z}} + \frac{DownloadableTaskLength_i}{DB_{CC_{j,z}}} * P\_RC_{CC_{j,z}} \tag{11}$$

where $P\_TS_{CC_{j,z}}$ is the power consumption to transmit task from jth cloud to zth cloud in unit time. $P\_RC_{CC_{i,z}}$ is the power consumption to receive the executed task result from zth cloud to jth cloud in unit time.

After that, power consumption to communicate with home cloud and chief scheduler is required which is calculated using Eq. (12).

$$P\_COMM\,(CCM)_{i,j} = \frac{MetaLength_i}{UB_{CCM_j}} * P\_TS_{CCM_j} + \frac{MapLength_i}{DB_{CC_j}} * P\_RC_{CCM_j} \tag{12}$$

where $P\_TS_{CCM_j}$ is the power consumption to transmit metadata of task from jth cloud to chief scheduler in unit time. $P\_RC_{CCM_j}$ is the power consumed to receive the mapped information of the task from chief scheduler to jth cloud in unit time.

Table 1 Various weight factors values

User type	Task type	User criteria	a	b	c
Primary	Real time	Low energy	0.09	0.1	0.14
Primary	Real time	Less time	0.09	0.14	0.1
Primary	Real time	Both	0.09	0.12	0.12
Primary	Non-real time	Low energy	0.11	0.07	0.09
Primary	Non-real time	Less time	0.11	0.09	0.7
Primary	Non-real time	Both	0.11	0.08	0.08
Secondary	Real time	Low energy	0.06	0.07	0.09
Secondary	Real time	Less time	0.06	0.09	0.07
Secondary	Real time	Both	0.06	0.08	0.08
Secondary	Non-real time	Low energy	0.08	0.04	0.06
Secondary	Non-real time	Less time	0.08	0.06	0.04
Secondary	Non-real time	Both	0.08	0.05	0.05

3.1 Priority Function Model for the Proposed Approach

Cloud users are divided into two categories—primary users and secondary users. Primary users of a cloud are the users who are registered to CSP, whereas secondary users are floating, unregistered users of a CSP. So, primary users receive uninterrupted service from cloud and are given more priority compared to secondary users.

Priority function model is a hybrid model which considers cost of task execution, network condition and power required for task execution as denoted in Eq. (13).

$$Prt_{i,j,l} = Deadline_i * \frac{\left[\left(a * Ct_{i,j,l}\right) + \left(b * FT_{i,j,l}\right) + \left(c * Power_{i,j,l}\right)\right]}{ExpertiseFactor_{j,k}} \quad (13)$$

where $Prt_{i,j,l}$ is the priority value of ith task on lth VM of jth CSP. a, b and c are weight value for cost, time and power, respectively. $ExpertiseFactor_j$, k is the expertise factor for jth cloud providing kth type of service. These weight factors make the priority model dynamic. Table 1 shows some weight values for cost, time and power. The proposed algorithm for task scheduling is given in Table 2.

4 Results and Discussion

The algorithm is simulated in MATLAB 2014b, and the results are compared with cost-deadline-based task scheduling in cloud computing [11] and energy aware scheduling of HPC tasks in decentralized cloud systems [17].

Table 2 Proposed task scheduling approach

Input consideration: Set of CSP. **Output consideration: (Task,VM,Status).**

Procedure:
Module – Task Information Collector
Step 1: Begin
Step 2:TT(i=1...n) = Collect task info from each cloud connected to the chief scheduler;
// TT contains columns like Task Name, user type, length, deadline, task type, service type and location. All information of a task is stored in appropriate column of the table.
Module – ResourceManager
Step 3: Cloud_Table (j=1...m) = Collect all information from each cloud.
//This Cloud_Table contains columns like Cloud_Name, Service Type, Location, Unit Size, Per Unit Cost and ExpertiseFactor.
End for
Step 4: For each j in cloud do
Step 5: VMT (l=1…p) = Collect all Information from VM_Info_Table.
Step 6: VM_MI = VM_MIPS * GTP;
// This VMT table contains columns like VM_Name, VM_MIPS, VM_MI, VM_TaskList, VM_Load,VM_Status, P_EXEC.
Done;
Module – Priority Generation
Step 7: For each j in Cloud_Table do
Step 8: For each i in TT do
Step 9: Calculate priority of each task using equation (13)
Done;
Done;
Module – Map Handler
Step 10: Priority Generator sends the priority list to map handler
Step 11: It assigns maps task to its corresponding VM
Step 12: End

4.1 *Profit Earned from Tasks*

In any sort of business, profit is one the prime aim. As CSP is also doing business, it also tries to maximize its profit. Figure 2 shows the comparison between [11, 17], and the proposed method showing an improvement of 32.14%. Profit is calculated with the help of Eq. (14).

$$Total\ Profit = \sum_{\substack{i=1 \\ j=1}}^{\substack{m \\ n}} Ch_{i,j,l} - \sum_{\substack{i=1 \\ j=1}}^{\substack{m \\ n}} Ct_{i,j,l} \tag{14}$$

where $Ch_{i,j}$ is the charge of ith task's execution on lth VM of jth cloud taken from user.

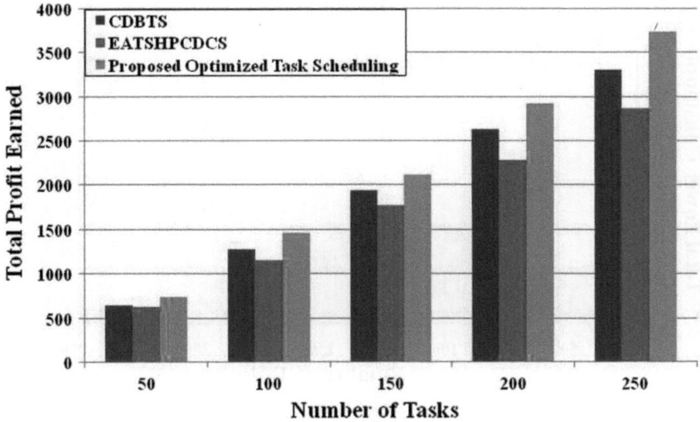

Fig. 2 Comparison of profit in multiple instances

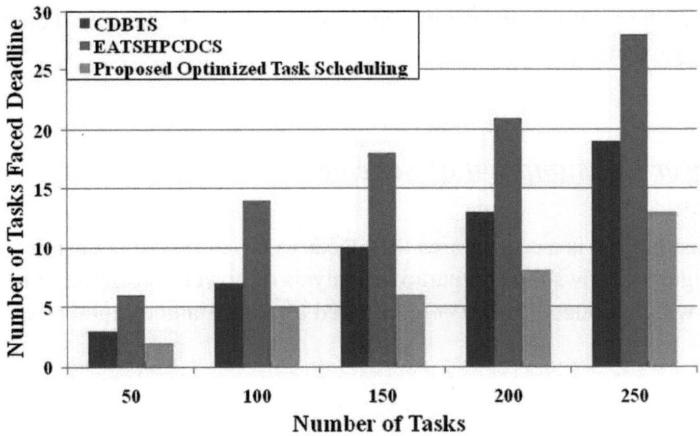

Fig. 3 Comparison of deadline faced by tasks in multiple instances

4.2 Deadline Faced by Tasks

To maintain good QoS, the system must ensure that task is completed before meeting deadline. When there are n numbers of tasks then number of deadline misses is determined by Eq. (15). Figure 3 shows the count of deadline faced by allotted tasks in multiple instances. There is a reduction of 31.57% tasks that faced deadline for the proposed approach

$$No.\,of\ Deadline\ misses = \left(\sum_{i=1}^{n} IT - \sum_{i=1}^{n} Tse\right) \tag{15}$$

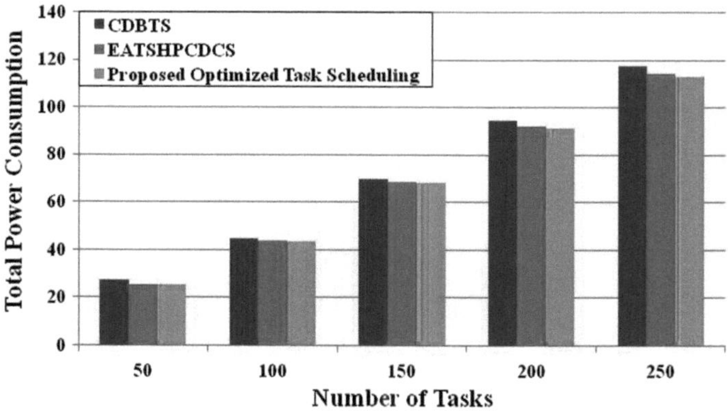

Fig. 4 Power consumed by VMs in multiple instances

where *IT* is the total count of incoming tasks and *Tse* is the total count of tasks successfully executed.

4.3 Power Consumption of Schemes

Power consumption is a critical area that needs to be addressed to design an efficient system. Figure 4 shows the comparative analysis of power consumption by different resources which reduced by 5.08% compared to other standard methods.

5 Conclusion

In cloud computing, resources are allocated virtually. This allotment is carried out by task scheduling algorithms. In real-world scenario, there are multiple available clouds with varying network configuration and storage space. So, for deploying an efficient task scheduler in real world requires the scheduler to consider multiple parameters while taking task scheduling decision. In this work, a well-structured hybrid task scheduling algorithm is designed that works in multi-cloud environment where multiple clouds are interconnected with each other and are managed by a chief scheduler. The proposed approach aims to provide services to users with low cost, in reduced time, thus decreasing the number of task facing deadline, and also takes into account the amount of power required to execute a task. Simulation of the proposed approach shows that the profit earned by CSP using our approach is 32.14% more compared to other established approaches. At the same time, tasks facing deadline are reduced by 31.57% using this approach.

Acknowledgements The authors are grateful to the TEQIP—III program of Maulana Abul Kalam Azad University of Technology, A World Bank project.

References

1. Mathew, T., Sekaran, K.C., Jose, J.: Study and analysis of various task scheduling algorithms in the cloud computing environment. In: International Conference in Advances in Computing, Communications and Informatics (ICACCI), pp. 658–664. IEEE, New Delhi, India (2014)
2. Roy, A., Midya, S., Majumder, K., Phadikar, S., Dasgupta, A.: Optimized secondary user selection for quality of service enhancement of two-tier multi-user cognitive radio network: a game theoretic approach. Comput. Netw. **123**, 1–18 (2017)
3. Mell, P., Grance, T.: The NIST definition of cloud computing. Natl. Inst. Stand. Technol. **53**(6), 50 (2009)
4. Midya, S., Roy, A., Majumder, K., Phadikar, S.: Multi-objective optimization technique for resource allocation and task scheduling in vehicular cloud architecture: a hybrid adaptive nature inspired approach. J. Netw. Comput. Appl. **103**, 58–84 (2018)
5. Armbrust, M., Fox, A., Griffith, R., Joseph, A.D., Katz, R., Konwinski, A., Zaharia, M.: A view of cloud computing. Commun. ACM **53**(4), 50–58 (2010)
6. Nagadevi, S., Satyapriya, K., Malathy, D.: A survey on economic cloud schedulers for optimized task scheduling. Int. J. Adv. Eng. Technol. **5**, 58–62 (2013)
7. Hazra, D., Roy, A., Midya, S., Majumder, K.: Distributed task scheduling in cloud platform: a survey. In: Smart Computing and Informatics, pp. 183–191. Springer, Singapore (2018)
8. Fang, Y., Wang, F., Ge, J.: A task scheduling algorithm based on load balancing in cloud computing. In: International Conference on Web Information Systems and Mining, pp. 271–277. Springer, Berlin, Heidelberg (2010)
9. Cao, Q., Wei, Z.B., Gong, W.M.: An optimized algorithm for task scheduling based on activity based costing in cloud computing. In: 3rd International Conference on Bioinformatics and Biomedical Engineering, pp. 1–3. IEEE (2009)
10. Garg, S., Govil, K., Singh, B.: Costbased task scheduling algorithm in cloud computing. Int. J. Res. Eng. Technol. **3**, 59–61 (2014)
11. Sidhu, H.S.: Cost-deadline based task scheduling in cloud computing. In: Second International Conference on Advances in Computing and Communication Engineering (ICACCE), pp. 273–279. IEEE (2015)
12. Van den Bossche, R., Vanmechelen, K., Broeckhove, J.: Cost-optimal scheduling in hybrid IAAS clouds for deadline constrained workloads. In: 3rd International Conference on Cloud Computing (CLOUD), pp. 228–235. IEEE (2010)
13. Rodriguez, M.A., Buyya, R.: Deadline based resource provisioning and scheduling algorithm for scientific workflows on clouds. IEEE Trans. Cloud Comput. **2**(2), 222–235 (2014)
14. Selvarani, S., Sadhasivam, G.S.: Improved cost-based algorithm for task scheduling in cloud computing. In: International Conference on Computational Intelligence and Computing Research (ICCIC), pp. 1–5. IEEE (2010)
15. Cheng, C., Li, J., Wang, Y.: An energy-saving task scheduling strategy based on vacation queuing theory in cloud computing. Tsinghua Sci. Technol. **20**(1), 28–39 (2015)
16. Alahmadi, A., Che, D., Khaleel, M., Zhu, M.M., Ghodous, P.: An innovative energy-aware cloud task scheduling framework. In: 8th IEEE International Conference on Cloud Computing (ICCC), pp. 493–500 (2015)
17. Alsughayyir, A., Erlebach, T.: Energy aware scheduling of HPC tasks in decentralized cloud systems. In: 24th Euromicro International Conference on Parallel, Distributed, and Network-Based Processing (PDP), pp. 617–621. IEEE (2016)

An Enhanced Post-migration Algorithm for Dynamic Load Balancing in Cloud Computing Environment

Anmol Bhandari and Kiranbir Kaur

Abstract Load balancing has been considered as one of the most important aspect of cloud computing in recent times. An increase in the number of users around the world has resulted in a large number of requests at a rapid rate. Researchers around the world have designed many algorithms to carry out the client's request at distributed cloud servers. Based on this, the cloud computing paradigm will automate configuration of servers in order to achieve efficient load balancing. Henceforth, selection of virtual machines has to be scheduled efficiently based on the load balancing algorithm. In this paper, a load balancing algorithm is proposed based on the availability of the VM. Specifically, the Availibility Index (AI) value is evaluated for every VM over a given period of time, and therefore a task is assigned to that machine based on the AI value. In order to validate the proposed model, it is compared with three famous load balancing algorithms are compared, namely Round Robin, Throttled and Active Monitoring. The performance of each algorithm was evaluated using CloudAnalyst. Simulation results show that the proposed algorithm is more efficient in load balancing over virtual machines as compared to other algorithms.

Keywords Cloud computing · Modified throttled · Virtual machine
Throttled algorithm · Round-robin algorithm · Active monitoring

1 Introduction

Cloud computing has been an innovative technology in recent times [1–3]. With the developments of cloud computing platform, resources in the form storage and computation are provided as a service and requirement of the user and guarantee them

A. Bhandari (✉) · K. Kaur
Department of Computer Engineering and Technology, Guru Nanak Dev University, Amritsar, India
e-mail: anmolbhandari60@gmail.com

K. Kaur
e-mail: kiran.dcse@gndu.ac.in

M. Chakraborty et al. (eds.), *Proceedings of International Ethical Hacking Conference 2018*, Advances in Intelligent Systems and Computing 811,
https://doi.org/10.1007/978-981-13-1544-2_6

by sticking to the service-level agreement (SLA) [4]. However, because of resource sharing, and the heterogeneous requirement of the users and platform independence, it results in inefficient utilization of resources, if the resources cannot be distributed properly. Moreover, another important problem for the cloud computing platform is balancing the load over various servers dynamically, for avoiding hot spot and enhancing resource utilization [5, 6]. Henceforth, dynamic and efficient allocation of cloud resources and meeting the requirements of the users become the open research problem (Fig. 1) [7].

In the recent time, development of virtualization technology has provided an efficient way of managing various dynamic resource requirements on the cloud computing platform [8]. Specifically, the job of the diverse user requirements and platform independence can be resolved efficiently by confining the required work in virtual machines and mapping it to each physically connected server, along with SLA [9, 10]. In addition to this, virtualization is capable of carrying out remapping between the virtual machine (VM) and physically allocated resources depending on the dynamic load change, so that optimal load balance can be effectively achieved. Because of these reasons, virtualization technology is being used in cloud computing with high efficacy. However, because of dynamic heterogeneity of resource requirement by users over cloud computing platform, virtual machines are required to readapt itself to the environment of cloud computing so that they can attain their optimal perfor-

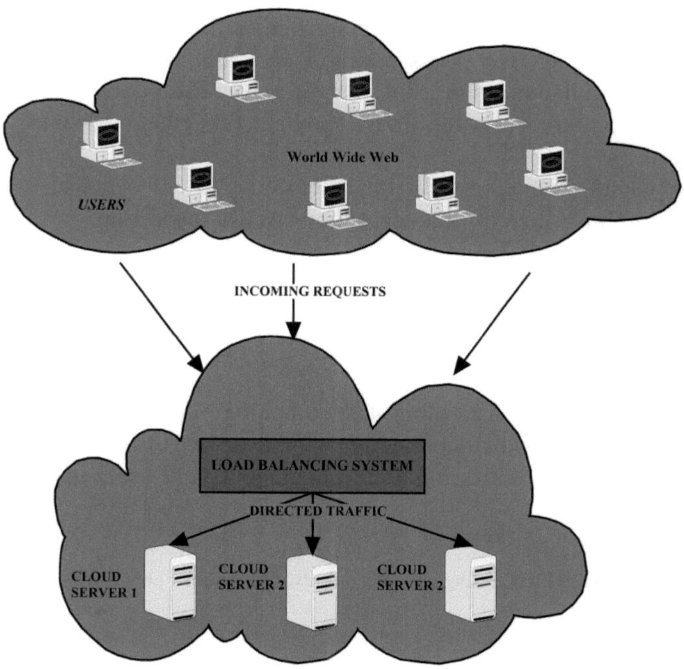

Fig. 1 Load balancing in cloud computing environment

mance by utilizing their allotted service and resource allocation to the fullest [11, 12]. Moreover, for the purpose of enhancing resource utilization, resources must be efficiently allocated and load balancing is to be guaranteed [13, 14]. Henceforth, in order to schedule VM resources for the realization of load balancing in cloud computing, enhancement of resource utilization becomes an important area of research [15]. However, in the current scenario of dynamic resource allocation over VM, users' resource requirements and fitness of the host machine are not considered during task transmission. Specifically, the fitness of the destination host in terms of available resources, such as RAM, and availability processors is not evaluated. Moreover, the time requirement with which host can complete the task is also not considered [5, 16]. Task transmission or migration consumes less time through the existing technique; however, the post-migration process is not considered.

Henceforth, in this paper, an enhanced load balancing technique is presented based on VM availability. Specifically, availability index is evaluated (AI), which is defined as the number of tasks assigned to a particular machine in a given span of time. The VM with less number of tasks assigned will be selected for the allocation of a new job. For the purpose of validation, three challenging post-migration algorithms named as round-robin (RR), throttled (TT), and active monitoring (AM) for dynamic load balancing on the cloud computing environment are considered. The major objectives for comparing these algorithms are (a) to minimize overall execution time associated with migration, (b) to enhance the reliability by ensuring check pointing in case of task failure, and (c) cost in terms of overhead required to be minimized.

The remaining paper is organized in the following sections. Section 2 provides an overview of the various load balancing algorithms. Section 3 reviews some of the important contributions in load balancing. The proposed model is discussed in Sect. 4. Experimental simulations are presented in Sect. 5. Finally, Sect. 7 concludes the paper with important discussion contribution in the current scenario.

2 Motivational Analysis

This section provides an overview of three post-migration algorithms that are considered for the current research.

2.1 Round-Robin Algorithm for Load Balancing

Round-robin algorithm [10, 17] for load balancing is counted as one of the most feasible techniques for dispensing user requests among a group of geographically distributed cloud servers. Starting from the first, it searches the list of servers in the group of servers and forwards a client request to every server turn-by-turn (Fig. 2). When the list of servers terminates, the load scheduler returns back and begins to

Fig. 2 Round-robin mechanism

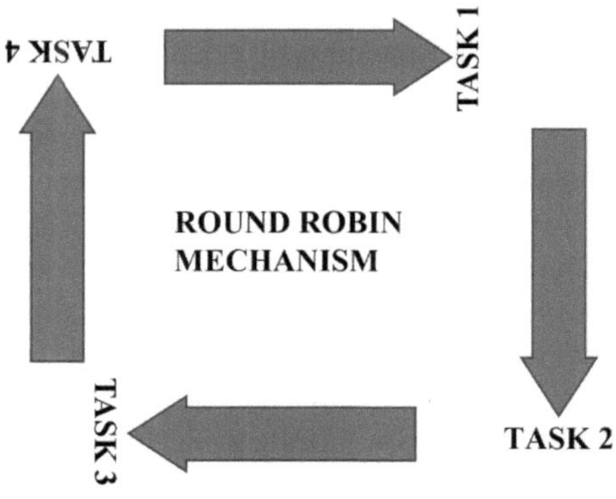

search the list again. In other words, it sends the next request to the first list of servers, then continuous to the second list, and so on.

The main advantage of this type of load balancing algorithm is that it is fairly easier to deploy. However, as part of its limitation, this algorithm does not always provide precise and effective distribution of user requests, because most of the load balancers used in this algorithm make an assumption that all cloud servers are similar in configuration: presently active, presently handling the similar load, and with the similar capability to store and compute. The advancement to round-robin algorithm is the two variants that are considered as additional factors which result in efficient load balancing.

Weighted Round-Robin (WRR)—WRR is a modified version of the round-robin scheduling algorithm in which a weight is assigned to each of the available cloud server nodes. The weight is a numerical number associated with a server based on a predefined criterion. The most acceptable criterion is the load-handling capacity of the server in real time. In other words, the highest number is allotted a server if it can receive a large number of user requests and vice versa. For instance, if a weight 5 is associated with a server node, then it is capable of receiving more requests in comparison with the other server which is assigned weight 2.

Dynamic Round-Robin (DRR)—It is another variant of the round-robin strategy in which a number is assigned dynamically to the server based on the real-time configuration of the server and its corresponding load-handling capacity. In other words, the associated weight is not static and is changed instantly if the server does not have sufficient resources.

2.2 *Active Monitoring-Based Load Balancing Algorithm*

Active monitoring-based load balancing algorithm [18] is another important load balancing algorithm that is available. The distinguishing feature of this algorithm

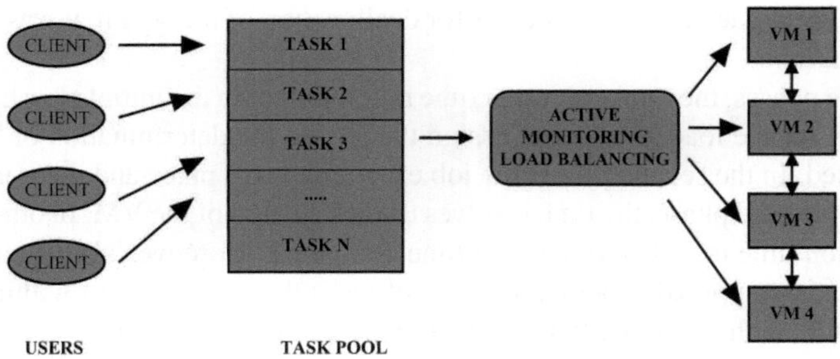

Fig. 3 Active monitoring-based load balancing technique

is that it maintains an information table which stores information about each of the VMs and the corresponding client generated data requests that are currently handled by it. The same is shown in Fig. 3. Moreover, when a user request arrives at the load balancer, it identifies a minimally loaded VM from the table. If multiple VMs are available, then the first selection is assigned the task. In addition to this, VM id is determined by the load balancer to keep the track of the server that has been assigned the current task. This information is again stored, and the information table is updated. The identified VM is returned to the data center controller, and an alert is sent for this new allocation.

2.3 Throttled Load Balancing Algorithm

Throttled load balancing algorithm [19, 20] is based on the conceptual framework of VM. In this algorithm, the client request that is generated is forwarded to the load balancer to determine the optimal VM which can handle the current request effectively and efficiently. Moreover, in this algorithm, various user operations are supported which must be considered during service provisioning as shown in Fig. 4.

In other words, an index table (IT) of various available VMs as well as their corresponding states is maintained by the load balancer, namely available or busy. The client or server initializes the load balancer by generating an appropriate request to the data center in order to find the suitable VM for service provisioning.

The data center forms a query of various requests arrived at the server for the provisioning of the VM. An optimal scan is performed by the load balancer from the top of IT list till an appropriate VM is located or IT is scanned completely. When the suitable VM is located, the data center assigns the job to the VM which is recognized by its VM id. Moreover, the load balancer is acknowledged by the data center upon determination of the VM. However, if the VM is not found, then -1 is returned to load balancer. The data center maintains the queue of the various requests within it, whenever a new request arrives. When the VM completes the assigned task, the request is forwarded to the data center for deallocation of the VM that was linked to the task earlier.

In three phases, the total execution time is computed. In the initial phase, VM will be waiting for the load balancer to assign the job, so the determination of VM time is computed. In the second phase, the job execution takes place and the total time is added. In the final phase, the time involves the destruction of the VM. In other words, deallocation time is added to the total time computed. Moreover, the throughput of the system is computed as a total number of tasks that are executed within a given span of time, without taking time for VM creation and destruction.

3 Related Work

In this section, we briefly summarize some of the important load balancing algorithms that have been developed by various researchers in the cloud computing environment. The main area of interest is the assignment of all incoming jobs among the available VM with minimal response time. As defined by most of the researchers, a process of ideally utilizing the resources by assigning the total job load to the individual systems of the unified system results in minimization of the response time of each job known as "load balancing."

Mondal et al. [21] have presented a stochastic hill-climbing algorithm for load balancing in the cloud computing environment. Authors have discussed stochastic hill climbing as one of many artificial intelligence approaches for dynamic load balancing. In the paper, a variant of stochastic model is discussed in which a random value is chosen which is mapped to a set of other values by making only minor changes to the original value. Moreover, the best element of the set is made as the next job. Authors have compared the results with conventional algorithms like round-robin and FCFS algorithms, and efficient results for the proposed model were achieved.

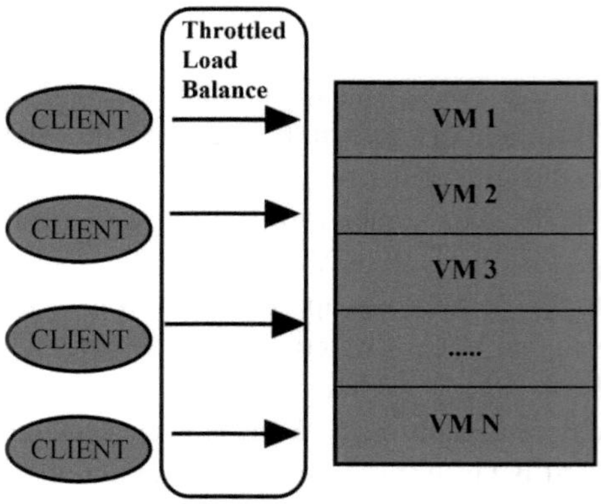

Fig. 4 Throttled load balancing algorithm

Dhinesh Babua and Venkata Krishna [22] have proposed an algorithm named as honey bee behavior-based load balancing (HBB-LB), which was focused on achieving efficient load balance across various VMs for maximizing the throughput of a large number of job operations. The presented algorithm discussed the preferences of job operations on the specific host machines in such a way that it minimizes the total amount of waiting time for the different tasks. In addition, to determine the validity of the proposed algorithm, a comparison is made with other prevailing load balancing and scheduling algorithms. The experimental results demonstrated that the proposed algorithm was efficient and more optimal.

Chhabra and Singh [14], proposed an optimal physical host with effective load balancing (OPH-LB) framework in which the service request of a client in infrastructure as a service (IaaS) architecture is modeled, by considering heterogeneous VMs. Initially, the OPH-LB approach involves the filtering of the identified hosts that satisfies the job requirement. Based on these qualified sets, a probabilistic model is applied, which is used to find the optimal host depending on its computing capability and performance. Furthermore, CloudSim simulation tool is used to analyze the performance of the proposed model. The results were compared with the state-of-the-art load balancing approaches, which showed that the proposed model was able to enhance the throughput, reduce the failure rate, and optimize cloud data centers.

Rahman et al. in [10] have presented an idea of load balancer as a service (LBaaS) in a cloud environment. In initial stages, load balancing problem, need of load balancing, and required characteristics in cloud computing have been focused by the authors and finally, and the load balancer as a service has been focused in cloud computing. The effectiveness of model was demonstrated in the form of various filed trials where effective results were registered.

Sharma et al. [23] presented different concepts of load balancing considering various throughput parameters. The presented research work provided a path to design a new algorithm by analyzing the behavior of conventional algorithms, on the basis of a specific parameter. In the end, authors have concluded that the behavior of static algorithms is feasible to understand as compared to other dynamic load balancing algorithms.

Sran and Kaur in [24] developed an efficient load balancer algorithm that on the basis of security thresholds performs the flow control of data traffic, for both static and dynamic environments, subject to the availability of VMs and network bandwidth. The authors have studied the conventional load balancing algorithms, such as round-robin, throttled, equally spread and biased random (ESBR), and have proposed a new algorithm which was able to suffice over the existing load balancing approach, by decrementing the overall waiting time and data processing time, and decreased the overall cost. Moreover, the presented algorithm provided data security in cloud server during the process of load balancing by indulging zero proof algorithm. In the proposed approach, the author utilizes VM migration policy in which on the basis of available resources, VMs are prioritized. Specifically, the algorithm checked if the utilization of CPU of a specific virtual machine (VM) was less than, equal to, or greater than 80%. Efficient results were obtained by the authors upon simulation.

4 Proposed Model of Load Balancing

The proposed model of the modified throttled algorithm is shown in Fig. 5. It mainly focuses on the number of incoming jobs that are currently been assigned to a particular virtual machine. In the proposed model, a modified throttled algorithm is presented, which is an extension of the conventional throttled load balancing technique. In the proposed technique, a table is maintained for every server which indicates the state of the machine in real time. The state of the machine is evaluated in terms of probability of availability of a particular VM, i.e., the availability index (AI). The AI value is defined as the number of tasks that have been directed to the particular VM in a given span of time. More value of AI indicates less availability. However, the minimal value of AI indicates more availability. Based on this AI value, each job request generated by the user is directed based on the minimal AI value. The detailed algorithm steps are shown ahead.

5 Experimental Simulations

This section provides a comprehensive overview of the various experimental implementations that was performed based on various load balancing policies of a cloud computing environment. The simulation was performed using a cloud computing

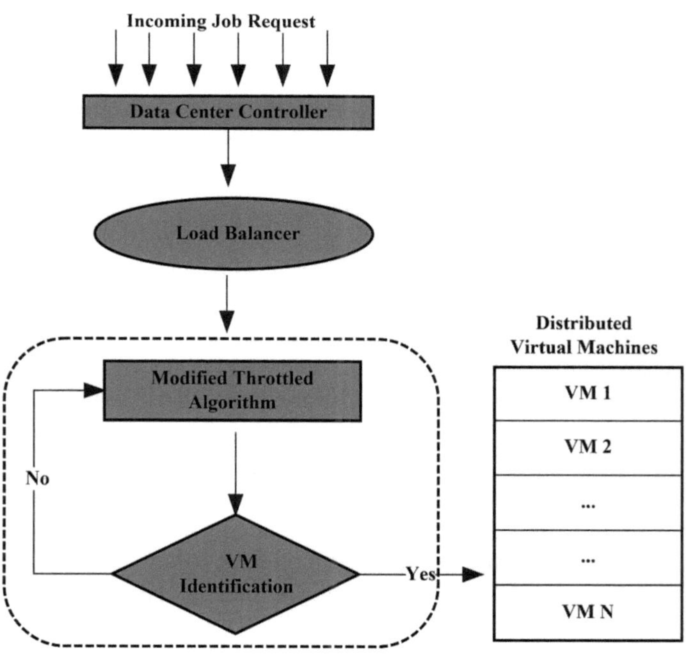

Fig. 5 Proposed modified throttled algorithm

Algorithm: Proposed Load Balancing Algorithm
Input: No of incoming jobs a1, a2…an Corresponding Availability Index VM: b1, b2…bn Output: All incoming jobs a1, a2…an are allocated one by one to the available VM based on b1, b2…bn Step 1: Proposed model maintains an index table of VMs and evaluates AI value for a given period of time for every VM. Step 2: Data Center at the cloud receives a new job request. Step 3: Data Center inquires the proposed load balancer algorithm for the next allocation. Step 4: Proposed Model starts with the VM with minimal AI value of VM. Step 5: The VM with minimal AI value is determined and its unique id is returned to the load balancer. Step 6: The proposed load balancer then directs the traffic to the corresponding VM

assessment tool, named as CloudAnalyst. CloudAnalyst [7] is most famous graphical user interface (GUI) based on CloudSim architecture. CloudSim is developed by the University of Melbourne that allows a computing-based modeling, simulation, and experimentation in an effective manner. Moreover, this tool allows multiple simulations with minimal change in the parameter for efficient assessment of the model. Moreover, CloudAnalyst allows users to set multiple data centers in the geographically distributed area so that simulations can be executed considering the practical aspects like network delay, latency, and throughput. In addition to this, parameters like number of users involved, number of processors, computing capability of the processor, its type, and storage can be easily altered by the users for effective simulations. Based on the simulations, the results can be formed graphically in terms of time and cost functions which are essential in the overall assessment of any algorithm.

5.1 Simulation Parameters

5.1.1 User Base

A user base model is a group of virtual users that in the simulating experiment is considered as a single unit. The main task of the user base is to spawn traffic for the overall simulation procedure. An individual user base may be made of thousands of users but is represented as a single unit, and its size is represented by the traffic that is generated in simultaneous bursts. It can be possible that a modeler represents a single user through user base, but in ideal conditions, a large number of users are represented by a user base for the efficiency of simulation. In the proposed simulation experiment, five user bases with a variable number of peak users for each user base are considered as shown in Fig. 6. Each user base is geographically distributed over five different regions. Moreover, each user base was configured to generate a data request of 60 per hour each of size 100 bytes per request. For service broker policy, dynamically reconfiguring router policy is adopted. A service broker is used to decide

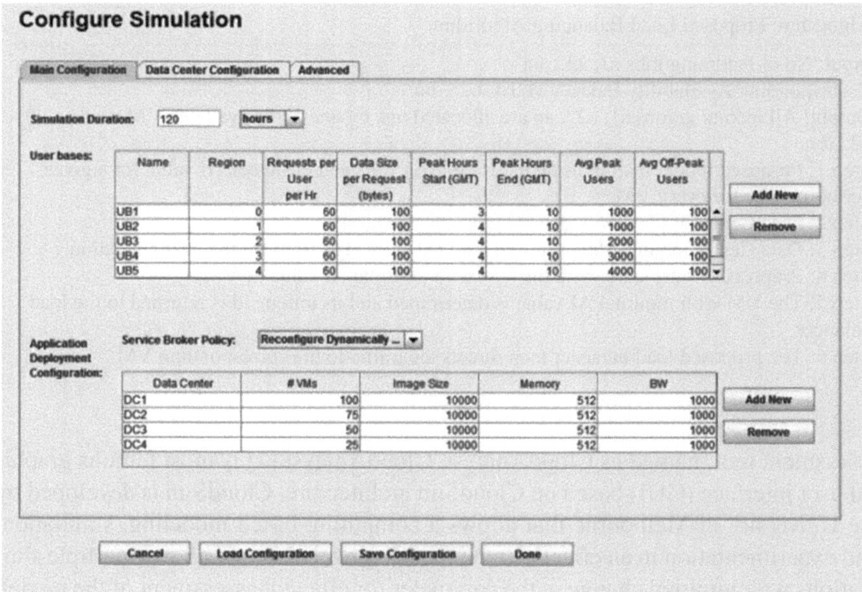

Fig. 6 User base configuration

which data center should be able to provide the service to the requests generated from each user base.

5.1.2 Data Centers

The data center is the most important component in the CloudAnalyst. Mapping of a single data center controller is done to a single CloudSim server simulator. Data center is capable of managing the data-related activities such as VM establishment, formation and termination of VM, and the requests of users that are gathered from user bases via the Internet are routed to the VMs. In the proposed study, four data centers are selected based on ×86 architecture running on Xen VMM of Linux operating system as shown in Fig. 6. Moreover, cost per VM/HR is remained same for all data centers in order to determine the overall average cost. Each data center is comprised of different hardware structures. For instance, DC with id 02 is shown in Fig. 7 with the corresponding configuration like memory, available bandwidth, and a number of processors.

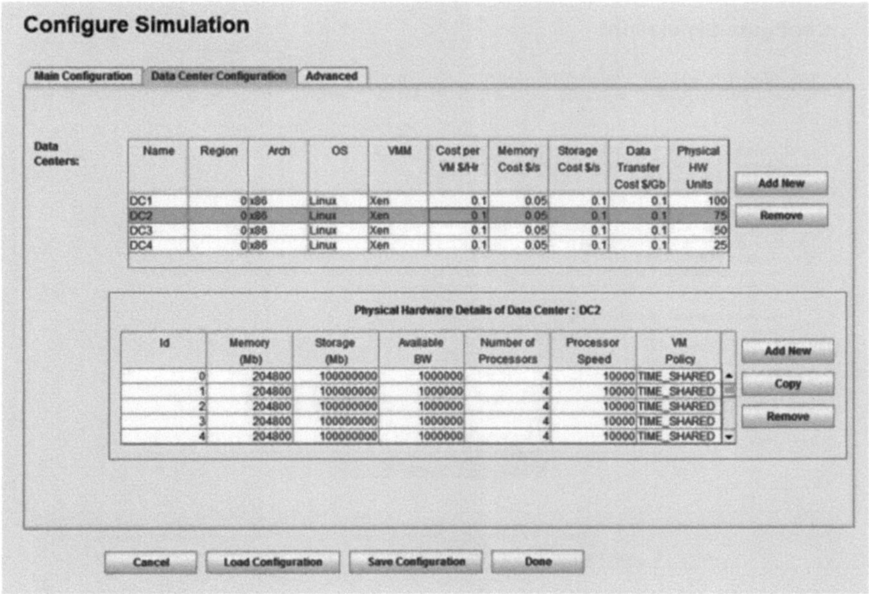

Fig. 7 Data center configuration

5.1.3 VM Load Balancer

The main responsibility of this component of CloudAnalyst is to assign the data load on various data centers according to the request that is generated by user bases. Various load balancing policies were selected once a time, and results are compared accordingly. The given policies included conventional round-robin algorithm, active monitoring, throttled, proposed modified throttled (Fig. 8).

5.1.4 Cloud App Service Broker

As mentioned in the earlier section, the main task of this component is to map the service broker for handling different data loads between the user bases and the data centers. The service broker is capable of adopting one among several routing algorithms, such as closest data center first, optimal response time first, and reconfiguring the load balance dynamically. The nearest one routes the data load to the geographically nearest data center based on network latency from the user database. On the other hand, when the efficiency of the data center is dropped down to a certain threshold, then dynamic reconfiguration is used, in which a load of that data center is again scattered equally among other available server nodes.

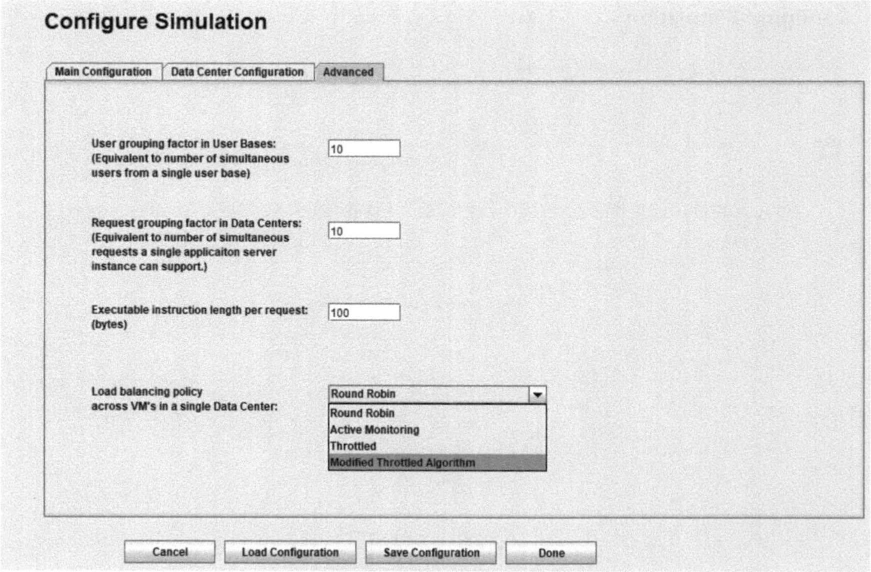

Fig. 8 VM load balancer module

Table 1 Average response time

User base	Round robin (MS)	Throttled	Active monitoring	Modified throttled
UB1	50.391	50.132	50.133	50.055
UB2	200.28	200.055	200.066	199.125
UB3	300.755	300.482	300.477	300.125
UB4	500.354	499.967	499.958	498.568
UB5	500.774	500.426	500.423	500.023
UB6	201.049	200.579	200.576	200.425

6 Results

After performing various simulations with variable parameters, results are computed and stored in Tables 1, 2, and 3. The predefined configuration for each load balancing policy has been used one after another. Depending upon this, the results are calculated for the estimating parameters like response time, request processing time, and a final cost function of providing the service as shown in Figs. 9, 10, and 11. Parameters such as average response time, data center service time, and total cost of different data centers have been taken into account for the analysis purpose.

Table 2 Average data center request servicing time

Data center	Round robin	Throttled	Active monitoring	Modified throttled
DC1	0.371	0.37	0.37	0.368
DC2	0.46	0.366	0.367	0.364
DC3	0.699	0.366	0.367	0.662
DC4	1.306	0.387	0.388	0.385

Table 3 Comparison of load balancing policies

Parameter	Round robin	Throttled	Active monitoring	Modified throttled
Overall response time	316.15	315.79	315.79	314.75
Data center processing time	0.71	0.37	0.37	0.36
Total cost	$4934.56	$4933.89	$4934.01	$4933.21

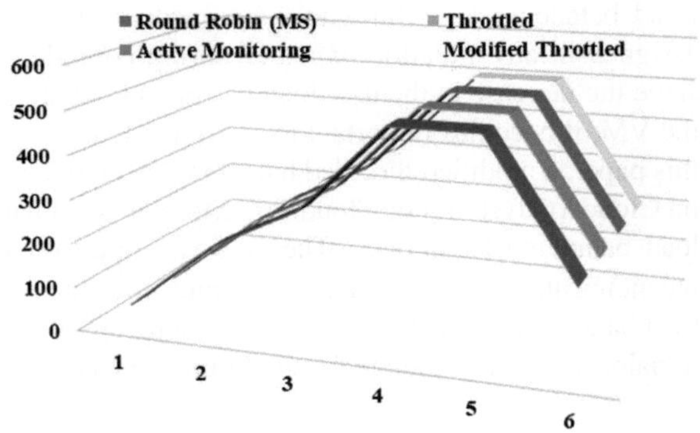

Fig. 9 Average response time results

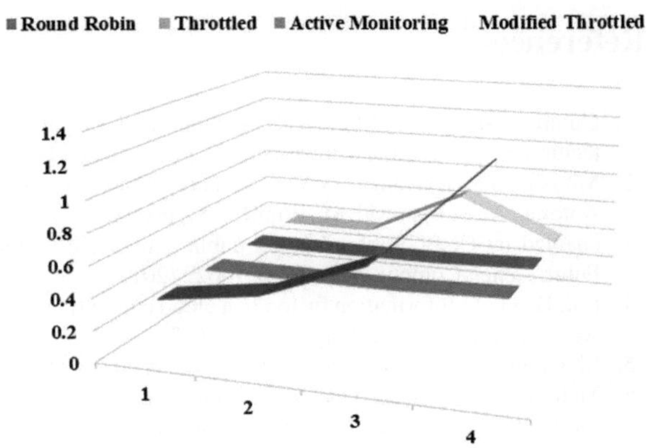

Fig. 10 Average data center request servicing time

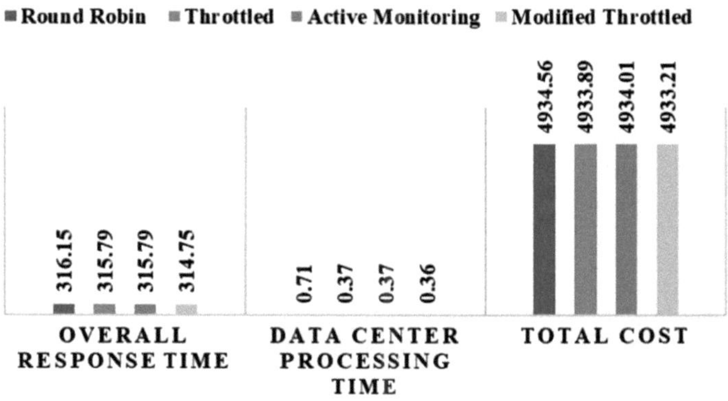

Fig. 11 Overall comparison of load balancing policies

7 Conclusion

Load balancing algorithms take into account the principle that workload can be assigned in any situation, whether during compile time or at runtime. Moreover, since the increase in the user load has put a load on the computational capability of the VM, it becomes important to distribute load to appropriate VM. Henceforth, in this paper, a modified throttled balancing technique is proposed and is implemented on CloudAnalyst tool of CloudSim. More, it is validated by comparing it with other load balancing techniques. The above comparison shows that the proposed load balancing algorithm is more effective and efficient as compared to other conventional load balancing algorithms. However, evaluation of appropriate load in real time still remains an open challenge for the future perspective.

References

1. Gubbi, J., Buyya, R., Marusic, S., Palaniswami, M.: Internet of Things (IoT): a vision, architectural elements, and future directions. Futur. Gener. Comput. Syst. **29**(7), 1645–1660 (2013)
2. Mirashe, S.P., Kalyankar, N.V.: Cloud computing. Commun. ACM **51**(7), 9 (2010)
3. Antonić, A., Marjanović, M., Pripužić, K., Podnar Žarko, I.: A mobile crowd sensing ecosystem enabled by CUPUS: cloud-based publish/subscribe middleware for the Internet of Things. Futur. Gener. Comput. Syst. **56**, 607–622 (2014)
4. Liu, B., et al.: Information fusion in a cloud computing era: a systems-level perspective. IEEE Aerosp. Electron. Syst. Mag. **29**(10), 16–24 (2014)
5. Moharana, S.S., Ramesh, R.D., Powar, D.: Analysis of load balancers in cloud computing
6. Mahmud, S., Iqbal, R., Doctor, F.: Cloud enabled data analytics and visualization framework for health-shocks prediction. Futur. Gener. Comput. Syst. (2015)
7. Wickremasinghe, B., Calheiros, R.N., Buyya, R.: CloudAnalyst: a CloudSim-based visual modeller for analysing cloud computing environments and applications. In: 2010 24th IEEE International Conference on Advanced Information Networking and Applications, pp. 446–452 (2010)

8. Hu, J., Gu, J., Sun, G., Zhao, T.: A scheduling strategy on load balancing of virtual machine resources in cloud computing environment. In: 2010 3rd International Symposium on Parallel Architectures, Algorithms and Programming, pp. 89–96 (2010)
9. Fang, Y., Wang, F., Ge, J.: A Task Scheduling Algorithm Based on Load Balancing in Cloud Computing, pp. 271–277. Springer, Berlin, Heidelberg (2010)
10. Rahman, M., Iqbal, S., Gao, J.: Load balancer as a service in cloud computing. In: 2014 IEEE 8th International Symposium on Service Oriented System Engineering, pp. 204–211 (2014)
11. Bagwaiya, V., Raghuwanshi, S.K.: Hybrid approach using throttled and ESCE load balancing algorithms in cloud computing. In: 2014 International Conference on Green Computing Communication and Electrical Engineering (ICGCCEE), pp. 1–6 (2014)
12. Busato, F., Bombieri, N.: A dynamic approach for workload partitioning on GPU architectures. IEEE Trans. Parallel Distrib. Syst. **28**(6), 1535–1549 (2017)
13. Neto, E.C.P., Callou, G., Aires, F.: An algorithm to optimise the load distribution of fog environments. In: 2017 IEEE International Conference on Systems, Man, and Cybernetics (SMC), pp. 1292–1297 (2017)
14. Chhabra, S., Singh, A.K.: ScienceDirect a probabilistic model for finding an optimal host framework and load distribution in cloud environment. Procedia Comput. Sci. **125**, 683–690 (2018)
15. Al Nuaimi, K., Mohamed, N., Al Nuaimi, M., Al-Jaroodi, J.: A survey of load balancing in cloud computing: challenges and algorithms. In: 2012 Second Symposium on Network Cloud Computing and Applications, pp. 137–142 (2012)
16. Xu, G., Pang, J., Fu, X.: A load balancing model based on cloud partitioning for the public cloud. Tsinghua Sci. Technol. **18**(1), 34–39 (2013)
17. Shreedhar, M., Varghese, G.: Efficient fair queuing using deficit round-robin. IEEE/ACM Trans. Netw. **4**(3), 375–385 (1996)
18. Bryhni, H., Klovning, E., Kure, O.: A comparison of load balancing techniques for scalable Web servers. IEEE Netw. **14**(4), 58–64 (2000)
19. Domanal, S.G., Reddy, G.R.M.: Load balancing in cloud computing using modified throttled algorithm. In: 2013 IEEE International Conference on Cloud Computing in Emerging Markets (CCEM), pp. 1–5 (2013)
20. Soni, G., Kalra, M.: A novel approach for load balancing in cloud data center. In: 2014 IEEE International Advance Computing Conference (IACC), pp. 807–812 (2014)
21. Mondal, B., Dasgupta, K., Dutta, P.: Load balancing in cloud computing using stochastic hill climbing—a soft computing approach. Procedia Technol. **4**, 783–789 (2012)
22. Dhinesh Babua, L.D., Venkata Krishna, P.: Honey bee behavior inspired load balancing of tasks in cloud computing environments. Appl. Soft Comput. **13**(5), 2292–2303 (2013)
23. Sharma, S., Singh, S., Meenakshi, S.: Performance analysis of load balancing algorithms. Int. J. Civ. Environ. Eng. **2**(2), 367–370 (2008)
24. Sran, N., Kaur, N.: Zero proof authentication and efficient load balancing algorithm for dynamic cloud environment. Int. J. Adv. Res. Comput. Sci. Softw. Eng. **3**(7), 2277–128 (2013)

Cryptanalysis and Improvement of Three-Factor-Based Confidentiality-Preserving Remote User Authentication Scheme in Multi-server Environment

Subhas Barman, Prantik Guha, Rituparna Saha and Soumil Ghosh

Abstract Lately, Ali–Pal addressed an improvement to Guo–Wen's scheme which proclaims to protect the anonymity of the user during remote authentication in a multi-server environment. But, the cryptanalysis of their scheme finds leakage of some sensitive information. Even, the scheme is not resilient to insider attack. In this paper, we address the problems and attempt to improve the security of the scheme. In addition, security of the proposed scheme is analyzed with the pi-calculus-based formal verification tool ProVerif. The proposed scheme is compared with other existing key exchange protocols reported in the literature with respect to computation and communication costs. We also prove that our proposed scheme provides mutual authentication and it is secured against various well-known attacks.

Keywords Multi-server environment · Remote authentication · ProVerif · Key exchange protocol

1 Introduction

In the era of modern technology, biometrics are used to either generate [1–3] or exchange a cryptographic key [4–6] for better network security. Now a user can access remote servers through a smart card in a public channel. Smart cards which contain the biometric data are vulnerable towards common attacks like stolen smart card, password update in the server without a secure channel, privileged insider attack, user impersonation attack, replay attack, and also offline password guessing attack. Multi-server environment provides a better solution as the user can communicate with any server by doing one-time registration. Mishra et al. [7] provided a secure and resilient scheme for a multi-server environment which was developed to deal with the user and server impersonation attack and stolen smart card attack of the previously available schemes. Later, Lu et al. [8] addressed the drawbacks of

S. Barman (✉) · P. Guha · R. Saha · S. Ghosh
Jalpaiguri Government Engineering College, Jalpaiguri 735102, West Bengal, India
e-mail: subhas.barman@gmail.com; bsubhas1980@gmail.com

M. Chakraborty et al. (eds.), *Proceedings of International Ethical Hacking Conference 2018*, Advances in Intelligent Systems and Computing 811,
https://doi.org/10.1007/978-981-13-1544-2_7

Mishra et al.'s scheme like forgery and server masquerading and lacks password forward secrecy. So they projected a scheme to overcome the issues. Lu et al. also pointed out the vulnerability towards replay attack and also incapability of password update phase. Then, they developed an improved authentication scheme. Thereafter, Chaudhry [9] found that Lu et al.'s scheme [8] prone to impersonation attack and is not facilitated to user anonymity. Similarly, Guo–Wen [10] proposed a more reliable and robust authentication protocol in a multi-server environment. However, this scheme is prone to fall prey to password and identity guessing attack, new smart card issue threat, user impersonation threat, known session key temporary information attack, and insider attack. To overcome these problems, Ali–Pal [11] came up with an enhanced and resilient three-factor-based confidentiality-preserving remote authentication scheme in multi-server environment. They addressed the pitfalls of previous schemes like new smart card issue attack, identity guessing attack, and known session key attack. Computation cost and estimated time are minimized in [11]. However, we find that there are still some threats to this scheme. For example, one of the random nonce can be computed from intercepted message. Moreover, Ali–Pal's scheme [11] is also vulnerable to insider attack.

We propose an improved scheme to surmount these drawbacks. Our scheme provides mutual authentication in multi-server environment. Moreover, in this paper, we use pi-calculus [12]-based formal verification tool ProVerif [13] to prove authentication and security of the proposed protocol.

2 Our Contribution

- We cryptanalysis of the Ali–Pal's scheme [11].
- We improved the scheme [11] to overcome the drawback and also add some new features. We simulate our scheme for the formal security analysis using ProVerif tool and show that proposed scheme is protected from different security attacks.
- We also compare communication cost and performance of the proposed scheme with other existing schemes.

3 Literature Review

Ali–Pal [11] addressed an improvement to Guo–Wen's [10] scheme. In this section, we reviewed the Ali–Pal's scheme [11]. The symbols and its meanings are given in Table 1.

Table 1 Meaning of notations

Notation	Description
RC	Registration center
ID_i, PW, BIO_i	Identity, password, biometrics of ith user U_i
$\mathcal{A}$	An attacker/adversary
SID_j, e_j, d_j	Identity, public key, private key of jth server S_j
h(.), H(.)	Hash function and Biohashing function
$\|\|$, $\oplus$	Concatenation and XOR operation
SK	Session key shared between U_i and S_j

3.1 Server Registration

S_j selects own identity SID_j and sends it to RC via a trustworthy channel. Then RC computes $X_j = h(d_j||SID_j)$ and transfers X_j, d_j to the server S_j via secure channel.

3.2 User Registration

In this phase, U_i selects own identity ID_i, password PW_i and imprints biometric BIO_i and then calculates $RPW_i = h(PW_i||H(BIO_i))$. Then U_i transfers ID_i, RPW_i to RC via trustworthy channel. Then RC computes $A_i = X_j \oplus h(RPW_i)$, $B_i = h(ID_i||RPW_i||X_j)$ and issues a smart card holding parameters $\langle A_i, B_i, h(.), H(.)\rangle$. RC finally sends smart card to U_i via secure channel.

3.3 Login and Authentication

U_i inserts smart card into a smart card reader and inputs own ID_i, password PW_i and imprints biometric BIO_i. Then smart card calculates $RPW_i = h(PW_i||H(BIO_i))$, $F_i = h(ID_i)$, $X_j' = A_i \oplus h(RPW_i)$ and $B_i' = h(ID_i||RPW_i||X_j')$. Now, if $B_i' = B_i$, U_i chooses a random nonce R_1 and computes $RPW_{ij} = h(RPW_i||SID_j)$, $M_1 = (F_i||R_1||SID_j)_j^e \mod n_j$, $M_2 = R_1 \oplus RPW_{ij} \oplus ID_i$ and $M_3 = h(RPW_{ij}||F_i||X_j|| R_1)$ and then M_1, M_2, M_3 is sent to S_j via a public channel. After getting the login message M_1, M_2, M_3 from U_i, S_j decrypts $(F_i||R_1||SID_j) = M_1^{d_j} \mod n_j$ and computes $RPW_{ij}' = M_2 \oplus R_1 \oplus F_i$ and $M_3' = h(RPW_{ij}||F_i||X_j||SID_j)$ and compares with M_3. If M_3' is equals to M_3, then S_j believes that U_i is legal; otherwise, the session is expired. Now, S_j Selects a random nonce R_2 and computes $M_4 = h(RPW_{ij}||R_1) \oplus R_2$, $SK = h(R_1||R_2||X_j||ID_i)$ and $M_5 = h(RPW_{ij}||SK)$ and transmits M_4, M_5 via a public channel. After receiving M_4, M_5, U_i computes $R_2' = M_4 \oplus h(RPW_{ij}||R_1)$, $SK' = h(R_1||R_2'||X_j||F_i)$ and $M_5' = h(RPW_{ij}||SK)$. If M_5' is not equals to M_5, then session is terminated. Otherwise, U_i believes on the legitimacy of S_j and mutual authentication holds.

4 Cryptanalysis of Ali–Pal's Scheme

In the Ali–Pal's scheme, ID_i, $H(BIO_i)$ are fixed for every communication initiated by U_i to any server S_k. X_j, R_1^j are varied for different servers (i.e., S_j and $j = 1, 2, \ldots$ and $S_k \neq S_j$). From received message, S_j can extract ID_i, R_1^j from message M_1^j and subsequently, S_j can computes $H(BIO_i)$ from message M_2^k, that is, $H(BIO_i = M_2^j \oplus R_1^j \oplus ID_i$. Now, S_j can act as an insider attacker and may try to know some information for communication of U_i with other server S_k. Attacker S_j intercepts login messages (say, M_2^k) from public channels. From the knowledge ID_i, $H(BIO_i)$ and publicly shared message M_2^k (shared by U_i with S_k), attacker can reveal R_1^k from M_2^k, that is, $R_1^k = M_2^k \oplus ID_i \oplus H(BIO_i)$. Moreover, Ali–Pal's does not consider the biometrics change phase.

5 Proposed Scheme

We present a three-factor-based authentication protocol. This can be used in multi-server environment. Our scheme consists of five phases (i) system setup, (ii) registration, (iii) login and authentication, (iv) password change, and (v) biometrics change.

5.1 System Setup

In this phase, the system setup is carried out following the similar process of Ali–Pal's scheme [11]. The detailed description is given below. Step 1: Registration center RC selects two large prime numbers, i.e., p_j and q_j for m servers where j=1 to m. After that RC computes $n_j = p_j \times q_j$, where $pj \neq q_j$.
Step 2: RC chooses $1 < e_j < \phi(n_j)$ where $\phi(n_j) = (p_j - 1)x(q_j - 1)$ and calculates d_j. Where $d_j = e_j - 1 \bmod \phi(n_j)$ and issues (e_1, n_1), (e_2, n_2), …, (e_m, n_m) as public key and $d_1, d_2, \ldots, d_m$ as private key.

5.2 Registration

This phase consists of two phases server and user registration.

1. Server Registration: Step 1. Server selects own identity SID_j and computes $C_j = h(SID_j)$. C_j is transferred to RC via secure channel.
   Step 2. After getting C_j, RC computes $X_j = h(d_j||C_j)$. Now RC transmits X_j, d_j to server via a secure channel.
2. User Registration: An user U_i can register by the following ways: Step 1. U_i selects an ID_i, password PW_i and imprints biometric BIO_i. Then U_i computes $RPW_i = h(PW_i||H(BIO_i)$ and $F_i = h(ID_i)$ and transfers F_i, RPW_i to RC via a secure channel.
   Step 2. Upon receiving F_i, RPW_i, RC computes $A_i = X_j \oplus h(RPW_i)$, $B_i = h(F_i||RPW_i||X_j)$ and stores the values A_i, B_i, $h(.)$, $H(.)$ into a smart card. Finally, RC transmits smart card to U_i via a secure channel. We elaborate the user registration phase in Table 2.

5.3 Login and Authentication

User U_i is authenticated by the smart card reader to access the remote server. Then the smart card reader sends a login message to the server S_j via a public channel. This phase is given Fig. 1. Detailed description is given below.

Step 1. U_i inserts smart card into a smart card reader and inputs own ID_i, password PW_i and imprints biometric BIO_i.Then smart card calculates $RPW_i = h(PW_i||H(BIO_i))$, $F_i = h(ID_i)$, $X_j' = A_i \oplus h(RPW_i)$ and $B_i' = h(ID_i|| RPW_i||X_j')$. Now, smart card compares B_i' with B_i. If B_i' is not equals to B_i, then U_i is rejected.

Step 2. Otherwise, U_i chooses a random nonce R_1 and computes $RPW_{ij} = h(RPW_i||SID_j)$, $M_1 = (F_i||R_1||SID_j)_j^e \bmod n_j$, $M_2 = R_1 \oplus RPW_{ij} \oplus ID_i$ and $M_3 = h(RPW_{ij}||F_i||X_j||R_1)$ and then M_1, M_2, M_3 is sent to S_j via a public channel.

Step 3. After getting the login message M_1, M_2, M_3 from U_i, S_j decrypts $(F_i||R_1|| SID_j) = M_1^{d_j} \bmod n_j$ and computes $RPW_{ij}' = M_2 \oplus R_1 \oplus F_i$ and $M_3' = h(RPW_{ij}||F_i||X_j||SID_j)$ and compares with M_3. If M_3' is equals to M_3, then S_j believes that U_i is legal otherwise the session is expired.

Step 4. Now, S_j Selects a random nonce R_2 and computes $M_4 = h(RPW_{ij}||R_1) \oplus R_2$, $SK = h(R_1||R_2||X_j||ID_i)$ and $M_5 = h(RPW_{ij}||SK)$ and transmits M_4, M_5 via a public channel.

Step 5. After receiving M_4, M_5, U_i computes $R_2' = M_4 \oplus h(RPW_{ij}||R_1)$, $SK' = h(R_1||R_2'||X_j||F_i)$ and $M_5' = h(RPW_{ij}||SK)$. If M_5' is not equals to M_5, then session is terminated. Otherwise, U_i believes on the legitimacy of S_j and mutual authentication holds.

User U_i **Server S_j**

Inputs ID_i, PW_i
Imprints BIO_i
Computes $RPW_i = h(PW_i || H(BIO_i))$
AND $F_i = h(ID_i)$
$X'_j = A_i \oplus h(RPW_i)$
$B'_i = h(F_i || RPW_i || X'_j)$
If B'_i B_i Then Rejects
Else selects R_1 randomly and computes
$RPW_{ij} = h(RPW_i || SID_j)$,
$M_1 = (F_i || R_1 || SID_j)^{ej} \bmod n_j$
$M_2 = R_1 \oplus RPW_{ij} \oplus F_i$
$M_3 = h(RPW_{ij} || F_i || X'_j || R_1)$
Transmits $<M_1, M_2, M_3>$ (via public channel) →

Decrypts $(F_i || R_1 || SID_j) = M_1^{dj} \bmod n_j$
$RPW'_{ij} = M_2 \oplus R_1 \oplus F_i$
$M'_3 = h(RPW'_{ij} || F_i || X_j || SID_j)$
If M'_3 M_3 then rejects
Else selects R_2 randomly and computes
$M_4 = h(RPW'_{ij} || R_1) \oplus R_2$
$SK = h(R_1 || R_2 || X_j || F_i)$
$M_5 = h(RPW'_{ij} || SK)$
← Transmits $<M_4, M_5>$ (via public channel)

Computes $R'_2 = M_4 \oplus h(RPW_{ij} || R_1)$
$SK' = h(R_1 || R'_2 || X'_j || F_i)$
$M'_5 = h(RPW_{ij} || SK')$
If M'_5 M_5 then rejects
Else *mutual authentication* holds.

Fig. 1 Login and authentication protocol

5.4 Password Change

Changing of password in a varied time interval is a good habit which incurs the security. If the user wants to change his/her password, he/she can easily do that through simple steps.

Step 1. U_i touches own smart card into a terminal, input his/her ID_i and PW_i and also imprints BIO_i. After that the smart card computes $RPW_i' = h(PW_i||H(BIO_i))$, $X_j' = A_i \oplus h(RPW_i')$, $B_i' = h(h(ID_i)||RPW_i'||X_j')$. Then it checks B_i' is equals to B_i or not. If it is true, it means that the input for the user U_i 's identification is authorized that means U_i is a authorized user for that smart card and then smart card allows the user U_i to change his/her password and asks to input new password PW_i^{new}. Otherwise rejects.

Step 2. Now smart card calculates $RPW_i^{new} = h(PW_i^{new}||H(BIO_i))$, $A_i^{new} = X_j' \oplus h(RPW_i^{new})$, $B_i^{new} = h(h(ID_i)||RPW_i^{new}||X_j')$. Finally, smart card replaces A_i, B_i with the new A_i^{new}, B_i^{new} and stored it into the smart card.

5.5 Biometric Change

Suppose any user wants to update the biometric data. Then, authentication of the user is done by the same way as discussed in the password change phase. If the authentication holds, smart card allows the user U_i to change his/her biometric with the new biometric and asks to input new biometric BIO_i^{new}. Otherwise rejects. Now, smart card calculates $RPW_i^{new} = h(PW_i||H(BIO_i^{new}))$, $A_i^{new} = X_j' \oplus h(RPW_i^{new})$, $B_i^{new} = h(h(ID_i)||RPW_i^{new}||X_j')$. Finally, smart card replaces A_i, B_i with the new A_i^{new}, B_i^{new} and stored it into the smart card.

6 Security Analysis

The security of the proposed scheme is analyzed with formal as well as informal security analysis. We have verified our proposed protocol using ProVerif simulator.

6.1 Formal Security Analysis

In order to prove the security of cryptographic protocols, ProVerif is a widely used formal verification tool [9, 12, 13]. In this section, we prove secrecy and authentication using ProVerif, because it is performed automatically and efficiently and can detect errors easily. ProVerif makes use of Dolev–Yao model [14] and supports many cryptographic primitives, including digital signature, symmetric and asymmetric encryption, hash function.

The user and the server communicate among themselves through a public channel, which is defined as below:

free Ch_Pub:channel.

The variables used in the protocol are defined as follows:

free IDi:bitstring.
free PWi:bitstring.
free BIOi:bitstring [private].
free RPWi:bitstring [private].
const dj:bitstring[private].
const nj:bitstring.
const ej:bitstring.
free SIDj:bitstring [private].
free Ai:bitstring [private].
free Bi:bitstring [private].
free X_j:bitstring [private].
free SK:bitstring [private].
free SK':bitstring [private].

The functions (xor(), exp(), mod(), mult(), and concat() represent exclusive-OR, exponent function, modulo operation, scalar multiplication, and string concatenation, respectively) used in the protocol are defined as follows:

```
fun h(bitstring):bitstring.
fun H(bitstring):bitstring.
fun xor(bitstring,bitstring):bitstring.
fun mod(bitstring,bitstring):bitstring.
fun exp(bitstring,bitstring):bitstring.
fun mult(bitstring,bitstring):bitstring.
fun concat(bitstring,bitstring):bitstring.
```

The algebraic properties of the functions are defined as below:

```
equation for all a:bitstring,b:bitstring; xor(xor(a,b),b)=a.
```

According to the protocol, the user Ui computes and sends $M1$, $M2$, $M3$ to the Server Sj and then waits until he receives $M4$, $M5$ from the Server Sj. So, the user Ui is defined as follows:

```
let UserUi=
let RPWi=h(concat(PWi,H(BIOi))) in
let Fi=h(IDi) in
let Xj'=xor(Ai,h(RPWi)) in
let Bi'=h(concat(Fi,concat(RPWi,Xj))) in
if (Bi'=Bi) then
new R1:bitstring;
let RPWij=h(concat(RPWi,SIDj)) in
let M1=mod(exp(concat(Fi,concat(R1,SIDj)),ej),nj) in
let M2=xor(R1,xor(RPWij,Fi)) in
let M3=h(concat(RPWij,concat(Fi,concat(Xj',R1)))) in
out(Ch_Pub,(M1,M2,M3));
in(Ch_Pub,(xM4:bitstring,xM5:bitstring));
let R2'= xor(xM4,h(concat(RPWij,R1))) in
let SK'=h(concat(R1,concat(R2',concat(Xj',Fi)))) in
let M5'=h(concat(RPWij,SK')) in
if (M5'=xM5) then 0.
```

According to the protocol, the server Sj receives `{M1,M2,M3}` from the user Ui, then computes and sends `{M4,M5}` to the user Ui. We can define the server Sj as follows:

```
let ServerSj=
let Xj=h(concat(dj,SIDj)) in
in(Ch_Pub,(xM1:bitstring,xM2:bitstring,xM3:bitstring));
new R1':bitstring;
new Fi':bitstring;
let RPWij'=xor(xM2,xor(R1',Fi')) in
let M3'=h(concat(RPWij',concat(Fi',concat(Xj,SIDj)))) in
if (xM3=M3') then
new R2:bitstring;
let M4=xor(h(concat(RPWij',R1')),R2) in
```

```
let SK=h(concat(R1',concat(R2,concat(Xj,Fi')))) in
let M5=h(concat(RPWij',SK)) in
out(Ch_Pub,(M4,M5))
else 0.
```

In order to ensure mutual authentication, we define events as follows:

```
event begin_UserUi(bitstring).
event end_UserUi(bitstring).
event begin_ServerSj(bitstring).
event end_ServerSj(bitstring).
```

The process can be defined by:

```
process ((!UserUi) | (ServerSj)
```

To verify mutual authentication and session key's security, we define the following queries:

```
query attacker(SK).
query attacker(SK').
query id:bitstring; event(end_UserUi(id)) ==> event(begin_UserUi(id)).
query id:bitstring; event(end_ServerSj(id)) ==> event(begin_ServerSj(id)).
```

When the above code is performed in ProVerif, we find that both the correspondence queries are true and both the (not)attacker queries are true, thus indicating that both the mutual authentication property and session key security are satisfied for our proposed scheme.

6.2 *Informal Security Analysis*

In this section, we elaborate informal security analysis of our scheme and prove that our protocol is able to protect from different types of security vulnerabilities.

1. Password and identity guessing attack: We assume an adversary can eavesdrop all communication messages M_1, M_2, M_3, M_4, M_5 and extract all information A_i, B_i from smart card. But still $\mathcal{A}$ is not able to calculate PW$_i$ and ID$_i$ from A_i.$A_i = X_j \oplus h(RPW_i)$, where $X_j = h(SID_j||d_j)$ and $RPW_i = h(PW_i||H(BIO_i))$. To calculate PW_i, $\mathcal{A}$ needs to know BIO_i, SID_j, d_j at one time which is infeasible in polynomial time. $\mathcal{A}$ is not able to calculate PW_i from $B_i = (F_i||RPW_i||X_j)$, where $F_i = h(ID_i)$. For computing PW_i, $\mathcal{A}$ has to know the parameters $H(BIO_i)$, ID_i, SID_j, d_j at one time which is impossible. $\mathcal{A}$ also cannot evaluate ID_i from B_i. $\mathcal{A}$ cannot obtain ID_i from M_1, M_2, M_3, and M_5 because of hash function where $M_1 = (F_i||R_1||SID_j)^{e_j} \text{mod} n_j$, $M_2 = R_1 \oplus RPW_{ij} \oplus F_i$, $M_3 = h(RPW_{ij}||F_i||X_j||R_1)$ $M_5 = h(RPW_{ij}||SK)$, $RPW_{ij} = h(RPW_i||SID_j)$ and $SK = h(R_1||R_2||X_j||F_i)$.
2. Impersonation attack: We assume an attacker $\mathcal{A}$ intercepts all communication messages, and then he modifies all messages and tries to imitate as a legal server or user. But, in our protocol it is not possible for some reasons like $\mathcal{A}$ cannot calcu-

late $M_1 = (F_i||R_1||SID_j)_j^e \bmod n_j$ where R_1 is a random nonce because A is unable to obtain ID_i. $M_2 = R_1 \oplus RPW_{ij} \oplus F_i$ and $M_3 = h(RPW_{ij}||F_i||X_j||R_1)$ where $RPW_{ij} = h(RPW_i||SID_j)$ and $RPW_i = h(PW_i||H(BIO_i))$. To compute M_2, $\mathcal{A}$ has to know BIO_i, ID_i, PW_i, and SID_j at same time, which is not possible. For calculating M_3, $\mathcal{A}$ has to know PW_i, ID_i, BIO_i and X_j which is not feasible. $M_4 = h(RPW_{ij}||R_1) \oplus R_2$ and $M_5 = h(RPW_{ij}||SK)$ where $SK = h(R_1||R_2||X_j||F_i)$. So, $\mathcal{A}$ has to know $PW_i, BIO_i, SID_j, X_j, ID_i$ at one time to calculate M_4 and M_5 which is not possible.

3. User untraceability attack: In this type of threat, an attacker intercepts two communication messages and tries to extract identity of user or server by matching values of each parameter. But, our protocol is able to protect this type of attack. In $M_1 = (F_i||R_1||SID_j)_j^e \bmod n_j$, user ID_i is secured using hash function and R_1 is a random nonce. So, value of M_1 is different in each session due to uniqueness property of R_1. $M_2 = R_1 \oplus RPW_{ij} \oplus F_i$ and $M_3 = h(RPW_{ij}||F_i||X_j||R_1)$ are also different in each session due to uniqueness property of R_1. Therefore, our protocol resists user untraceability attack.
4. Replay attack: Our protocol resists replay attack by using random nonce R_1 and R_2.
5. Insider attack: Our scheme is not vulnerable to insider attack because user U_i sends $RPW_i = h(PW_i||H(BIO_i))$ to RC. So, an insider of system cannot obtain

Table 2 Security features comparison

SF	Guo–Wen [10]	He–Wang [15]	Wen et al. [16]	Li et al. [17]	Irshad et al. [18]	Ali–Pal [11]	PS
A1	Yes	Yes	Yes	No	Yes	Yes	Yes
A2	Yes	No	No	No	No	No	Yes
A3	Yes	No	Yes	No	No	Yes	Yes
A4	Yes	No	Yes	Yes	No	Yes	Yes
A5	No	Yes	Yes	Yes	Yes	Yes	Yes
A6	No	No	No	Yes	Yes	Yes	Yes
A7	Yes	Yes	No	No	Yes	No	Yes
A8	No	Yes	Yes	Yes	Yes	No	Yes
A9	No	Yes	Yes	Yes	No	Yes	Yes
A10	Yes	Yes	Yes	Yes	Yes	Yes	Yes
A12	Yes	Yes	Yes	Yes	Yes	Yes	Yes
A13	Yes	Yes	Yes	Yes	Yes	Yes	Yes

SF security features, PS proposed scheme, A1 be proof against password guessing attack, A2 facilitating user anonymity, A3 be proof against user impersonation attack, A4 be proof against server impersonation attack, A5 be proof against replay attack, A6 be proof against session key temporary information attack, A7 be proof against user untraceability attack, A8 be proof against privileged insider attack, A9 be proof against identity guessing attack, A10 forward secrecy, A12 be proof against smart card theft attack, A13 session key verification password because of hash function. Though attacker guesses PW_i but still he/she is unable to validate password without knowledge of biometrics BIO_i.

6. Known session key temporary information attack: In our scheme, attacker cannot compute session key $SK = h(R_1||R_2||X_j||F_i)$ with the knowledge of random nonce R_1 and R_2. Because, SK also depends on X_j and F_i.
7. Smart card stolen attack: Suppose attacker gets the smart card of an user and extracts parameters $A_i = X_j \oplus h(RPWi)$ and $B_i = (F_i||RPW_i||X_j)$, where $RPW_i = h(PW_i||H(BIO_i))$, $X_j = h(SID_j||d_j)$ and $F_i = h(ID_i)$. But, attacker is unable to calculate ID_i from F_i and PW_i from A_i, B_i.
8. Forward secrecy: Our scheme facilitates forward secrecy property. With the knowledge of a session key, attacker is not able to compute other session key.

We compare our scheme along with other schemes with respect to different security attacks and given in Table 2.

7 Performance

In this section, we compare our scheme with other existing schemes based on communication cost and estimated time.

7.1 Communication Cost

In Table 3, we represent comparison of communication cost of our scheme with respect to other existing schemes. Here, we assume lengths of ID_i, PW_i, random nonce and hash functions are 160 bits. e_j, d_j are 1024 bits and symmetric encryption, decryption is of 512 bits for each.

7.2 Estimated Time

To calculate estimated time, we have used the following notations, *Th*: time complexity of hash function, *Ts*: symmetric encryption or decryption, *Te*: modular exponentiation, and *Tm*: point multiplication of elliptic curve. We calculate estimated time in seconds. The time complexity of our scheme is $(28\text{Th} + 2\text{Te}) = 28 \times 0.0005 + 2 \times 0.522 = 1.058$. The comparison of computation time of our scheme with other scheme is given in Table 3.

Table 3 Estimated time comparison and communication cost

PC	Li et al. [17]	Ali–Pal [11]	Irshad et al. [18]	He–Wang [15]	Wen et al. [16]	PS
CCRP	8Th	4Th	5Th + 1Ts + 1Tm	3Th	4Th	4Th
CCLAP	20Th + 4Te	16Th + 2Te	17Th + 5Ts + 10Tm	23Th + 8Tm	12Th + 10Ts	16Th + 2Te
CCPCP	6Th	10Th	7Th + 1Ts + 1Tm	2Th	4Th	8Th
TCC	34Th + 4Te	30Th + 2Te	29Th + 7Ts + 12Tm	28Th + 8Tm	20Th + 10Ts	28Th + 2Te
ET	2.105	1.059	0.8232	0.5186	0.097	1.058
CC	2688	1664	2784	3360	4032	1664

PC performance comparison, *PS* proposed scheme, *CC* communication cost, *CCRP* computation cost of registration phase, *CCLAP* computation cost of login and authentication phase, *CCPCP* computation cost of password change phase, *TCC* total computation cost, *ET* estimated time

8 Conclusion

In this paper, we found some faults of Ali–Pal's scheme and overcome the drawbacks of the same scheme. We use ProVerif to verify the security of our scheme. Communication cost and estimated time of our scheme are comparatively better than other schemes. In our scheme, a legal user can change his/her password and biometrics without help of server's involvement.

References

1. Barman, S., Chattopadhyay, S., Samanta, D.: Fingerprint based symmetric cryptography. In: 2014 International Conference on High Performance Computing and Applications (ICHPCA), Bhubaneswar, pp. 1–6 (2014). https://doi.org/10.1109/ICHPCA.2014.7045306
2. Barman, S., Samanta, D., Chattopadhyay, S.: Approach to cryptographic key generation from fingerprint biometrics. Int. J. Biom. **7**(3), 226–248 (2015)
3. Barman, S., Samanta, D., Chattopadhyay, S.: Fingerprint-based crypto-biometric system for network security. EURASIP J. Info. Secur. **3** (2015). https://doi.org/10.1186/s13635-015-0020-1
4. Barman, S., Chattopadhyay, S., Samanta, D.: An approach to cryptographic key exchange using fingerprint. In: Mauri, J.L., Thampi, S.M., Rawat, D.B., Jin, D. (eds.) Security in Computing and Communications. Communications in Computer and Information Science SSCC 2014, vol. 467. Springer, Berlin (2014)
5. Barman, S., Chattopadhyay, S., Samanta, D.: An approach to cryptographic key distribution through fingerprint based key distribution center. In: 2014 International Conference on Advances in Computing, Communications and Informatics (ICACCI), New Delhi, pp. 1629–1635 (2014). https://doi.org/10.1109/ICACCI.2014.6968299

6. Barman, S., Chattopadhyay, S., Samanta, D., Panchal, G.: A novel secure key-exchange protocol using biometrics of the sender and receiver. Comput. Electr. Eng. **64**, 65–82 (2017)
7. Mishra, D., Das, A.K., Mukhopadhyay, S.: A secure user anonymity-preserving biometric-based multi-server authenticated key agreement scheme using smart cards. Exp. Syst. Appl. **41**(18), 8129–8143 (2014)
8. Lu, Y., Li, L., Peng, H., Yang, Y.: A biometrics and smart cards based authentication scheme for multi-server environments. Secur. Commun. Netw. **8**(17), 3219–3228 (2015)
9. Chaudhry, S.A.: A secure biometric based multi-server authentication scheme for social multimedia networks. Multimed. Tools Appl. **75**, 12705 (2016). https://doi.org/10.1007/s11042-015-3194-0
10. Guo, D., Wen, F.: Analysis and improvement of a robust smart card based-authentication scheme for multi-server architecture. Wirel. Pers. Commun. **78**(1), 475–490 (2014)
11. Ali, R., Pal, A.K.: Three-factor-based confidentiality-preserving remote user authentication scheme in multi-server environment. Arab. J. Sci. Eng. **42**(8), 3655–3672 (2017). https://doi.org/10.1007/s13369-017-2665-1
12. Abadi, M., Fournet, C.: Mobile values, new names, and secure communication. In: Proceedings of the 28th ACM SIGPLAN-SIGACT Symposium on Principles of Programming Languages, pp. 104–115. ACM, New York (2001)
13. Abadi, M., Blanchet, B., Comon-Lundh, H.: Models and proofs of protocol security: a progress report. In: Computer Aided Verification, vol. 5643, pp. 35–49. Springer, Heidelberg (2009)
14. Dolev, D., Yao, A.C.: On the security of public key protocols. IEEE Trans. Inf. Theory **29**(2), 198–208 (1983)
15. He, D., Wang, D.: Robust biometrics-based authentication scheme for multi-server environment. IEEE Syst. J. **9**(3), 816–823 (2015)
16. Wen, F., Susilo, W., Yang, G.: Analysis and improvement on a biometric-based remote user authentication scheme using smart-cards. Wirel. Pers. Commun. **80**(4), 1747–1760 (2015)
17. Li, X., Niu, J., Kumari, S., Liao, J., Liang, W.: An enhancement of a smart card authentication scheme for multi-server architecture. Wirel. Pers. Commun. **80**(1), 175–192 (2015)
18. Irshad, A., Sher, M., Nawaz, O., Chaudhry, S.A., Khan, I., Kumari, S.: A secure and provable multi-server authenticated key agreement for TMIS based on Amin et al. scheme. Multimed. Tools Appl. (2016). https://doi.org/10.1007/s11042-016-3921-1

Part III
Session 1B: Cryptography

Bi-symmetric Key Exchange: A Novel Cryptographic Key Exchanging Algorithm

Shekhar Sonthalia, Trideep Mandal and Mohuya Chakraborty

Abstract The most sensitive part of any crypto graphical system is ensuring the security of key exchange. Classical crypto graphical algorithms make use of one-way complex mathematical functions, to encode the key. On the other hand, Quantum crypto graphical systems make use of Q-bits or photons to encode the key. These methods are useful to secure the key but they come with a lot of trade-offs. Classical one is complex and requires a lot of mathematical calculations and is easy to use as it does not involve much hardware but with quantum computing on the rise, it is a piece of cake for an eavesdropper to break the key and hamper the security. Quantum Cryptography ensures that safety by not allowing the eavesdropper to access the data without hampering the key. However the hardware requirements make it inaccessible. In this paper a novel algorithm of key exchange that involves the best of both quantum and classical worlds has been proposed. It is called Bi-Symmetric Key Exchange. Simulation result and subsequent performance analysis show the efficiency of this algorithm over existing key exchange algorithms with respect to average computational time for key generation.

Keywords Quantum cryptography · Classical cryptography · Photons · Q-bit
Bi-symmetric · Security

S. Sonthalia (✉) · T. Mandal · M. Chakraborty
Department of Information Technology, Institute of Engineering and Management, Salt Lake, Kolkata, West Bengal, India
e-mail: sonthalia1996@outlook.com

T. Mandal
e-mail: mandal.trideep1435@gmail.com

M. Chakraborty
e-mail: mohuyacb@iemcal.com

M. Chakraborty et al. (eds.), *Proceedings of International Ethical Hacking Conference 2018*, Advances in Intelligent Systems and Computing 811,
https://doi.org/10.1007/978-981-13-1544-2_8

1 Introduction

Classical cryptology is the science of rendering information unintelligible to unintended parties. Various methods are used to encrypt and decrypt information. Classical cryptology relies on a key, which may be a mathematical function, alphanumeric string or even a matrix which is pre-decided and recorded either on paper or in digital format. Quantum physics has introduced us to qubits, or quantum bits. They lose their state when any attempt to measure their state is made. Their state can be read only once, irrespective of whether intended data can be read or not. This has sparked interest in using this counter-intuitive probabilistic nature to transmit data securely. Existing cryptographic protocols are either strictly classical or purely based upon quantum physics [1, 2]. In this paper we have proposed a new algorithm which makes use of best of both the worlds to provide a better cryptographic key exchange protocol and is called Bi-Symmetric Key Exchange.

The organization of the paper is as follows. After the introduction in Sect. 1, brief overview of existing cryptographic key exchange protocols has been given in Sect. 2. Section 3 holds description, algorithm, simulation results and performance analysis of Bi-Symmetric Key Exchange. Section 4 concludes the paper with some highlights on future works in this area.

2 Overview of Existing Cryptographic Key Exchange Protocols

2.1 BB84 Protocol

In 1984 Charles Bennet and Gilles Brassard developed the most famous quantum key distribution protocol known as BB84 [3, 4]. The system makes use of polarized photons to transmit information. Information is transmitted through fiber-optic cables or free space. According to the design, none of these channels need to be secure. Photons follow Heisenberg's Uncertainty Principle [5] and thus the state of a photon cannot be determined without affecting its state. If an eavesdropper tries to intercept the data transmission, he/she cannot do it without changing the state of the transmitted photons. As a result, presence of an eavesdropper increases error percentage in the data stream. BB84 protocol encodes the information in non-orthogonal states. Table 1 shows the encoding scheme of BB84.

Table 1 BB84 encoding scheme

Basis	0	1
+	$\uparrow$	$\longrightarrow$
$\times$	$\nearrow$	$\searrow$

In this protocol, Sender decides upon a basis, where he denotes two of four polarization states of a photon by binary zero and the other two polarization states by binary one, as shown in Table 1. Sender then sends the encoded string to the Receiver over public channel. Receiver having no idea about the basis that Sender has encoded in chooses a convention of his own. Thus, he may detect the photon correctly or incorrectly. In other words, there is a 50% chance that Receiver decodes the information sent by Sender correctly and a 50% chance that he interprets it erroneously. Nevertheless, Receiver continues to interpret the string using his own basis. If an eavesdropper tries to understand the states of the photons, he would have to detect their state, thus changing their original state. This would result in 50% drop in integrity of the transmitted information. Now when Receiver would try to understand the data sent to him, he would have only 50% chance to interpret the data.

Now, Receiver tells Sender about the basis used by him/her over the public channel. If his basis is compatible with the photon generated by Sender, then Receiver responds with "MATCH", else replies "DISCARD". The photons for which both of them have used the same basis is converted to its equivalent binary 0 or 1, as shown in Table 1.

Then, Receiver calculates his match percentage. If the match percentage is above 50%, the generated key is used for encrypting data for further communication. On the other hand, if match percentage is below 50%, it implies the presence of an Eavesdropper in the channel. Sender and Receiver discard the present key and proceed to generate a new key until they have a better match percentage. Figure 1 shows the workflow diagram of BB84.

2.2 *Diffie-Hellman-Merkle Protocol*

Diffie-Hellman-Merkle key exchange (DHM) [6, 7, 8] is one of the first public-key protocols as designed by Ralph Merkle and named after Whitfield Diffie and Martin Hellman. It is a method of securely exchanging cryptographic symmetric keys over a public channel. Both Sender and Receiver share the same key which is used to encrypt data for future transmissions. This algorithm is based upon discrete logarithm problem solving, which takes way too long to be calculated by the fastest classical computing system [9, 10, 11].

In DHM protocol, Sender and Receiver decide beforehand the multiplicative group of integers modulo c, and d, which is a primitive root modulo c. According to the design, c is a prime number. Values of c and d are chosen such that the shared key lies between 1 and $c - 1$. Now, Sender chooses a secret number, say a. Receiver too chooses a secret number, say b.

Sender sends to Receiver

$$J = d^a mod\, c \tag{1}$$

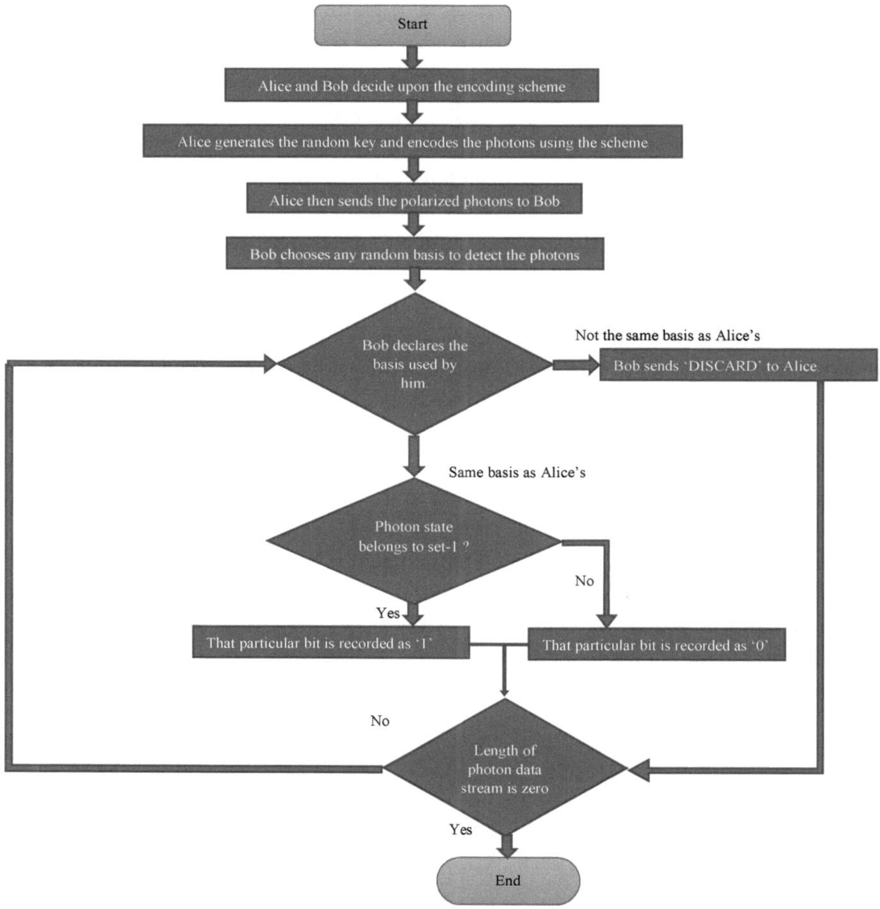

Fig. 1 BB84 flowchart

Receiver sends to Sender

$$K = d^b mod\ c \tag{2}$$

Sender determines

$$S = K^a mod\ c \tag{3}$$

Receiver determines

$$S = K^b mod\ c \tag{4}$$

Both Receiver and Sender now share the same key S, as under mod c,

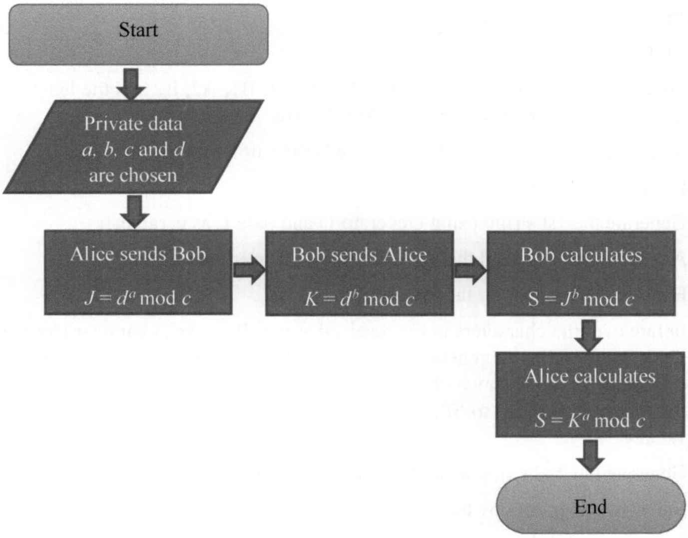

Fig. 2 Diffie-Hellman-Merkle protocol flowchart

$$K^a mod\ c = d^{ab} mod\ c = d^{ab} mod\ c = K^b mod\ c \tag{5}$$

This key 'S' can now be used sending encrypted data over the public channel. The workflow diagram of Diffie-Hellman-Merkle protocol is shown in Fig. 2.

3 Bi-symmetric Key Exchange: Proposed Cryptographic Key Exchange Algorithm

Bi-symmetric Key Exchange protocol improves upon BB84 and Diffie-Hellman-Merkle by using some of their key features along with its own. In this section we have proposed our algorithm for key distribution using concepts from both the protocols. Instead of going in with quantum bits or photons we rely upon classical bits. Tables 2, 3 and 4 demonstrate the algorithm of this protocol.

3.1 Description

The system starts with the creation of a universal set U, which is a set of symbols to be used to encode the private key by generating random strings. For the sake of explanation, we have considered a universal set U, consisting of lowercase and uppercase alphabets and numbers as given in Eq. (6).

Table 2 Algorithm for main function

Steps	Actions
Step 1.	First define the universal set U, private sets A1, B1, A2, B2 and the length L of the string which is to be generated for key transfer
Step 2.	Ask the user to choose his/her role as a Sender or Receiver
Step 3.	If user chooses Sender
Step 4.	Generate the 1st string using Generator() and pass L as parameter
Step 5.	Ask the Sender to send this string to Receiver over public channel
Step 6.	Enter the string sent by the Receiver
Step 7.	Iterate over the characters in the received string. Pass each character along with the sets corresponding to generation of 1st key. If Set_Check() returns '1', print '0' or '1' depending on set with which the function was called. If Set_Check() was called with the set corresponding to '0', print '0' and vice versa. This string of '0' and '1' is the 1st key
Step 8.	Generate the 2nd string using Generator() and pass L as parameter
Step 9.	Enter the string sent by the Receiver
Step 10.	Iterate over the characters in the received string. If the character in the generated and received string belong to the same set (the sets corresponding to generation of 2nd key), store 'M' in an array. Else, store 'D' in the array
Step 11.	Send this array to Receiver
Step 12.	For the characters lying in the same set, print the number corresponding to that set to generate the 2nd key
Step 13.	Else if the user chooses Receiver
Step 14.	Generate the 1st string using Generator() and pass L as parameter
Step 15.	Enter the string sent by Sender
Step 16.	Iterate over the characters in the received string. If the character in the generated and received string belong to the same set (the sets corresponding to generation of 1st key), store 'M' in an array. Else, store 'D' in the array
Step 17.	Send this array to the Sender
Step 18.	For the characters lying in the same set, print the number corresponding to that set to generate the 1st key
Step 19.	Generate the 2nd string using Generator() and pass L as parameter
Step 20.	Enter the string sent by Sender
Step 21.	Iterate over the characters in the received string. Pass each character along with the sets corresponding to generation of 2nd key. If Set_Check() returns '1', print '0' or '1' depending on set with which the function was called. If Set_Check() was called with the set corresponding to '0', print '0' and vice versa. This string of '0' and '1' is the 2nd key
Step 22.	Append both the keys to generate the final key

Table 3 Algorithm for generator function

Steps	Actions
Step 1.	Take length L as an argument
Step 2.	Apply the correct seeding for random generator function
Step 3.	Select a random character from the universal set of characters and store it in an array
Step 4.	Repeat the previous step until the length of array created is not equal to L

Table 4 Algorithm for Set_Check function

Steps	Actions
Step 1.	Accept a character and a character array as parameters
Step 2.	If the character exists in the character array, return '1'
Step 3.	Return '0' as a default case

$$U = \{ABCDEFGHIJKLMNOPQRSTUVWXYZ\ abcdefghijklmnopqrstuvwxyz\ 0123456789\} \tag{6}$$

Before the sender and receiver start the exchange of keys, they decide upon four private sets which are subsets of the universal set U. These four sets, by design, would be disjoint in nature and their union would contain fewer elements than the universal set. Let us name these sets as A_1, B_1, A_2 and B_2.

$$\{A_1, A_2, B_1, B_2\} \subset U \tag{7}$$

$$A_1 \cap B_1 = \phi \tag{8}$$

$$A_2 \cap B_2 = \phi \tag{9}$$

The size and symmetry of the above four sets would be limited only by the above stated requirements and the users' needs. These four sets would be used in pairs of two, the first pair, A_1 and B_1 will be used first to generate the first private key (Key 1) and the other pair, A_2 and B_2 will be used to generate the second private key (Key 2). So, at any given instant only one of the two pairs will be in use as the private set for generation of the key (be it Key 1 or Key 2).

These sets are similar to the polarization conventions used in the BB84 protocol. In BB84 protocol, two out of the four photon states signify binary zero and the other two states represent binary one (as shown in Table 1). Similarly, in our system, the sender and the receiver will decide upon which of the above two sets (i.e. A_1 and B_1) would denote to binary zero, making the other denote to binary one. This is done for generating the first key, similarly for the second key, one of the two sets (i.e. A_2 and B_2) would denote to binary zero, making the other denote to binary one.

Let us assume that in our system set A_1 and A_2 represent binary 0, and set B_1 and B_2 represent binary 1. All of this is done privately and is user dependent. Once Sender and Receiver have decided on the private sets, they can start the exchange of keys. Instead of transmitting any photons over free space, both of them will generate a string of X characters on their systems. This string of X characters will be generated over the Universal Set U, where the characters can be repeated. Let Sender be the sender and Receiver the receiver. Sender then will transmit her string to Receiver. Let i be the position of the ith element in both Sender's and Receiver's string. Receiver upon receiving the string, checks if the ith character in his string belongs to the same set containing Sender's ith character. Since they are generating the first key (Key 1) so Receiver will be checking the first pair of sets, A_1 and B_1. If both the ith elements belong to the same private set (either A_1 or B_1) then Receiver prints 'M' which stands for "MATCH". If both the characters don't belong to the same set or are not lying in the private set A_1 or B_1, he prints 'D' which stands for "DISCARD". In this way Receiver goes through all the characters of Sender's string and generates his own string of X characters containing 'M' and 'D' which is known as "MATCH-DISCARD" string. This string is sent over to Sender. Now both Sender and Receiver have their own string of X characters and the "MATCH-DISCARD" string which is also of X characters. They then determine the characters in their string which are at the same position as that of 'M' in the "MATCH-DISCARD" string. If their character belongs to A_1, they record it as binary 0, else they record it as binary 1. This string of 0's and 1's forms the first key (Key 1).

Now Sender and Receiver interchange their roles. Sender becomes the receiver and Receiver becomes the sender. They repeat the above process with sets A_2 and B_2 as their private sets. In this way they end up generating the second key (Key 2). The final private key is produced by applying any complex one-way mathematical function between Key 1 and Key 2 (over here we are appending both the keys).

Similarities with BB84 protocol. Bi-symmetric key exchange protocol is inspired from two aspects of BB84 protocol. First the basis and second the Match-Discard verification process. BB84 used four non-orthogonal polarization states of a photon as the smallest package of information and were used to transmit information from Sender to Receiver. Two of these states were mapped to binary 0 and the other two to binary 1, forming the basis. Our system uses a set of 62 characters comprising of lower case alphabets, upper case alphabets and numbers. Each of these 62 characters is the smallest package of information, and a combination of these characters is used to generate strings and private sets. The private sets A1, B1, A2 and B2 function in a way similar to basis in BB84 protocol. The strings are used to exchange data among the Sender and Receiver. This exchanged data is verified using the private sets through a process similar to the Match-Discard verification process of BB84 protocol.

Similarities with Diffie-Hellman-Merkle protocol. In contrast to BB84 protocol, where data transmission is directed from Sender to Receiver, Diffie-Hellman-Merkel protocol allows exchange of data among Sender and Receiver over a public channel to generate a symmetric key. Bi-symmetric key exchange protocol is inspired from this exchange of data among Sender and Receiver, as it allows bi-directional transfer of data to generate two symmetric keys, which are then used to generate the final key.

Table 5 String size and average key size (Key 1 or Key 2) comparison for a private set of 15 elements

Length of string used	100	200	300	400	500	600	700	800	900	1000
Average length of key generated	25	36	53	72	92	117	128	132	155	185

3.2 Bi-symmetric Key Exchange Flowchart

Figure 3 shows the workflow of Bi-Symmetric key exchange algorithm.

3.3 Algorithm of Bi-symmetric Key Exchange Protocol

See Tables 2, 3 and 4.

3.4 Performance Analysis

Figure 4 shows the plot of average computational time for key generation. It indicates that computational time required to generate a key from a string of different lengths, based on pre-specified private sets is more or less constant, with a difference of about 500 μs. In Table 2, we get an idea of the size of generated keys for different lengths of strings used for key generation. Looking at Table 2 and Fig. 1, we can conclude that generation of a key for practical use takes a time of 0.5–1 ms only. This is in contrast with other algorithms where time taken to generate a key rises exponentially with increase in size of keys. This low variance in the time taken for generation of larger keys and non-exponential nature of increase in the time required to generate larger keys gives the user flexibility to generate keys of any length, based on his/her security requirements. They wouldn't have to think twice before increasing the size of their key as the increase in computation time is considerably less when compared to other protocols. The overall time required to generate keys of 150–190 characters is around a millisecond. As a result, Sender and Receiver have the flexibility to generate a new key in real time. This would eliminate the need to wait too long or pause exchange of information to secure the communication with a new key (Table 5).

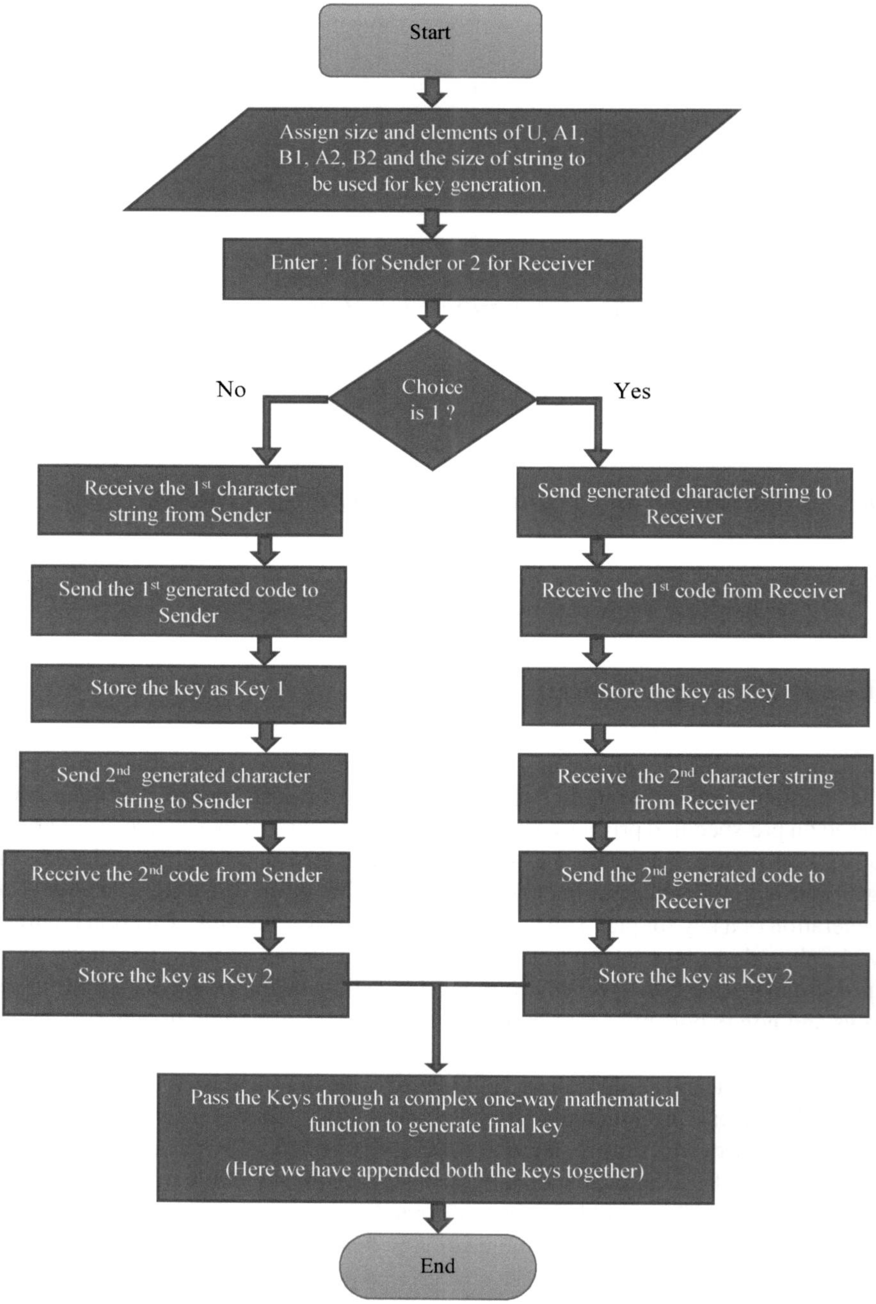

Fig. 3 Bi-symmetric key exchange flowchart

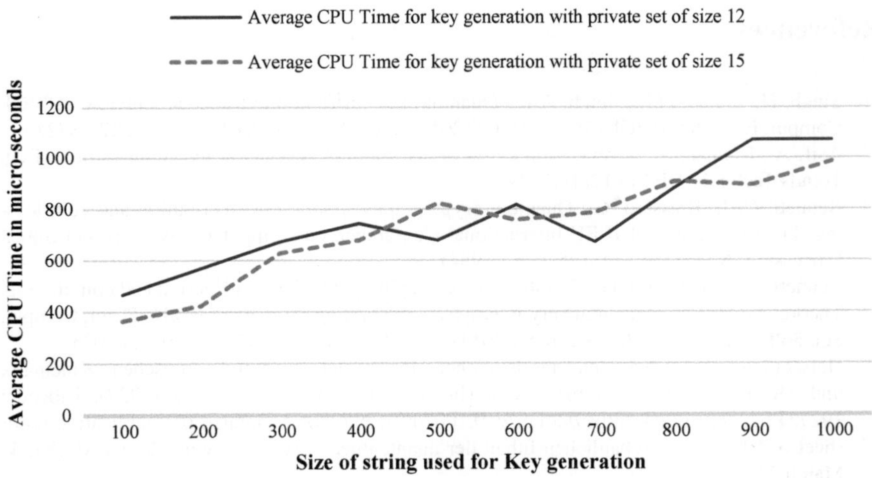

Fig. 4 The plot of average computational time required to generate key on the user's system vs the length of string used to generate the key by the user

4 Conclusion

BB84, in spite of being a robust quantum key exchange protocol has a few major flaws which make it impractical for current use. The sender needs to have access to a quantum laser setup for generating the photons and the receiver needs to detect their polarization, both of which add to hardware cost. Moreover, the transmission of these photons takes place through optical fibers or free space [12]. Whereas Bi-symmetric Key exchange protocol improves upon these flaws by eliminating the need of any dedicated hardware setup for generation and transmission of keys. As a result, this protocol can be used to secure transmissions between mobile and low powered devices too. The time required for generation of larger keys doesn't increase exponentially, which gives the user the freedom to generate key as per his security requirements. Additionally, the user can choose any number system for implementation of this protocol for improved encryption. Here we made use of binary system where the key generated consists of both 0 and 1. Since both the Universal Set U and the private sets are user defined, he/she can choose any number system like quaternary, octal, or decimal etc. This makes our system flexible and less prone to attacks, the required changes for which would be implemented by us based on the user's demands. To sum up, Bi-Symmetric Key Exchange is a capable protocol which takes in better features of existing cryptographic algorithms and produces a powerful yet simple way of securing today's key exchange.

References

1. Singh, H., Gupta, D.L., Singh, A.K.: Quantum key distribution protocols: a review. IOSR J. Comput. Eng. (IOSR-JCE) **16**(2), 01–09 (2014). e-ISSN: 2278–0661 p-ISSN: 2278-8727
2. Asif, A., Hannan, S.: A review on classical and modern encryption techniques. Int. J. Eng. Trends Technol. (IJETT) **12**(4) (2014)
3. Bennett, C.H., Brassard, G.: Quantum cryptography: public key distribution and coin tossing. In: Proceedings of IEEE International Conference on Computers, Systems and Signal Processing, New York, vol. 175, p. 8. (1984)
4. Bennett, C.H., Brassard, G.: Quantum cryptography: public key distribution and coin tossing. Theoretical aspects of quantum cryptography—celebrating 30 years of BB84. Theor. Comput. Sci. **560**(Part 1), 7–11 (04 December 2014). https://doi.org/10.1016/j.tcs.2014.05.025
5. Heisenberg, W.: Über den anschaulichen Inhalt der quantentheoretischen Kinematik und Mechanik. Zeitschrift für Physik (in German), **43**(3–4), 172–198 (1927). Bibcode: 1927ZPhy…43..172H, https://doi.org/10.1007/bf01397280. Annotated pre-publication proof sheet of Über den anschaulichen Inhalt der quantentheoretischen Kinematik und Mechanik, March 21, 1927
6. Merkle, R.C.: Secure communications over insecure channels. Commun. ACM. **21**(4), 294–299 (1978). https://doi.org/10.1145/359460.359473. Accessed August 1975; Revised September 1977
7. Diffie, W., Hellman, M.: New directions in cryptography (PDF). IEEE Trans. Inf. Theory **22**(6), 644–654 (1976). https://doi.org/10.1109/TIT.1976.1055638
8. Hellman, Martin E.: An overview of public key cryptography (PDF). IEEE Commun. Mag. **40**(5), 42–49 (2002). https://doi.org/10.1109/MCOM.2002.1006971
9. Garzia, F.: Handbook of Communications Security, p. 182. WIT Press, (2013). ISBN: 1845647688
10. Buchmann, J.A.: Introduction to Cryptography, 2nd ed. Springer Science & Business Media, pp. 190–191 (2013). ISBN: 1441990038
11. Barbulescu, R., Gaudry, P., Joux, A., Thomé, E.: A heuristic quasi-polynomial algorithm for discrete logarithm in finite fields of small characteristic. In: Advances in Cryptology—EUROCRYPT 2014. Proceedings 33rd Annual International Conference on the Theory and Applications of Cryptographic Techniques. Lecture Notes in Computer Science, Copenhagen, Denmark, vol. 8441, pp. 1–16 (2014). https://doi.org/10.1007/978-3-642-55220-5_1. ISBN: 978-3-642-55220-5
12. Ojha, V., Sharma, A., Goar, V., Trivedi, P.: Limitations of practical quantum cryptography. Int. J. Comput. Trends Technol. (2011). ISSN: 2231-2803

DNA Cryptography-Based Secured Weather Prediction Model in High-Performance Computing

Animesh Kairi, Suruchi Gagan, Tania Bera and Mohuya Chakraborty

Abstract This paper discusses the design of a DNA cryptography-based secured weather prediction model by the use of supercomputing or cluster type computing environment. The model is based on Markov's chain. The supercomputer clusters are mainly required to run high-resource and time-demanding applications which a single computer cannot run. Use of supercomputers ensures faster and efficient computational power. High-performance computing (HPC) can be used to build a centralized file server for the Web and can easily process the information with its high processing speeds. A weather prediction system generally involves a large amount of past data to be processed over for an efficient prediction of the future weather. This paper lays emphasis on the use of Markov's chain to develop a weather prediction model which depends on a larger set of input data types and can be implemented on a HPC system environment for a lesser computational time and higher accuracy. A flexible algorithm is proposed for weather prediction named averaged transit prediction (ATP) algorithm here. This model has been further integrated with a novel DNA cryptography-based algorithm named decimal bond DNA (DBD) algorithm for secured transmission of data between different processors of HPC. The simulated results on test bed formed by connecting five nodes in parallel mode forming supercomputing environment and having a performance of 0.1 Tflops gave predicted temperature, humidity, and wind speed for three different days with an accuracy of 85–95%.

A. Kairi (✉) · S. Gagan (✉) · T. Bera · M. Chakraborty
Department of Information Technology, Institute of Engineering & Management, Salt Lake, Kolkata, West Bengal, India
e-mail: animesh.kairi@iemcal.com

S. Gagan
e-mail: suruchigagan62@gmail.com

T. Bera
e-mail: taniabera008@gmail.com

M. Chakraborty
e-mail: mohuyacb@iemcal.com

M. Chakraborty et al. (eds.), *Proceedings of International Ethical Hacking Conference 2018*, Advances in Intelligent Systems and Computing 811,
https://doi.org/10.1007/978-981-13-1544-2_9

Keywords HPC · Weather forecasting model · Markov's chain · Weather prediction · Cluster computing · Numerical weather prediction · DNA cryptography

1 Introduction

1.1 HPC

It is a computer which has very high computation power, means the computing ability of a HPC is very high as compared to a general working computer [1]. Supercomputers were introduced back in the 1960s since then there has been a lot of development in the overall performance and efficient use of this technology. HPC is the future generation computer architecture which allows the large database and software to move to a larger data centers [2, 3]. This technology has completely revolutionized the computing environment. Cloud computing has been majorly enhanced by the advent of this technology, moreover, the HPC can provide numerous number of services [4, 5] which single computer was not capable of. HPC is majorly used to solve higher capability problem leading to the solution of a very large single problem in the least time required.

An HPC basically involves a very large number of computer nodes connected to a master node with the help of high-speed switch to send and receive data among itself. The master node is the main controller of every slave node, and it monitors the data being shared inside all the nodes (see Fig. 1).

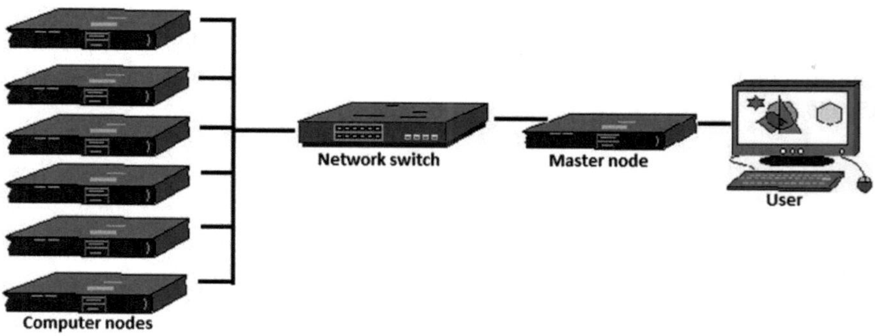

Fig. 1 HPC setup

1.2 Weather Forecasting Model

Weather forecasting or numerical weather prediction (NWP) [6] generally involves the use of mathematical equations on Earth models to calculate the future weather conditions based on the previous weather conditions. The predictions are generally categorized into two types: (i) short-term predictions and (ii) long-term predictions.

Short-term predictions are fairly more accurate as compared to that of long-term predictions. Here we have discussed short-term prediction model.

1.3 Markov's Chain

Markov's chain is a stochastic process which means a future dataset can be predicted depending completely on the present dataset [7]. Through this method, next set of possible events can be predicted depending completely on the previous event where the probability of happening of the next event is known. It is modeled by finite state machines where the state space is defined, and each state space is connected by a probability function which constitutes a random walk throughout the system.

Due to its simplicity, Markov's chain has many applications for real-world models. It plays a crucial role in the weather prediction models used worldwide [8]. Calculation is carried out by the common matrix method. The changes of the states present in the system are known as transitions, and the associated probabilities are called transition probabilities. To move from one state to another, we follow the transition matrix. As the probabilities may change rapidly, it is inaccurate to say that the prediction is 100% correct (see Fig. 2).

The organization of the paper is as follows. After the introduction in Sect. 1, the proposed weather prediction model called averaged transit prediction (ATP) is presented in Sect. 2. Section 3 describes the implementation of the model in HPC. Section 4 provides the actual physical implementation and simulation result in test bed. In Sect. 5, integration of the DNA-based secured algorithm called decimal bond DNA (DBD) with ATP is described. Section 6 concludes the paper with future scope.

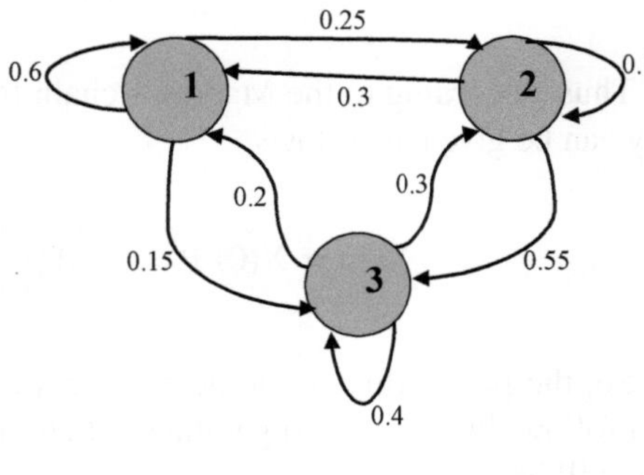

Fig. 2 A state diagram illustrating Markov's chain

Table 1 Transition table between two weather conditions

From	To	Probability
Sunny	Sunny	0.9
Sunny	Cloudy	0.1
Cloudy	Cloudy	0.5
Cloudy	Sunny	0.5

2 Markov's Chain Implementation for Weather Prediction

A simple model can demonstrate how Markov's chain can be used to predict weather for a short term. The different transition probabilities are stored in the transition matrix. Consider two weather conditions, namely sunny and cloudy. Table 1 shows the transition table.

The table shows that the probability of the next day being sunny is 90% and cloudy is 10% if the present day is sunny. The probability of the next day being sunny or cloudy if the present day is cloudy is 50% and 50%, respectively, which gives us the following transition matrix.

$$P = \begin{bmatrix} 0.9 & 0.1 \\ 0.5 & 0.5 \end{bmatrix}$$

Here the first row represents the sunny probabilities, and the 2nd row represents the cloudy probabilities [9]. The calculation of transition matrix plays a crucial role in the overall accuracy of weather prediction.

2.1 *Predicting the Weather on the Subsequent Days*

Let the weather on 1st day be sunny. Then a matrix X(0) will consists of two entries, that is, 1 for sunny and 0 for cloudy. So, we get the matrix as follows:

$$X(0) = [1\ 0]$$

Thus, according to the Markov's chain formula, the weather on the forthcoming day can be given as follows:

$$X(1) = X(0)\ P = \begin{bmatrix} 1 & 0 \end{bmatrix} \begin{bmatrix} 0.9 & 0.1 \\ 0.5 & 0.5 \end{bmatrix} = \begin{bmatrix} 0.9 & 0.1 \end{bmatrix}$$

So, the probability of the next day being sunny is 90% and cloudy is 10%. We can follow this process to get the predictions for more days. The weather after this day will be:

$$X(2) = X(1)\ P = \begin{bmatrix} 0.9\ 0.1 \end{bmatrix} \begin{bmatrix} 0.9\ 0.1 \\ 0.5\ 0.5 \end{bmatrix} = \begin{bmatrix} 0.86\ 0.14 \end{bmatrix}$$

Thus, we get a generalized formula that can be given as follows:

$$X(n) = X(n-1)P$$
$$X(n) = X(0)\ P^n$$

where n is the nth day from the present.

2.2 *Proposed Averaged Transit Prediction (ATP) Algorithm*

It can be realized that the calculation of the transition table or matrix is the most vital part of prediction through this method. So, to make the model efficient a transition table with the most accurate probability distribution is required.

In order to calculate the transition table, we have to be taken into account various parameters of weather, viz. temperatures, humidity, rainfall, visibility, wind speed, pressure. A table consisting of different weather conditions with their varying parameters must be made which will be considered as the pivot or standards. This means that a fixed value of the parameters must be set for all the different weather conditions to be taken into account so that one can differentiate between two or more weather conditions.

Now, all the parameter values must be stored in a database (this may be obtained from a weather station) for a duration of at least 2 days. Starting from the 1st day, the standard deviation of each of the parameters from the pivot values must be taken so as to decide the transition matrix which will give the transition probabilities from one weather to another. This process of calculation of transition matrix must be done on each of the previous day's data to get multiple transition matrices. These matrices will be then averaged to get a final transition matrix which will be used for the prediction of the forthcoming days.

The overall function of this method is to increase the efficiency in the calculation of short-term prediction which can only be increased by enhancing the accuracy of the of transition table calculation in Markov's chain. A care should always be maintained while calculating the transition matrix that the sum of all the probabilities in a row must every time add to 1.

3 HPC Implementation for Weather Prediction

The above method proposed is rather complex for a single location. In case of calculations required, the calculations for multiple locations would take significant amount of time as the process involves storing, managing, and manipulating a large amount of dataset. This cannot be carried out with a single machine, so a computer with extraordinary computational power is required to carry out the given task.

Apart from just the calculations, the graphical rendering or the simulation of the weather conditions would require a very high processing machine [10]. The involvement of HPC will divide the process over multiple nodes so that different nodes can handle data from different locations. A two-way communicating HPC link should be fixed so that in case of failure, the data is preserved in at least one of the systems. Weather prediction was not possible before 1950 because of low processing speed. After the advancement of technology and introduction of HPC, weather prediction is much more easy, reliable and efficient. The benefit of HPC is that it would process all the data in parallel, so similar amount of time would be required in prediction of multiple locations.

HPC enables larger data access, and by using the above-proposed method over an HPC system would result in a fairly accurate prediction of the weather. Not only short-term predictions, nowadays supercomputers have become so powerful that the long-term predictions can be taken care of easily. The computing process in case of HPC differs mainly on its two types, that is, (i) grid computing and (ii) cluster computing [11].

Use of other technologies such as machine learning and deep learning can be easily applied over the datasets for more accurate and reliable result [3]. The use of artificial neural network (ANN) has paved a new path in the overall prediction and is also aware to a little extent of the abrupt changes in the weather conditions [12, 13].

The proposed model can have an n * n transition matrix with n number of parameters to compute. A supercomputer can handle multiple parameters and on priority basis render useful parameters to give out an accurate transition matrix suitable for prediction.

4 Physical Implementation of the Model

The test bed setup of an HPC system by the use of five high processing machines (Intel Pentium CPU G3240@3.10 GHz dual core) connected through a single network switch (CISCO SF95D-08). The HPC connections are totally handled by the python language package MPI (message passing interface) [14], which enables the machine to communicate with each other.

The weather parameters have been obtained from Wunderground Web site [15], which provides the data from weather stations throughout the world. It also supplies data to giant platforms like Google. All the data of different parameters are called via an API to the system and saved in the database, and the above-mentioned method is implemented over the data values to predict the weather condition for the coming next three days. In this implementation, the master node is the main data conveyer and is the only user interface. All the other nodes communicate solely to the master node, and their progress is not traceable. The overall physical implementation of this

Table 2 Data table of actual weather values

Days in February 2018	6 February (partly cloudy)	7 February (clear sunny day)	8 February (partly cloudy)	9 February (scattered clouds)	10 February (scattered clouds)
Average temperature (C)	24.5	24.5	25	24.5	25
Average humidity (%)	62.25	52.75	54.25	61.25	56.75
Average wind speed (km/hr)	5.5	5	5	4.75	6.75

model is not very tough and can be easily duplicated over a large scale for larger coverage of locations.

4.1 Real-Life Prediction Simulation

The actual weather data of February 6 and 7, 2018 for the region of Kolkata was fed to the system. The 6th day was marked as partly cloudy and 7th day as clear sunny day according to a trusted source [16]. We give the averaged data as input to the computer to predict the weather of the upcoming days, that is, February 8–10, 2018. Table 2 shows the actual weather values.

The averaged data of February 6 and 7, 2018 were taken to calculate the transition table and provide prediction for 8, 9, and 10 February. The upcoming days according to this data were predicted to be partly cloudy, and the limits of the prediction for each parameter are provided in Fig. 3. The values thus predicted correspond closely to the real values of the day. The efficiency of this algorithm lies in the range of (85–95)% for a period of one-week prediction.

4.2 Application of DNA Cryptography

As there is communication of data involved between separate nodes (an interconnected network), the protection of the large amount of data becomes a concern for the admin. We use DNA cryptography to solve this problem.

DNA cryptography as the name suggests helps in preserving or hiding data in the form of DNA sequences. It requires the conversion of each of the alphabets in the file in combinational form of the four nitrogen bases, namely A (Adenine), G (Guanine), C (Cytosine), and T (Thymine), present in the DNA sequence present in human body (see Fig. 4).

DNA cryptography is one of the newest forms of cryptography which provides efficient and speedy protection of data while using less storage and computational power. The benefit of DNA cryptography is that it provides the decrypted data without

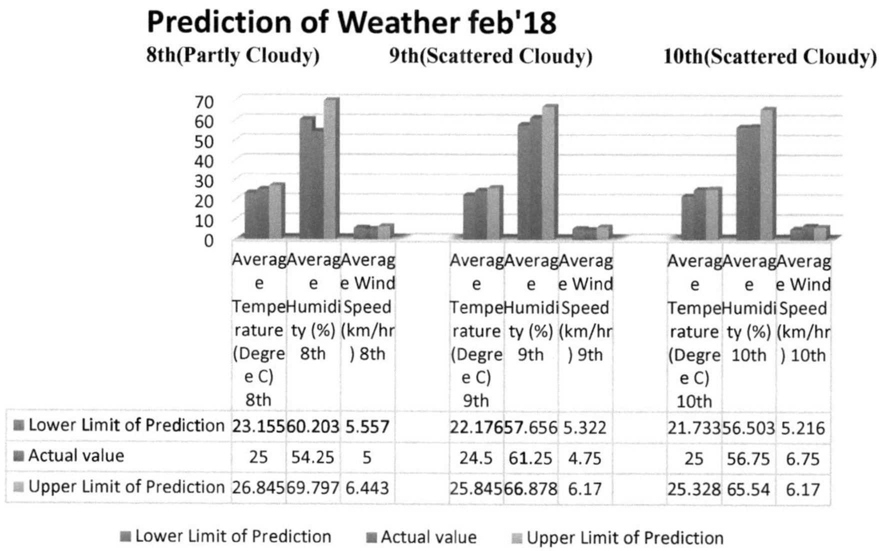

Fig. 3 Prediction diagram (using mentioned algorithm)

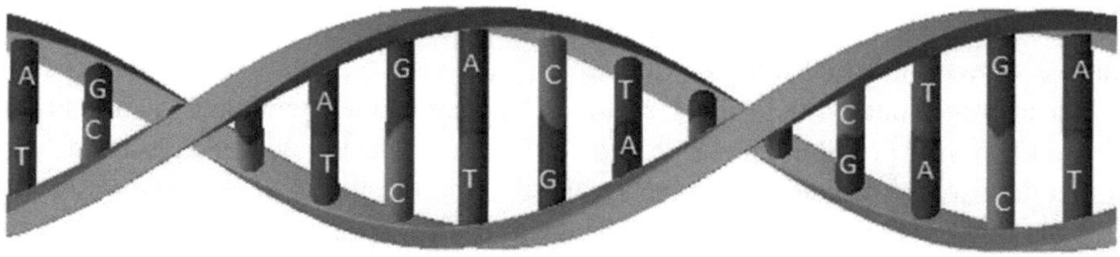

Fig. 4 A DNA strand

uncovering the entire encrypted message for respective users [17]. The traditional encryption methods involve application of long public and private keys and complex mathematical calculations which were tough to crack for the previous generation computers, but the latest technological advancements in computing would lead to these methods remaining futile.

DNA cryptography is still in infancy period and would require a bit more time to cultivate completely. Various researches all around the world have led to the development of different algorithms based on DNA cryptography, namely symmetric and asymmetric cryptography systems, using methods such as DNA sequencing and triple stage DNA cryptography.

For the purpose of this research work, we have used a novel cryptography algorithm to convert the data in database file (consisting vital information) into sequences of DNA that means the database files (.db file systems) are encrypted to DNA files (.db.dna file systems) [18] using the proposed DBD algorithm. These encrypted files are then transferred to another node through the network and then decrypted by the respective user node. This ensures that the integrity and confidentiality of the data present in the database remain safe, and even if the data is stolen during the process of communication through the network, the hacker would not be able to decipher the original contents of the database.

5 Proposed Decimal Bond DNA (DBD) Algorithm

5.1 Encryption

Consider a message M = "sunny" (5 characters) stored in the database.

We now use a DNA encode library which consists of the keywords 'A', 'G', 'C', 'T' mapped so as to give every character present on the keyboard a definite combination of the keywords.

Such as 's' → 'GCCT', 'u' → 'GACT', 'n' → 'ACCT', and 'y' → 'TAGC'.

As we have four keywords available, which can be repeated, a total of $4^4 = 256$ characters can be mapped.

So the message in example becomes M' = 'GCCTGACTACCTACCTTAGC' (20 characters).

Now, we use a padding library to conceal the message more which can be done by the use of padding table given below (see Table 3).

Using this padding table, we further encode each character of the message to get M'' = 'GCTACTAGCTAGTAGCGCTAAGCTCTAGTAGCAGCTCTAGCTAG-TAGCAGCTCTAGCTAGTAGCTAGCAGCTGCTACTAG' (80 characters).

Now, a binary base table can be used to convert the following message into binary information as shown in Table 4.

Using this binary table we convert the above message in binary sequences to give:

Table 3 Padding table

Keyword	Pad
A	AGCT
G	GCTA
C	CTAG
T	TAGC

Table 4 Binary table

Keyword	Binary value
A	00
G	01
C	10
T	11

Table 5 Bond table

Bond	Code
A-T	0
C-G	1

M''' = '01101100101100011011000111000110011011000001101110110001110001100001101110110001101100011100011000011011101100011011000111000 11011000110000110110110110010110001' (160 characters).

Now this message will again be padded using two of the keywords (which will be passed on as key to the host), as in a real DNA strand 'A' makes bond with 'T' and 'C' makes a bond with 'G'. So these bonds will be used to pad the message M''' as shown in Table 5.

In any of the two cases either 'A-T' or 'C-G' bonds, the first member will be padded at the starting end of the message and the second member of the bond will be added to the end of the message.

For example, if the bond 0, that is, 'A-T' bond is taken, then the message becomes, M'''' = '**AM'''T**', that is:
M'''' = '**00**01101100101100011011000111000110011011000001101110110001 110001100001101110110001101100011100011000011011101100011011000111000110110001100001101101101100101100011**11**' (164characters).

This binary message will lastly be converted to its decimal equivalent considering two bits at a time to give final encoded message:
M(F) = '01230230123013012123001232301301201231301230130120123210 12301301230120123123023013' (82 characters)

This M(F) is the final encoded message which is needed to be sent to the receiver. Here in this case, it is the slave nodes. All the weather data present in the database file will be encoded in this format (file.db.dna) and will be shared among the nodes where the receiver nodes will ultimately decode the file.

5.2 Decryption

The receiving nodes will be sent the data as well as the decryption key which can be given as:

DECRYPTION KEY—BOND CODE | PAD VALUE A | BINARY CODE A | PAD VALUE G | BINARY CODE G | PAD VALUE C | BINARY CODE C| PAD VALUE T | BINARY CODE T |

For this case which is '0AGCT00GCTA01CTAG10TAGC11'
The process of decryption starts by first converting the binary equivalent of each number of the encrypted message which will give M''''.

The bond pad will be removed according to the code value provided in the decryption key (M''').

The binary table code provided with the key will be used to get the message M''. Then the padding will be removed via the pad information provided in the key to get M'. Lastly, the 4 character encode library will be used to finally give the original message.

The receiver only requires to keep the encode library only, and all the other information is self-contained in the decryption key. By the change in just the binary table, the machine can create random encrypted messages for the different nodes to decode.

6 Conclusion

In this paper, implementation and result analysis of a DNA cryptography-based weather prediction model have been given. Simulated results in terms of various weather parameters like humidity, rainfall, and wind speed in test bed supercomputer show accuracy of the weather prediction in the range 85–95%. The efficiency of the proposed method is higher for short-term predictions and is less for long-term predictions (efficiency will decrease for long-term predictions). This model can be implemented for a small region to mentor the overall climatic region with passage of time. The proposed method is ignorant to abrupt change in weather, which can be rectified by changing the method of creation of transition table. The proposed DBD algorithm may be analyzed further to increase the overall performance of the model.

Various researchers have devised various methods of weather predictions using newer technologies [19] which can be modified to act on HPC to increase throughput. With the advancement in technology, the methods of prediction of weather might be modified, but the concept of usage of HPC for weather predictions will flourish.

References

1. https://en.wikipedia.org/wiki/Supercomputer
2. Ateniese, G., Pietro, R.D., Mancini, L.V., Tsudik, G.: Scalable and efficient provable data possession. In: Proceedings of SecureComm'08, pp. 1–10 (2008)
3. Rittinghouse, J.: HPC: Implementation, Management, and Security. Amazon Book Stores (2009)
4. Miller, M.: HPC: web-based applications that change the way you work and collaborate online. Online Journal (August 2008 Issue)
5. Curtmola, R., Khan, O., Burns, R., Ateniese, G.: MRPDP: multiple-replica provable data possession. In: Proceedings of ICDCS'08, pp. 411–420 (2008)
6. https://en.wikipedia.org/wiki/Numerical_weather_prediction
7. https://brilliant.org/wiki/markov-chains/
8. https://en.wikipedia.org/wiki/Markov_chain
9. https://en.wikipedia.org/wiki/Examples_of_Markov_chains
10. https://www.altair.com/c2r/ws2017/weather-forecasting-gets-real-thanks-high-performance-computing.aspx
11. Gu, Y., Grossman, R.L.: Sector and sphere: the design and implementation of a high performance data cloud. UK (2008)

12. Beltrn-Castro, J., Valencia-Aguirre, J., Orozco-Alzate, M., Castellanos-Domnguez, G., Travieso-Gonzlez, C.M.: Rainfall forecasting based on ensemble empirical mode decomposition and neural networks. In: Rojas, I., Joya, G., Gabestany, J. (eds.) Advances in Computational Intelligence. Lecture Notes in Computer Science, vol. 7902, pp. 471–480. Springer Berlin, Heidelberg (2013)
13. Abhishek, K., Kumar, A., Ranjan, R., Kumar, S.: A rainfall prediction model using artificial neural network. In: 2012 IEEE Control and System Graduate Research Colloquium (ICSGRC), pp. 82–87, July 2012
14. http://mpi4py.scipy.org/docs/
15. https://www.wunderground.com/
16. https://www.timeanddate.com/weather/india/kolkata/historic
17. http://securityaffairs.co/wordpress/33879/security/dna-cryptography.html
18. https://pypi.python.org/pypi/file2dna
19. Hernandez, E., Sanchez-Anguix, V., Julian, V., Palanca, J., Duque, N.: Rainfall prediction: a deep learning approach. In: Conference paper, April 2016

A Novel Approach of Image Steganography with Encoding and Location Selection

Debalina Ghosh, Arup Kumar Chattopadhyay and Amitava Nag

Abstract Steganography is a well-known technique of data hiding. The confidential pieces of information are concealed within cover media like image, audio, video such that it does not arouse any attention of eavesdroppers to scrutinize the object and that is the main advantage of steganography over conventional cryptography methods. Steganography techniques are increasingly used for audios and images. Least significant bit modification (LSB) is the most popular method for steganography. In this paper, we propose a novel secure LSB modification scheme, that can be used to hide a secret image with a few cover images. First, we encode the secret image using simple bitwise XOR operations to break strong correlation between adjacent pixels. Then the location selection algorithm has been carried out to find a particular position from least significant four bits to hide a secret bit (from secret image). We have experimented with the proposed method using MATLAB and tested on a grayscale secret image and two grayscale cover images.

Keywords Steganography · LSB · XOR · Cryptography · Grayscale image
Secret image

1 Introduction

Securing the confidentiality of digital data has a great importance in our daily life. As the usage of Internet is increasing, the need to secure secret information is also increasing. Along with this requirement, the number of data hiding techniques [1] is

D. Ghosh (✉) · A. K. Chattopadhyay
Institute of Engineering and Management, Salt Lake, Kolkata, West Bengal, India
e-mail: debalinag1986@gmail.com

A. K. Chattopadhyay
e-mail: ardent.arup@gmail.com

A. Nag
Central Institute of Technology, Kokrajhar, India

M. Chakraborty et al. (eds.), *Proceedings of International Ethical Hacking Conference 2018*, Advances in Intelligent Systems and Computing 811,
https://doi.org/10.1007/978-981-13-1544-2_10

also increasing. Watermarking, cryptography, and steganography are some of them. Each of these techniques has their own advantages and disadvantages.

Watermarking [2] is a very efficient data hiding technique. In this technique, noise tolerant signals are used. These noise tolerant signals are embedded in digital media for security purpose. The disadvantage comes with the introduction of these noise tolerant signals because sometimes it is impossible to extract the digital media at the receiver end.

In cryptography [3, 4], the secret data is encrypted by a key but any unauthorized access of the key can hamper the security of the secret data. If an intruder gets the key, then easily original data could be recognized by deciphering the ciphertext.

Steganography is a technique of hiding information within some digital cover media like text, image, audio, video. Steganography [5] consists of two Greek words—"*stego*" means "*cover*" and "*grafia*" means "*writing*"; defining as "*covered writing*." Steganography hides the existence of the secret data as the secret data is hidden in a carrier file. So the existence of secret data will remain unknown to the observer [6]. There are various image steganography techniques. Some are based on spatial domain, and in some cases, we need to consider pixel values in binary format. In spatial domain, the steganographer modifies the secret data and the cover medium which involves encoding at the level of the LSBs. LSB is the least significant bit in a series of numbers in binary [7]. For example in the binary number: 101100010, the least significant bit is the far right 0. In the LSB-based steganography to embed the secret data, the least significant bits of the pixel values in the cover images are used. Cover images are needed to hide secret data. Generally, cover images are of 8-bit gray level or color image [8]. Other popular techniques for image steganography are (a) **Discrete cosine transform** (DCT) [9] is a mathematical transformation that takes a signal or image and transforms it from spatial domain to frequency domain. So it can separate an image into high-, middle- and low-frequency components. For JPEG compression, DCT coefficients are used. And (b) **Discrete wavelet transform** (DWT) [10] in which the wavelet coefficients of the cover image are modified to embed the secret message. We have considered LSB-based steganographic technique in this paper. LSB-based methods manipulate the least significant bit (LSB) planes by directly replacing the LSBs of the cover image with the secret message bits. LSB methods typically achieve high capacity. Different LSB-based steganography methods are proposed in [11–15]. In the proposed scheme, the hiding of a secret image in one or more cover images has been explained where the original data has been encoded first. Then encoded data of the secret image has been embedded in such a way that undistorted recovery of the image is possible. Here for steganography a modified LSB technique has been used. Since the data has been encoded first and then hidden, such that the security has also increased. So, to achieve higher security two levels have been introduced in the approach. In the first level, all the bits of the secret image have been encoded, and in the second level, steganography has been carried out.

Rest of the paper is divided into five major parts; in Sect. 2, we have the discussion of the previous work done in the domain of steganography. Section 3 comprises the proposed algorithm. Section 4 contains the experimental results and analysis of the result. Section 5 concludes the paper.

2 Related Study

In the proposed scheme, we first encode the image to break the correlation between the adjacent pixels. Our encoding scheme utilizes only bitwise XOR operations between consecutive bits in all the pixels of the secret image. For steganography, we have used LSB technique very similar to [16] proposed by Pathak et al.

2.1 Brief Discussion on Audio Steganograpy Scheme Based on Location Selection [16]

It considers a secret text T_s to be hidden in a cover audio file A_c (16-bit .wav format). Traditional LSB scheme converts the secret text T_s in binary format. Then, each bit will be replacing the bit at LSB position of each sample of the cover audio sequentially. Unlike traditional LSB, this scheme selects a specific bit from least significant eight bits of a sample where the secret binary bit will be inserted. Hence, it enhances the security of the stego-audio. The steps are as follows:

Embedding of secret. Consider the encrypted secret text T_s which is in binary format with m number of bits. Store the audio samples from the cover audio file A_c into an array $SAMPLES[]$.
Step 1: for $i = 1$ to m.
Step 1.1: Consider $SAMPLES[i]$ as cover sample.
Step 1.2: Compute the decimal equivalent of first (MSB) three bits as j.
Step 1.3: Insert the secret bit $T_s[i]$ at j position from LSB of $SAMPLES[i]$.
Step 2: Store the modified $SAMPLES[i]$ for ($i = 1$, to, m) into stego-audio file A_{stego}. Keep the rest of the samples in A_{stego} same as original cover A_c.
Step 3: Transfer the stego-audio file A_{stego} on public channel.

Extraction of secret. The encrypted secret message can be extracted if the number of bits in the secret text m is known to the receiver. The receiver performs the following actions to reveal the secret inside the stego-image.
Step 1: Store the audio samples from the stego-audio file A_{stego} into an array $SAMPLES[]$.
Step 2: for $i = 1$ to m.
Step 2.1: Consider $SAMPLES[i]$ as stego-sample.
Step 2.2: Compute the decimal equivalent of first (MSB) three bits as j.
Step 2.3: Extract the secret bit b_i from jth position from LSB of $SAMPLES[i]$.
Step 3.0: Combine the m bits as binary sequence as $b_m b_{m-1} \ldots b_2 b_1$ and regenerate the secret encrypted text T_s.

3 Proposed Method

In the proposed scheme, the inputs are the grayscale secret image ($n \times n$) and m grayscale cover images ($s \times r$). In this scheme, the secret image will be first encoded and then LSB-based steganography has been used to hide the secret image in cover images. For steganography, a modified LSB with location selection has been used. The process for encoding and embedding is as shown in Fig. 1, whereas extraction and decoding are as shown in Fig. 2.

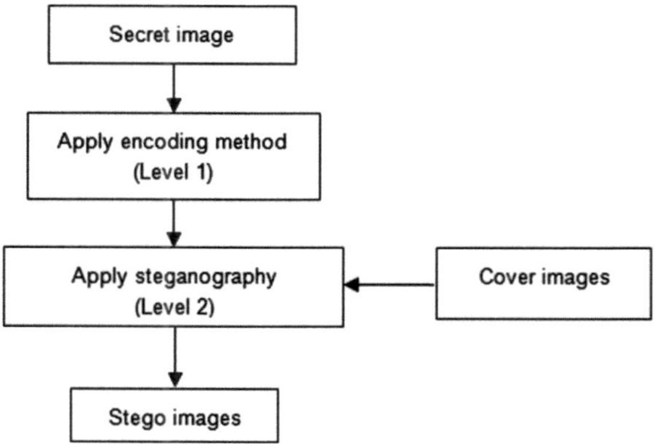

Fig. 1 Block diagram presenting embedding process

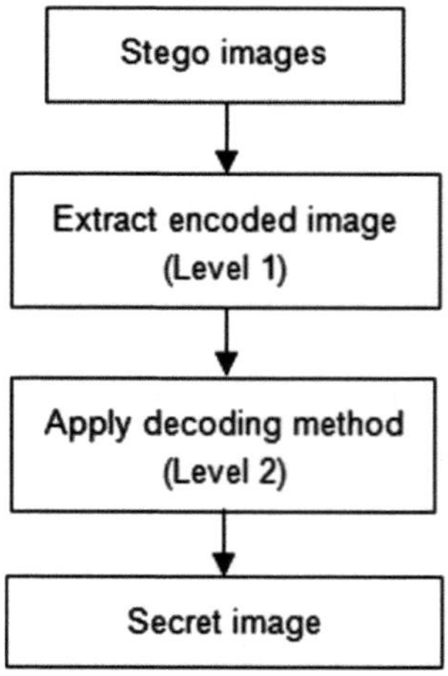

Fig. 2 Block diagram presenting extraction process

3.1 Embedding and Encoding Phase

1. Let I_s be the original secret image of $n \times n$ pixels. If the original image having $n \times k(k < n)$ pixels, then use sufficient padding to make it $n \times n$ pixels.
2. Now $n \times n \times 8$ bits of the secret image I_s will be embedded separately in each pixel (byte) of m cover images by using the method of encoding and steganography. If each cover image contains $s \times r$ pixels, then we need m cover images to embed $n \times n \times 8$ bits of secret image where $m = \left\lceil \frac{n \times n \times 8}{r \times s} \right\rceil$.
3. Now for each bit b_i; $(i = 1$ to $n \times n \times 8)$ from the secret image I_s, each bit will be modified in the following manner:
   Suppose b_i is the original bit and t_i is the modified one where $i = 1$ to $n \times n \times 8$ then,
   $t_1 = b_1 \;\; if \;\; i = 1$
   $t_i = t_{i-1} \oplus b_i \;\; if \;\; i > 1 \; and \; i \leq n \times n \times 8.$
4. For steganography, we have m different cover images. As $n \times n \times 8$ bits to be concealed into $m \times s \times r$ bytes and number of bits from secret image is equal to total number bytes in cover images, we need to insert one bit in each byte (or pixel).
   For each bit t_i from secret encoded image and each byte (pixel) P_i from cover images (where $i = 1$ to $n \times n \times 8$) compute as follows:
   (a) Convert P_i to its 8-bit binary representation $\{p_8 p_7 p_6 p_5 p_4 p_3 p_2 p_1\}_2$.
   (b) Compute the decimal value for $\{p_8 p_7\}_2$ as d_i.
   (c) Insert the secret bit t_i at position $(d_i + 1)$ of the binary representation of P_i. The t_i will replace either p_1 or p_2 or p_3 or p_4.
5. By modifying specific one of the four least significant bits of each pixel of cover images, we achieve m stego-images.

3.2 Extraction and Decoding Phase

We first extract the secret bits from the stego-images, then construct the encoded secret image. After that apply decoding algorithm to retrieve the actual secret image.

1. For each byte (or pixel) P_i $(i = 1$ to $m \times r \times s)$ from m stego-images extracts the secret bits t_i as follows:
   (a) Convert P_i to its 8-bit binary representation $\{p_8 p_7 p_6 p_5 p_4 p_3 p_2 p_1\}_2$.
   (b) Compute the decimal value for $\{p_8 p_7\}_2$ as d_i.
   (c) $t_i = p_{d_i+1}$.
2. After extracting $n \times n \times 8$ bits, reconstruct the encoded secret image.
3. For each bit t_i, $i = 1$ to $n \times n \times 8$ from encoded secret image compute the decoded bit b_i as follows:

$b_1 = t_1 \; if \; i = 1$
$b_i = b_{i-1} \oplus t_i \; if \; i > 1 \; and \, i \leq n \times n \times 8$

4. Now, from $b_i(i = 1$ to $n \times n \times 8)$ generate the secret image I_s.

4 Experimental Results

This section deals with the experimental results of the proposed method using MATLAB R2012a. For experiment, we have considered two grayscale cover images (mandril_gray.tif and cameraman.tif) of dimension 256 × 256 and one grayscale secret image (lena.tif) of dimension 128 × 128. Since we are going to embed 128 × 128 × 8 bits of secret image in 256 × 256 pixels cover images, so we need $m = \frac{128 \times 128 \times 8}{256 \times 256} = 2$.

The secret image and two cover images are shown in Fig. 3a–c. After encoding, the encoded image is shown in Fig. 3d. By embedding the encoded secret bits, we get two stego-images as shown in Fig. 4a, b. The encoded image extracted from the stego-images is shown in Fig. 4c. The final decoded secret image is shown in Fig. 4d.

(a) Secret Image

(b) Cover Image-1

(c) Cover Image-2

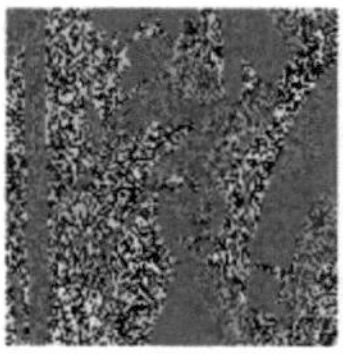

(d) Encoded image before embed

Fig. 3 Secret image (**a**), cover images (**b**), (**c**) and encoded image before embed (**d**)

(a) Stego Image-1

(b) Stego Image-2

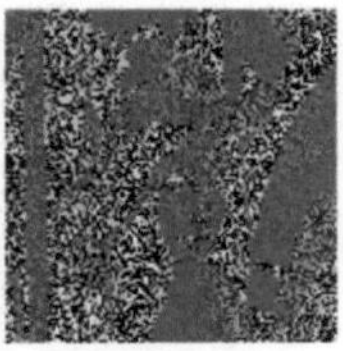

(c) Extracted encoded image

(d) Final decoded image

Fig. 4 Stego-images (**a**), (**b**), extracted encoded image (**c**), and final decoded image (**d**)

5 Analysis of Results

The cover images—mandril_gray.tif and cameraman.tif—are shown in Fig. 3b, c, whereas the corresponding stego-images are shown in Fig. 4a, b. The histograms of cover-image1 and stego-image1 (for mandril_gray.tif) are shown in Fig. 5a, c. Similarly, the histograms of cover-image2 and stego-image2 (for cameraman.tif) are shown in Fig. 5b, d. The stego-image and the cover image are compared to verify the quality of the obtained stego-image in the proposed scheme as follows.

5.1 Mean Square Error (MSE)

The mean square error between the cover image $g(x, y)$ and stego-image $\hat{g}(\hat{x}, \hat{y})$ can be represented as:

$$MSE = \frac{1}{M \times N} \sum_{n=1}^{M} \sum_{m=1}^{N} [\hat{g}(n, m) - g(n, m)]^2$$

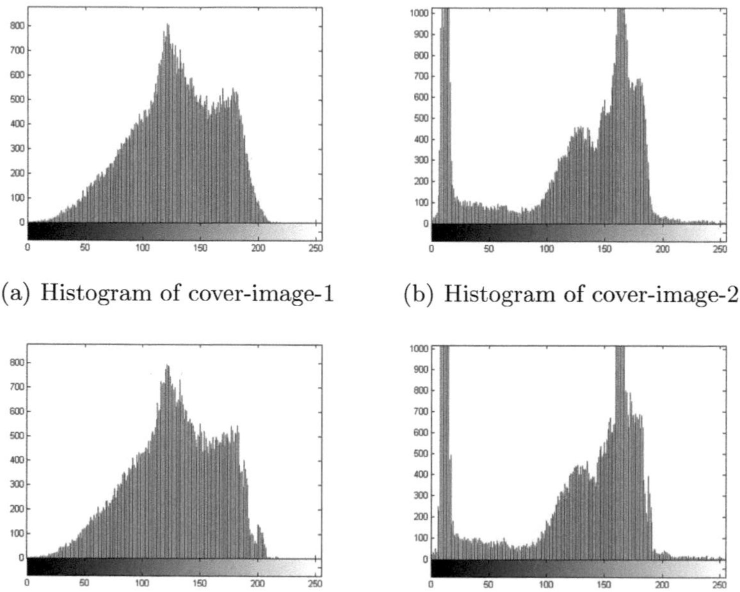

(a) Histogram of cover-image-1 (b) Histogram of cover-image-2

(c) Histogram of stego-image-1 (d) Histogram of stego-image-2

Fig. 5 Comparison of histogram of cover images and stego-images

The MSE values computed for mandril_gray.tif and cameraman.tif are 3.02 and 2.90, which is low considering other traditional steganography schemes means very little distortion induced to the stego-images by our algorithm.

5.2 *Peak Signal-to-Noise Ratio (PSNR)*

The quality of the image better represented by PSNR (as MSE having strong dependency image intensity scaling which does not effect PSNR). The PSNR can be calculated from MSE as:

$$PSNR = 10\log_{10}\frac{S^2}{MSE}$$

where S is the maximum pixel value and result is measured in decibels (dB). The PSNR values computed for mandril_gray.tif and cameraman.tif are 43.36 and 43.55 dB, which is high considering other traditional steganography schemes means high quality of stego-images.

5.3 *Structural Similarity Index Metric (SSIM)*

SSIM index is used to find dissimilarities between two images. SSIM index value is within the range from 0 to 1. A value 0 presents two images that are all dissimilar and 1 means the images are identical. If two images are X and Y, the SSIM is defined as:

$$SSIM(X, Y) = \frac{(2\mu_X \mu_Y + C_1)(2\sigma_{XY} + C_2)}{(\mu_X^2 + \mu_Y^2 + C_1)(\sigma_X^2 + \sigma_Y^2 + C_2)}$$

where μ_X and μ_Y are the mean intensity of X and Y;
σ_X^2 and σ_Y^2 are the variance of X and Y;
σ_{XY} is the covariance between X and Y;
$C_1 = (k_1 L)^2, C_2 = (k_2 L)^2$ are two variables to stabilize the division with weak denominator and L is the dynamic range of the pixel values chosen as $L = 255$. The value of $k_1(\ll 1)$ and $k_2(\ll 1)$ are chosen as $k_1 = 0.01, k_2 = 0.03$.

SSIM index between cover image and stego-image calculated for mandril_gray.tif is 0.9879 and cameraman.tif is 0.9568, which implies that the stego-images are almost similar of the cover images.

6 Conclusion

In this paper, we have proposed a steganography scheme to conceal a secret image within multiple cover images. In proposed work, we have considered grayscale images only. But the scheme can be easily extended for color images, if we repeat the algorithm for each of the three color planes (RGB images). Before hiding, the secret image has encoded first using bitwise XOR operations (computationally low-cost operation) and then the encoded bits are inserted into multiple cover images. We have used a modified LSB technique which will select a bit position out of least significant four bits of a pixel of a cover image. For this scheme, we may need more than one cover image as each single bit from the secret image after encoding will be inserted in one byte (or one pixel) of the cover image. So, if the cover images are of less dimensions, then we may need more than one cover images. Then from those stego-images, we first extract the encoded secret image and decode it to regenerate the original secret image.

References

1. Fridrich, J.: Applications of data hiding in digital images. In: 5th International Symposium on Signal Processing and Its Applications (ISSPA). IEEE, Brisbane, Queensland, Australia (1999)
2. Nin, J., Ricciardi, S.: Digital watermarking techniques and security issues in the information and communication society: the challenges of noise. In: 27th International Conference on Advanced

Information Networking and Applications Workshops, pp. 1553–1558. IEEE, Barcelona, Spain (2013)
3. Kumari, S.: A research paper on cryptography encryption and compression techniques. Int. J. Eng. Comput. Sci. **6**(4), 20915–20919 (2015)
4. Huang, Q., Wong, D.S., Yang, G.: Heterogeneous signcryption with key privacy. Comput. J. **54**(4), 525–536 (2011)
5. Petitcolas, F.A.P., Anderson, R.J., Kuhn, M.G.: Information hiding a survey. Proc. IEEE **87**(7), 1062–1078 (1999)
6. Cachin, C.: An information-theoretic model for steganography. Inf. Comput. **192**(1), 41–56 (2004)
7. Kaur, N., Behal, S.: A survey on various types of steganography and analysis of hiding techniques. Int. J. Eng. Trends Technol. **11**(8), 388–392 (2014)
8. Samidha, D., Agrawal, D.: Random image steganography in spatial domain. In: International Conference on Emerging Trends in VLSI. Embedded System, Nano Electronics and Telecommunication System (ICEVENT), pp. 1–3. IEEE, Tiruvannamalai, India (2013)
9. Sheidaee, A., Farzinvash, L.: A novel image steganography method based on DCT and LSB. In: 9th International Conference on Information and Knowledge Technology (IKT), pp. 116–123. IEEE, Tehran, Iran (2017)
10. Surse, N.M., Vinayakray-Jani, P.: A comparative study on recent image steganography techniques based on DWT. In: International Conference on Wireless Communications. Signal Processing and Networking (WiSPNET), pp. 1308–1314. IEEE, Chennai, India (2017)
11. Wang, R.-Z., Lin, C.-F., Lin, J.-C.: Hiding data in images by optimal moderately-significant-bit replacement. Electron. Lett. **36**(25), 2069–2070 (2000)
12. Chan, C.-K., Cheng, L.M.: Hiding data in images by simple LSB substitution. Pattern Recognit. **37**, 469–474 (2004)
13. Karim, S.M.M., Rahman, M.S., Hossain M.I.: A new approach for LSB based image steganography using secret key. In: 14th International Conference on Computer and Information Technology (ICCIT), pp. 286–291. IEEE, Dhaka, Bangladesh (2011)
14. Sapra, P.S., Mittal, H.: Secured LSB modification using dual randomness. In: International Conference on Recent Advances and Innovations in Engineering (ICRAIE), pp. 1–4. IEEE, Jaipur, India (2016)
15. Blue, J., Condell, J., Lunney, T.: Identity document authentication using steganographic techniques: the challenges of noise. In: 28th Irish Signals and Systems Conference, pp. 1–6. IEEE, Killarney, Ireland (2017)
16. Pathak, P., Chattopadhyay, A.K., Nag, A.: A new audio steganography scheme based on location selection with enhanced security. In: First International Conference on Automation, Control, Energy and Systems (ACES), pp. 1–4. IEEE, Hooghy, India (2014)

Image Encryption Using Pseudorandom Permutation

Tapan Kumar Hazra, Kishlay Raj, M. Sumanth Kumar, Soummyo Priyo Chattopadhyay and Ajoy Kumar Chakraborty

Abstract A simple, fast, and dynamic image encryption scheme is proposed in this paper based on pixel shuffling. The primary idea applied for the purpose lies in designing of dynamic pseudorandom permutation map using simple divide and conquer method that will introduce diffusion among strongly correlated image pixels. As a result, the visual information content is completely lost. Data encryption is to hide and conceal the information content present in the secret data. Data encryption differs from data hiding, where the prime objective is to conceal the secret data. Using the technique of reversible data encryption, we can get back the original data content out of the encrypted message, without loss of any original information. The decryption process can be applied only at the authentic receiver end. Thus, we can transfer the encrypted data through any communication channel because it has completely lost its resemblance to the original data. Though theoretically it is possible to break such security, it does not appear feasible by any known practical means as it appears as random noise to any unintended recipient. To incorporate second fold of security, some proper data hiding methods can be used to embed the encrypted data. In this paper, we have presented a new visual data encryption technique by randomly shuffling the pixels within the image. The newly formed encrypted image using this technique completely loses the original characteristics. Our experimental results show that the proposed method can achieve the claim.

Keywords Circular left shift · Circular right shift · Cryptography
Divide and conquer · Horizontal encryption · Image encryption
Pseudorandom permutation · Vertical encryption

T. K. Hazra (✉) · K. Raj · M. Sumanth Kumar · S. P. Chattopadhyay · A. K. Chakraborty
Department of Information Technology, Institute of Engineering & Management, Y-12, Salt Lake Electronics Complex, Sector-V, Kolkata, India
e-mail: tapankumar.hazra@iemcal.com; tapankumarh@yahoo.com

M. Chakraborty et al. (eds.), *Proceedings of International Ethical Hacking Conference 2018*, Advances in Intelligent Systems and Computing 811,
https://doi.org/10.1007/978-981-13-1544-2_11

1 Introduction

It is Internet and dawn of the digital natives that have made this world a small village with advanced information and communication system for the society. The whole of the global system is intensifying with the quantity of data that is stored and transmitted. Unfortunately, this ease has also raised new challenges concerning the security and protection of important information and data against unauthorized access. Increase in interconnectivity, growth of networks, and number of users and decentralization has increased system vulnerability. So the much-needed security of information and communication system involves the protection and confidentiality of the system and the data to be transmitted and stored.

Steganography and cryptography are two different information and data hiding techniques. Steganography hides messages inside some other digital media while cryptography protects the content of messages for secure communication in the presence of third parties (called adversaries) [1–4]. Sometimes both reapplied in succession [5, 6]. The word steganography is derived from Greek, and it means "covered writing", the art of hiding information in ways that prevent detection. There can be many ways to hide information. To embed secret data, either every byte of cover media are selected sequentially, or bytes are selectively chosen from insignificant regions that draw less attention of viewer [5]. After location selection in cover media, a straight message insertion may encode every byte. Messages may also be scattered following some SCAN pattern throughout the cover data [6]. Least significant bit (LSB) insertion is one of the common methods of data hiding.

Image encryption techniques have been increasingly studied to satisfy the demands for secure image transmission over the network. Traditional image encryption techniques which are nonrandom and static in nature are more vulnerable to be hacked with possibility of finding similarity in the encoded image. In most cases, the encoded message that comes out is same every time we apply it on the same image. In this paper, we present an image encryption technique which is both dynamic and robust. It generates a new random image with set of keys, every time we apply it on an image, making it pseudorandom and at the same time difficult to analyze or to decode. We have found that the encryption technique discussed here creates new random images, out of which none of them carry any resemblance with the original image and seem to be pure noise.

In the present paper, we have considered the visual information in a grayscale image and propose a new method by shuffling pixel information randomly within the image. The technique breaks the array of pixels of the image into different subsets using divide and conquer technique, applied in a way similar to merge sort. Then a circular rotation is performed on every subset after dividing the array with the help of a random key generated. This way of shuffling pixel information within the image dimension gives some sort of pseudorandom permutation and also we can get back the original image exactly using generated key.

The organization of the paper is as follows: Sect. 2 focusses on existing works on image encryption, Sect. 3 describes the proposed encryption and decryption algorithm with illustrations, Sect. 4 focusses on result and discussion, and Sect. 5 concludes the paper.

2 Existing Works on Image Encryption

Digital image is a two-dimensional array of pixel intensities, and those intensity values are highly correlated for plain image. Image encryption techniques primarily disrupt the normal interpretation of visual information represented by those intensity values. There are various methods to achieve this. Pseudorandom permutation is applied to shuffle spatial locations of pixels within image boundaries. Using pseudorandom permutation (PRP), we can generate permutation cipher that is successfully applied on text or images to perform encryption [1–3]. The main limitation is this cipher is vulnerable to statistical attack, and security increases with the length of the key. In case of text file encryption, slightly different techniques are applied [7, 8].

Image encryption based on chaos theory is very popular and efficiently encrypt images. Limitations of PRP-based image encryption are mostly overcame [9–17].

Image encryption using principle of optics is also applied [18]. There is also some research work that focusses on security enhancement in addition to standard RSA algorithm [19, 20].

2.1 Pseudorandom Permutation

In the present work, we have rearranged the information using divide and conquer and circular right shift technique which creates a new randomly rearranged image of the original image, every time we apply it. The proposed technique is not truly random and generates a set of keys, or else it would become impossible to get back the original data.

We applied the proposed technique on image data which are intensity level of pixels. We treat the image as a simple 2-D array with pixels intensity values. This technique works on each line of pixels treating them as simple 1-D array.

At each step in the process, we break the array in two equal halves (ignore the middle element in case of odd number of elements). We swap the positions of each of the elements to the other side. Now we do the right shift of the each of these two groups for which we generate two random numbers and perform the right shift with these values. The randomly generated numbers are stored in the key set where it will be used in the future to retrieve the information back. The steps are repeated recursively until we break them down to the level of individual pixels.

3 Proposed Algorithm

A digital image is a 2-D array of pixel intensity values. The proposed algorithm consists of two parts: encryption and decryption. Encryption process is done by shuffling the elements horizontally row-wise as well as vertically column-wise. We call these as horizontal encryption and vertical encryption, respectively. Exactly inverse operations are done in decryption.

3.1 Encryption Process

This technique works on each row of pixels, treating them as simple 1-D array. We applied the discussed technique on an image data to the level of pixels. We have rearranged the information using divide and conquer and circular right shift technique which creates a different randomly rearranged image of the original image, every time we apply it.

At each step in the process, we break the array in two equal halves (ignore the middle element in case of odd number of elements). We swap the positions of each of the elements to the other side. Now we do the right shift of the each of these two groups for which we generate two random numbers and perform the right shift with these values. The randomly generated numbers are stored in the key set where it will be used in the future to retrieve the information back. The steps are repeated recursively until we break them down to the level of individual pixels.

For example, let us take an array of 10 elements consisting of numbers from 0 to 9, and each encryption algorithm steps are specified and illustrated with the help of these dummy elements.

0 1 2 3 4 5 6 7 8 9

0 1 2 3 4 5 6 7 8 9

Step 1: Break into two equal parts (divide).

0 1 2 3 4	5 6 7 8 9

Step 2: Swap two parts

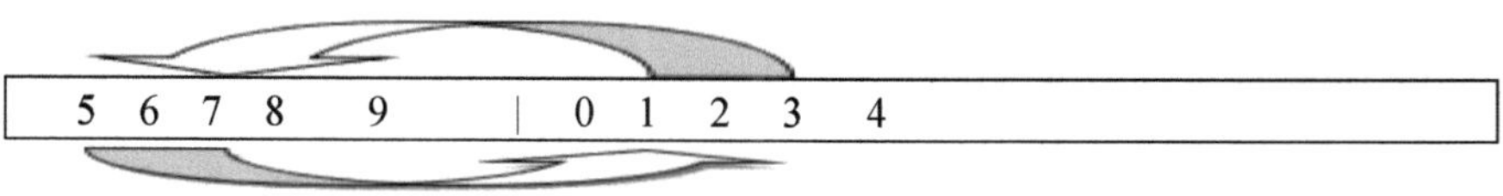

Step 3: (a) Generate two random numbers
Let random number 1 is 4 and random number 2 is 1

(b) Perform Circular Right Shift on each half by respective random number.

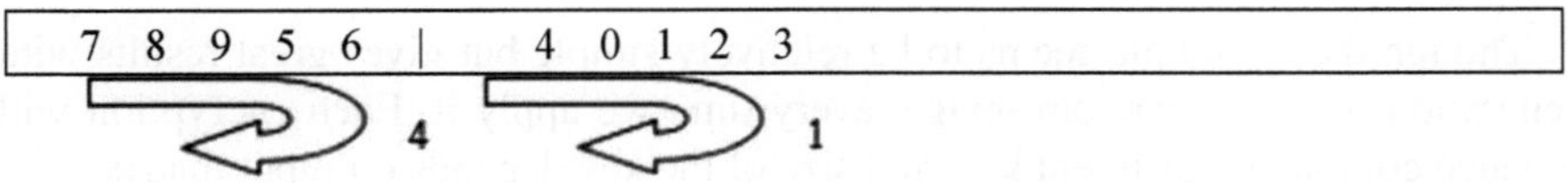

Step 4: Break each part into two sub parts and apply steps 1 through 3 recursively until number of elements in each part is two.

Further illustrations for successive recursive calls are shown in sub-steps and associated diagrams as follows:

Step 1:

78|9|56 | 40|1|23

Step 2:
Swap each sub-part

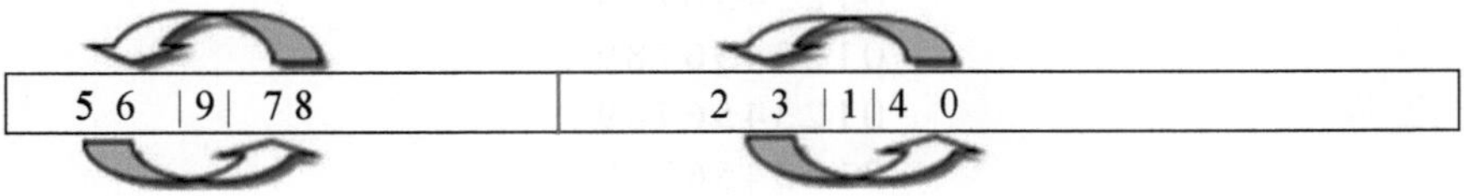

Step 3:
Generate random numbers for each part:
For first part, random number 1 is 1 and random number 2 is 1.
For second part, random number 1 is 0 and random number 2 is 1.
Perform circular right shift

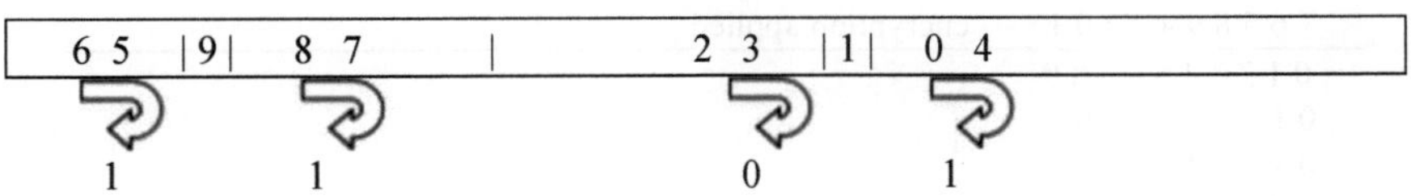

Now stop the recursive calls as the number of elements in no exceeding 2. So the final encrypted array obtained is

6 5 9 8 7 2 3 1 0 4

And the randomly generated key is generated as **4 1 1 1 0 1**.

This encryption technique is applied on each row of the 2-D array of pixels treating them as separate 1-D array.

Now, the same technique can also be applied to the vertical line, i.e., column-wise, treating them as another form of 1-D array.

We can apply the technique both horizontally and vertically (every time getting a more complex shuffled image), and therefore, we need to remember the sequence and do the exact reverse to get back the original image.

Though the technique seems to be relatively simple but gives great results with generation of a new random image, every time we apply it. Each encryption will produce completely different key, and size of the key depends on input matrix.

This technique is also very flexible as well because any user who wants to encrypt can create his/her own sequence of horizontal and vertical encryption which will have a relatively different technique to decode (the exact reverse sequence which only the user knows). So it can be easily customized and still remains utterly difficult to decode.

To understand its functioning more deeply, let us take an example of a simple 10×10 matrix

0 1 2 3 4 5 6 7 8 9
0 1 2 3 4 5 6 7 8 9
0 1 2 3 4 5 6 7 8 9
0 1 2 3 4 5 6 7 8 9
0 1 2 3 4 5 6 7 8 9
0 1 2 3 4 5 6 7 8 9
0 1 2 3 4 5 6 7 8 9
0 1 2 3 4 5 6 7 8 9
0 1 2 3 4 5 6 7 8 9
0 1 2 3 4 5 6 7 8 9

Now we apply to above-discussed method on the first row of matrix and that becomes

7 6 5 8 9 4 3 2 0 1 <--- encryption applied	
0 1 2 3 4 5 6 7 8 9	others rows are not disturbed
0 1 2 3 4 5 6 7 8 9	
0 1 2 3 4 5 6 7 8 9	
0 1 2 3 4 5 6 7 8 9	
0 1 2 3 4 5 6 7 8 9	
0 1 2 3 4 5 6 7 8 9	
0 1 2 3 4 5 6 7 8 9	
0 1 2 3 4 5 6 7 8 9	
0 1 2 3 4 5 6 7 8 9	

With the key 2 0 0 1 0 1 for row 1, which were again generated randomly so the permutation came out to be different from that of the previous case. We now apply the same technique to all the rows of the given matrix.

	Row
7 6 5 8 9 4 3 2 0 1	1
9 8 7 5 6 0 4 3 1 2	2
6 5 9 7 8 0 4 3 1 2	3
6 5 9 7 8 0 4 3 2 1	4
8 7 6 5 9 1 0 4 3 2	5
6 5 9 7 8 1 0 4 2 3	6
8 9 7 5 6 3 4 2 0 1	7
9 5 8 7 6 0 1 4 3 2	8
6 7 5 8 9 0 1 4 2 3	9
8 9 7 6 5 2 3 1 0 4	10

We call this part of encryption as horizontal encryption. Horizontal encryption jumbles up all the rows among themselves. We also keep the keys generated for encryption in 2-D array form so that we can distinguish keys for different rows easily. So, the list of keys in 2-D array form is given below.

Keys	
2 0 0 1 0 1	for Row1
0 4 0 1 0 1	for Row2
3 4 0 1 0 1	for Row3
3 4 0 1 0 1	for Row4
1 3 0 0 0 0	for Row5
3 3 0 1 0 1	for Row6
0 0 1 1 1 1	for Row7
4 3 1 0 1 0	for Row8
2 3 1 1 1 1	for Row9
0 1 1 0 1 0	for Row10

To make our encryption more secure, we apply the same method on the vertical lines, i.e., on the columns of the encrypted array and treating each column as single 2-D array. We call the method of applying encryption on each vertical lines, i.e., columns as vertical encryption. Separate key is produced dynamically for the process.

So, we observe that just after two series of encryption one after other the encrypted matrix has lost its resemblance totally with the original matrix. Unlike other transposition ciphering processes, the proposed process is dynamic and random, and localized transposition effect is absolutely removed for large 2-D array.

To get back the original matrix, we have to perform the decryption technique in the exact reverse manner that we choose to encrypt it. In this situation, we first performed horizontal encryption and then the vertical encryption on it. So we will have to first perform the vertical decryption on it and then the horizontal decryption.

When we perform the vertical decryption, we get back the matrix which was formed after the horizontal encryption was applied to the original matrix so as second step we perform horizontal decryption and get back the original matrix.

```
Input: Original Matrix
Step1: Horizontal encryption (on rows)
Step2: Vertical encryption (on columns)
Step3: Vertical decryption
Step4: Horizontal decryption
```

Any other sequence of decryption other than the exact reverse of the original process will not be able to recover the original matrix.

This also means the encryption technique can be easily customized by user making it more secure, because the sequence will be known only to the original key. For example, if the encryption was performed via three successive horizontal encryptions, only three horizontal decryptions will be able to decrypt it with the corresponding keys.

```
Input: Original Matrix
Step1: Horizontal encryption (1)
Step2: Horizontal encryption (2)
Step3: Horizontal encryption (3)
Step4: Horizontal decryption (key set 3)
Step5: Horizontal decryption (key set 2)
Step6: Horizontal decryption (key set 1)
```

3.2 Decryption Process

We take the encrypted array 6 5 9 8 7 2 3 1 0 4 and break it down to the lowest level.

```
Step 1: Break into two parts
```

6 5 9 8 7	2 3 1 0 4

```
Again break it into sub parts
```

6 5	9	8 7	2 3	1	0 4

Now, after reaching the lowest level, we will start using the key (randomly generated keys which were generated during the time of encryption). Key: 4 1 1 1 0 1.

We will start using the keys in exact reverse fashion and do circular left shift and then swap the elements.

6 5 | 9 | 8 7 | 2 3 | 1 | 0 4

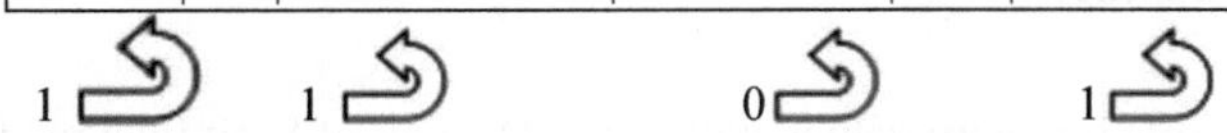

After the circular left shift, the array becomes

5 6 | 9 | 7 8 | 2 3 | 1 | 4 0

Step 2: Swap each subpart

7 8 |9| 5 6 | 4 0 | 1 | 2 3

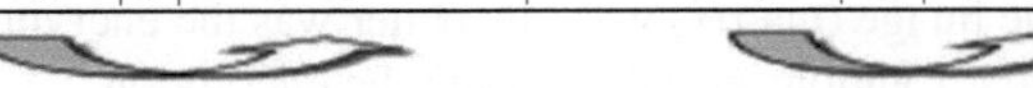

After swapping, the array becomes

7 8 9 5 6 | 4 0 1 2 3

Key: 4 1 1 1 0 1, and the part yet to used (using the reverse order) are 4 1

7 8 9 5 6 | 4 0 1 2 3

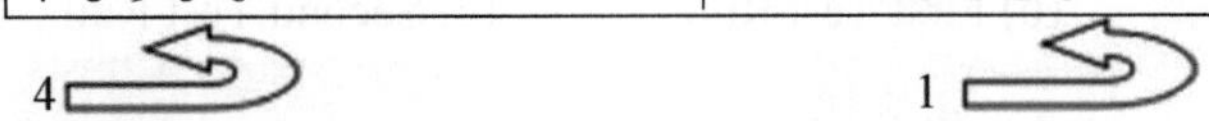

After the shift the array becomes

5 6 7 8 9 | 0 1 2 3 4

Performing the swap operation:

5 6 7 8 9 | 0 1 2 3 4

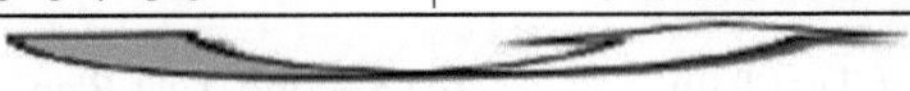

Decrypted array becomes

0 1 2 3 4 | 5 6 7 8 9

So, the shuffled elements are placed back to the original position correctly.

4 Results and Discussion

A digital image records pixels intensity values within a 2-D array of 8-bit unsigned integers. We applied the proposed encryption technique to the image and observed that the image has totally lost its resemblance to the original image. First encryption is applied on all the rows (horizontal lines). After encryption is completed on all the horizontal lines, we applied it to all the columns (vertical lines).

Even with the simplest permutation of horizontal and vertical encryption, the results were astounding. The technique is applied on Lena and Cameraman and results are shown in Figs. 1 and 2. Two different encryption results are shown, and various encryption quality parameters indicate that the proposed process is dynamic. We also found that the bigger the image (matrix) was, the better was the encryption because of the bigger right circular rotations possible during the encryption.

Necessary security analysis was carried out on plain image and encrypted images using the new algorithm, and simulation results show that encryption and decryption are good, and the algorithm demands good security and robustness. Table 1 represents entropy and PSNR values computed at various stages of encryption. Plain image Lena has entropy 5.332146228071736, and cameraman has entropy 5.037729765949852.

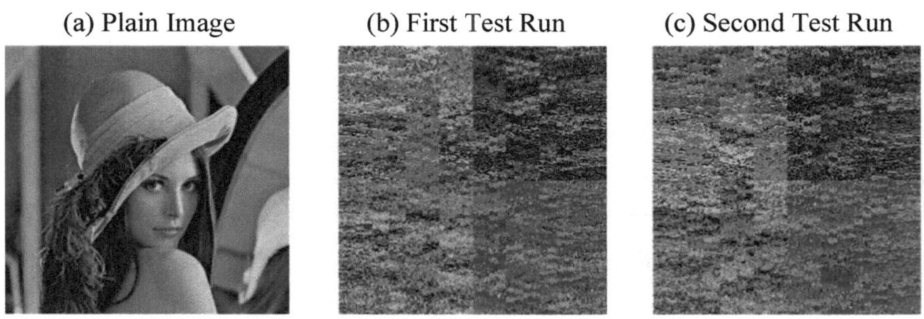

Fig. 1 Lena plain image and two encrypted image in two test runs

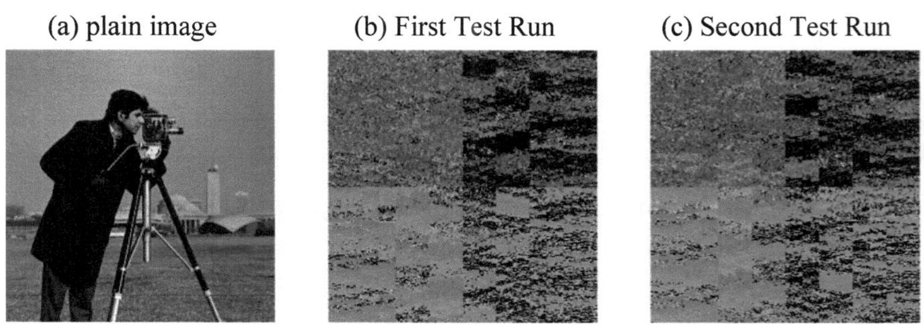

Fig. 2 Cameraman plain image and two encrypted image in two test runs

Table 1 Entropy and PSNR of encrypted images

Selected image	Reference image	PSNR of selected image	Entropy of selected image
Encrypted Lena, Fig. 1b	Lena plain image, Fig. 1a	−36.928817	5.332146228071736
Encrypted Lena, Fig. 1c	Lena plain image, Fig. 1a	−36.943271	5.332146228071736
Encrypted Cameraman, Fig. 2b	Cameraman plain image, Fig. 2a	−39.636006	5.037729765949852
Cameraman Second encrypt in Fig. 2c	Cameraman plain image, Fig. 2a	−39.673154	05.037729765949852

Table 2 Pearson's correlation coefficient

Selected image	Horizontal correlation coefficient	Vertical correlation coefficient
Plain image Lena, Fig. 1a	0.9071539397408981	0.9746624329061755
Encrypted Lena, Fig. 1b	0.3902039586852926	0.164632642344182
Encrypted Lena, Fig. 1c	0.4032467558705724	0.294682842482829
Plain image Cameraman, Fig. 2a	0.9433007313117971	0.8533010814343324
Encrypted Cameraman, Fig. 2b	0.49107879669398824	0.08674830224963347
Encrypted Cameraman, Fig. 2c	0.5031422316833177	0.32548714931357287

Pearson's correlation coefficient is computed using Eq. (1) and presented in Table 2 to determine the degree of correlation between adjacent pixels in both horizontal and vertical directions for all images. It is clearly observed that pixels of plain image are highly correlated, but the same is lost when encrypted by the proposed algorithm.

$$r = \frac{\frac{1}{n}\sum_{i=1}^{n}(x_i - \bar{x})(y_i - \bar{y})}{\sqrt{\left(\frac{1}{n}\sum_{i=1}^{n}(x_i - \bar{x})^2\right)\left(\frac{1}{n}\sum_{i=1}^{n}(y_i - \bar{y})^2\right)}}, \quad \text{where } \bar{x} = \frac{1}{n}\sum_{i=1}^{n} x_i \text{ and } \bar{y} = \frac{1}{n}\sum_{i=1}^{n} y_i \tag{1}$$

Here, $(x_{i,}\ y_i)$ refers to the ith pair of vertically/horizontally adjacent pixels. Considering an image of dimensions $(a \times b)$, the value of n is set as, $n = (b - 1) * a$, in order to calculate the correlation coefficient for the horizontally adjacent pixels and $n = (a - 1) * b$.

Two of the most common techniques to evaluate the strength of image encryption algorithms, especially with respect to differential attacks, are NPCR (number of pixel changing rate) and UACI (Unified Average Changing Intensity). NPCR and

Table 3 Correlation coefficient

Referred images	Inter correlation coefficient	NPCR (%)	UACI (%)
Figure 1a, b	−0.07608793056431	99.5254516601563	22.5088560814951
Figure 1a, c	−0.079675319256034	99.5269775390625	22.5448907590380
Figure 1b, c	0.085660634687198	99.3209838867188	20.3426226447610
Figure 2a, b	−0.183095212288852	99.5651245117188	31.0831346698836
Figure 2a, c	−0.1933	99.6002197265625	31.3529818665748
Figure 2b, c	0.268571041672525	97.7081298828125	19.8884253408395

UACI are used to quantify the influence of change of pixel positions in the plain image. NPCR measures the number of pixel change rate, whereas UACI measures the average intensity of difference between the cipher image and the plain image.

The respective values of NPCR and UACI for all cases are computed using Eqs. (2) and (3) and are provided in Table 3. A high NPCR/UACI value indicates a high resistance to differential attacks.

$$NPCR : N\left(Cipher^{o}, Cipher^{c}\right) = \sum_{i,j} \frac{D(i,j)}{T} \times 100\,\% \tag{2}$$

$$UACI : u\left(Cipher^{o}, Cipher^{c}\right) = \sum_{i,j} \frac{[Cipher^{o}(i,j) - Cipher^{c}(i,j)\}}{F \times T} \times 100\,\% \tag{3}$$

where

$$D(i,j) = \begin{cases} 0, \ if\ Cipher^{o}(i,j) = Cipher^{c}(i,j) \\ 1, \ if\ Cipher^{o}(i,j) \neq Cipher^{c}(i,j) \end{cases}$$

The encryption algorithm proposed in the paper has the ability to resist brute-force attack. Encryption key is totally randomly generated.

Estimation of key size: Let the number of elements in one row/column is n. The size of the key involved in that row/column is decided by the number of random numbers involved to perform rotations. Every time the array is divided into two equal parts, skipping middle element in case of odd size, and random rotation is performed if length of subpart is more than one. So there are two random numbers at first stage, 2^2 in the 2nd stage, and so on until kth stage such that $\lfloor \frac{n}{2^k} \rfloor = 2$ i.e. $k = \lfloor \log_2 n \rfloor - 1$. So, the size of key = 2 + 2^2 + ⋯ + 2^{k} = 2^{k+1} – 2. For 2-D array, same key size is applied for each row/column. So exponential time is required for brute-force attacks.

With the key, the attacker also needs to know the proper sequence (horizontal or vertical) in which the encryption was made.

5 Conclusion

The proposed shuffle-based image encryption technique produces a dynamic result for each run of the application due to the inclusion of random number to achieve pseudorandom permutation of image pixels positions. Although the histogram remains same and computed entropy varies marginally, security can be breached for known images. But the proposed technique promises many advantages like large key space which is dynamic, and the process is fast. So the proposed technique may be a potential step in hybrid cryptosystem to overcome all existing limitations.

References

1. Hazra, T.K., Bhattacharyya, S.: Image encryption by blockwise pixel shuffling using modified Fisher-Yates shuffle and pseudorandom permutations. In: 2016 IEEE 7th Annual Information Technology, Electronics and Mobile Communication Conference (IEMCON), Vancouver, BC, pp. 1–6 (2016). https://doi.org/10.1109/iemcon.2016.7746312
2. Chowdhury, S.R., Hazra, T.K., Chakroborty, A.K.: Image encryption using pseudo random permutations. Am. J. Adv. Comput. **1**(1) (2014). http://dx.doi.org/10.15864/ajac.v1i1.2
3. Mallik, S., Saha, P., Singha Roy, A., Sinha, R., Hazra, T.K., Chakraborty, A.K.: Image hiding using zigzag and spiral traversal algorithms. Am. J. Adv. Comput. **1**(1) (2014)
4. Hazra, T.K., Chowdhury, S.R., Chakraborty, A.K.: Encrypted image retrieval system: a machine learning approach. In: 2016 IEEE 7th Annual Information Technology, Electronics and Mobile Communication Conference (IEMCON), Vancouver, BC, pp. 1–6 (2016). https://doi.org/10.1109/iemcon.2016.7746351
5. Hazra, T.K., Anand, A., Shyam, A., Chakraborty, A.K.: A new approach to compressed image steganography using wavelet transform. IOSR J. Comput. Eng. **17**(5), 53–59 (2015)
6. Hazra, T.K., Samanta, R., Mukherjee, N., Chakraborty, A.K.: Hybrid image encryption and steganography using SCAN pattern for secure communication. In: 2017 8th Annual Industrial Automation and Electromechanical Engineering Conference (IEMECON), Bangkok, pp. 370–380 (2017). https://doi.org/10.1109/iemecon.2017.8079625
7. Hazra, T.K., Ghosh, R., Kumar, S., Dutta, S., Chakraborty, A.K.: File encryption using Fisher-Yates shuffle. In: 2015 International Conference and Workshop on Computing and Communication (IEMCON), Vancouver, BC, pp. 1–7 (2015). https://doi.org/10.1109/iemcon.2015.7344521
8. Hazra, T.K., Mahato, A., Mandal, A., Chakraborty, A.K.: A hybrid cryptosystem of image and text files using blowfish and Diffie-Hellman techniques. In: 2017 8th Annual Industrial Automation and Electromechanical Engineering Conference (IEMECON), Bangkok, pp. 137–141 (2017). https://doi.org/10.1109/iemecon.2017.8079577
9. Li, C.: Cracking a hierarchical chaotic image encryption algorithm based on permutation. Sig. Process. **118**, 203–210 (2016)
10. Arroyo, D., Li, C., Li, S., Alvarez, G., Halang, W.A.: Cryptanalysis of an image encryption scheme based on a new total shuffling algorithm. Chaos, Solitons Fractals **41**(5), 2613–2616 (2009)
11. Pisarchik, A.N., Zanin, M.: Image encryption with chaotically coupled chaotic maps. Phys. D **237**(20), 2638–2648 (2008)
12. Chen, G.R., Mao, Y., Chui, C.K.: A symmetric image encryption scheme based on 3D chaotic cat maps. Chaos, Solitons Fractals **21**, 749–761 (2003)
13. Kocarev, L.: Chaos-based cryptography: a brief overview. IEEE Circ. Syst. Mag. **1**(3), 6–21 (2001)

14. Li, S.J., Zheng, X., Mou, X.Q., Cai, Y.L.: Chaotic encryption scheme for real-time digital video. Real-Time Imag. **VI**, 149–160 (2002)
15. Shujun, L., Xuanqin, M., Yuanlong, C.: Pseudo-random bit generator based on couple chaotic systems and its applications in stream-cipher cryptography. In: Progress in Cryptology—INDOCRYPT 2001, pp. 316–329 (2001)
16. Ye, R.: A novel chaos-based image encryption scheme with an efficient permutation-diffusion mechanism. Optics Commun. **284**(22), 5290–5298 (2011)
17. Jolfaei, A., Mirghadri, A.: Image encryption using chaos and block cipher. Comput. Inf. Sci. **4**(1), 172 (2010)
18. Hazra, T.K., Kumari, N., Monica, Priya, S., Chakraborty, A.K.: Image encryption and decryption using phase mask over sinusoidal single and cross grating. In: 2016 IEEE 7th Annual Information Technology, Electronics and Mobile Communication Conference (IEMCON), Vancouver, BC, pp. 1–6 (2016). https://doi.org/10.1109/iemcon.2016.7746317
19. Mustafi, K., Sheikh, N., Hazra, T.K., Mazumder, M., Bhattacharya, I., Chakraborty, A.K.: A novel approach to enhance the security dimension of RSA algorithm using bijective function. In: 2016 IEEE 7th Annual Information Technology, Electronics and Mobile Communication Conference (IEMCON), Vancouver, BC, pp. 1–6 (2016). https://doi.org/10.1109/iemcon.2016.7746304
20. Sheikh, N.U., Hazra, T.K., Rahman, H., Mustafi, K., Chakraborty, A.K.: Multi-variable bijective mapping for secure encryption and decryption. In: 2017 8th Annual Industrial Automation and Electromechanical Engineering Conference (IEMECON), Bangkok, pp. 338–345 (2017). https://doi.org/10.1109/iemecon.2017.8079619

Authentication of Diffie-Hellman Protocol Against Man-in-the-Middle Attack Using Cryptographically Secure CRC

Nazmun Naher, Asaduzzaman and Md. Mokammel Haque

Abstract Diffie-Hellman key exchange (DHKE) protocol, which is also known as exponential key exchange protocol, is one of the practical ways of generating a common secret key between two communicating parties. But this protocol itself is a non-authenticated protocol; hence, the protocol is exposed to man-in-the-middle (MITM) attack. An attacker can easily hijack sender's public value. Attacker calculates his own public value and sends this value to the receiver instead of sending the original value. Attacker does the same thing when receiver replies back to the sender. After this exchange, attacker can decrypt any messages sent by both of the communicating parties. In this paper, a simple authentication mechanism is developed based on the cryptographically secure version of well-known cyclic redundancy check (CRC). A cryptographically secure CRC is capable of detecting both random and malicious errors where the CRC divisor polynomial is randomly generated and secret. A common CRC divisor polynomial is generated for both of the communicating parties. The system is capable of generating cryptographically secure random numbers which are different in every session. Here the length of the divisor polynomial for CRC must be large. In our proposed system, cryptographically secure CRC is combined with the Diffie-Hellman algorithm for checking whether the public value of the sender is changed by an attacker. MITM attack is detected successfully by using only one securely and randomly generated secret nonzero divisor polynomial of cryptographically secure CRC. The length of public keys to be sent in the Diffie-Hellman protocol and modified system are also compared to show the overhead is negligible.

N. Naher (✉) · Asaduzzaman · Md. M. Haque
Department of Computer Science and Engineering, Chittagong University of Engineering and Technology, Chittagong 4349, Bangladesh
e-mail: nazmunsonia403@gmail.com

Asaduzzaman
e-mail: asad@cuet.ac.bd

Md. M. Haque
e-mail: mokammel@cuet.ac.bd

M. Chakraborty et al. (eds.), *Proceedings of International Ethical Hacking Conference 2018*, Advances in Intelligent Systems and Computing 811,
https://doi.org/10.1007/978-981-13-1544-2_12

Keywords Man-in-the-middle attack · Message authentication · Diffie-Hellman key exchange protocol · Cryptographically secure cyclic redundancy check (CRC)

1 Introduction

In the present era of high technology, environment network security is an important aspect of system admiration. Without having to be physically present, the network allows people to access geographically distant resources remotely. Hence, there can happen many unpredictable incidents, i.e., hacking, information loss, unauthorized access, and misuse. Cryptology is the study of designing a system which will ensure to keep the four aspects of modern cryptology such as confidentiality, integrity, authentication, and non-repudiation of the information in an organized and systematic way. It is mainly used to protect information that is sent over a channel which is insecure. Cryptography can be divided into two sections depending on the nature of key used in the symmetric key cryptography and asymmetric key cryptography. In a symmetric key cryptosystem, same key is used by both sender and receiver for encryption and decryption, respectively. On the other hand, in an asymmetric key cryptosystem, two different keys are used for encryption and decryption mechanism. Symmetric key cryptography is generally very fast and ideal for encrypting a large number of data. Security of the symmetric key encryption depends on the key exchange protocol. In 1976, two researchers at Stanford University, Diffie and Hellman presented a key exchange protocol. This protocol can proceed over public communication channels [1]. Diffie-Hellman (DH) protocol is still extensively used in the present days. But one of the major problems of this protocol is that it is not self-authenticated, so man-in-the-middle attack can be possible in DH protocol. In this paper, some currently available solutions are described in Sect. 2 with some of their shortcomings. But we proposed a system totally in a different approach to provide authentication in a simple manner but effectively which will prevent MITM attack on DH protocol.

Cyclic redundancy check is extensively used as a safeguard of data in transmission channels. It prevents errors occurred randomly in the communication channel [2]. The basis of our proposed system is the use of division modulo a random and secret polynomial over GF (2) for the purpose of authentication of the protocol. Cyclic redundancy check (CRC), which is most of the time used as a random error detection mechanism in the communication channel, is cryptographically varied here. A satisfactory level of security without losing reliability can be guaranteed, if the conventional CRC is made cryptographically secure. The main concept is to make the generator polynomial of CRC variable and secret [3, 4]. It is also computationally very simple and secure than other hash function when the simple CRC is converted to cryptographically secure CRC.

In this paper, Sect. 2 discusses the related works on this problem. Section 3 represents the proposed system, and Sect. 4 explains the experimental results and their explanation. This section also includes secrecy analysis of the proposed system and comparison of the existing and modified proposed system. And finally, Sect. 5 represents the conclusion and future works.

2 Related Works

In 1976, Diffie and Hellman proposed their algorithm for symmetric key encryption system to exchange shared key between two communicating parties. But their system is not self-authenticated, so it is possible to attack both of the communicating parties by an attacker [1]. Here we will discuss some existing solution to prevent this attack. In the study [5], they proposed a secure system against MITM attack based on Geffe's generation of binary sequences and server to handle the communication. But server has to handle all user tables with a large number of entities, and both of the communicating parties have to send their private key to the server. In this study [5], they also discussed some problem of other authentication systems.

In another study [6], a biometric-based sender authentication system was developed using speech. But speech can be changed by acoustic surrounding and transduction device such as microphone.

2.1 Diffie-Hellman Key Exchange Protocol

Diffie-Hellman key exchange (DHKE) is one of the primitive concepts of public key cryptosystem. This protocol allows two communicating parties to share a common secret key over insecure communication mediums without meeting in advance. The process of this protocol supposes that Alice and Bob have different private keys, and they have to agree upon two relatively prime numbers p, g, and then each of them uses the obtained information to calculate the public keys. As a result, both of Alice and Bob obtained the shared key without sending their private keys through the channel [7]. Steps in Diffie-Hellman key exchange protocol are described below:

- Alice and Bob have to agree on two large numbers p and g, where p is a prime number, and p and g are public.
- Alice picks a large number a and keeps it secret; similarly Bob picks his secret key b.
- Alice sends a message to Bob containing (p, g, g^a mod p).
- Bob sends a message to Alice containing (p, g, g^b mod p).
- After receiving Bob's message, Alice performs the following computation:

  $(g^b \bmod p)^a \bmod p$, which yields $g^{ab} \bmod p$.

- After receiving Alice's message, Bob performs the same computation:

  $(g^a \bmod p)^b \bmod p$, which yields $g^{ab} \bmod p$.

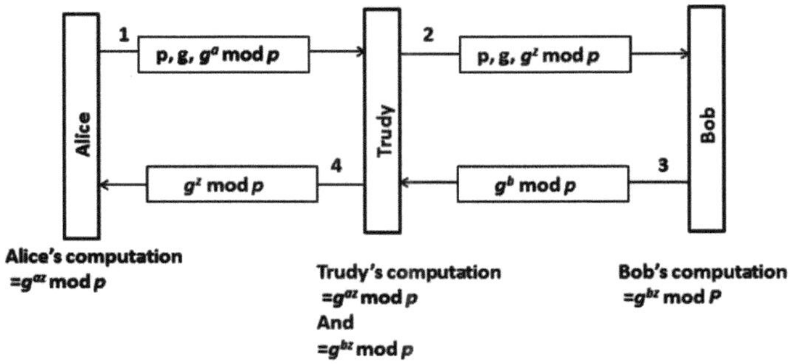

Fig. 1 Man-in-the-middle attack scenario [8]

Hence, the secret key of Alice and Bob is g^{ab} mod p, using which they will communicate securely [8].

The Diffie-Hellman key exchange protocol is exposed to man-in-the-middle (MITM) attack. An attacker can interpose Alice's public value and send his own calculated public value to Bob. When Bob replies back his public value, attacker again replaces it with his own public value and sends it to Alice. Attacker and Alice thus agree on one shared secret key. Attacker also agrees on another shared key with Bob. At the end of the session, any messages sent by Alice or Bob can be seen and modified by the attacker before encrypting them again with the key and sending them to the receiver. This attack is possible because any of the communicating parties are not authenticated in DH protocol. A man-in-the-middle attack can succeed only if a middleman or attacker can intercept each of the parties to their satisfaction as expected from the legitimate parties (see Fig. 1).

2.2 *Cryptographically Secure CRC*

Authentication can be done simply by using cryptographically secure CRC. The concept of cryptographically secure CRC is to use secret and random CRC generator polynomial. Steps involved in the sender side of cryptographically secure CRCs are as follows:

- Firstly, choose a message M(x) of m bit.
- Randomly generate the polynomial q(x) from a set. This set consists of all possible polynomials which has degree n over GF(2), where q(x) must be a nonzero polynomial, i.e., n > 1.
- Then M(x) is multiplied by x^n. Here n is the degree of generator polynomial q(x). Then the checksum is calculated and appended to the message (i.e., CRC (M)) for sending to the other end.

The CRC decoding or the receiver side check steps are as follows:

- Get the secure random polynomial q(x) form the server which is same as the sender's polynomial.
- Divide the message CRC(M) modulo q(x), where CRC(M) is the received message and q(x) is generator polynomial.
- Finally, the coefficients of the resulting remainder and the received CRC check bits are compared.

Any mismatch in the receiving side points out the occurrence of an error [9].

3 Proposed System

The fundamental target of the system proposed is to authenticate the sender of Diffie-Hellman protocol, so that the attacker's interception will be failed. Attacker could not be able to change the public value of any parties without being detected.

Here the server performs as a trusted third party. Alice and Bob are two communicating parties. The main concept of the server is hired from trusted third party (TTP). In the concept of TTP, both of the communicating parties use this trust to make a secure interaction between them [10]. TTP are common in most of the commercial transactions and in cryptographic transactions as well as cryptographic protocols. Steps in the modified system are described as follows:

- A trusted server and two clients are implemented who are capable of generating secure random number.
- Alice requests a session key from the server. The request message involves two information: Who is the sender and who is the receiver. They have identified the sender's and receiver's ID which is autogenerated after connecting to the server.
- Server then runs its generation algorithm to get the secure random and variable length nonzero generator polynomial. Server then replies to Alice with n-bit generator polynomial.
- Server also sends a message to Bob which consists of two parts: A token and n-bit generator polynomial. Token has also two parts: the sender id and the receiver id.
- Then Alice will run her generation algorithm gen() which will generate a large prime number p and another prime number g (relative prime of p). g is also chosen randomly. This generation algorithm also generates a random number a, which is the secret key for Alice. Generated p, g and a by the generation algorithm p, ganda are random and different in every session. Then Alice calculates the following things:

$$A = g^{a} mod\ p \tag{1}$$

$$r = A \cdot x^{n} \tag{2}$$

$$z = r\ mod\ q \tag{3}$$

$$CRC(A) = r \oplus z \tag{4}$$

- Alice sends a message to Bob for start communication which is in the format (p, g, CRC(A)) where p and g are the generated prime and relative prime, respectively. And CRC(A) is the calculated CRC value of the public key A.
- Now, Bob runs his generation algorithm which generates a random large secret key b, which less than p.
- Bob calculates the following things:

$$B = g^b mod\ p \tag{5}$$

$$r = B \cdot x^n \tag{6}$$

$$z = B\ mod\ q(x) \tag{7}$$

$$CRC(A) = r \oplus z \tag{8}$$

- Bob replies back to Alice with a message in format CRC(B). Here CRC(B) is the CRC of public message B calculated by Bob.
- Alice runs reverse CRC to get the public key of Bob. She divides CRC(B) modulo q(x) and then compares the coefficients of the resulting remainder with the CRC check bits received from the sender. If the result matches, then she will calculate:

$$B = CRC^{-1}(B) \tag{9}$$

$$K = B^a mod\ p \tag{10}$$

- Bob also runs reverse CRC to get the public key received from Alice. He divides the received message CRC(A) modulo q(x) and compares the coefficients of the resulting remainder with the CRC check bits that are received from the sender. If the result matches, then he will calculate:

$$A = CRC^{-1}(A) \tag{11}$$

$$K = A^b mod\ p \tag{12}$$

At the end of the algorithm, both parties gain the same shared secret key without any interception of attacker which is actually as follows:

$$K = g^{ab} mod\ p \tag{13}$$

Using this shared secret key, both of the communicating parties are able to communicate with each other securely (see Fig. 2).

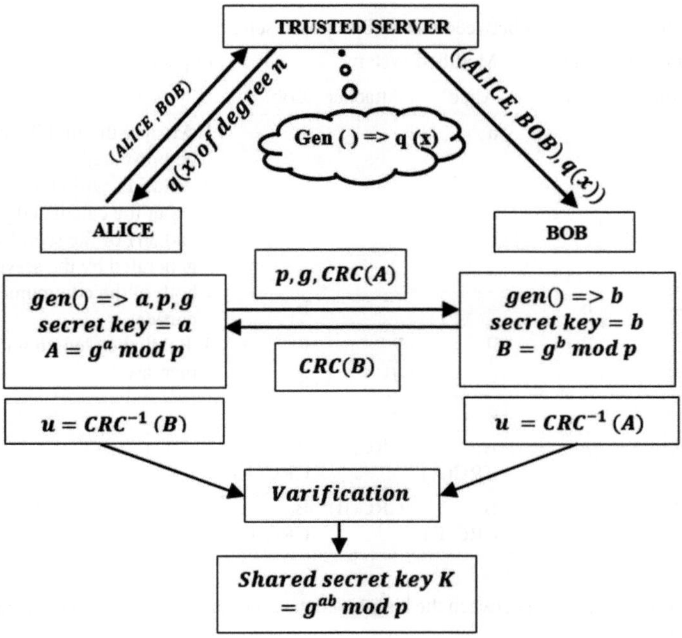

Fig. 2 System architecture

A simple example of the authentication can be represented as follows for simplification:

1. Sender (Client1) → Sends request to the server to communicate with Client2.
2. Server randomly generates 10011001 and sends the number to both clients (Client1 and Client2).
3. Client1 → Generates random secret key, a = 00100000, p = 11101111 and g = 00010111. He also calculates $g^a \bmod p$ = 0 1 0 0 1 0 1 1 (7 bits), the CRC value of 01001011 using 10011001 is 0010010110001111 (14 bits) and sends (11101111, 00010111, 0010010110001111) to Client2.
4. Receiver (Client2) → Checks the validity of the received message by CRC check bits. After that, he calculates reverse CRC using the same secret random polynomial 10011001 and retrieve the main public value 0 1 0 0 1 0 1 1.
5. Client2 → Generates random secret key, b = 00011000. He calculates g^b mod p = 0 1 1 0 0 1 0 1 (7 bits), the CRC value of 01100101 using 10011001 is 0011001011111110 (14 bits) and sends 0011001011111110 to Client1.
6. Client1 do the same thing as step 4.

The algorithm of authenticated DH protocol which is proposed has obviously some difference from the non-authenticated version of this protocol. There has been created some variation in our system from Diffie-Hellman algorithm which obviously will

Table 1 A short comparison between DH and proposed scheme

Diffie-Hellman algorithm			Modified system			Impact
Alice	Attacker	Bob	Alice	Attacker	Bob	
a	–	–	a	–	–	1. Man-in-the-middle attack can be prevented 2. Attacker will know nothing about the calculated public key and any of the secret key generated by the server and both of the communicating parties 3. Public key length will be increased
–	–	b	–	–	b	
p	P	p	p	p	p	
g	g	g	g	g	g	
–	–	–	q	–	q	
A	A	A	A, CRC(A)	CRC(A)	A, CRC(A)	
B	B	B	B, CRC(B)	CRC(B)	B, CRC(B)	

Table 2 A short comparison between the length of data to be sent in the DH protocol and modified version (bits)

Generated key length (bits)					Diffie-Hellman public key length (bits)		Modified Diffie-Hellman hashed public key length (bits)	
A	b	p	g	q	Alice	Bob	Alice	Bob
5	3	6	4	6	6	4	11	8
4	5	6	2	8	5	6	12	13
3	3	8	5	8	4	6	11	13
4	2	9	4	7	7	8	13	14
6	5	8	5	8	7	7	14	14
5	4	8	6	4	7	6	10	9
5	5	7	4	8	6	6	13	13
5	4	8	6	7	6	6	12	14

create some impacts and extra overheads on the existing protocol. These variations, impacts, and overheads are shown theoretically in Table 1.

Table 1 shows that here only one overhead is: The length of the data to be transmitted during this protocol will be increased. But it is not a problem here, and actually, it creates no overhead at all because the application of DH protocol is in SSL/TLS (secure socket layer/transport layer security). In SSL/TLS data, unit is called record with maximum payload field or the maximum length of data 14 k or 16,384 bytes. Hence, generated overhead will be negligible here. Public key length of DH and modified DH is compared in Table 2.

The client–server communication will be unicast communication like DHCP (Dynamic Host Configuration Protocol) server. Every client IP will be stored in the server's table. So, there is no need to store any extra entity in the server table which saves both memory and time. And there is no need to share the secret key with the server. Server also does not need to store the generated polynomial by itself. Unicast communication is used for all network processes in which a private or unique resource is requested. The client unicast a request message to the server for generator, as the server IP is known to him. After receiving the request, server will unicast the keys to the client. There is no need to store any secret keys in the server table.

4 Performance Analysis

4.1 Security Analysis

Security of the proposed system depends on the attacker's ability to get the generator polynomial. If he could figure out the secret generator polynomial, then he will be able to change the data. But in our system, we showed that attacker will not be able to get the polynomial because the server will generate a different polynomial in every session. And for brute force search, the probability of finding the polynomial is 2^n, where n is the length of the polynomial (generating capability of the server). So, for secure communication n should be larger, i.e., 80 bit or more, so that the probability will be greater than or equal 2^{80}, which is non-negligible in secrecy analysis of any cryptographic methods.

Attacker fails to get the shared secret key (see Fig. 3). He requested to the server acting like he is Alice, but he is given a secret key which is $q'(x)$ or he randomly generates $q'(x)$ which is different from $q(x)$. Attacker gets CRC(A) from Alice, then he calculates $CRC'(A)$ using his own generated secret key a' and $q'(x)$. He sends $CRC'(A)$ to Bob instead of CRC(A). But Bob can recognize the change because he performs the authentication using $q(x)$, which is same as Alice's calculation. Attacker also replaces CRC(B) by $CRC'(B)$ using a' and $q'(x)$. But none of his changes are succeeded without being detected by using only one secure random polynomial.

The example given before in Sect. 3 can be extended for showing attacker's activity:

1. Attacker changes Client1's message 0010010110001111 by 0001000111101110 using his own secret key 00001010 and polynomial 01010101 because he doesn't know the secret generator polynomial.
2. After that, he sends (11101111, 00010111, and 0001000111101110) to client2.
3. After receiving the message, Client2 can recognize the change because the calculated checksum by 00010111 and 10011001 are not same.
4. Client2 sends an attack notification to Client1 and the server.

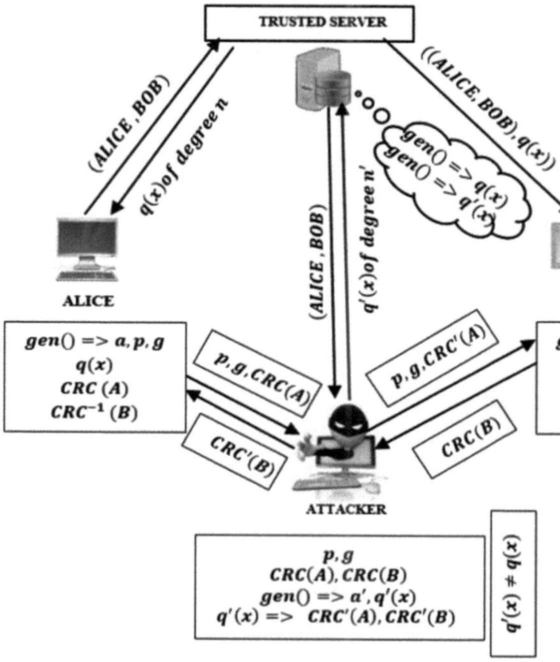

Fig. 3 Security against attack

So, attacker can't change the value of any keys without being caught. There is no option to steal any key from the server because the server need not store any secret or public key of any communicating parties. And it is not possible to get that how much bits are appended after the key because every time the length of the generator polynomial provided by the server is different.

4.2 Comparison of Existing and Modified System

Authentication can be done simply by using cryptographically secure CRC. From the Table 2, we can see that in the modified algorithm public key size is increased depending on the length of CRC polynomial and the public key length in the existing Diffie-Hellman protocol. So it becomes tough to guess by the attacker that what the value of the public key is in the modified algorithm. And here we can see that the appended value after the key is not fixed, it is changed in every session.

So for man-in-the-middle attack, the attacker must have to gain the generator polynomial or guess it. But guessing is not so when the length of the polynomial is larger. And the probability that the attacker can find the polynomial is 2^n where n is the length of the generator polynomial. But the increasing length will not create any overhead in the channel because TLS/SSL where DH protocol is used has a larger unit of data called record which has 16,384 bytes or 14 kB payload field. The length is handled here in bits or some bytes but needs not to be more than 1 kB.

5 Conclusion

In the present era, Diffie-Hellman key exchange protocol is playing a very vital role in information technology and security. It is the most widely used protocol. The Diffie-Hellman algorithm can be efficiently authenticated using cryptographically secure CRC and a trusted server. Implementation of this algorithm is very easy to understand because the implementation of cryptographically secure CRC is most likely as the conventional CRC except the generator polynomial is programmable (random) and secret. And the responsibilities of the server are reduced. Here server's job is to provide only a secure random divisor polynomial of CRC which is different in every session but not to store them. There is no need to send any other secret or public key values to the server by the communicating parties. All the private or secret key values should be random, large, and different in every session. Hence, man-in-the-middle attack can be detected simply.

As a future work, our proposed system can be altered in order to contribute a better secure system and also can be used in the particular fields where it is applicable. We are also very much interested to study the comparative analysis of our proposed system with other existing systems in order to enhance the performance and security of the system.

References

1. Diffie, W., Hellman, M.: New directions in cryptography. IEEE Trans. Inf. Theory **22**, 644–654 (1976)
2. Baylis, J.: Cyclic codes. In: Error-correcting Codes, pp. 141–159 (1998)
3. Dubrova, E., Näslund, M., Selander, G., Lindqvist, F.: Message authentication based on cryptographically secure CRC without polynomial irreducibility test. Cryptogr. Commun. **10**, 383–399 (2017)
4. Krawczyk, H.: LFSR-based hashing and authentication. In: Advances in Cryptology—CRYPTO, pp. 129–139 (1994)
5. Khader, A., Lai, D.: Preventing man-in-the-middle attack in Diffie-Hellman key exchange protocol. In: 2015 22nd International Conference on Telecommunications (ICT) (2015)
6. Laser, J., Jain, V.: Enhanced security mechanism in public key cryptosystems using biometric person authentication. In: 2016 International Conference on Computation of Power, Energy Information and Communication (ICCPEIC) (2016)
7. Kumar, C.K., Jose, G.J.A., Sajeev, Suyambulingom, C.: Safety measures against man-in-the-middle attack in key exchange. Asia Research Publishing Network (ARPN) J. Eng. Appl. Sci. **7**, 243–246 (2006)
8. Tanenbaum, A., Tanenbaum, A.: Computer Networks, 4th edn. Prentice Hall PTR, Upper Saddle River, NJ (2003)

9. Shamir, A.: Identity-based cryptosystems and signature schemes. In: Advances in Cryptology: Proceedings of CRYPTO 84. Lecture Notes in Computer Science, vol. 7, pp. 47—53 (1984)
10. Trusted third party. https://en.wikipedia.org/wiki/Trusted_third_party, Accessed 11 Aug 2017

Multiple RGB Image Steganography Using Arnold and Discrete Cosine Transformation

Diptasree Debnath, Emlon Ghosh and Barnali Gupta Banik

Abstract The aim of this paper is to establish a new method for RGB image steganography which can hide more than one RGB secret image in a single RGB cover image. To do so, discrete cosine transformation is used here in a blockwise manner. The cover image is divided into three-color channels (Red, Green, and Blue) and each channel is further divided into 8×8 blocks. From each block, one bit can be selected to replace with one-pixel value of a secret. Now according to the block size, more than one bit can be manipulated as after applying DCT on a block, mid-band and lower band coefficients have considerably less impact on the overall image. If 3 bits are chosen, then for each block those 3 bits will be replaced by the values from 3 respective secret images. Hence each channels of the cover will contain the same color channel values of the secrets. Just following these steps in reverse order extraction can easily be done and the steganography method will be blind i.e. there will be no need for original cover image while extracting the secrets. Now to enhance the security, before embedding, the secret image is encrypted using Arnold transformation. The proposed method has been thoroughly tested using different color images and the results of these experiments have been scrutinized through various quality metrics, which prove its effectiveness.

Keywords Data privacy · Information security · Discrete cosine transformation Image quality

D. Debnath (✉) · E. Ghosh · B. Gupta Banik
Department of Computer Science & Engineering, St. Thomas' College of Engineering & Technology, Kolkata, India
e-mail: diptasree.debnath@gmail.com

E. Ghosh
e-mail: emlonghosh@gmail.com

B. Gupta Banik
e-mail: barnali.guptabanik@stcet.ac.in

M. Chakraborty et al. (eds.), *Proceedings of International Ethical Hacking Conference 2018*, Advances in Intelligent Systems and Computing 811,
https://doi.org/10.1007/978-981-13-1544-2_13

1 Introduction

The aim of steganography is to hide any secret information in a cover media and transmit it through a common channel such that its existence remains hidden for any third party. In case of image steganography, this cover media is an image. Image steganography can be further classified where the secret information is an RGB color image; this method is popularly known as RGB steganography, which is the main concern of this paper.

2 Literature Survey with Related Knowledge

Using discrete cosine transformation or least significant bit method, a secret text can be hidden in an RGB cover image with vast capacity, but it is not clear how the same technique can be used for hiding more than one RGB secret images [1]. Secret data can be hidden in an RGB image file using quantized DCT coefficients. This method also has a huge capacity to hide data, but how it can be effective in hiding an RGB image is left to discover [2]. Hence in both the cases, how multiple images can be hidden as secret data for respective algorithms, is needed to be found.

While using randomization to hide secret images, the main constraint of the algorithm is the binary secret images [3]. Multiple secret images can be hidden using discrete wavelet transformation. However, in that case, the scope for increasing the capacity is less and the secret has to be a grayscale image, not an RGB image [4].

Using 3-SWT technique effectively, multiple RGB image steganography can be achieved, but the extraction process becomes non-blind which is a major disadvantage of the process [5].

When LSB substitution method along with artificial neural network and advanced encryption standard is applied to hide secret images in an RGB cover image, each color plane contains only one secret image [6]. Hence, at most three images can be hidden in the cover across the three image planes. However, further capacity improvement is not possible, as well as, this method is prone to RS steganalysis attack.

While hiding multiple RGB images in a single container, steganographic key generation is an effective method with good capacity, and it can be implemented by several methods as mentioned in [7]. Using RSA and 3-DWT algorithm, multiple RGB images and secret text messages can be hidden in a single RGB image efficiently with optimized security [8]. Since the perceptual quality of this method is very low, there is scope for a new method which is blind and delivers a high-quality stego image keeping the standard of extracted secrets intact.

A. **Discrete Cosine Transform (DCT)**

Discrete cosine transform is one of the most important techniques used in the field of image steganography. Now if the changes are made directly on the cover pixel,

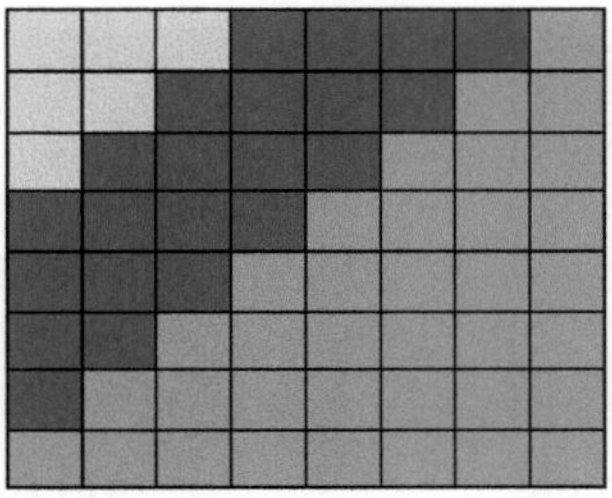

Where
RED=> Mid frequency coefficients

YELLOW=> Lower frequency coefficients

GREEN=> Higher frequency coefficients

Fig. 1 8 * 8 DCT block

possibility of them being visible is high. So, to solve this, DCT has been applied to the image. DCT transforms a block of pixel values into its frequency domain.

Figure 1 shows 8 × 8 DCT block, which can be further divided into three parts. The values lying in the higher band affect the image most. So, any change in those places is easily detectable. However, mid and lower frequency positions do not have much effect on the image, and changes can be made in these places without it being detected [9]. DCT can be calculated as follows:

$$DCT(i,j) = \frac{1}{\sqrt{2N}} C(i)C(j) \sum_{x=0}^{N-1} \sum_{y=0}^{N-1} pixel(x,y) \cos\left[\frac{(2x+1)i\pi}{2N}\right] \cos\left[\frac{(2y+1)j\pi}{2N}\right] \tag{1}$$

where $C(x) = \frac{1}{\sqrt{2}}$ if x is 0 else 1 if x>0

B. **RGB Image**

Images can be of different formats like binary, grayscale, and color images. A color image (also known as RGB image) is a combination of three grayscale images for each of the three color channels red, green, and blue, respectively. All these color channels are individually represented as m * n matrices. When these color channels are combined, the resultant picture is formed having the size of m * n * 3. To manipulate an individual bit of RGB image, one needs to make those changes separately on the individual color channels.

C. **Arnold Transformation**

One of the most important aspects of image steganography is to keep the secret image as secure as possible. However, if somehow the secret is extracted from the cover, there must exist some way to keep the secret message secure from outsider even in that circumstance. This is where Arnold's transformation is used.

Arnold transform is a periodic scrambling method. It changes the position of the pixel values by encoding a few processes iteratively. As the pixel value of a position is iteratively changed by following the same process, after N times, the pixel value must return to its initial position. So, to encrypt, this N needs to be found out. In

(a) Original Image (b) After Arnold (c) After Reverse Arnold

Fig. 2 Application of Arnold transform

the encrypting process, 'N1' iterations are made. Thus, the picture at this point is scrambled and can not be read. However, knowing the value of N and N1, the process can be executed on the picture (N–N1) times again and thus the initial pixel positions can be restored.

After applying this Arnold encryption algorithm, one gets the output as a scrambled or an encrypted image which is meaningless. Hence, the original message is secure. Being periodic in nature, one can inverse the Arnold effect, i.e., by the cyclic process, the original secret can be retrieved back from the encrypted image. As this process can be applied to only grayscale images, Arnold transform must be applied to all the three channels of RGB image [10].

The formulae used to perform Arnold transformation is given in Eq. (2)

$$\begin{bmatrix} x' \\ y' \end{bmatrix} = \begin{bmatrix} 1 & 1 \\ 1 & 2 \end{bmatrix} \begin{bmatrix} x \\ y \end{bmatrix} \bmod 1 \tag{2}$$

And the inverse Arnold transformation is executed using Eq. (3).

$$\begin{bmatrix} x \\ y \end{bmatrix} = \begin{bmatrix} 2 & -1 \\ -1 & 1 \end{bmatrix} \begin{bmatrix} x' \\ y' \end{bmatrix} \bmod 1 \tag{3}$$

Figure 2 shows how Arnold transformation and reverse Arnold transformation affect an image.

Arnold Transform Algorithm:

```
Let 'in' be the input image matrix and 'iter' be an
integer parameter denoting the number of iterations.
Let m, n be the size of the image.

STEP 1: Check if the input is two-dimensional or not as
this can be performed only on two-dimensional matrices
corresponding to the images. If not 2 dimensional then
display error.
```

```
STEP 2: Check for the square matrix. This process is
for a square matrix only. If not square display error.
STEP 3: Create a matrix of 'm' zeros say 'out'
STEP 4: n = n - 1
STEP 5: for j = 1 to 'iter' do
STEP 6: for y = 0 to n do
STEP 7: for x = 0 to n do
STEP 8: p = [1 1; 1 2] * [x; y]
STEP 9: out[mod(p(2), m) + 1, mod(p(1), m) + 1) = in(y + 1,
x + 1)
STEP 10: End yth and xth loop
STEP 11: in = out
STEP 12: END
```

Reverse Arnold Transform Algorithm:

```
Let 'in' be the input image matrix and 'iter' be an
integer parameter denoting the number of iterations.
  Let m, n be the size of the image

STEP 1: Check if the input is two-dimensional or not
as this can be performed only on 2-dimensional matrices
corresponding to the images. If not 2 dimensional then
display error.
STEP 2: Check for the square matrix. This process is
for a square matrix only. If not square display error.
STEP 3: Create a matrix of 'm' zeros say 'out'
STEP 4: n = n − 1
STEP 5: for j = 1 to 'iter' do
STEP 6: for y = 0 to n do
STEP 7: for x = 0 to n do
STEP 8: p = [2 −1; −1 1] * [x; y]
STEP 9: out[mod(p(2), m) + 1, mod(p(1), m) + 1) = in(y + 1,
x + 1)
STEP 10: End yth and xth loop
STEP 11: in = out
STEP 12: END
```

3 Proposed Method

Using this method, the embedding and extraction of multiple RGB secret images into one RGB cover image are successfully achieved without secret being detectable and without any color data loss. Here DCT and Arnold transform have been used to improve the security of the secret.

A. Embedding Procedure

Here the cover and all the secrets are RGB images. An RGB image is made of three different color components (red, green, and blue) which are represented in an m * n * 3 matrix format. The cover is divided into three m * n image planes where each plane represents each color. Similarly, the secret is also divided into its components. After this, each cover component is divided into several blocks such that each block contains one modified component of the secret image. To make the process more secure, each component of the secret is scrambled using Arnold transformation. Now, DCT is applied to convert the cover image into its frequency domain. Then changes made in the lower frequency coefficients as that will not have much effect on the cover. Then the value of the alpha factor is set. As the DCT value of the cover is much less as compared to the pixel values of the secret, it is necessary to modify the secret's pixel values to an extent such that it is comparable to the original cover's DCT values. This modification is done using the alpha factor [11]. Then the modified pixel value is embedded or placed in a position where the cover image will be affected the least. It is observed that for a block of elements, there exist more than one element—whose value when changes, do not affect the cover. Therefore, these places can be used to embed the values of another secret image. (e.g., if an 8 * 8 block is taken, more than one position e.g., (8, 8), (7, 8) and (6, 8) do not affect the cover much. Hence their values can be modified.) So, three separate secret images can be hidden by embedding the values of each secret on one position of all the blocks for each color channel. After this, IDCT is used on each block to get the pixel values back from the frequency domain. Now the modified color channels of the cover contain the embedded values representing the same color channels of multiple secrets. Now the three channels are combined to get a m * n * 3 stego which is an RGB image like the cover. Thus, successful implementation of multiple RGB image steganography is done.

Embedding Algorithm

STEP 1: Read the cover and the secrets (all RGB)
STEP 2: Divide the cover and the secrets into the 3 color channels each
STEP 3: Divide the cover image into a number of blocks depending upon the size of the secrets to be embedded
STEP 4: Apply Arnold Transformation for respective channels of each secret
STEP 5: Apply DCT on each of the blocks of the cover
STEP 6: Set the alpha factor
STEP 7: Place the manipulated pixel value of each secret at one of the least affected positions of the block
STEP 8: Apply IDCT on the block

STEP 9: Repeat step 4 to step 8 for each channel of the cover image
STEP 10: Combine the 3 color channels to get the stego image

B. **Extraction Procedure**

Unlike most of the extraction techniques, this algorithm requires only the stego and the alpha factor to retrieve the original secrets, i.e.; the extraction process is blind. This makes the method more secured as transmission of the cover image is not required.

Initially, the stego image is divided into its three color channels, and each component is furthermore divided into several blocks same as the size of the secret to be retrieved from the stego. To each block, DCT has been applied. Now modified pixel values of the respective secret images are retrieved from each block, and using the alpha factor, the encrypted pixel values for the secrets are calculated and saved in an m * n matrix format. Now on completion of this process, for each color channel, three m * n matrices of pixel values are received for each secret image. These are the scrambled secret images' color channels which were formed using Arnold transformation to make the process more secure. By periodically using reverse Arnold transformation on each color channel of each secret, original pixel values of the secrets' color channels are achieved. Finally, the respective three color channels of each secret image are combined to get the RGB secret images back using blind extraction process.

Extraction Algorithm

STEP 1: Read the stego and split it into 3 color channels
STEP 2: Divide the channels into a number of blocks same as that of the size of the secret
STEP 3: Apply DCT to the block
STEP 4: Set the alpha factor same as embedding algorithm
STEP 5: Take the value of the modification position for each secret to calculate the original pixel value using the alpha factor and save the value on an m * n matrix
STEP 6: Apply reverse Arnold on each color channel
STEP 7: Repeat step 2 to step 6 for all the 3 color channels of each RGB secret image
STEP 8: Combine the respective channels for each of the secrets to get the RGB secret images back from the Stego.

4 Result and Quality Analysis

Peak Signal-to-Noise Ratio (PSNR)
Peak signal-to-noise ratio is a ratio between the signal values and noise content of an image when compared to a reference image. This is one of the most frequently used quality analysis methods as it is very effective in measuring the perceptual transparency. PSNR is mathematically defined as follows:

$$PSNR = 20\log_{10}(\frac{Max_I}{\sqrt{MSE}}) \quad (4)$$

where MSE is the mean squared error. Here, Max_i is the maximum pixel value of the cover image.

Structural Similarity Index (SSIM)
Structural similarity index is a comparison of similarity between an ideal and a modified image. SSIM has a value ranging between 0 and 1 where 1 indicates 100% similarity and 0 implies completely unrelated image. Therefore, it is obvious that the ideal embedding process must have a higher value of SSIM. It is mathematically computed as follows:

$$SSIM(x, y) = \frac{(2\,\mu_x\,\mu_y + C_1)(2\,\sigma_{xy} + C_2)}{(\mu_x^2 + \mu_y^2 + C_1)(\sigma_x^2 + \sigma_y^2 + C_2)} \quad (5)$$

where μ_x, μ_y, σ_x, σ_y, and σ_{xy} are the local mean, standard deviation, and cross-covariance of images x and y, respectively. In Tables 1 and 2, test results of various cover, secret, and stego image combinations have been shown with the help of aforesaid quality metrics.

Embedding Capacity
The embedding capacity depends on the block size of cover image. If the block size is small, then there are few bit positions lying in upper and mid-band frequency regions hence, there would be less place for manipulation. When the block size is large, the number of bits with less significance are more. Hence, more number of positions can be manipulated. Now, if there are total ‘x’ number of bits in the cover image qualified for manipulation and ‘n’ number of secret images at hand, then the average number of bits that can be used to hide each and individual secret will be x/n.

For example, if the cover size is 512 * 512 and a block size of 8 * 8 is chosen, then in each block there are at least three bits from the mid-frequency region where secret image bits can be embedded. Now all those bits can be used to embed one secret, or those can be distributed to hide three 64 * 64 secrets. Thus, the size of the secret image can be controlled. Embedding capacity can be calculated as follows:

$$\text{Embedding capacity} = (64 * 64 * 3)/(512 * 512) = 4.6875\%$$

Table 1 Quality comparison between cover and stego image

Cover Image	**1st Secret Image**	**2nd Secret Image**	**3rd Secret Image**	**Stego Image**
			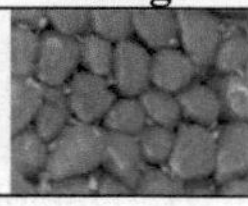	
Quality Analysis between Cover and Stego Images				PSNR:81.2670 SSIM:0.9120

Quality Analysis between Cover and Stego Images				PSNR:82.5030 SSIM:0.8808

Quality Analysis between Cover and Stego Images				PSNR:79.5327 SSIM:0.7963
	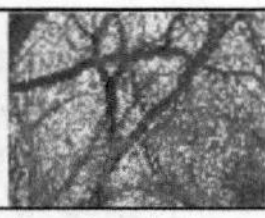			
Quality Analysis between Cover and Stego Images				PSNR:81.6959 SSIM:0.8562
				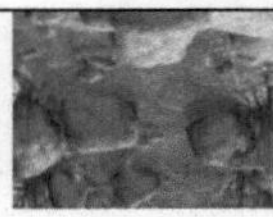
Quality Analysis between Cover and Stego Images				PSNR:83.7494 SSIM:0.8890

Comparison with Existing Technique

In [6], another method was discussed using which multiple RGB images can be hidden into a single cover image. The proposed method fulfills the same objective, but the results are better. There exist a few differences such as authors of [6] used AES, whereas the proposed method uses Arnold transformation for encryption. The changes made in the original cover in [6] are in the spatial domain, therefore, the secrets are binarized, whereas in the proposed method, the changes are made in the frequency domain hence, the original pixel value of the secret is hidden and is retrieved later. The proposed method also has minimum data loss. Comparison of these two methods w.r.t. PSNR values obtained for the stego images are given in Table 3.

Table 2 Quality comparison between embedded versus extracted secret image

Cover Image	Secret Image	Extract-ed Image	Secret Image	Extract-ed Image	Secret Image	Extract-ed Image
					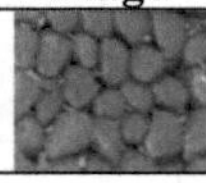	
Quality Analysis	**PSNR**: 69.8380 **SSIM**: 0.9424		**PSNR**: 71.5690 **SSIM**: 0.9890		**PSNR**: 79.7822 **SSIM**: 0.9853	
					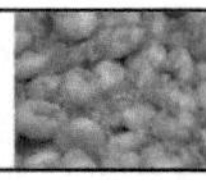	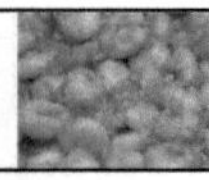
Quality Analysis	**PSNR**:65.6042 **SSIM**:0.9126		**PSNR**:67.9318 **SSIM**:0.9461		**PSNR**:70.0954 **SSIM**:0.9248	
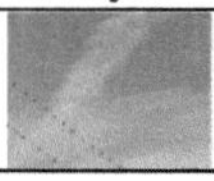						
Quality Analysis	**PSNR**:81.4202 **SSIM**:0.9890		**PSNR**:80.8951 **SSIM**:0.9948		**PSNR**:80.6798 **SSIM**:0.9846	
			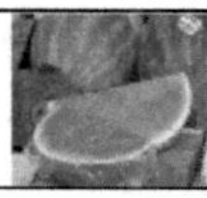	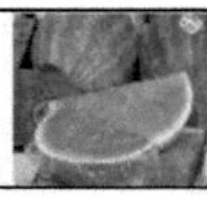	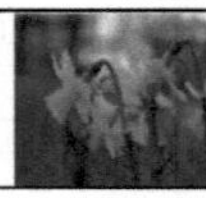	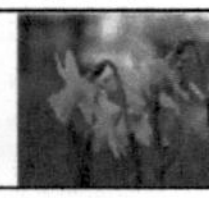
Quality Analysis	**PSNR**:80.7842 **SSIM**:0.9950		**PSNR**:80.2735 **SSIM**:0.9753		**PSNR**:84.4300 **SSIM**:0.9666	

Quality Analysis	**PSNR**:82.2703 **SSIM**:0.9831		**PSNR**:82.7293 **SSIM**:0.9869		**PSNR**:85.2370 **SSIM**:0.9825	

Table 3 Quality comparison between stego images of proposed versus existing method

Stego Image	PSNR of Proposed Method	PSNR of Existing Method
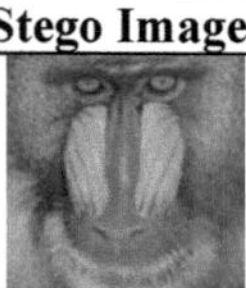	82.6935	70.786
	83.8223	71.349
	81.6900	69.673

5 Conclusion

In this research article, it has been shown that three secret RGB images can be hidden in one RGB color image, and all the secret images can be successfully retrieved without the help of the original cover image. Both the stego image and extracted secret images have high quality in terms of PSNR and SSIM values. Hence, successful implementation of blind multiple RGB image steganography can be achieved by this proposed method. The capacity can be further increased without compromising the high quality of the stego image as more bits are available to be manipulated. One limitation of this algorithm is the size of the secret image due to its inverse relation with the block size of the cover. The security of this algorithm can be further improvised by using different encryption algorithms.

References

1. Singh, Y.K., Sharma, S.: Image steganography on gray and color image using DCT enhancement and RSA with LSB method. https://doi.org/10.1109/inventive.2016.7830106 © IEEE
2. Bansal, D., Chhikara, R.: An improved DCT based steganography technique. Int. J. Comput. Appl. **102**(14). https://doi.org/10.5120/17887-8861
3. Das, P., Kushwaha, S.C., Chakraborty, M.: Data hiding using randomization andmultiple encrypted secret images. https://doi.org/10.1109/iccsp.2015.7322892 © IEEE
4. Hemalatha, S., Dinesh Acharya, U., Renuka, A., Kamath, P.R.: A secure image steganography technique to hide multiple secret images. https://doi.org/10.1007/978-1-4614-6154-8_60 © Springer
5. Tripathi, D., Sharma, S.: A robust 3-SWT multiple image steganography and contrast enhancement technique. https://doi.org/10.1109/inventive.2016.7823256 © IEEE
6. Aynapur, D., Thenmozhi, S.: A secure steganography approach of multiple secret images using ANN. Int. J. Recent Trends Eng. Res. **02**(04), 468–473 (2016)
7. Koptyra, K., Ogiela, M.R.: Key generation for multi-secret steganography. https://doi.org/10.1109/icissec.2015.7371013 © IEEE
8. Sharma, S., Sejwar, V.: QR code steganography for multiple image and text hiding using improved RSA-3DWT algorithm. Int. J. Secur. Appl. **10**(7), 393–406 (2016)
9. Shinu, V.L.: Mid band DCT coefficients based steganography. Int. J. Sci. Eng. Technol. Res. ISSN 2319-8885, **03**(48), 9838–9842 (2014)
10. Shankar, S.S., Rengarajan, A.: Data hiding in encrypted images using Arnold transform. ICTACT J. Image Video Process. **07**(01). ISSN: 0976-9102 (Online) https://doi.org/10.21917/ijivp.2016.0194
11. Dey, N., Roy, A.B., Dey, S.: A novel approach of color image hiding using RGB color planes and DWT. Int. J. Comput. Appl. **36**(5). https://doi.org/10.5120/4487-6316

Part IV
Session 2A: Modeling and Simulation

Brain–Computer Interface-Based Fear Detection: A Self-defense Mechanism

Rheya Chakraborty, Arup Kumar Chattopadhyay, Animesh Kairi and Mohuya Chakraborty

Abstract In this paper, brain–computer interface (BCI)-based fear signal detection and subsequent self-defense system has been presented. Self-defense is a countermeasure that involves protecting the health and well-being of oneself from detriment by others including human beings, animals. The system aims at designing an automated alert mechanism that operates involuntarily by taking into consideration the biological signals of a human being without the knowledge of the victim. This device is known as Silent Alert Self-Defense System (SiLERT). It is a small device that may be embedded in a cap, which monitors the human heartbeat rate and brainwaves to detect the fearful condition of a person when he is in danger. Upon detection of fear signals, the system automatically dials and sends emergency alert information including the location of the user via GPS to some predefined mobile numbers silently without the knowledge of the victim and attacker for help. The system has been designed and implemented using heartbeat and brain sensors along with a microcontroller to do the necessary steps. Real-time experimental results for two cases performed on two persons show the normal as well as fear state of mind. The GSM module attached to the system which automatically sends alert to the predefined mobile numbers is clearly shown experimentally.

Keywords Brain–computer interface · Brainwaves · Fear signal · Heartbeat Automatic · Involuntary · Self-defense · Silent alert

R. Chakraborty (✉)
Department of Electronics & Communication Engineering, Institute of Engineering & Management, Salt Lake, Kolkata, India
e-mail: crheya97@gmail.com

A. K. Chattopadhyay · A. Kairi · M. Chakraborty
Department of Information Technology, Institute of Engineering & Management, Salt Lake, Kolkata, West Bengal, India
e-mail: arup.chattopadhyay@iemcal.com

A. Kairi
e-mail: animesh.kairi@iemcal.com

M. Chakraborty
e-mail: mohuyacb@iemcal.com

M. Chakraborty et al. (eds.), *Proceedings of International Ethical Hacking Conference 2018*, Advances in Intelligent Systems and Computing 811,
https://doi.org/10.1007/978-981-13-1544-2_14

1 Introduction

A brain–computer interface (BCI) is an interface that directly communicates between brain signals and an external device. BCIs are often directed at researching, mapping, assisting, augmenting, or mending human cognitive or sensory-motor functions by monitoring brainwaves [1, 2].

Self-defense is the use of bodily strength to counter a sudden threat of violence either armed or unarmed. There are a large number of parameters on which chances of success depend, viz. hardness of the attack, mental and physical swiftness of the defender, strength of the defender. Self-defense may be accomplished by using licensed arms, martial arts, handheld portable alarms that generate high-pitch sounds to attract passerby, etc. The defender may either be strong or may be weak being elderly, disabled, child, sick, etc. If the defender is strong enough to react swiftly while in danger, then the above-mentioned mechanisms work well. However, if the defender is weak or the situation happens so abruptly that the defender may not get enough time to use the defense mechanism, then protection becomes a complete failure. With the rising crimes in every country where children, females, elderly persons become victims of human threat of violence, this paper presents a novel portable brain–computer interface for the detection of fear, based on biological signals like heartbeat rate and EEG signals of the brain to detect fear condition of a person in case of any danger and sends alert information through the GSM link to predefined mobile numbers for help. This device, which performs involuntarily and automatically, is known as Silent Alert Self-Defense System (SiLERT). In order to reduce false alarms, the owner of the device should avoid wearing this when he feels that he is in a safe environment or while enjoying because under extreme happiness and excitement, the pulse rate increases which activates the process.

1.1 Background Study

The speed of the heartbeat measured by the number of contractions of the heart per minute (bpm) is known as heart rate, which can differ under diverse situations based on the body's physical needs. It is usually found to be equal or close to the pulse measured in the peripheral point of the body. Activities that usually provoke the change in normal heart rate are physical exercise, sleep, anxiety, stress, illness, and ingestion of drugs. According to several studies, the normal heart rate of an adult human being is found to be 60–100 bpm. While tachycardia explains the condition of fast heart rate, defined as 100 bpm at rest, bradycardia is the slow heart rate condition, below 60 bpm at rest. When the heart beats in an irregular pattern, it is referred to as the condition named arrhythmia. Rhythm refers to the type of heartbeat. Normally, the heart beats in a sinus rhythm. Each electrical impulse is generated by the SA node which results in a ventricular contraction, or heartbeat. There are various abnormal electrical signals among which some are normal variants while the rest are potentially dangerous. Some electrical rhythms do not generate a heartbeat and are the cause of sudden death. This paper takes into consideration the trigger which occurs during conditions of excitement which are represented by increased heart rates. This trigger in turn activates the next part of the device which detects the required brainwaves for detection of fear.

The brain signals play a vital role behind the occurrence of every emotion like relaxed state, attention, alertness, fear [3]. Neural oscillation, generally known as brainwave, is the rhythmic or repetitive brain activity in the central nervous system. Neural tissue is capable of generating oscillatory activity in many ways, which are driven either by mechanisms within individual neurons or by interactions between the neurons. For the individual neurons, oscillations can appear in two different ways, i.e., either as oscillations in membrane potential or as rhythmic patterns of action potentials, after which oscillatory activation of post-synaptic neurons is produced. At the level of neural ensembles, macroscopic oscillations can be produced when synchronized activity of large numbers of neurons occur. This process can be observed in an electroencephalogram (EEG) that is an electrophysiological monitoring method to record the electrical activity of the brain. It measures voltage fluctuations obtained from the ionic current within the neurons of the brain. In clinical terms, EEG is referred to as the recording of the brain's spontaneous electrical activity over a period of time, as obtained from multiple electrodes placed on the scalp [4].

In order to find the fear condition, a thorough knowledge of the brainwaves and their meanings must be understood. Researchers have found that fear is established unconsciously in the amygdala region situated in the brain. A new study suggests that brainwave oscillations of a particular frequency may serve to create a "fearful" brain state, generating conditioned behaviors associated with fear. Previous studies have shown that such responses to conditioned terror depend on an interaction between brain regions called the dorsal medial prefrontal cortex (dmPFC) and basolateral amygdala (BLA). The following frequency ranges of brainwave oscillations are associated with the various human emotions (see Fig. 1).

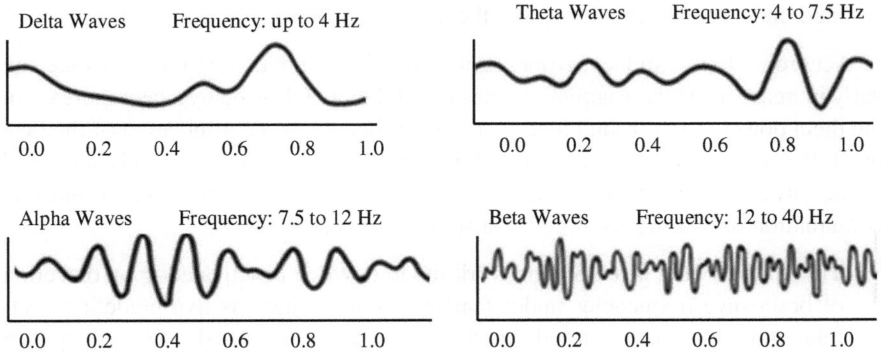

Fig. 1 Brainwave frequencies and their nature

A GSM module is basically a GSM modem connected to a circuit with different types of outputs taken from the breadboard, viz. TTL output for microcontrollers to send messages to mobile numbers through GSM link using the cellular network.

The organization of the paper is as follows. Section 1 holds the introduction with a background study of the proposed work. Sect. 2 contains the basic concept of the proposed system called SiLERT with the necessary flow diagrams and the respective explanations. Section 3 describes the block diagram representation of the hardware device and experimental setup. Section 4 shows the experimental result analysis. Section 5 highlights conclusion to this paper with few focus points on imminent works and the future scope of the proposed work.

2 Basic Concept of SiLERT

2.1 Overview

This research work relates to the design of an automated involuntary hardware system for self-defense that would monitor the heart rate as well as the electrical signals sent by the brain during fearful or tensed conditions. This would activate a GSM module to dial predefined mobile numbers, thus sending alert messages simultaneously locating the victim via GPS. Figure 2 depicts the basic idea of SiLERT [5].

The system basically consists of three components.

1. The first module is a microcontroller-based heartbeat sensor that monitors the heart rate in the form of pulses found at specific regions of the body. In this device, the sensors are attached behind the ears. This system is capable of detecting any increase in heartbeat rate above the threshold value.
2. The second module is a microcontroller-based system consisting of brainwave sensors which activate whenever the human heart rate goes above the normal threshold value, to monitor brainwave frequencies that are expected to be generated under fearful condition, i.e., theta, and beta waves.

According to research experiments, it is found that threatening images evoked an early increase in the theta activity in the occipital lobe, followed by a later increase in the theta power in the frontal lobe. A left lateralized de-synchronization of the beta band, the wave pattern associated with motor behavior, also consistently appeared in the threatening conditions. Thus, fear can be detected by obtaining an unusual combination of beta waves along with some theta waves.

3. The third module is a GSM network module which is activated upon detection of brainwave frequencies under fearful condition that has five predefined cellular mobile numbers stored which automatically dials and sends emergency information to these mobile numbers indicating danger [6].

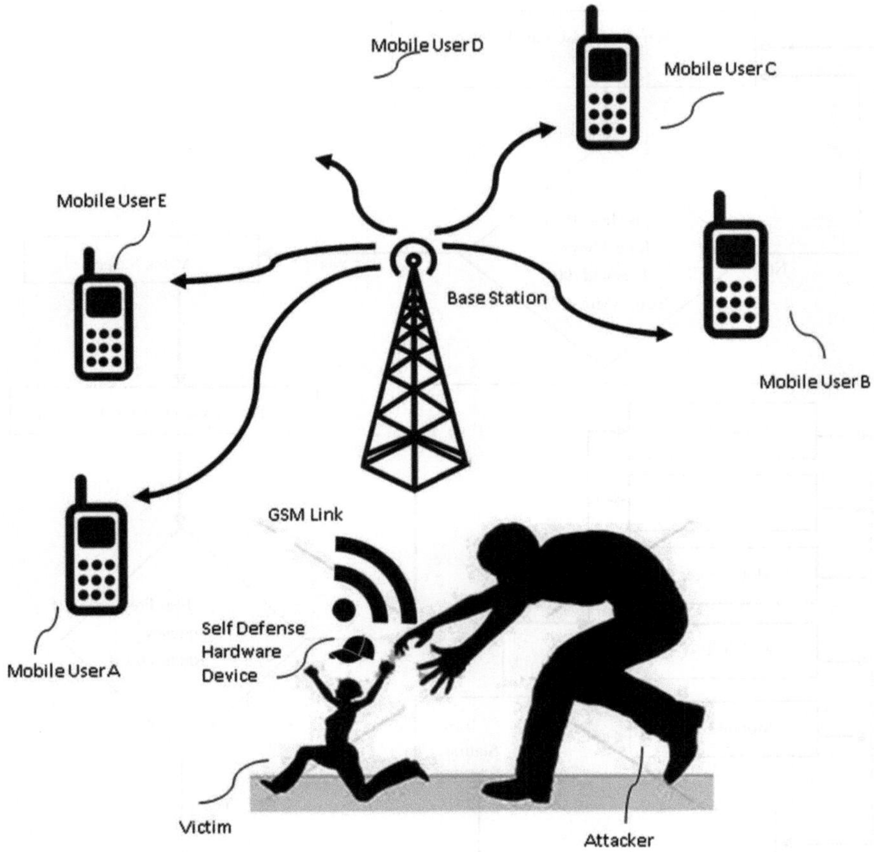

Fig. 2 Basic idea of SiLERT

The entire self-defense system is designed in the shape of cap so that the owner can wear the device very easily without being recognized by the attacker so as to obtain the heartbeat rate and brainwave frequencies involuntarily.

2.2 *Work Flow of the Self-defense Hardware Device*

The flowchart of SiLERT is represented in Fig. 3. The figure indicates that pulse rate detector has to be activated whenever a person wants self-defense. If the heart rate goes above the threshold value, attack is suspected but not confirmed. This is because human heart rate may go above normal value whenever a person is excited or extremely happy. Once the attack is suspected, the brainwave detector gets activated and monitors the frequencies. If the fear frequencies of various brainwaves like

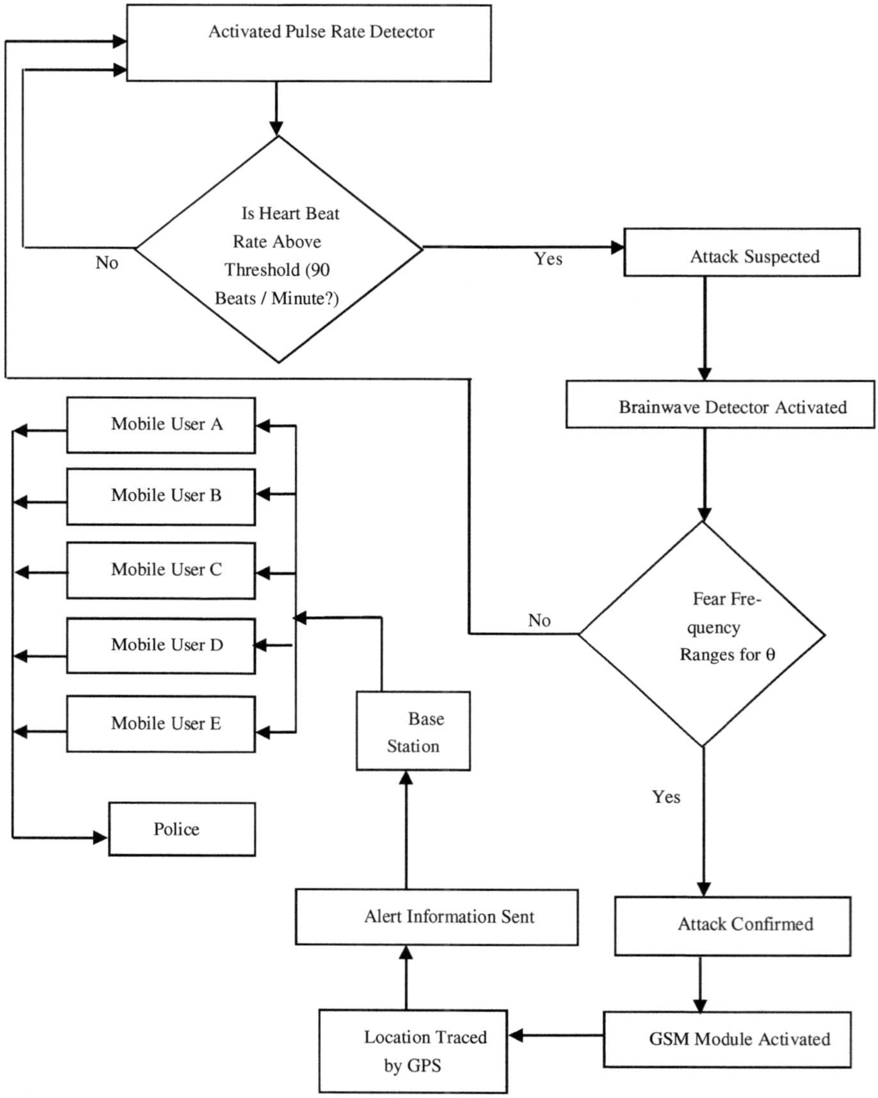

Fig. 3 Work flow of the self-defense hardware device

alpha, beta, gamma and theta are detected, then attack gets confirmed. On confirmation of attack, the GSM module gets activated, location is traced by GPS, and alert information is sent to all the predefined mobile numbers through base station of the service provider which can then contact law and enforcement for the safety of the victim [7].

3 Experimental Setup

3.1 Heartbeat Rate Monitor

Figure 4 shows the block diagram representation of the heartbeat rate module, which is designed using a heartbeat sensor (SEN11574) attached to the back of the ear to detect heartbeat.

This sensor module consists of an IR pair which actually detects heartbeat from the blood. Heart pumps the blood in body which is called the heartbeat. On occurrence of heartbeat, the blood concentration in our body changes. A voltage or electrical pulse may be generated by using the above change. A microcontroller (Arduino UNO) has been attached to the heartbeat sensor module that controls the whole process of the system like reading pulses from heartbeat sensor module, calculating heart rate, and sending this data to a threshold detector to detect whether the heartbeat rate is above normal value or not. Push button is used to activate the module. The flowchart of the algorithm used to detect pulse rate is given in Fig. 5.

3.2 Brainwave Monitor and Integrated GSM Module

Figure 6 shows the block diagram representation of brainwave monitor integrated with GSM module. Upon detection of increased heartbeat rate, the next module that gets activated is the brainwave monitor [8]. The brainwave monitor has been designed using brain sensors made up of EEG electrodes [9, 10]; amplifier to amplify weak brain signals; and GSM module (SIM900) which is integrated with the microcontroller supports communication in 900 MHz band and sends desired alert SMS. We have used SIM900 as most of the mobile network providers in India operate in the 900 MHz band.

The detailed circuit of brainwave amplification consisting of instrumentation amplifier (AD620), operational amplifiers (LM324N), resistors, capacitors, microcontroller (Arduino UNO), and power supply is given in Fig. 7.

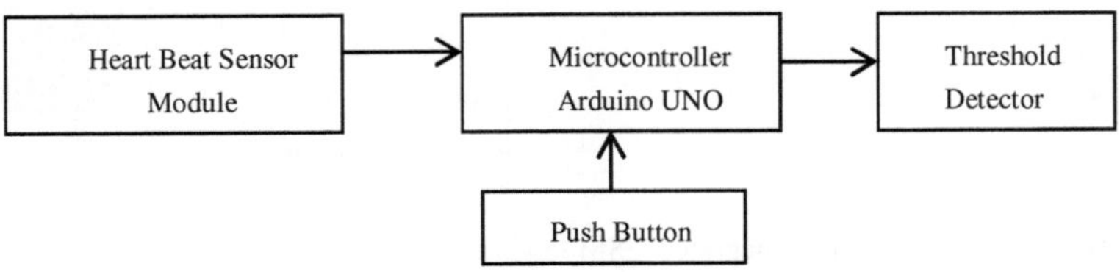

Fig. 4 Heartbeat rate monitor

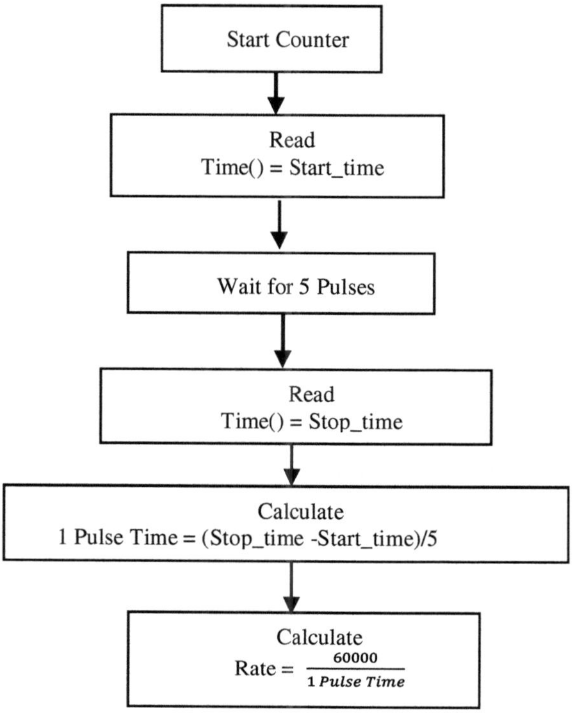

Fig. 5 Work flow of the algorithm to calculate heartbeat rate

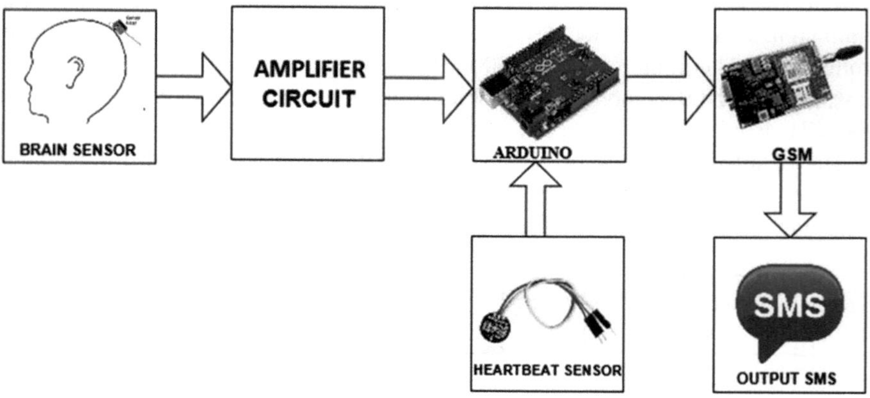

Fig. 6 Brainwave monitor and integrated GSM module

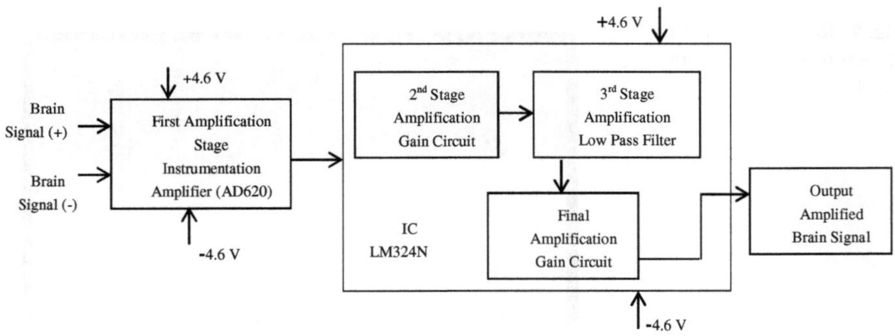

Fig. 7 Brainwave amplifier

4 Experimental Result Analysis

The experiment to indicate fear detection in humans was conducted in the college laboratory at the institute premises. The experiment was performed on two different persons for normal case as well as fear state of mind. Experimental results obtained in Arduino UNO were conducted on two persons for normal as well as fear state of mind. Figure 8 shows the real-time FFT plotting of brainwaves for a period of 1 min. The circled portion of the plot where both beta and theta waves have been detected determines fear.

Figure 9 shows the real-time reception of alert messages on detection of fear signals in a predefined mobile number (+919934493642) on February 24, 2018, during the time interval 14:48–14:50. Location may be obtained by attaching GPS with the system.

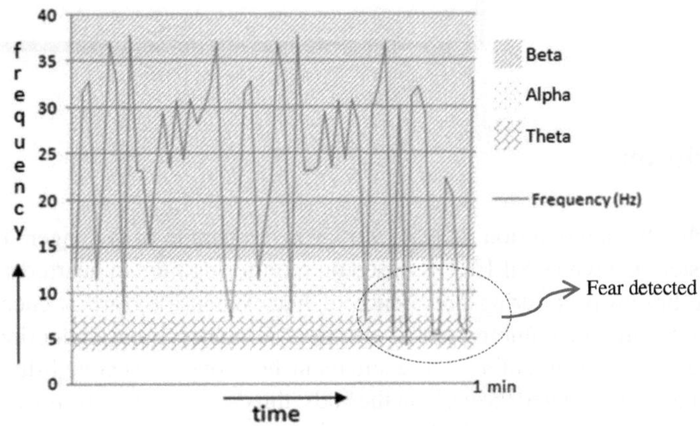

Fig. 8 FFT of real-time brainwaves plotted in Arduino UNO determining fear

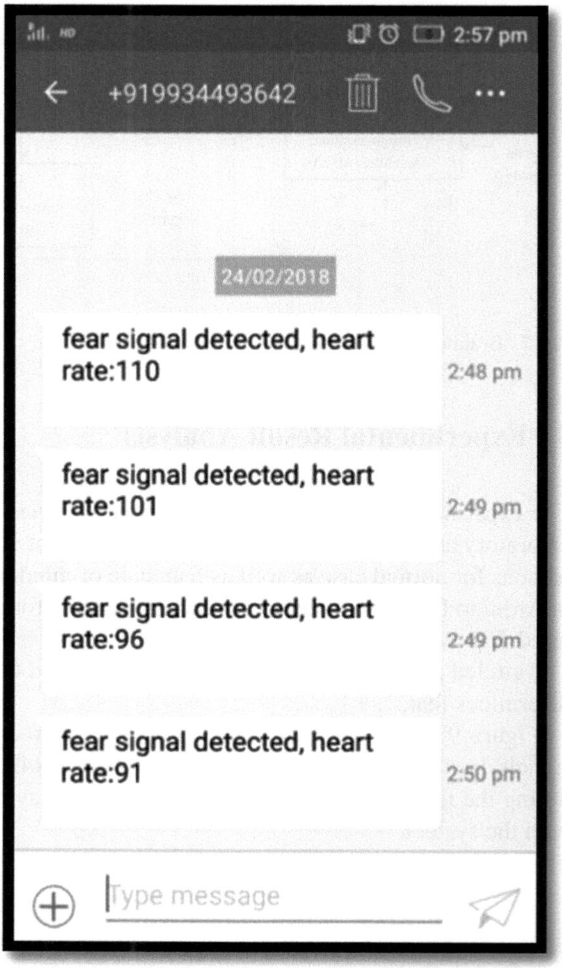

Fig. 9 Real-time reception of alert messages through GSM link

5 Conclusion

Fear is such a human emotion which occurs when a person is in danger. This self-defense system known as SiLERT is based on monitoring human heartbeat rate and appropriate brainwaves during prevalence of fear. When a hazard is encountered, our brain sends unique combination of electrical impulses to our endocrine system through the nerves. Adrenaline, the excitement hormone, is secreted during such situations that are circulated throughout the body, thus sending us alert signals. During any kind of excited situations, the heart rate increases due to the fact that the heart starts pumping at a much faster rate so as to supply oxygen to the various parts of the body. So by monitoring the heartbeat rate and brainwaves, fear may be detected.

The most significant concern in this research work is the design of a small size device in the form of a cap that may be worn by the victim for detection of fear. Experimental results performed on two different persons for two different cases including normal and fear state of mind confirm the success of sending the alert information to predefined mobile numbers through GSM link. On receiving the emergency information, proper safeguard of the victim is guaranteed. This device promises freedom of movement of common men, local population, and above all assures the nation a safe surrounding.

6 Compliance with Ethical Standards

6.1 Research Involving Human Participation

The experiment of the work described in this paper was conducted in the college laboratory at the institute premises and was performed on two teachers of the college with their written consent in the ethics approval form of the institute.

6.2 Research Involving Mobile Device

The present work also involved the use of a SIM card. The mobile number used in this connection is that of a student of the college whose written consent has been taken in the ethics approval form of the institute prior to the experiment.

References

1. Krucoff, M.O., Rahimpour, S., Slutzky, M.W., Edgerton, V.R., Turner, D.A.: Enhancing nervous system recovery through neurobiologics, neural interface training, and neurorehabilitation. Front. Neuroprosthetics **10**, 584 (2016); J. **2**(5), 99–110. https://doi.org/10.3389/fnins.2016.00584.5186786.pmid28082858.author
2. Donchin, E., Spencer, K.M., Wijesinghe, R.: The mental prosthesis: assessing the speed of a P300-Based brain-computer interface. IEEE Trans. Rehabil. Eng. **8**(2), 174–179 (2000)
3. Bozinovski, S., Sestakov, M., Bozinovska, L.: Using EEG alpha rhythm to control a mobile robot. In: Proceedings of IEEE Annual Conference of Medical and Biological Society, pp. 1515–1516, New Orleans (1988)
4. Leuthardt1, E.C., Schalk, G., Wolpaw, J.R., Ojemann, J.G., Moran, D.W.: A brain–computer interface using electrocorticographic signals in humans. J. Neural Eng. 63–71 (2004). stacks.iop.org/JNE/1/63, https://doi.org/10.1088/1741-2560/1/2/001
5. Lazar, S.W., Bush, G., Gollub, R.L., Fricchione, G.L., Khalsa, G., Benson, H.: Functional brain mapping of the relaxation response and meditation. Neuroreport (2000)

6. Dorfer, C., Widjaja, E., Ochi, A., Carter, O.S.I., Rutka, J.T.: Epilepsy surgery: recent advances in brain mapping, neuroimaging and surgical procedures. J. Neurosurg. Sci. **59**(2), 141–155 (2015)
7. Sriranjini, R.: GPS and GSM based self defense system for women safety. J. Electr. Electron. Syst. (2017). https://doi.org/10.4172/2332-0796.1000233
8. Johnson, N.N., Carey, J., Edelman, B.J., Doud, A., Grande, A., Lakshminarayan, K., He, B.: Combined rTMS and virtual reality brain-computer interface training for motor recovery after stroke. J. Neural Eng. **15**(1) (2018) © IOP Publishing Ltd
9. Rajesh Kannan, V., Joseph, K.O.: Brain controlled mobile robot using brain wave sensor. IOSR J. VLSI Signal Process. (IOSR-JVSP) 77–82. e-ISSN: 2319–4200, p-ISSN : 2319–4197. www.iosrjournals.org; International Conference on Emerging Trends in Engineering and Technology Research, pp. 77–82
10. Ying, R., Weisz, J., Allen, P.K.: Grasping with your brain: a brain-computer interface for fast grasp selection. In: Bicchi, A., Burgard, W. (eds.) AG 2018 Robotics Research. In: Springer Proceedings in Advanced Robotics 2. Springer International Publishing, pp. 325–340. https://doi.org/10.1007/978-3-319-51532-8_20

Modelling and Simulation of Proton Exchange Membrane Fuel Cell for Stand-Alone System

Rajesh Singla

Abstract Fuel cell system is an unconventional energy source that can be used in various stand-alone applications. The fuel cell gives an unstabilized voltage that is exclusively unacceptable for segregated applications. The primary objective of this study is to design an appropriate power conditioning unit that comprises of DC-DC converter stages along with DC-AC inverter. The capacitance and resistance are dependent on the proton exchange membrane fuel cell and cause electrical effects due to the behavioral changes of the output voltage of the fuel cell stack. This article discusses the electrical parameters of dynamic model of proton exchange membrane fuel cell. Its dynamic model was related to the boost converter-averaged dynamic model that is obtained by using the mathematical model of the boost converter circuit. This circuit keeps the output voltage of the converter constant and is being fed into the inverter to rectify the voltage, and a filter is also used to eliminate harmonics in the AC signal.

Keywords Proton exchange membrane fuel cell · Electrochemical impedance spectroscopy · DC-to-AC inverter · DC-to-DC converter · Filter and PWM

1 Introduction

Many traditional methods for the production of electricity such as burning fossil fuels were used for last many decades despite having various kinds of destructive environmental impacts. For instance, such methods increased the global warming with the rise in emission of greenhouse gases [1, 2]. Such factors make fuel cell (FC) technology a satisfactory choice for various external utilization.

Polymer exchange membrane fuel cell (PEMFC) is a type of fuel cell (FC), which proves its supremacy over other types of fuel cells in terms of characteristics like high power density, low temperature, and fast response. Despite this, factors like

R. Singla (✉)
Dr. B. R. Ambedkar NIT Jalandhar, Jalandhar, India
e-mail: rksingla1975@gmail.com

M. Chakraborty et al. (eds.), *Proceedings of International Ethical Hacking Conference 2018*, Advances in Intelligent Systems and Computing 811,
https://doi.org/10.1007/978-981-13-1544-2_15

tremendous cost, limited lifespan, and awful authenticity of PEMFCs have limited the extensive applications of FC in real-time world [3, 4]. The most prominent drawbacks of FCs include load current imbalance and maximal load swings which further sometimes become the reason for voltage swings and power issues. Such issue is resolved by using proper power conditioning unit (PCU). The PCU is required to process the unprocessed power output of FCs so that it becomes utilizable.

DC-to-DC converters are used whenever an averaged voltage is needed as the output. Inverter is extensively used for backup generation of electricity for analytical loads like computers in various offices, life support systems in hospitals, powerhouses, and also in transportation workstations, and communications systems [5]. This study focuses on modelling an isolated application where a PEMFC is considered as a basic energy source.

The output voltage of PEMFC is in a fluctuating DC form with the variations in the load. The prime and fluctuating 48 V DC input source models FC, and the boost DC-to-DC converter regulates the output of the PEMFC to 72 V.

The ideal standard potential of a PEMFC is 1.229 V (25 °C and 1 atm) with liquid water product. The permanent voltage losses happening in the FCs decrease the actual FC potential from the equilibrium point.

2 Losses in Fuel Cell

FC losses are credited to following grades: (i) activation overvoltage, (ii) ohmic overvoltage, and (iii) concentration overvoltage [3].

2.1 *Activation Overvoltage*

The activation overvoltage is the result of electron transfer need and breaking-forming of chemical bonds at anode as well as cathode. The activation overvoltage happens on both anode and cathode of the FC.

The hydrogen oxidation reaction at the anode is too fast, whereas oxygen reduction process at the cathode is significantly steady in speed [6]. There hence, potential loss because of the loss of activation is overcome by the cathode reaction.

2.2 *Ohmic Overvoltage*

The ohmic overvoltage is because of the polymer membrane resistance offered to the protons transfer and the resistance of both the electrodes: cathode and anode along with collecting plate offered to electron transfer [6, 9].

2.3 *Concentration Overvoltage*

The concentration overvoltage emerges as a result of concentration changes in the reactants because of their consumption during the reaction. Such losses lead to instant potential drop at greater currents [6, 7, 9].

3 Electrochemical Impedance Spectroscopy

Electrochemical impedance spectroscopy is a very useful technique in the characterization of the behavior of electrochemical cells. The data obtained from this model can be fitted to an electric circuit model that mimics chemical processes or models [10].

4 Cell Terminal Voltage

From the combination of all the voltage drops related to all losses mentioned above, the operating voltage of the FC is represented as follows:

$$\text{Vfc} = \text{E} - \text{Vact} - \text{Vohm} - \text{Vconc} \tag{4.1}$$

where E is the open circuit potential E. The calculated potential Vfc shows potential of a single FC [5, 7–9]. Total stack voltage is calculated from the product of single cell voltage and the total number of cell in stack.

$$\text{Vst} = \text{n} \times \text{Vfc} \tag{4.2}$$

5 PEM Fuel Cell Electrical Circuit Model

Electrical circuit model consists of the membrane cell resistance Rm in series with a parallel combination of double-layer capacitance Cdl representing the membrane electrode interface and an impedance of faradic reaction consisting of charge transfer resistance Rct and specific electrochemical element of diffusion Zw as shown in Fig. 1 [10].

The characteristics of the above circuit when 40 cells are stacked together are shown in Fig. 2. It draws 1.11 A current when the values of C1 and C2 are considered negligible. The output voltage of the fuel cell is unregulated with change in the load (current) voltage, so we need a power converter circuit to correct this problem.

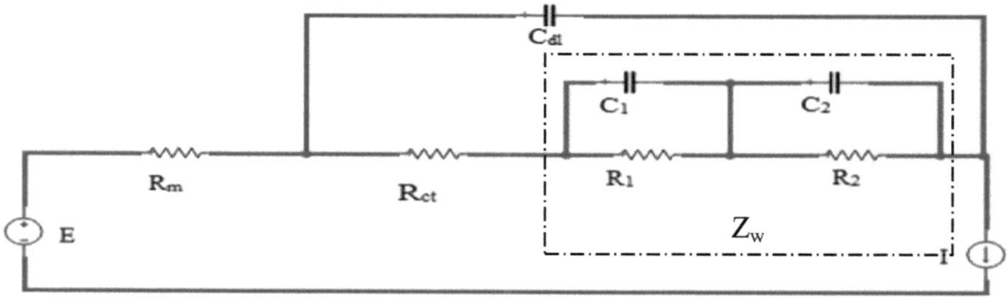

Fig. 1 Circuit of PEMFC

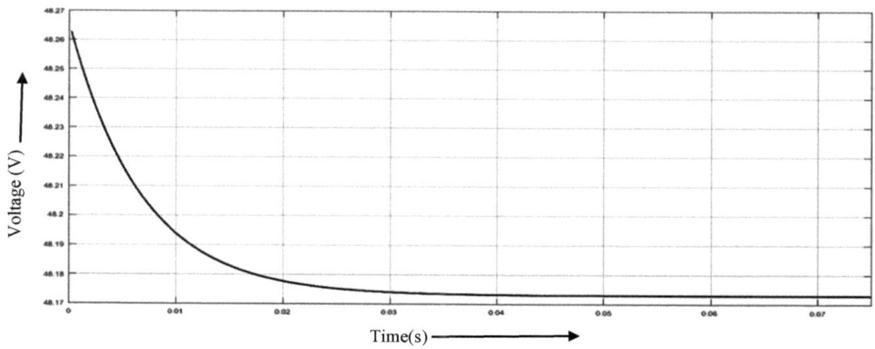

Fig. 2 Voltage drops across fuel cell

6 Power Conditioning Unit

This unit of power conditioning transforms the raw energy into utilizable energy for various applications.

While the DC output voltage in the fuel cell is not regulated, its updating is required for practical application. The power conditioning unit maintains the harmonics at an acceptable level. The conditioning system converts the gross energy into usable energy for all applications.

The FC output is an unregulated DC voltage that gets transformed to DC voltage administered by boost converter. The potential procured post filtration gets fed to DC-to-AC inverter. As per the load, the PCU will be brought into use to take out the needed current from the FC. The load current will be an explicit wave and will depend on frequency, the size of the inductor and the capacitor. Other than this, the inverter interfaces the FC from the grid power system which is used to create a mesh with the help of voltage or current in a suitable phase, frequency, and magnitude. PCU structure along with DC-to-DC pulse converter and DC-to-AC inverter stage is depicted in Fig. 3. There are various transitional phases shown in Fig. 3 that constitute filters for dampening of current harmonics as well as undesired voltages at the DC-to-DC converter and the DC-to-AC inverter output [2, 4].

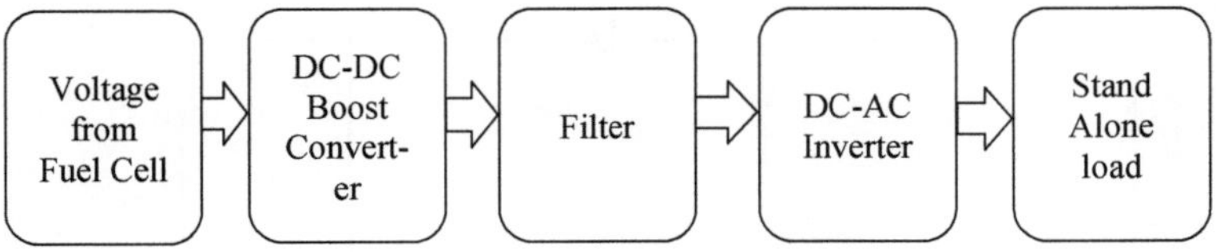

Fig. 3 Power conditioning blocks for PEMFC

6.1 The Levels of PCU

The FC output is a fluctuating DC voltage that gets varied along the variations in load. The boost converter is further transformed into a controlled DC voltage. Inverter acts as a network interface [2, 4].

6.2 DC-DC Boost Converter

As the load current increases, the FC voltage will be reduced, so the voltage of the unregulated terminals cannot be precisely linked to the DC bus. The DC-AC inverter cannot be used for stand-alone applications.

As a result, the FC stack is used in the linear work area for designing the converter (due to the strength of the internal components), as the use of the FC in nonlinear region damages the lamination.

Working of Boost Converter. When switch S is closed for time duration t1, the inductor current rises, and energy gets stored in the inductor. If the switch S is opened during time t2, the energy stored in the inductor is transferred to the load through the diode D, and the current of the inductor drops. When the switch S is open, the capacitor is charged by the voltage stored in the inductor. Therefore, the capacitor gets fully charged for use. The unsteady high-frequency switching voltage in the boost converter with 33% duty cycle becomes the required stable voltage [11, 12] (Figs. 4, 5, 6).

6.3 Designing of Boost Converter

A boost converter is designed to provide an output of 48 V from a 72 V source. The load is 80 W. The voltage and current ripple must be less than 0.5%. Specify the duty ratio, switching frequency, values, and ratings of each of the components. All the components used are assumed in their ideal states [11, 13, 14].

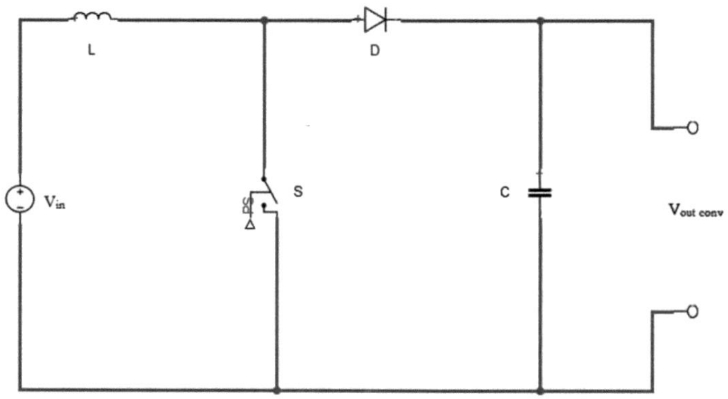

Fig. 4 Boost converter circuit

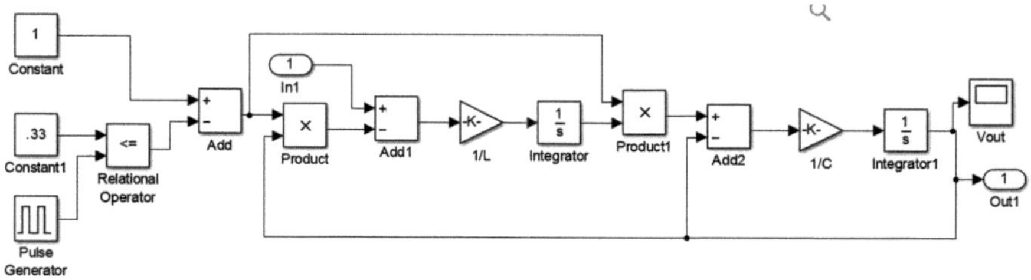

Fig. 5 Mathematical model of boost converter

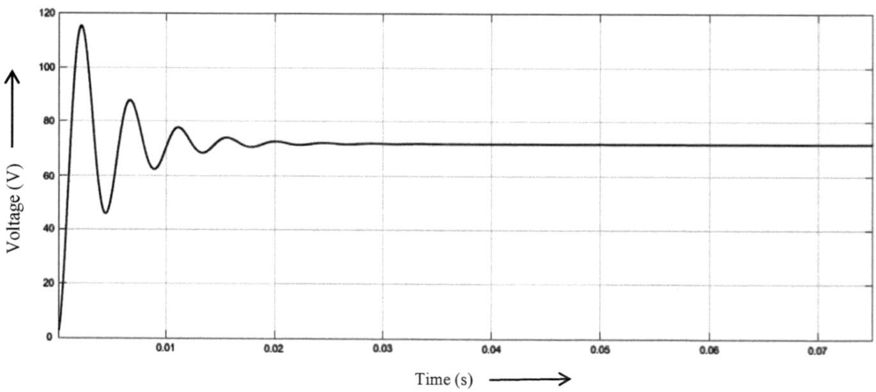

Fig. 6 Characteristic of boost converter

Given:

$$V_{in} = 48\,V$$
$$V_0 = 72\,V$$

$$\Delta V_0 \le 0.5\% V_0$$

Design Assumptions:

i. The switching frequency is assumed to be 25 kHz.
ii. Inductor current is assumed to be continuous.

$$Io = \frac{P}{V_0} = \frac{80}{72} = 1.11\,\text{A}$$

If D is the duty cycle of boost converter, then

$$\frac{V_0}{V_{in}} = \frac{1}{1-D};$$
$$\text{D} = 0.33$$
$$Lmin = \frac{D(1-D)}{2f}R;$$
$$Lmin = 48\,\mu\text{H}$$
$$\text{Let L} = 150\%\text{ of } Lmin$$
$$\text{L} = 1.5 * 48 = 72\,\mu\text{H}$$
$$\text{C} \ge \frac{D}{R\frac{\Delta V}{V}f}R \ge 163\,\mu\text{F}$$
$$\therefore \text{C} = 165\,\mu\text{F}$$

7 Filter

The line filter reduces the high-frequency harmonic content of the line current that was caused by the switched operation of the voltage source inverter (VSI). Factors like filter cost, size, and application purpose are considered for the selection purpose [13].

LC filter can be used for minimization of harmonics at output side of the inverter.

7.1 Calculation of L and C

The inductive part of the low pass filter is designed on the basis of the allowable current ripple at switching frequency on the grid side [13, 14].

$$\text{L} = \frac{V_{DC}}{4f_S\Delta i}$$

where VDC is the DC voltage, Δi is the value of the current ripple, and fs is switching frequency.

For calculation of capacitor,

$$C = \frac{\Delta i}{8 f_S \Delta V_0}$$

where ΔV_0 = voltage ripple.

8 PWM Generation

PWM techniques are identified by constant amplitude pulses. Their width is to be modulated for obtaining output voltage control along with reduction of its harmonic content [13]. Here, three different PWM techniques generally used are single pulse modulation, multiple pulse modulation, and sinusoidal pulse width modulation (carrier-based PWM technique). Here, the only carrier-based PWM technique is explained which is used in this paper.

8.1 Carrier-Based PWM

During this process, a triangular carrier wave gets compared to the reference sinusoidal voltage with the specified output frequency. When the reference signal is larger as compared to carrier signal, the upper switch gets on while the lower switch gets off, else the other way. Whenever reference signal is larger compared to carrier signal, the switching signal begins while it stops when carrier signal is larger compared to reference signal [13]. The functional value of switching signals seems sinusoidal where every pulse width is associated with amplitude of reference wave. The changes in amplitude of reference wave regulate output voltage of inverter. Mathematical model and characteristics of the PWM are shown in Fig. 7 and Fig. 8, respectively.

9 Single-Phase DC-to-AC Inverter

The inverters play a crucial role in converting DC to AC using electrical equipment like induction motor drive, uninterruptible power supply (UPS), and automatic voltage regulator (AVR). The basic aim to use the inverter is constructing a sinusoidal output voltage, which regardless of the kind of load creates stable and smooth waveform. Figure 9 shows the circuit diagram of the single-phase inverter. The single -phase complete bridge inverter comprises two identical arms consisting of four

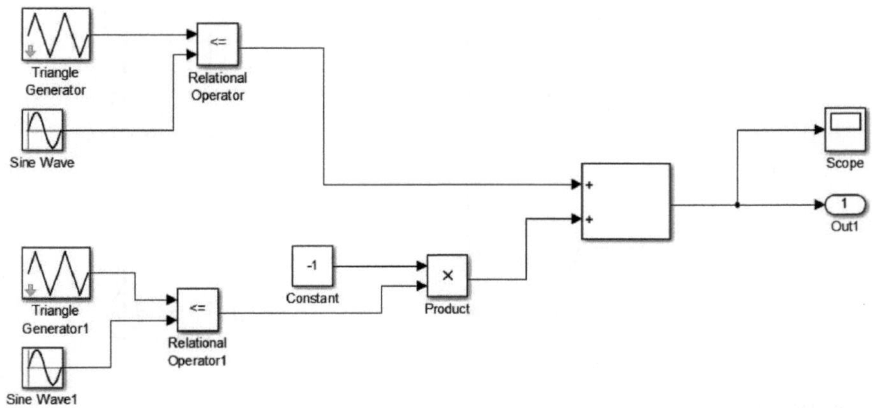

Fig. 7 Mathematical model of PWM

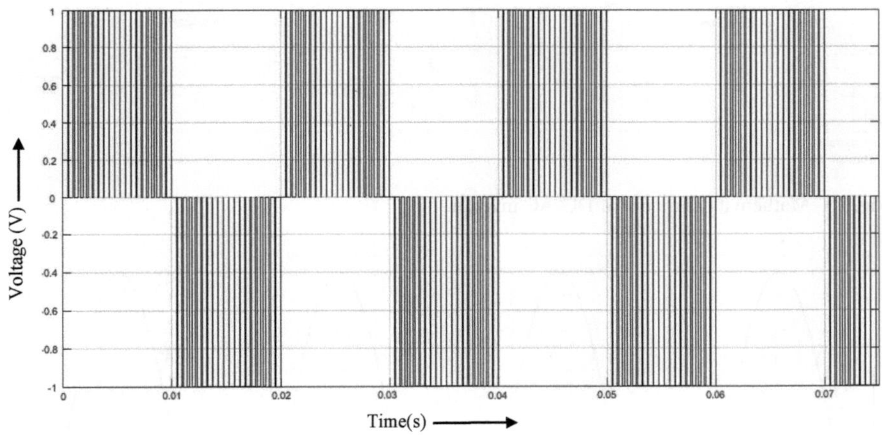

Fig. 8 PWM signal generator output

switches (S1, S2, S3, S4) along with four antiparallel diodes (D1, D2, D3, D4). On switching on S1 and S2, the output voltage becomes DC bus voltage +Vdc, and likewise on switching on S4 and S3, the output voltage becomes −Vdc [11, 12]. Figure 10 shows the mathematical model of DC-AC inverter. Figure 11 shows the sinusoidal output response of the DC-to-AC inverter.

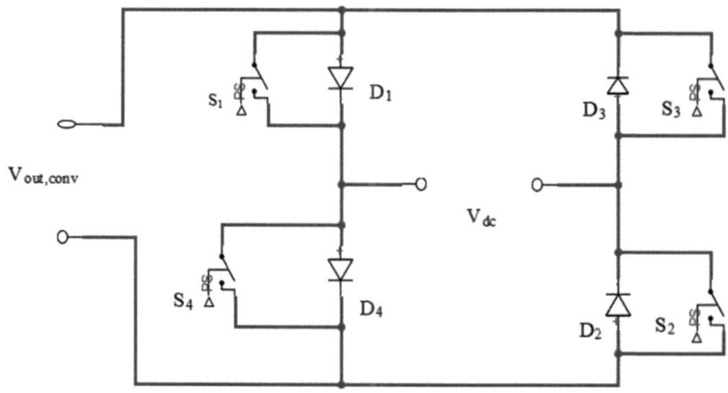

Fig. 9 DC-AC inverter circuit

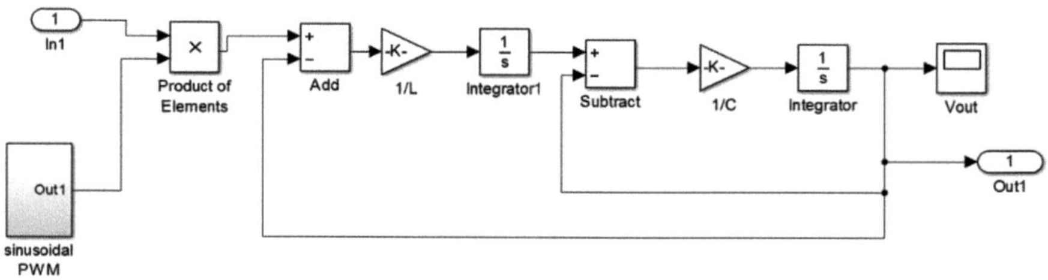

Fig. 10 Mathematical model of DC-AC inverter

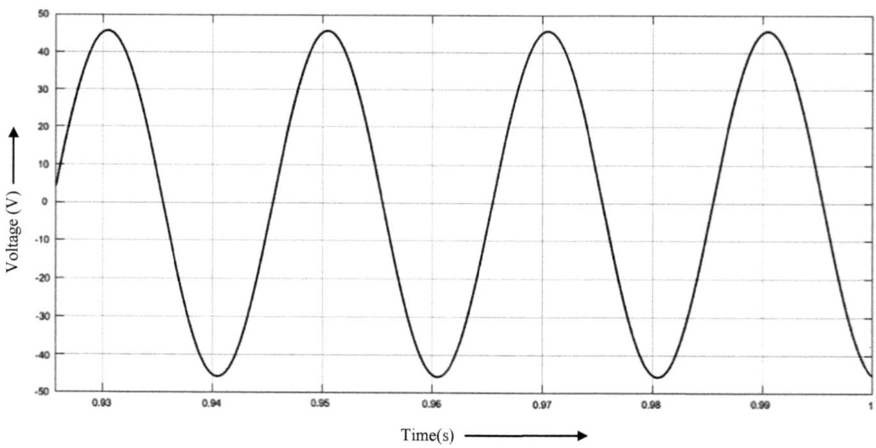

Fig. 11 Characteristic of inverter

10 Conclusion

This study presents a Simulink model for the PEMFC power system consisting of PEMFC stack, DC-to-DC converter, PWM-based DC-to-AC inverter. Modelling was done by adopting and deriving a mathematical model that explained the behavior of the fuel cell system. The PEMFC stack was modelled while counting the irreversibility. The performances of the fuel cells voltage are analyzed for a particular current range. Switching signals were input to the inverter model that were generated by the PWM.

References

1. Itoh, J.I., Hayashi, F.: Ripple current reduction of a fuel cell for a single-phase isolated converter using a DC active filter with a center tap. IEEE Trans. Power Electron, 550–560 (2010)
2. Lee, S.H., Song, S.G., Park, S.J., Moon, C.J., Lee, M.H.: Grid connected photovoltaic system using current-source inverter. Sol. Energy, 411–419 (2008)
3. Larminie, J., Dicks, A.: Fuel Cell Systems Explained, 2nd edn. SAE International and Wiley Ltd, New York (2003)
4. Wang, C., Nehrir, M.H., Gao, H.: Control of PEM fuel cell distributed generation systems. IEEE Trans. Energy Convers. 586–595 (2006)
5. Liu, G., Liu, Y., Qi, Z.: Single-phase sinusoidal inverter based on fuzzy PID control for small wind power system. In: 2nd International Conference on Computer Science and Network Technology (ICCSNT), pp. 625–638 (2012)
6. Na, W.K., Gou, B.: Feedback-linearization-based nonlinear control for PEM fuel cells. IEEE Trans. Energy Convers. 179–190 (2008)
7. Iqbal, M.: Simulation of a small wind fuel cell hybrid energy system. Renew. Energy, 511–522 (2003)
8. Rakht Ala, S.M., Ghaderi, R., Ranjbar, A., Fadaeian, T., Nabavi, A.: Current stabilization in fuel cell/battery hybrid system using fuzzy-based controller. In: Presented at the IEEE Conference on Electrical Power & Energy, Canada (2009)
9. Na, W.K., Gou, B., Diong, B.: Nonlinear control of PEM fuel cells by exact linearization. IEEE Trans. Ind. Appl. 1426–1433 (2007)
10. Fardoun, A.A., Hejase, H.A., Al-Marzouqi, A.: Electric circuit modelling of fuel cell system including compressor effect and current ripples. Int. J. Hydrogen Energy, **42**(2), 1558–1566 (2017)
11. Rakhtala, S.M., Shafiee Roudbari, E.: Fuzzy PID control of a stand-alone system based on PEM fuel cell. Electr. Power Energy Syst. **78**, 576–590 (2016)
12. Utkin, V., Wenguang, Y., Longya, X.: Sliding mode pulse width modulation. In: American Control Conference, ACC'07, pp. 4530–4550 (2007)
13. Dave, M., Vyas, S.R.: Simulation and modelling of single phase dc-ac converter of solar inverter. Int. Res. J. Eng. Technol. (IRJET) **02**, 2225–2236 (2015)
14. Bimbhra, P.S.: DC-DC Power Converters Power Electronics. ISBN: 817409279X (Edition 2012)

Hardware Realization of Power Adaptation Technique for Cognitive Radio Sensor Node

S. Roy Chatterjee, J. Chowdhury and M. Chakraborty

Abstract Prototype developments of cognitive radio sensor nodes (CRSNs) need to minimize the utilization of hardware and power consumption as they have an inherent limitation in terms of transmission power consumption, communication capabilities, processing speed, and memory resources. In this paper, a power-sharing algorithm based on game theory is implemented in FPGA for an embedded wireless system in which both Primary User (PU) and CRSN operate simultaneously and a dedicated hardware unit takes the decision about the power transmission for both PU and CRSN. Hardware architecture is designed in Verilog hardware description language in Vivado Design Suite 2015.3 using IEEE 754 floating point format with 64-bit double-precision. The hardware module analyzed in real time in DIGILENT ZED BOARD (xcz-7z020 clg484-1) using integrated logic analyzer shows computational time and computational power of 4.55 μs and 9 mw, respectively. Comparative performance analysis of the hardware and MATLAB simulation shows that the former provides less computing power to CRSN compared to the simulated value. However, number of iteration varies in simulation for the distance of the node from the base station whereas it is almost constant in the hardware module.

Keywords Cognitive radio sensor node · FPGA · Hardware architecture · Power control

S. Roy Chatterjee (✉) · J. Chowdhury
Netaji Subhash Engineering College, Kolkata, India
e-mail: rcswagata@gmail.com

J. Chowdhury
e-mail: jayantachowdhury32@gmail.com

M. Chakraborty
Department of Information Technology, Institute of Engineering & Management, Salt Lake, Kolkata, West Bengal, India
e-mail: mohuyacb@iemcal.com

M. Chakraborty et al. (eds.), *Proceedings of International Ethical Hacking Conference 2018*, Advances in Intelligent Systems and Computing 811,
https://doi.org/10.1007/978-981-13-1544-2_16

1 Introduction

Cognitive radio sensor network is an extension of conventional wireless sensor network (WSN). Here, conventional sensor nodes are replaced by the cognitive radio sensor nodes (CRSNs). These nodes sense the surrounding RF environment for detecting PUs' transmission pattern and select either spectrum holes or any of the PUs' bands for their signal transmission for opportunistic spectrum access [1, 2]. Incorporation of cognitive radio (CR) may help in overcoming the spectrum scarcity of the WSN; however, CR needs to perform extra functions like spectrum sensing, power control that may appear as burden for the tiny sensor node. Spectrum sensing techniques are utilized for finding out the presence of spectrum holes that are not used by the PUs. CR units send information through these spectrum holes in overlay mode of communication. They also transmit information simultaneously with the PU under the interference temperature limit in underlay mode of communication. It requires highly efficient power adaptation technique to avoid the interference with the PU. Power adaptation and spectrum sensing are the imperative tasks with respect to the performance of the CR. However, these tasks also raise the delay which is not desirable in time-bound emergency situations. Dedicated hardware units for these tasks may help in parallelism in hardware to speed up the process, and also the overhead of software applications like decoding and interpretation of software commands can be avoided that in turn speed up the process [3–5]. Unification of PU and CRSN in system-on-chip (SOC) implementation may further improve the speed and minimize the resource constraint. At the stage of hardware resource allocation for the PU and CRSN, a particular hardware module may be dedicated for computation of power for both PU and CRSN, considering the present condition of the RF environment. These factors lead us to design parametric reconfiguration hardware architecture for power control unit for the network architecture where CRSN has given high priority in emergency time-bound situation and PU agrees to compromise QoS up to a certain level. The hardware module iteratively changes the computed parameters' values to optimize the transmission power for both PU and CRSN. Hardware architecture is designed in Verilog hardware description language using IEEE 754 floating point format with 64-bit double-precision. The performance of the designed hardware module is analyzed in real time in DIGILENT ZED BOARD (xc-7z020 clg484-1) using integrated logic analyzer.

After the introduction in Sect. 1, the network architecture with unification of PU and CRSN and corresponding power control technique is discussed in Sect. 2. The detailed description of the designed hardware architecture followed by FPGA board implementation with port assignment is given in Sect. 3. Section 4 provides comparative analysis of the hardware implementation result with the MATLAB simulation result. Section 5 highlights future works and concludes the paper.

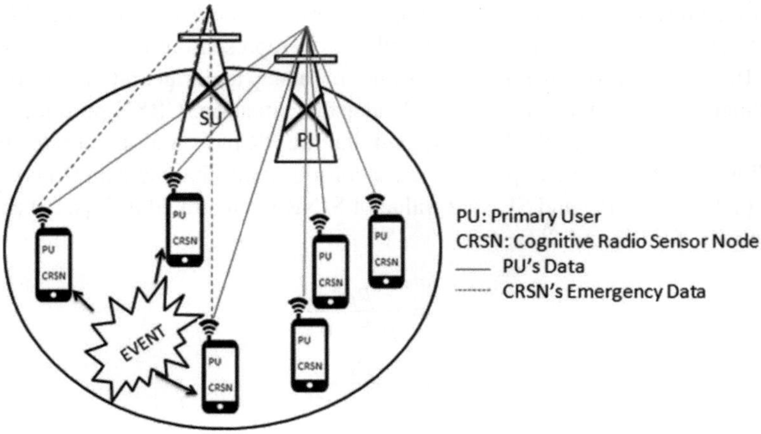

Fig. 1 Proposed network architecture

2 Previous Work

In [6] we proposed a network architecture in which PU and SU are interconnected into a single device as depicted in Fig. 1. In the network, PU has the registration of using the resources whereas SU only shares the resources in the emergency situation with the joint decision with the PU. In the application domain, a GSM/CDMA user may act as PU whereas CRSN may operate in SU mode. In the emergency time-bound situation like the leakage/presence of poisonous gas in the surrounding, CRSN operates in underlay mode and sends information simultaneously with the PU. So PU shares its resources with the CRSN in the emergency situation compromising its QoS. The CRSN solely shares the spectrum of its own PU without the need for sharing the power with the other CRSNs. As a consequence, the main objective is to maximize the QoS of the integrated system in the emergency situation under the constraint of interference. In normal operation, PU has its own power control strategy to maximize its QoS, but in emergency environment we proposed Modified Nash Bargaining Power Control (MNBPC) scheme to comprise of its QoS up to a certain threshold value, sustaining the link connection with the base station to maximize the CRNS's QoS from its threshold.

It in turn provides reliable transmission of emergency data without establishing a new network for the CRSN. Single-hop communication instead of multihop communication from the CRSN to base station is considered for minimization of time delay. However, MNBPC may be applied for node-to-node transmission in multihop communication as well.

MNBPC is based on Nash Bargaining Game Theory considering signal-to-interference-plus-noise ratio (SINR) as QoS. Game theory has been recognized as an important tool in modeling and analyzing the cognitive interaction process. References [7–9] have discussed the design methodologies of power control algorithms based on game theory. Let p_p, p_s represent the transmit power of the PU and CRSN, respectively, of the integrated system with γ_p, γ_s representing their SINR, respectively. Both PU and SU operate on the same frequency and transmit the information simultaneously from the same device. Under this situation, CRSN produces maximum interference to the PU. It is assumed that the PR base station operates for SU as well with added cognitive property. As a consequence, channel gain (h) is considered equal for both PU and SU. The value of SINR is illustrated in Eqs. (1) and (2), respectively, for PU and SU.

$$\gamma_p = \frac{p_p h}{p_s h + \sigma^2} \tag{1}$$

$$\gamma_s = \frac{p_s h}{p_p h + \sigma^2} \tag{2}$$

where σ^2 is the noise variance.

The interference from other SUs under different PUs with different frequencies does not come into consideration here. The power level of PU is decremented from the maximum power P_{max} in each iteration step to enhance the power of CRSN; however, it maintains threshold value of SINR ($\gamma_{p(min)}$). Thus, the CRSN gains power from the PU to increase its QoS from the minimum threshold value ($\gamma_{s(min)}$). The MNBPC model based on Nash Bargaining is formulated with the utility function as shown in Eq. (3b) for power control optimization problem that decides the transmit powers $p = (p_p, p_s)$. Equations (4a) and (4b) represent the constraints in deciding the power of both PU and SU.

$$p = \underset{(p_{i,\ldots,}\, p_2)}{max} \sum_{i=1}^{N} U_i(p_{i,\ldots,}\, p_2) \tag{3a}$$

$$p = (p_{i,\ldots,}\, p_2)\left(\sum_{i=1}^{N} log(\gamma_i(\mathrm{p}) - \gamma_{i,\mathrm{min}})\right) \tag{3b}$$

$$\text{subject to } P_p + P_s \le P_{max} \tag{4a}$$

$$\gamma_i \ge \gamma_{i,min}, \quad i = 1, 2 \tag{4b}$$

The iterative method for evaluating the optimal power for both PU and CRSN using Lagrange multiplier (μ and λ) is represented by Eqs. (5a)–(5e).

$$p_p^{(i+1)} = \left[\frac{\gamma_p \mathrm{min}}{\upsilon_p^{(i)}} - \frac{\alpha_p}{\mu_p^{(i)} \frac{\partial \gamma_p^{(i)}}{\partial p_p^{(i)}} - \lambda^{(i)}}\right]_{p_p^i}^{+\mathrm{P_{th}}} \tag{5a}$$

$$\upsilon^i = \frac{\gamma_p^i}{P_p^i} \tag{5b}$$

$$\mu^{(i)} = \mu^{(i-1)} - c_t\left(\gamma_p^{(i)} - \gamma_{pmin}\right) \tag{5c}$$

$$\lambda^{(i)} = \left[\lambda^{(i-1)} - c_t\left(p_{max} - \sum_{j=1}^{2} p_j^{(i)}\right)\right]_0^+ \tag{5d}$$

$$p_s^{(i+1)} = P_{max} - p_p^{(i+1)} \tag{5e}$$

Equation (5d) implies that there is no need to update the value of λ as $\lambda^{(t)} = [\lambda^{(t-1)}]$. PU starts the iteration from a large value of SINR, so it is essential to set the value of μ between zero and one for fast convergence. Equation (5a) implies that the power of PU decremented and symbol $[x]_{p_p^i}^{+P_{th}} = x$ if power is greater than the power P_{th} corresponding to $\gamma_{p,\min}$ and $x = p_p^i$ if the power is less than P_{th}.

3 Hardware Architecture and FPGA Board Implementation

The entire hardware architecture and state diagram of the power-sharing unit are represented in Fig. 2 and Fig. 3, respectively. It consists of several dedicated circuit module for specific calculation for the ease of complexity. The key controller unit 'CU' is designed based on FSM for smooth operation of the entire hardware circuit. It generates several chip select signals for activation of particular hardware unit. The entire hardware unit is activated by the master control signal 'rst_n'. Different calculation stages are assigned as the different states of the controller unit as shown in figure. The states are selected accordingly with the receiving of the handshaking signals that are generated by the dedicated units and feed to the control unit. The processing of controller is terminated by 'comp signal' with logic one. A floating point conditional subtractor 'COND SUB' is designed for calculation of initial value of the PU's power for given value of maximum power and sleep power of the CRSN. It also acts as iterative subtractor for the calculation of CRSN's power during iteration with updated value of PU's power. When the 'flag' signal is at logic one, the conditional subtractor operates on the updated value of Pp and Pmax and calculates new value of Ps, whereas for logic zero, it evaluates initial value of Pp for starting the iteration process. In the architecture, the hardware module 'SINR CALC' is utilized for evaluating SINR in each round of iteration and comparing with threshold value of it. It is activated by the chip select signal 'cs2' with logic one.

If SINR value is greater than threshold value of it, correspondingly, Pp and Ps take the new values and the output 'comp' is set to logic zero otherwise iteration ends with the previous values of Pp and Ps. A handshaking signal 'rdy_si' is generated by it to acknowledge the control module after completing the specified job. It is designed with floating point divider, floating point adder, and comparator circuits. The variable V in Eq. (5b) and the Lagrange variable μ are required for calculating the updated value of Pp. As they are sensitive to the iteration, dedicated modules are designed for updating their values for the sake of ease and isolation. The hardware module 'VI CALC' for calculation V is activated with the chip select signal 'cs3',

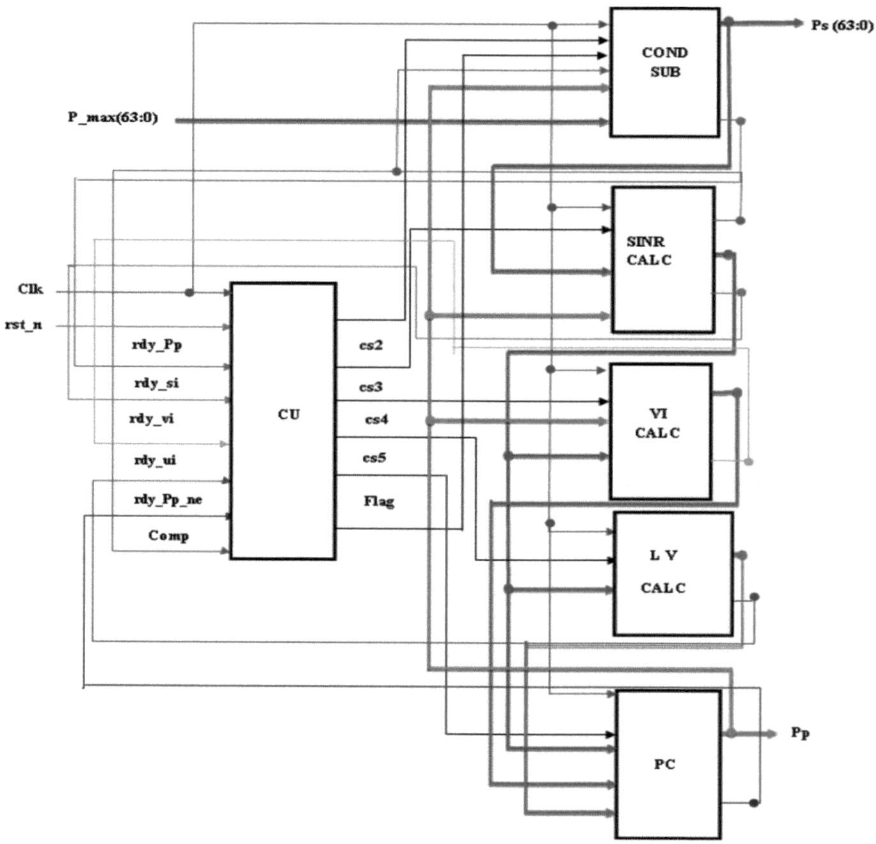

Fig. 2 Designed hardware architecture

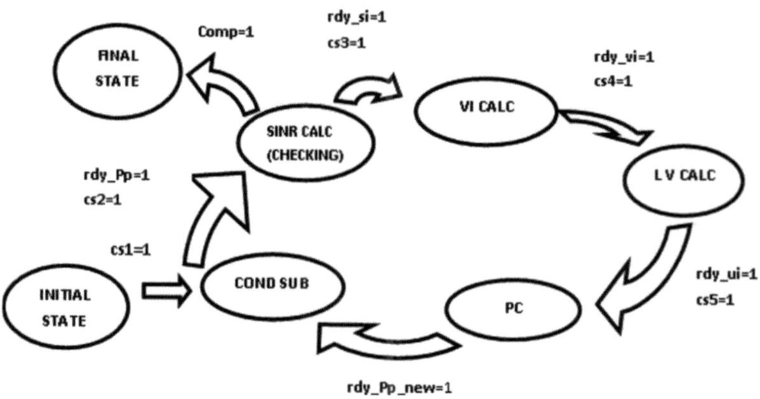

Fig. 3 State diagram of the hardware module

Fig. 4 Implementation in FPGA ZED BOARD

Table 1 Utilization of hardware resources

Resources in ZED BOARD	Utilization	Available	Utilization (%)
LUT	23,192	53,200	43.59
FF	38,189	106,400	35.89
IO	130	200	65.00
BRAM	4	140	2.86

and it sends the acknowledgment signal 'rdy_vi' to provide the control module feedback about the completion of the calculation. The hardware module 'LV CALC' is designed for calculating the value of Lagrange variable μ with floating point subtractor and multiplier in each round of iteration. It is activated by the chip select signal 'cs4' and generates the signal 'rdy_ui' to acknowledge the control module. The hardware module 'PC' is responsible for the calculation of the updated Pp. This module generates the maximum delay and timing constraint issues as it contains the maximum number of circuits. It is activated by the chip select signal 'cs5' and generates the acknowledgment 'rdy_Pp_new', and feeds to the control module for starting the next round of the iteration.

Figure 4 depicts the experimental setup for evaluating the performance of the designed hardware architecture in real time on FPGA evaluation board named '*Xilinx ZynqTM-7000 All Programmable SoC* (*AP SoC*)' ZED BOARD. The detail features are available in [10]. The computed values of Pp and Ps are in IEEE 754 format 64-bit double-precision. As a consequence, a large number of output ports of the FPGA board are required for examination of the optimum value of Pp and Ps. To reduce the complexity of the routing of 128-bit output, it is split into sixteen 8-bit signals and observed using onboard LED. The hardware resources of the ZED BOARD utilized by the implemented algorithm are represented in Table 1.

Table 2 Requirement of clock cycle of the hardware modules

Hardware module	Required clock cycle
CU	1
COND SUB	2
LV CALC	4
VI CALC	2
SINR CALC	4
PC	7

It was observed that 43% of LUT, 35% of FF, and 2% of BRAM have been utilized. The detailed the clock cycle requirement for each hardware module is illustrated in Table 2. It is observed that 'PC' module requires highest number clock cycle as it contains large number of circuits.

4 Comparative Analysis of the Results

Figure 5 illustrates the optimal power allocation to CRSN by MNBPC in MATLAB simulation and FPGA by varying the distance of the node from the base station. Fig. 5 shows that hardware circuit provides less power to the CRSN than MATLAB simulation. This is due to the fact that the initial value of power called sleep power of CRSN implemented in FPGA has been taken as 0.007 W. The figure depicts that for simulation results, MNPC allows the power-sharing between PU and CRSN up to 1000 m from the base station for the 1 W maximum allotted power to the system, whereas hardware module is effective up to around 900 m for the same maximum power and the CRSN is staying in the sleep mode.

Figure 6 illustrates that the power requirement/consumption in FPGA board for the computation of the optimum power is 9 mw, and it is almost constant with the variation of the distance of the node from the base station.

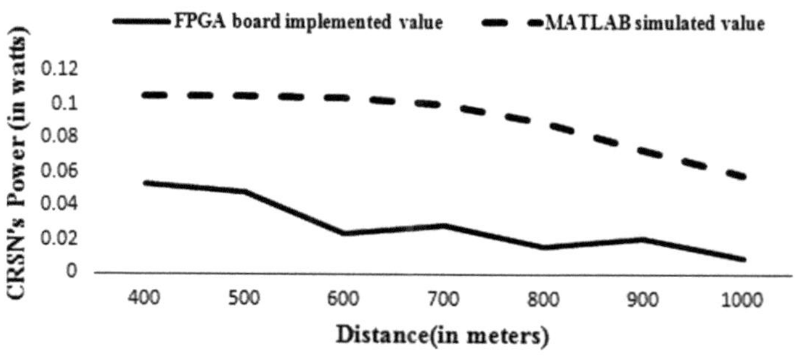

Fig. 5 Variation of optimal power of CRSN with the distance of the system from the base station

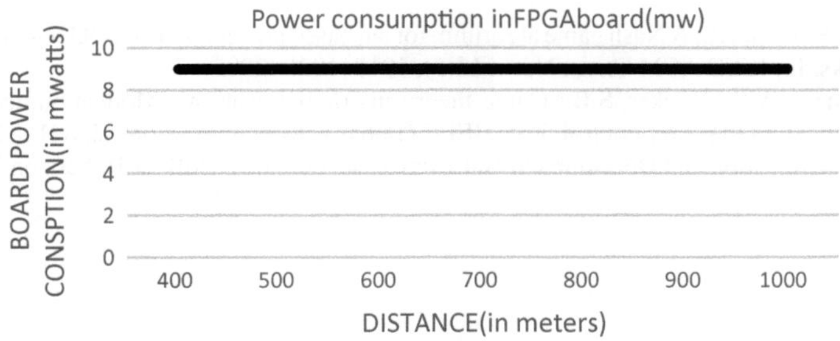

Fig. 6 Variation of power consumption in FPGA board with the distance of the system from the base station

5 Conclusion

The designed hardware architecture of the proposed algorithm efficiently computes the transmission power for both PU and SU for different distances of the system from the base station. So this may be effective for designing the dedicated hardware unit for the integrated system, minimizing the space complexity, and it enables the CRSN to work effectively in underlay mode. It also minimizes hardware overhead of the CRSN as the algorithm requires minimum hardware resources. Future works are focused on the minimization techniques of power consumption without compromising output performance.

References

1. Akan, O.B., Karli, O.B., Ergul, O.: Cognitive radio sensor networks. IEEE Netw. 34–40 (2009)
2. Prakash, P., Lee, S.R., Noh, S.K., Choi, D.Y.: Issues in realization of cognitive radio sensor network. Int. J. Control Autom. **7**, 141–152 (2014)
3. Das, S., Mukhopadhyay, S.: SoC FPGA implementation of energy based cooperative spectrum sensing algorithm for cognitive radio. In: 6th International Conference on Computers and Devices for Communication (CODEC), pp. 16–18, Dec 2015
4. Srinu, S., Sabat, S.L., Udgata, S.K.: FPGA implementation of cooperative spectrum sensing for cognitive radio networks. In: Second UK-India-IDRC International Workshop on Cognitive Wireless Systems (UKIWCWS), pp. 13–14, Dec 2010
5. Lotze, J., Fahmy, S.A., Noguera, J.: Development framework for implementing FPGA-based cognitive network nodes. In: Global Telecommunications Conference, 2009. GLOBECOM 2009. IEEE 30 Nov–4 Dec 2009
6. Chakraborty, M., Roy Chatterjee, S., Ray, S.: Performance evaluation of nash bargaining power sharing algorithm for integrated cellular phone system. In: 2016 IEEE Uttar Pradesh Section International Conference on Electrical, Computer and Electronics Engineering (UPĆON), Dec 9–11, 2016, pp. 125–131
7. Yang, C.G., Li, J.D., Tian, Z.: Optimal power control for cognitive radio networks under coupled interference constraints: a cooperative game-theoretic perspective. IEEE Trans. Veh. Technol. **59**(4), 1696–1706 (2010)

8. Koskie, S., Gajic, Z.: A Nash game algorithm for sir-based power control in 3G wireless CDMA networks. In: IEEE/ACM Trans. Netw. **13**(5), 1017–1026 (2005)
9. MacKenzie, A.B., Wicker, S.B.: Game theory in communications: Motivation, explanation, and application to power control. Proc. IEEE Global Telecommun. Conf. **2**, 821–826 (2001)
10. Zynq™ Evaluation and Development Hardware User's Guide (ZedBoard), Version 2.2, 27 Jan 2014

Driven by the Need for a Reliable and Cost-Effective LED Driver

Ajanta Dasgupta, Avijit Bose, Shamik Guha, Sourup Nag, Subham Mukherjee and Sumalya Saha

Abstract An LED driver is an electrical device that helps in the regulation of power to an LED or to an array of LEDs. The LED driver adapts to the requirements of the LED, by supplying a continuous quantity of power to the LED when its electrical properties vary. LEDs are as a matter of fact sensitive to the voltage used to power them (i.e., the current changes a lot with a little change in voltage). To avert the LED from becoming unstable due to voltage changes, an LED driver is necessary. An LED driver is a self-contained power supply which has outputs that are affianced to the electrical characteristics of the LED or to the array of LEDs. LED drivers may offer to dim by means of pulse width modulation circuits and may have different channels for separate control of different LEDs or LED arrays. Without the most suited driver, the LED may become hot and unstable, therefore, cause poor performance or failure. This paper presents a novel approach to the design of a LED Driver using various power transistors, power MOSFETs, operational amplifier and feedback circuits. The experimental results demonstrate the power delivered to the LED at different supply voltages upholding the performance of the driver circuit proposed.

Keywords LED · Driver · Power · Transistor · OP-AMP · Mosfet · Timer

1 Introduction

The preferable method for regulating the current in LEDs is to drive the LED array with a constant-current source. The constant-current source reduces greatly the changes in current due to changes in forward voltage, which translates into a

A. Dasgupta (✉) · A. Bose
Department of Information Technology, Institute of Engineering & Management, Kolkata, India
e-mail: ajanta.dasgupta10@gmail.com

S. Guha · S. Nag · S. Mukherjee · S. Saha
Department of Electronics & Communication Engineering, Institute of Engineering & Management, Kolkata, India

M. Chakraborty et al. (eds.), *Proceedings of International Ethical Hacking Conference 2018*, Advances in Intelligent Systems and Computing 811,
https://doi.org/10.1007/978-981-13-1544-2_17

constant LED brightness. The semiconductor junctions in an LED (producing the light) require very specific power to function properly. If the voltage supplied to the LED is lower than what is required, very small amount of current flows through the junction, and that results in low light output and poor performance. If the voltage is too high, a large amount of current flows and the LED can be severely damaged by overheating or fail completely (by thermal runaway). It is important to configure an appropriate LED driver for each application to maintain the optimum performance and reliability of the LED array. This is a practical problem that is faced. Different types and models of LEDs have separate voltage requirements even though they may share same current specifications. Also, the voltage that is required to achieve the proper current varies with temperature, the composition of chemicals in the LED and others. Hence, LEDs with the same specifications may have some variations in the exact level of voltage required for proper operation. However, it is desirable to regulate the current flowing through each LED as per the current rating in order to maximize the illumination and life of each LED. To do this, some type of true power regulation is required.

Till date, many circuits have been proposed to solve this problem. In [1], a power efficient high-power LED driver which can regulate the average current in the output accurately by integrating the voltage error in between the reference and sensing voltage was proposed. In [2], a linear constant current source without chip in inductor and capacitor was presented. The driver circuit consisted of a diode bridge, a resistor and a controller integrated circuits (IC). The controller IC shapes the output current according to the input voltage and drives the LED. In [3] Op-amps are used to control the LEDs in parallel configuration and to make a temperature independent I_{REF} by means of compensating the mobility and threshold voltage variations. In [4] No sensing resistor is used in the circuit but it extracts LED-current information from the output capacitor of the driver. This circuit thus formed for sensing is applied in a buck–boost LED driver. Cheng et al. [5] uses a dual boost power factor correction or PFC, an AC-DC converter with a half-bridge-type LLC DC-DC resonant converter. Two inductors which are present inside the dual boost converter sub-circuit are designed to function in discontinuous-conduction mode (DCM) for the purpose of achieving input current shaping.

Life of the battery is crucial in portable applications. For LED drivers to be useful, it has to be efficient. Measuring efficiency of an LED driver differs from that of a typical power supply. Efficiency of the system is defined as the ratio of output power to the input power. Using an LED driver, we are not interested in output power. The amount of input power required to generate the desired LED brightness is important. We can easily get it by taking the ratio of power in the LEDs to the input power. We mean that the power dissipated in the current-sensing resistor contributes to the power lost in the supply.

The following equation shows that smaller current sensing voltages contribute to higher-efficiency LED drivers

$$Efficency = \frac{\mathrm{P(LED)}}{P(LED) + P(Supply\ Losses) + P(Current_{Sense})} \quad (1)$$

A supply with a lower current sensing voltage is more efficient regardless of input voltage or LED current. With everything else being equal, a lower reference voltage can significantly improve efficiency and extend battery life.

Thus, we see that in every LED lighting system, we need an efficient driver to ensure smooth working of the system. As the world gets "smarter," the number of LED-using devices increases. This rising demand for LED drivers is leading to a more competitive market and hence the need for a reliable, high efficiency, cost-effective driver arises.

2 Related Works

According to [1], boost–buck converters are commonly used in LED driver circuits. In [1], the boost–buck main circuit and some related characteristics have been discussed. It is found that the system has many advantages like high efficiency, fast response and strong anti-interference. Also, it sustains good stability after analyses and simulations of its working dynamic characteristics. In the paper, the authors proposed a power-efficient high-power LED driver which can regulate the average current in the output accurately by integrating the voltage error in between the reference and sensing voltage. The performance of the driver is tested by time domain simulation. The final results are showing that the preset current 350 mA is equal to the average output current. The driver can function in different input voltages like 6, 12 and 24 V. When different power sources are applied, the LED current gets curved. Thus, the driver can provide good regulation and also protects MOS switches.

In [2], a linear constant current source without chip in inductor and capacitor has been presented. The driver circuit consisted of a diode bridge, a resistor and a controller integrated circuits (IC). The controller IC shapes the output current according to the input voltage and drives the LED. The HHNEC 1 μm 700 V process was used to design and implement the controller IC. The experimental result was found to show high PF of 97.93% and efficiency of 90.41% at 220 V RMS value of input voltage.

Also in [3], a new circuit for obtaining a constant current in parallel LED (Light Emitting Diodes) strings was proposed. This method not only achieves high sustainability and decreases susceptibility to temperature changes in all the strings but also reduces the internal and external resistances of the circuit. In this circuit, the LED current has a very small correlation with its forward voltage (VF). Also, this circuit matches the current with that of a multi-LED parallel string and maintains the accuracy by using operational amplifiers to control the LEDs. Therefore, it reduces the consumption of power considerably. The I_{REF} which is obtained in this circuit is almost not dependent on the temperature. This has been achieved by the mutually compensating the mobility and threshold voltage variations. Furthermore, the variations due to the temperature coefficient of resistance have been checked.

For lighting purposes [4], has opined that a high-power LED is driven at a current of 350 mA and a sensing resistor provides feedback for the current regulation of the LED. Thus an IR drop is added at the output branch. This limits the power efficiency as the current through the LED is large and tends to increase. A power-efficient LED current sensing circuit has been proposed in this paper. No sensing resistor is used in the circuit but it extracts LED-current information from the output capacitor of the driver. This circuit thus formed for sensing is applied in a buck–boost LED driver. Measurement results indicate that there is a power-conversion efficiency of 92% and power reduction in sensing has been reduced to almost 90%.

A unique single-stage, high-power-factor LED driver has been proposed in [5] for street-lighting applications. The LED driver which has been presented uses a dual boost power factor correction or PFC, an AC-DC converter with a half-bridge-type LLC DC-DC resonant converter. Two inductors which are present inside the dual boost converter sub-circuit are designed to function in discontinuous-conduction mode (DCM) for the purpose of achieving input current shaping. The switching losses of two power switches and two output rectifier diodes are lowered by the AC-DC resonant driver so that the circuit efficiency is increased. The proposed driver apart from being cheap also has a sufficiently high power factor and comes with low levels of input current harmonics. A prototype driver to power a 144 W LED lamp with a 110 V input voltage is developed and tested. The results are satisfying hence the circuit is feasible.

3 Working Principle

Here in Fig. 1, the circuit has been made in such a way that the current through the LEDs remains the same in spite of the change in the applied voltage. In order to do this, a power MOSFET (N-channel) (IRFZ44NS) has been used so that the gate voltage which is applied by the feedback part of the circuit can control the channel of the MOSFET and thus can control the current. A P-N diode (1N3064) has been connected across the drain and the source of the MOSFET in order to decrease the power drop across the MOSFET and hence to increase the efficiency. The other power MOSFET (IRFZ44) and a power transistor (TIP29AG) have been used to further control the current when the first MOSFET reaches its saturation point. The R2 (15 Ω) resistor is the current sensing resistor by changing the value of which we can vary the amount of constant current through the LEDs according to their wattage, max current specification, etc.

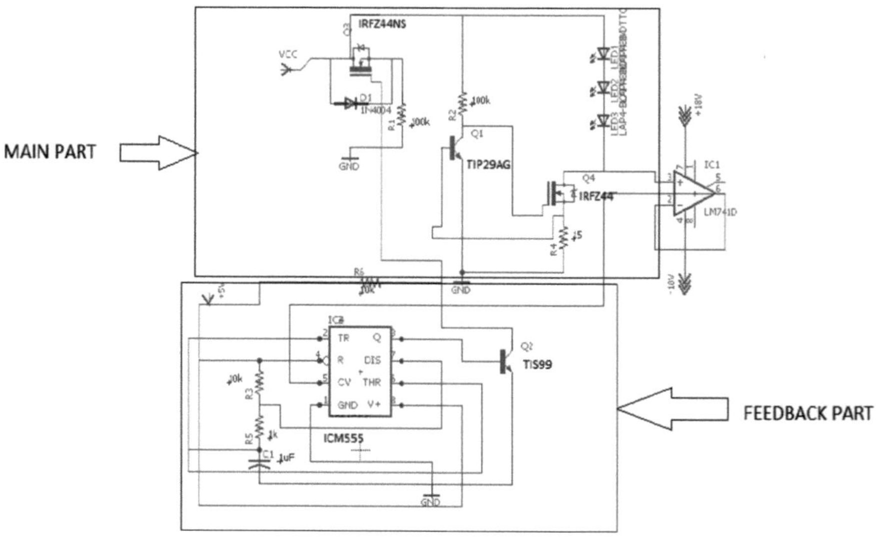

Fig. 1 Circuit diagram

The feedback part of the circuit helps in load regulation i.e. in case the number of LEDs changes slightly the circuit tends to keep the current through the LEDs constant. It is achieved by giving input pulses to the gate of the MOSFET Q3 i.e., by changing the duty cycle of the pulses(pulse width modulation). When the current through the load tends to increase i.e., the voltage drop across the load increases, the duty cycle of the input pulses is decreased to keep the current constant and when the load current decreases the duty cycle is increased proportionately. Here an OP-AMP buffer has been used to isolate the main and the feedback circuit. ICM555 timer IC has been used in a stable mode controlled by the load voltage in order to execute Pulse Width Modulation (PWM). The PWM pulses are fed back to the gate terminal of the MOSFET Q3 by a high power fast switching transistor (TIS99).

There has been 3 separate power supplies in the circuit, one is for driving LEDs and other two are used for powering OP-AMP and 555 timer IC respectively.

4 Experimental Details

Table 1 provides a detailed description of the experiment performed to test the driver circuit efficiency for different number of loads, i.e., different number of LEDs, viz. 3, 6 and 9.

Table 1 Driver circuit efficiencies for different amount of loads

Load (no. of LEDs)	Supply voltage (V)	Power loss through resistors, transistors and MOSFETs (mW)	Power consumed by the LEDs (mW)	Efficiency (%)
3	11	24.638	167.70	**87.19**
	12	44.12	169.75	79.37
	13	62.791	171.538	73.20
	14	80.263	173.158	68.32
	15	98.33	174.645	64.00
	16	116.192	175.998	60.23
	17	134.389	177.23	56.87
6	21	25.606	362.657	**93.40**
	22	46.973	364.42	88.58
	23	66.477	366.11	84.63
	24	85.637	367.73	81.11
	25	104.498	369.23	78.00
	26	129.009	370.717	74.18
	27	152.918	372.09	70.87
	28	171.162	373.50	68.57
	29	192.111	374.73	66.10
	30	211.204	375.94	64.02
9	31	23.034	433.58	**94.95**

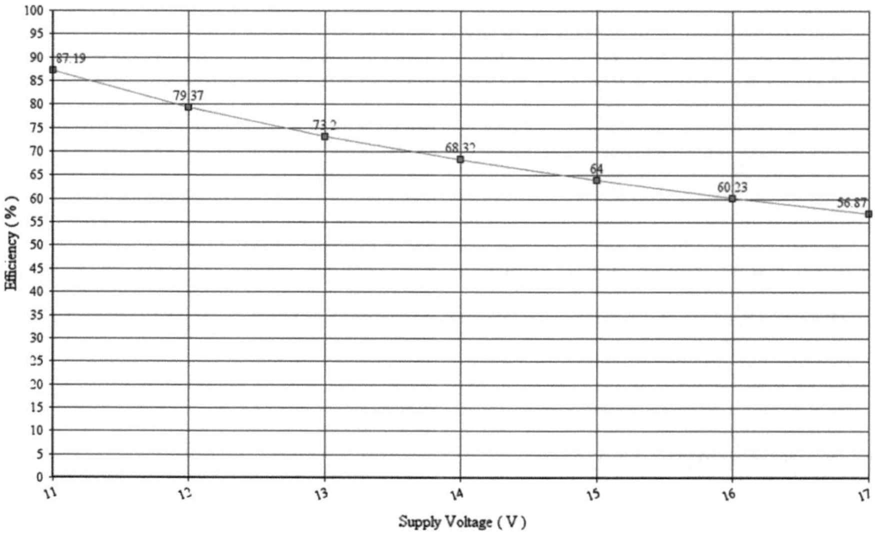

Fig. 2 Efficiency voltage curve for 3 LED array

It shows that as the supply voltage (V) is increased, for a particular load, the power consumption by the peripheral devices increases heavily. The major amount of this power is consumed by the 2 MOSFETs used in the circuit. This trend of increasing power consumption goes on as we keep on increasing the V, making us reach the highest efficiency at the start itself.

Again, when the load is changed, i.e. increased to 6 from 3, the trend repeats itself. The highest efficiency is achieved for the lowest supply voltage for that load. We get an efficiency of 93.40% using 6 LEDs at 21 V. The efficiency then gradually decreases, leading to the conclusion that the load must be altered to make the design more efficient at those high voltages.

Finally, when the load is changed to an array of 9 LEDs, we obtain an efficiency of 94.95% at 31 V. It is interesting to note that the power consumption of the MOSFETs (and the other peripheral devices) is least at this condition, and hence it is safe to conclude that for a system of 9 LEDs, the proposed circuit works at an overwhelming efficiency of **94.95%**.

Figure 2 depicts the variation of efficiency with the change in supply voltage for different loads.

From the graph we see that with an increase in supply voltage, the efficiency of the circuit decreases for a particular load.

Finally, we compare the highest efficiencies reached for each system, graphically.

Figure 3 gives us a clear picture that the maximum efficiency is reached at 31 V for a 9 LED array.

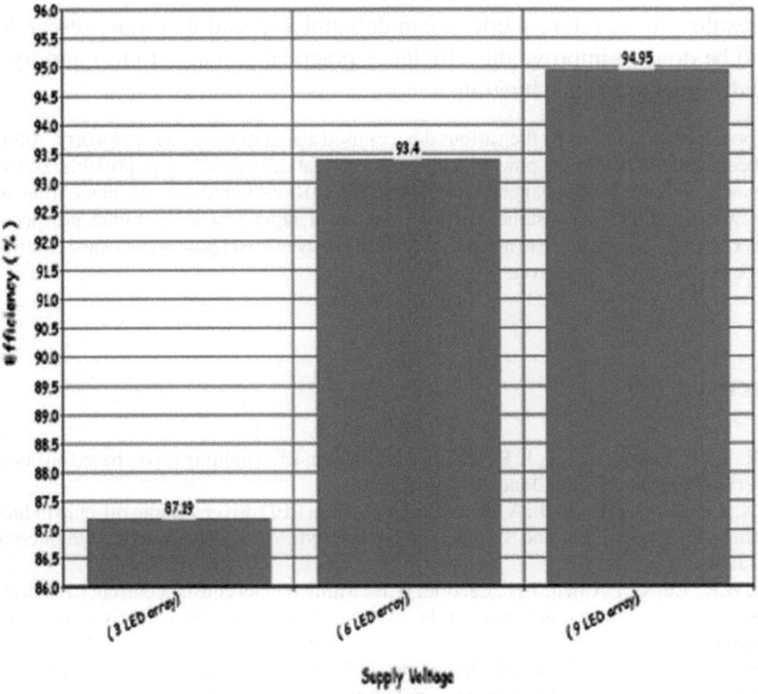

Fig. 3 Comparisons of highest efficiencies of three different LED arrays

5 Conclusions

On comparing the graphs and the experimental results, it can be concluded that the system works best for an array of 9 LEDs, giving an efficiency of 94.95%. Also, it is noted that for a particular load, the efficiency decreases with increase in supply voltage (giving the highest efficiencies at 11 V, 21 V and 31 V, respectively, for the three sets of LEDs).

This developed circuit can find effective uses in industrial/outdoor lighting and residential lighting, automotive interior or tail lights, street lights and in several other domains where efficient lighting is required with lesser power consumption. LED drivers can also find immense applications in several embedded systems and real-time IoT projects, which are at present, pretty much in the heart of the market.

6 Future Works

The LED driver circuit designed is expected to work with the desired efficiency, with minor variations under practical circumstances. The primary aim that always stays is to increase the efficiency and hence make the driver more practically usable with better results. So, as future scope, it can definitely be said that more research work needs to be done to improve this efficiency practically, that is to test and try under several different practical situations.

The power consumptions by the various devices used also remains a cause of worry. Attempt has been made to reduce the power intake by peripheral devices as less as possible. But still, the power MOSFETs (N-channel) (IRFZ44NS) and IRFZ44, and also the power transistor (TIP29AG) and TIS99 consume a certain amount of input power. As future prospective, further attempts can be made to use devices which use even lesser power and hence contribute to an even more efficient driver.

References

1. Xu, R., Li, Y., Zhong, L., Liu, J.: Research of an efficient LED lighting driver based on boost-buck converter. Published Online June 2014 in SciRes
2. Wu, X., Leng, Y., He, L., Xi, J.: A linear constant current LED driver without off-chip inductor and capacitor. In: New Circuits and Systems Conference (NEWCAS), 2015 IEEE 13th International, 7–10 June 2015
3. Liou, W.R., Lin, C.Y., Chen, T.H., Lacorte, W.B.: Multichannel constant current LED driver with temperature and voltage compensation. In: Communications, Circuits and Systems (ICCCAS), Chengdu, 28–30 July 2010
4. Leung, W.Y., Man, T.Y., Chan, M.: A high-power-LED driver with power-efficient LED-current sensing circuit. In: Solid-State Circuits Conference, 2008 ESSCIRC (2008)
5. Cheng, C.A., Chung, T.Y., Yang, F.L.: A single-stage LED driver for street-lighting applications with high PF. In: 2013 IEEE International Symposium on Industrial Electronics (ISIE), 28–31 May 2013

Part V
Session 2B: Network Security

SGSQoT: A Community-Based Trust Management Scheme in Internet of Things

Rupayan Das, Moutushi Singh and Koushik Majumder

Abstract Internet of things (*IoT*) has appeared as a current and vigorous research area where security in terms of trust management plays a major role. *IoT* network is established with some heterogeneous devices like smart sensor, smartphone, laptop. Apart from group communication, every smart device communicates with its nearest fog node in order to store data or share secret information. *Fog node* acts as a substitute of *cloud service provider* (*CSP*) or *cloud server* (*CS*) and helps *CS* by sharing overhead with it. Because of its distributiveness and openness in deployment, *IoT* network suffers from insecurity and constrains in terms of energy and memory. In this paper, we propose a community-based trust management architecture by considering *self-trust* (*SLT*), *social trust* (*ST*), *green trust* (*GT*), *and QoS trust*. The existing schemes in this platform have not considered all the trust management attributes together. *Self*, *green*, *social, and QoS trust* (*SGSQoT*) enables *IoT* devices as well as *IoT* network to fight against most of the attacks in *IoT* environment. The *SGSQoT* scheme is lightweight in terms of energy and memory consumption. Arithmetical calculations and extensive simulation have been done in order to find the trust value.

Keywords IoT · Fog · Cloud · Trust · QoS · Green

R. Das (✉)
Department of CSE, University of Engineering & Management, Jaipur, Rajasthan, India
e-mail: rupayan11@gmail.com

M. Singh (✉)
Department of IT, Institute of Engineering & Management, Kolkata, India
e-mail: moutushisingh01@gmail.com

K. Majumder
Department of CSE, Maulana Abul Kalam Azad University of Technology, BF 142 Sector 1, Salt Lake, Kolkata 700064, India
e-mail: koushikzone@yahoo.com; koushikwbutcse@gmail.com

M. Chakraborty et al. (eds.), *Proceedings of International Ethical Hacking Conference 2018*, Advances in Intelligent Systems and Computing 811,
https://doi.org/10.1007/978-981-13-1544-2_18

1 Introduction

IoT has appeared as a current research area where security in terms of trust management plays a major role [1]. *IoT* network consists of a number of devices like smart sensor, smartphone, laptop. The *IoT* system connects the physical world into the cyber world via radio frequency, sensors, and smart objects. The sensing capability of *IoT* device helps to extract, track, and monitor the environmental information. Cisco circulated the idea of *fog computing* to enable real-time applications on billions of connected devices, already connected in *IoT*. *Fog computing* is also known as *edge computing*, which offers open application and supports developers to produce their own applications and connectivity at the edge of the network [2]. The *fog node* acts as a substitute of *CSP* and shares the overhead [2, 3]. In *IoT* system, every smart device has a communication with its nearest *fog node* for storing and sharing information. Because of its distributive and open nature in deployment, *IoT* system faces insecurity and constrains in terms of energy and memory. The paper is arranged in the following manner: literature survey in Sect. 2, proposed scheme in Sect. 3, *SGSQoT* protocol design in Sect. 4, result and simulation in Sect. 5, and finally conclusion in Sect. 6.

2 Literature Survey

Various trust management schemes are there to determine trust in each level of network hierarchy with decision objectives like trust relationship and decision (*TRD*), privacy presentation (*PP*), data perception trust (*DPT*), data transmission and communication trust (*DTCT*), data fusion and mining trust (*DFMT*), self-trust (*ST*), and green trust (*GT*). The authors of [3] discussed trust management in *WSN* as well as *CSP* and *CS* levels by focusing on data integrity during sensing and processing. They have also maintained data privacy. Paper [4] indicates two important areas like trust bias minimization and application performance maximization in order to address the performance issue of trust management protocol design for *MANET*. To demonstrate the effectiveness of this approach, the authors integrated social trust and QoS trust to identify the best trust aggregation settings. In paper [5], the authors proposed a layered architecture which includes perception logical layer, mark recognition logical layer, decision control logical layer, and a trusted interface logical layer. This architecture can afford trusted and secure data transmission in *WSN*. The work is essentially focused on the physical perception layer and the sensory part of the network layer of the *IoT* system. However, it supports the trust management objectives regarding *DPT* and *DTCT*. The architecture in [6] consists of trusted user module, trusted perception module, trusted terminal module, trusted network module, and agent module. The modules mainly focused on *DPT* and *DTCT*.

Reference [7] reflected an *IoT* system architecture that suggests security in terms of system, network, and application with respect to basic security requirements by

Table 1 Comparison of inclusion of trust objectives for various papers

Paper title	TRD	PP	DPT	DTCT	DFMT	ST	GT
[3]	×	√	√	√	√	×	×
[4]	√	×	√	√	×	×	×
[5]	×	×	√	√	×	×	×
[6]	×	×	√	√	×	×	×
[7]	×	×	√	√	×	×	×
[8]	√	√	×	√	×	×	×
[9]	√	×	×	×	×	×	×
[10]	√	×	×	×	×	×	×
[11]	√	×	×	×	×	×	×
[12]	×	×	×	×	×	×	×
[13]	×	√	×	√	×	×	×

considering *DPT* and *DTCT*. Paper [8] proposed some functional components in order to enhance the trust. The components cover essential functions like identity management, key management, and finally trust and reputation management. The scheme ensures data integrity, confidentiality, service trust, and privacy of users and considers *TRD*, *PP*, and *DTCT*. The authors of [9–11] proposed community-based trust management scheme in *IoT*. The schemes are scalable and energy efficient. Intimacy of social trust is one of the vital parameters that should always be updated to deal with the total number of interaction and time taken to complete each interaction. The authors of [12] deal with user security by computing trust based on service classification. The penalty and authentication history are also considered. Security in each level of *IoT* is dealt in paper [13]. It considers lightweight cryptographic mechanism, authenticity, and sensor data security in physical layer, data encryption and secure communication in network layer, and secure multi-party computation, secure cloud computing and anti-virus for data processing, authentication and key management, security education and management, and privacy preservation in the application layer (Table 1).

3 Proposed Work

SGSQoT deals with four types of trust concepts: *self-trust* (*SLT*), *green trust* (*GT*), *social trust* (*ST*), *and QoS trust*. Along with this peer recommendation also comes into the field when node's trust information is unavailable.

A.3. Self-trust (*SLT*) Calculation by an IoT Device

The self-trust (*SLT*) is calculated by a node in the following way and is shared with other devices within the network.

IoT Data Processing Trust: Data processing trust deals with sensory data from environment and other devices. The device itself calculates data processing trust $\left(T_X^{SELF\_P}\right)$.

$$T_X^{SELF\_P} = \frac{100 * N_D}{R_D} \tag{1}$$

where N_D =correct data, R_D =combination of both correct and erroneous data. So, $R_D = N_D + E_D$; where E_D = erroneous data. $N_D = N_{SD} + N_{XD}$, where N_{SD} =correct sensor data and N_{XD} =correct data received from other nodes. $E_D = E_{SD} + E_{PD} + E_{XD}$, where E_{SD} =erroneous sensor data, E_{PD} =erroneous data after processing, and E_{XD} =erroneous data received from other nodes.

IoT Data Privacy Trust: The privacy trust is about the sensed data from the environment or received data from *IoT* devices accessed by others. This privacy trust mainly depends on the latest version of standard security protocol (*SSPVL*) used by *IoT* devices. We have assumed that number of data (*RD*) accessed by intruder is equal to false count (*FC*).

$$T_X^{SELF\_PV} = \begin{cases} 100 \; when \; F_C = 0, \; SSPV_L = 1 \\ 0 \qquad Others \end{cases} \tag{2}$$

IoT Data Transmission Trust: The transmission trust depends on successful data transmission among the nodes. Let S_T be the successful communication with respect to high feedback status $(FB_{I-X} = 1)$ and high acknowledgment status $(ACK_{I-X} = 1)$ from other nodes ({I – X}) and F_T =unsuccessful event due to no feedback or no acknowledgment.

$$T_X^{SELF\_DT} = \begin{cases} \frac{100 * S_T}{S_T + F_T} \; for \; all \; FB_{I-X} = 1 & \\ ACK_{I-X} = 1, & X \notin \{I - X\} \\ 0 & Others \end{cases} \tag{3}$$

Combining Eqs. (1)–(3), we get the combined self-trust.

$$T_X^{SELF} = T_X^{SELF\_P} + T_X^{SELF\_PV} + T_X^{SELF\_DT} \tag{4}$$

B.3. Green Trust (*GT*)

Green trust is also known as environmental trust, which deals with the characteristics of network. Newly deployed *IoT* devices are compared with the behavior of the network in order to check whether the devices are well fitted or not. The *GT* of node *X* can be calculated as follows:

$$T_{X\_GREEN} = T_{X\_GREEN}^{LT} + T_{X\_GREEN}^{RT} \tag{5}$$

Lifetime Trust (***LT***): The lifetime trust analyzes the relationship between the lifetime of deployed *IoT* network and the *IoT* network lifetime. Energy and memory consumption is the primary concern of *IoT* network. We assume that the matching and non-matching numbers of the *IoT* network lifetime of each service from *IoT* device to FOG node in the history recorded by trust center are *IL* and *FL*, respectively. The *LT* of node *X* shown by trust center is:

$$T^{LT}_{X\_GREEN} = \frac{100 * IL}{IL + FL} \tag{6}$$

Response Trust (***RT***): It seeks whether the response time of the deployed *IoT* network matches *IoT* network response time or not. The response time of *IoT* network is uncertain due to various factors like device dies, heterogeneity, and bad weather. The matching number (*IM*) and non-matching number (*NM*) of the *IoT* network response time of each service from *IoT* node to FOG node are taken from the history of fog server. *RT* of node *X* is:

$$T^{RT}_{X\_GREEN} = \frac{100 * IM}{IM + NM} \tag{7}$$

C.3. Node-Level Direct Trust

This trust consists of *social trust* and *QoS* trust.

Social Trust (***ST***): Social trust indicates the behavior of the nodes within the community. This trust is determined by intimacy, honesty, and social similarities. Device *X* calculates the social trust of Device *Y* as follows:

$$T^{ST}_{X,Y} = T^{IN}_{X,Y} + T^{HO}_{X,Y} + T^{SM}_{X,Y} \tag{8}$$

Intimacy: *ST* can be calculated in terms of intimacy. Intimacy measures how many times and how longer the devices are communicated with each other. Device *X* calculates the intimacy trust $\left(T^{IN}_{X,Y}\right)$ of Device *Y* as follows:

$$T^{IN}_{X,Y} = \left(\frac{100 * IT_{X,Y}}{IT_{X,Y} + IT_{X,I}}\right) + IT_{X,Y} * \left(\frac{100 * \sum_{J=1}^{J=IT_{LS}} T_{X_J,Y_J}}{T}\right) \tag{9}$$

$IT_{X,Y}$ No of interaction between Device *X* and *Y*,
$IT_{X,I}$ No of interaction between Device *X* and *I* ($Y \notin I$),
T_{X_J,Y_J} Interaction duration between *X* and *Y*,
T Total interaction duration with *X* and *Y* and *X* and *I*, where *I* can be any device except *Y*

Honesty: In case of community-based system, honesty is an important trust parameter to find social trust. Device *X* determines honesty of Device *Y* by the number of successful events and number of valid data (V_{DATA}) sent by *Y*. No of successful events and valid data are calculated within the time frame of $[T - T, T+T]$. $T - T$

indicates the past experience of X on Y, and $T+T$ indicates ongoing scenario with T interval. Here, two control factors are used: One is to control the increasing rate of successful events $\left(S_{X,Y}\right)$, and another one is for data tempering factor ($\Delta_{TF} \neq 0$, due to congestion impairment). Device X calculates the intimacy trust $\left(T_{X,Y}^{HO}\right)$ of Device Y as follows:

$$T_{X,Y}^{HO} = 100 * \left[\int_{T-\delta T}^{T+\delta T} \left(\frac{S_{X,Y}}{S_{X,Y} + U_{X,Y}}\right)\left(\frac{1}{\sqrt{1 + S_{X,Y}}}\right)\delta T + \int_{T-\delta T}^{T+\delta T} \left(\frac{V_{DATA}}{V_{DATA} + I_{DATA}}\right) * \left(\frac{1}{\Delta_{TF}}\right)\delta T\right] \quad (10)$$

Social Similarities:

Acquaintance List and *Fan Club* (*AF*): Each *IoT* device shares its own acquaintance list and fan club with other *IoT* devices. If Device X finds the friends and members similar with Device Y, then X generates a trust based on the percentage of similarities. Let X be P% similar to Y's acquaintance list and Q% similar to Y's fan club members, and then similarity trust is:

$$T_{AF}^{SM} = \begin{cases} \text{Trustworthy} & P + Q \succ 60\% \\ \text{Ignorance} & 60\% \succ P + Q \succ 40\% \\ \text{Untrustworthy} & P + Q \prec 40\% \end{cases} \quad (11)$$

Route Similarity (*RS*): Every *IoT* device shares its own route discovery list (to whom the device sends the request to get connected), complete route, number of hopes, and end device with its communicating devices. If X finds the route similarity with Y, then we can say that Y is trustworthy in terms of route similarity. The route similarity trust of Y observed by X as follows:

$$T_{RS}^{SM} = \frac{100 * R_S}{R_S + N_S} \quad (12)$$

Route similarity percentage is provided, where N_S = route not similar and R_S = route is similar. The values of R_S and N_S range within [0, 1] and $R_S + N_S = 1$. When $R_S = 1$, $N_S = 0$, then Y is trustworthy; when $R_S = 0$, $N_S = 1$, then Y is untrustworthy. Otherwise, the decision from T_{RS}^{SM} is uncertain hence we can ignore.

QoS Trust: *QoS* trust describes the trust on *QoS* of nodes within the community. The *QoS* trust is found by many parameters like protocol compliance and energy. Device X calculates the *QoS* trust of Y as follows:

$$T_{X,Y}^{QoS} = T_{X,Y}^{PC} + T_{X,Y}^{EN} \quad (13)$$

Protocol Compliance: Protocol compliance deals with the prescribed protocol execution sequence of a node. For instance, if Y follows sequence of protocol execution P_E, then protocol compliance trust of Y by X is calculated as:

$$T_{X,Y}^{PC} = \frac{100 * P_E}{T_E} \quad (14)$$

T_E = total protocol execution sequence and P_E = protocol execution sequences.

When $T_E = P_E$, then Y fully follows protocol execution sequence and trustworthy in terms of protocol compliance trust.

When $T_E \succ P_E \geq \frac{T_E}{2}$, then Y partially follows protocol execution sequence and the trust status is uncertain in terms of protocol compliance.

And when $\frac{T_E}{2} \succ P_E \geq 0$, then Y is not trustworthy.

Energy: Device X measures the energy trust $T^{EN}_{X,Y}$ of Y in terms of total number of acknowledgment (N_{Y_ACK}) and feedback (N_{Y_FB}) against total number of data (N_{X_DATA}) by X as follows:

$$T^{EN}_{X,Y} = \frac{N_{Y\_ACK} \cup N_{Y\_FB}}{N_{X\_DATA}} \tag{15}$$

D.3. Node-Level Indirect Trust (Peer Recommendation)

Let an *IoT* community be composed of n distinct nodes. Each node maintains a trust value for all other nodes. Whenever a node requires peer recommendation, it will send a request to all member nodes except for the untrusted one. Peer recommendation is possible within the mobility time period $[0,\ T_{mob}]$. Let j nodes are trusted in a community. Then, node X calculates the trust value of node Y as:

$$T^{R}_{X,Y} = \frac{W_{I,Y} R_{I,Y} T_{I,Y}}{T_{X,I} + T_{I,Y}}[0,\ T_{mob}] \tag{16}$$

$T^R_{X,Y}$ = indirect trusts of node X on Y and $T^R_{I,Y}$ = direct trust of node I on Y

$W_{I,\,Y}$ = remembrance factor of I node on Y node, where $I \epsilon\ n - X - Y$, $W \epsilon\ [0, 1]$

$R_{I,Y}$ = reputation points of I node on Y node, where $I \in n - X - Y, R \in [0, 10] T_{mob}$ = mobility time.

$$T_{X_I,Y} = Avg\left\{T_{X_1,Y}, T_{X_2,Y}, T_{X_3,Y}, \ldots, T_{X_{n-2},Y}\right\}$$

$$T_{X_I,Y}\begin{cases} \neq 0 & \text{for old } IoT \text{ node} \\ = 0 & \text{for newly joined } IoT \text{ node} \end{cases}$$

4 *SGSQOT* Protocol Design

Our proposed architecture consists of a community of *IoT* nodes and fog nodes. Every *IoT* node communicates with each other for group communication and also with its nearest fog node. Fog node acts as server and mitigates the overhead of *CS*. The two-layer architecture is shown in Fig. 1. In lower layer, *IoT* devices are deployed, and in the upper layer, Fog servers are there. *SGSQoT* deals with lower tier. The *IoT* devices calculate its own trust value called self-trust using Eqs. (1)–(5).

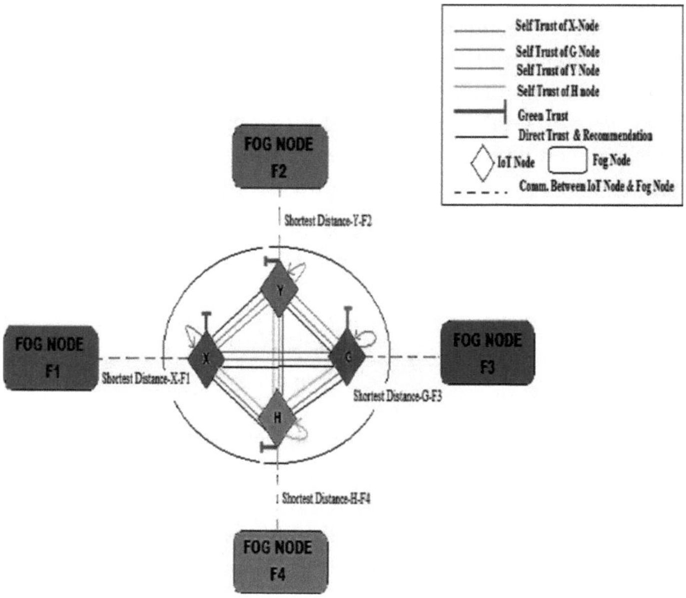

Fig. 1 Trust management architecture (*SGSQoT*) (*node level*)

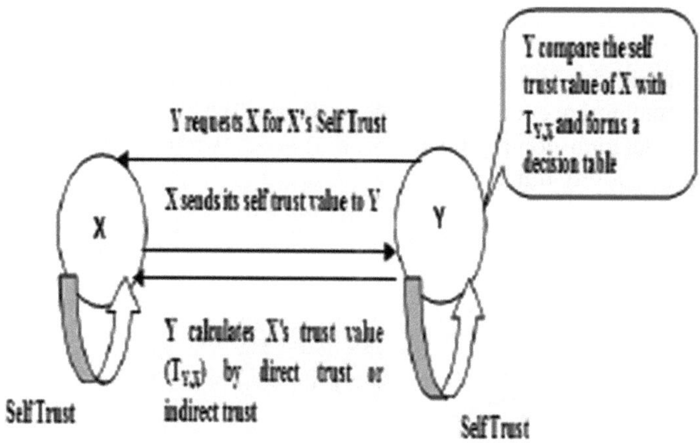

Fig. 2 Flow chart—a case when *Y* checks the trustworthiness of *X*

They share *SLT* with all other nodes with whom they are connected. *IoT* nodes also calculate *GT* using Eq. (5) to check whether the *IoT* network environment is well suited or not. In the next phase, each node calculates the trust value of other nodes within the community in terms of *ST* and *QoS* trust. After that, the *ST* value received from other nodes is compared with direct trust value (*ST* and *QoS* trust) or indirect trust value based on peer recommendation. The flow chart is given in Fig. 2, and Table 2 designates the decision table.

Table 2 Decision table to check vulnerability of node X

Case	Condition check	Comment	Probable attack
Case 1	$T_X^{SELF} \geq Th$ && $(T_{Y,X}^{ST} + T_{Y,X}^{QoS})\ U\ (T_{Y,X}^{R}) \geq Th$, $T_{X\_GREEN} \geq Th$	Node X is trustworthy and well suited in the network	Nil
Case 2	$T_X^{SELF} < Th$ && $(T_{Y,X}^{ST} + T_{Y,X}^{QoS})\ U(T_{Y,X}^{R}) \geq Th$, $T_{X\_GREEN} \geq Th$	Node is not trustworthy	Ballot stuffing attack prone (provide good reputation)
Case 3	$T_X^{SELF} < Th$ && $(T_{Y,X}^{ST} + T_{Y,X}^{QoS})\ U(T_{Y,X}^{R}) < Th$, $T_{X\_GREEN} \geq Th$	Node is not trustworthy	Bad mouthing attack prone
Case 4	$T_X^{SELF} > Th$ && $(T_{Y,X}^{ST} + T_{Y,X}^{QoS})\ U(T_{Y,X}^{R}) < Th$, $T_{X\_GREEN} \geq Th$	Node is not trustworthy	Self-promoting attack prone
Case 5	$T_X^{SELF} > Th$ && $(T_{Y,X}^{ST} + T_{Y,X}^{QoS})\ U(T_{Y,X}^{R}) \geq Th$, $T_{X\_GREEN} < Th$	Uncertain and node is not well suited	Uncertain

5 Simulation and Result

The simulation for *SGSQoT* is carried out in MATLAB. The trust value calculations are done with respect to *SLT*, *GT*, direct, and indirect trust using the above formulas. Simulation graph of recommendation trust is shown in Fig. 3. The simulation is done on 36 *IoT* nodes (one community) with 1 h duration. We assume that working area is 1000 * 1000 m. At the time of simulation, all the trust values introduced above are considered. The overall simulation graph considering all the nodes (36) is shown in Fig. 4. During the simulation, threshold (*Th*) values for *SLT*, direct trust (*ST* and QoS trust), and *GT* are 150, 250, and 100, respectively, on the basis of data range and data pattern. The trust value states for trustworthy nodes, untrustworthy nodes, and uncertain nodes are shown in Figs. 5, 6, and 7, respectively. In our proposed architecture, three types of communication take place and they are: communication during *SLT* exchange, direct trust calculation, and indirect trust calculation which is recommendation based. So if there exist *I* number of *IoT* devices, then maximum number of communication will be: $C_{Max} = \frac{3I(I-1)}{2}$ and we assume that each communication takes 1 kJ energy. So total energy for communication will be $E_{Comm} = \sum_{I=2}^{I=f} \frac{3I(I-1)}{2}$ KJ where f is a finite number. On the other hand, each trust value calculation takes certain computational power. Let us assume that for executing arithmetic formula (*af*) energy needed is 1 kJ. Now each *I* number of nodes calculates *SLT* (3*af*), *GT* (2*af*), *ST* (6*af*), *QoS* trust (2*af*), and indirect trust calculation (1*af*). Total *af*s are 15. So 15I KJ energy consumed by *I* number of nodes.

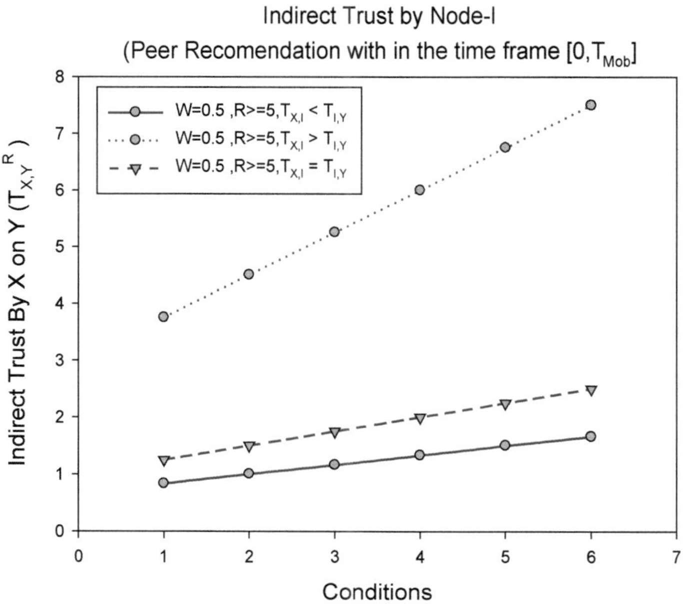

Fig. 3 Indirect trust by X on Y

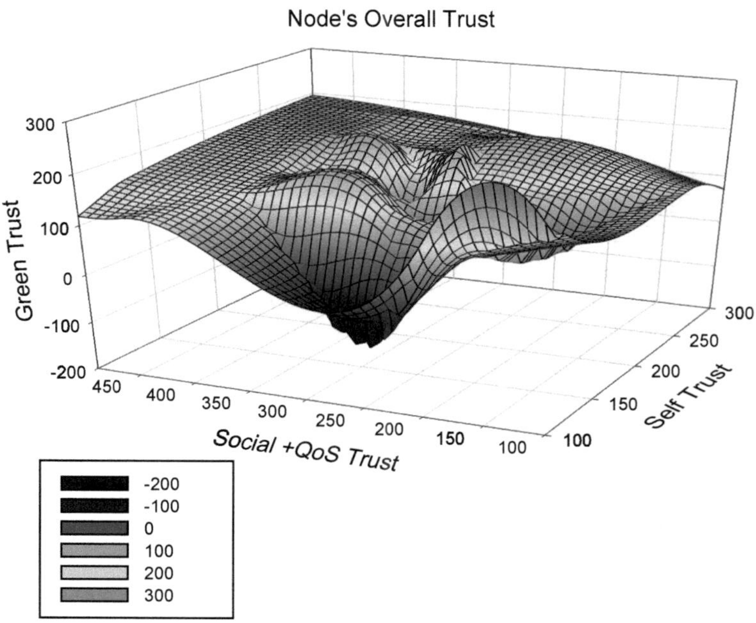

Fig. 4 Trust value states of all the nodes together

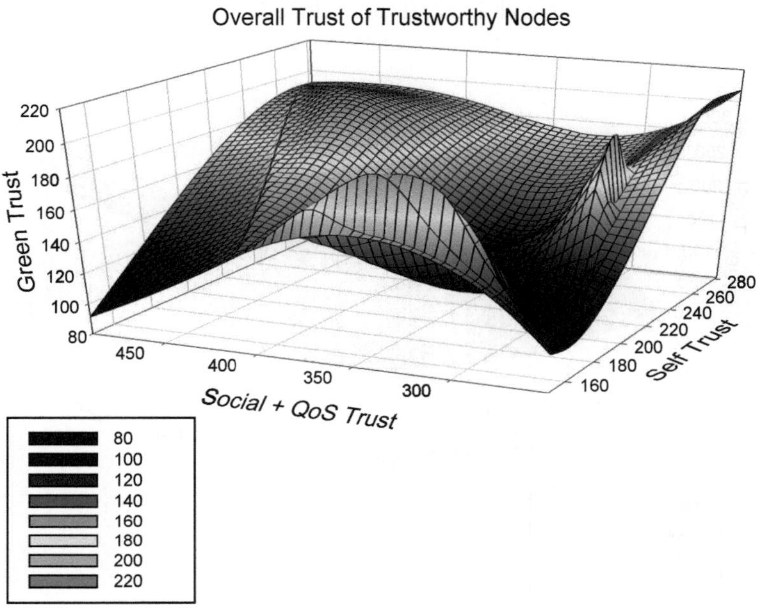

Fig. 5 Trust value state of trustworthy nodes

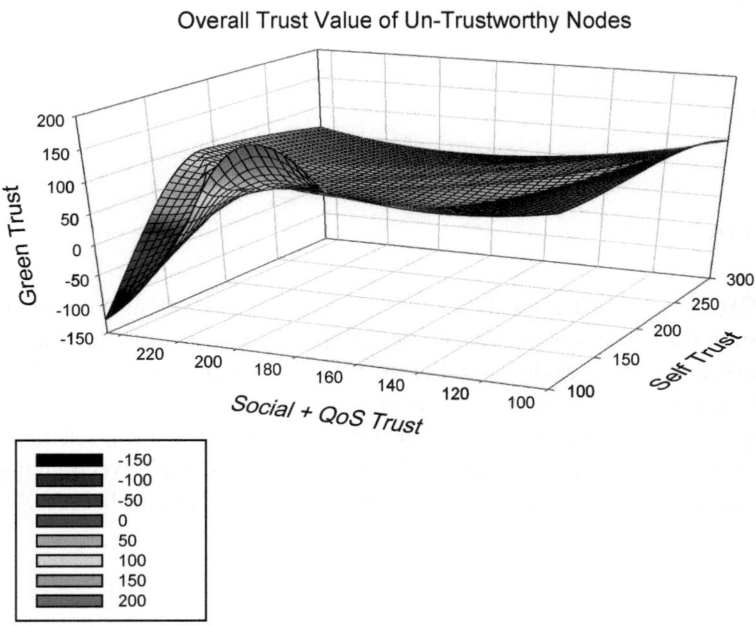

Fig. 6 Trust value state of untrustworthy nodes

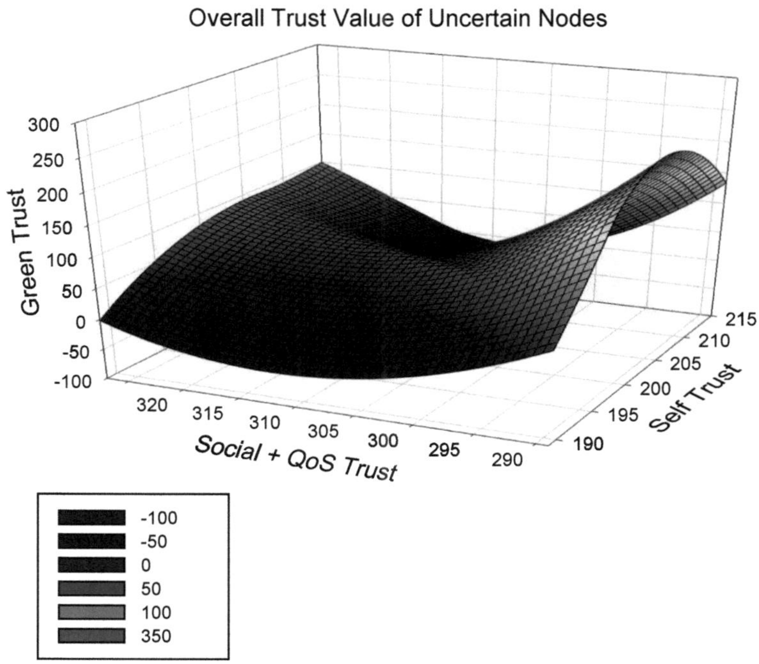

Fig. 7 Trust value state of uncertain nodes

Hence, total energy: $E_T = \left(\sum_{I=2}^{I=f} \frac{3I(I-1)}{2} + \sum_{I=2}^{I=f} 15I\right)$KJ. The energy graph is shown Fig. 8.

6 Conclusion

Our *SGSQoT* scheme deals with *SLT*, *GT*, *ST*, *and QoS trust* along with peer recommendation trust. The scheme tries to mitigate the drawbacks of existing techniques and satisfies the trust relationship. For the trust calculation, *DPT*, *PP*, *DFMT*, and *DTCT* are considered. Our protocol mainly focuses on community of *IoT* devices and *fog nodes*. In the future, we will develop new trust management schemes in *IoT* community and expand it to *IoT fog node* level and *fog node CSP* level for comparing the efficiency in terms of memory and energy with the existing techniques.

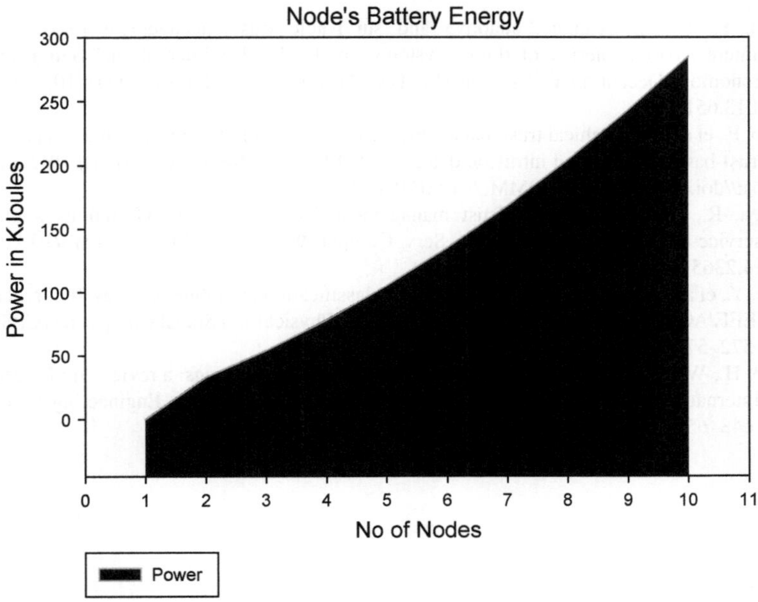

Fig. 8 Energy consumption for our scheme

References

1. Madakam, S., Ramaswamy, R., Tripathi, S.: Internet of things (IoT): a literature review. J. Comput. Commun. **3**, 164–173 (2015). https://doi.org/10.4236/jcc.2015.35021
2. Lee, W., et al.: A gateway based fog computing architecture for wireless sensors and actuator networks. In: 18th International Conference on Advanced Communication technology (ICACT), Kore (2016). ISBN 978-89-968650-6-3
3. Zhu, C., Nicanfar, H., Leung, V.C.M., Yang, L.T.: An authenticated trust and reputation calculation and management system for cloud and sensor networks integration. IEEE Trans. Inf. Forensics Secur. **10**(01) (2015). https://doi.org/10.1109/tifs.2014.2364679
4. Chen, R., et al.: Trust management in mobile ad hoc networks for bias minimization and application performance maximization. Ad Hoc Netw. **19**, 59–74 (2014)
5. Zhou, Q., Gui, F., Xiao, D., Tang, Y.: Trusted architecture for farmland wireless sensor networks. In: Proceedings of IEEE 4th International Conference on Cloud Computing Technology and Science (CloudCom), Taiwan, pp. 782–787 (2012). https://doi.org/10.1109/CloudCom.2012.6427496
6. Li, X., Xuan, Z., Wen, L.: Research on the architecture of trusted security system based on the internet of things. In: Proceedings of International Conference on Intelligent Computation Technology and Automation (ICICTA), pp. 1172–1175 (2011)
7. Ning, H., et al.: Cyber entity security in the internet of things. Computer **46**(4), 46–53 (2013)
8. Gessner, D., Olivereau, A., Segura, A.S., Serbanati, A.: Trustworthy infrastructure services for a secure and privacy-respecting internet of things. In: Proceedings of IEEE International Conference on Trust, Security and Privacy in Computing and Communications, pp. 998–1003 (2012)

9. Bao, F., et al.: Scalable, adaptive and survivable trust management for community of interest based internet of things systems. In: IEEE 11th International Symposium on Autonomous Decentralized Systems (ISADS), Mexico (2013). https://doi.org/10.1109/ISADS.2013.6513398
10. Bao, F., et al.: Hierarchical trust management for wireless sensor networks and its applications to trust-based routing and intrusion detection. IEEE Trans. Netw. Serv. Manag. **9**(2) (2012). https://doi.org/10.1109/TCOMM.2012.031912.110179
11. Chen, R., Guo, J., Bao, F.: Trust management for SOA based IoT and its application to service composition. IEEE Trans. Serv. Comput. **9**(3) (2016). https://doi.org/10.1109/tsc.2014.2365797
12. Liu, Y., et al.: A trust model based on service classification in mobile services. In: Proceedings of IEEE/ACM International Conference on Cyber, Physical and Social Computing (CPSCom), pp. 572–577 (2010)
13. Suo, H., Wan, J., Zou, C., Liu, J.: Security in the internet of things: a review. In: Proceedings of International Conference on Computer Science and Electronics Engineering (ICCSEE), pp. 648–651 (2012)

A Novel Trust Evaluation Model Based on Data Freshness in WBAN

Sanjoy Roy and Suparna Biswas

Abstract Wireless sensor networks foray into a promising field of remote healthcare applications nowadays, where the patients can be monitored remotely through wireless body area network (WBAN). The wireless medical sensors are deployed on the subject's body to closely monitor the person's vital body signs and transmit the data to the intended medical service providers. But medical sensor networks like WBAN are prone to potential security issues due to the sensitiveness of medical data and the exposure of data through wireless channel. The traditional cryptographic approaches and secure routing processes put overheads on the relatively light low-capacity nodes of the wireless body area network nodes. In this paper, a trust-based security model is proposed which will consider the data freshness factor and based on that a malicious or non-eligible node can be detected. A soft security mechanism like trust-based security model evaluates each nodes trust values that direct the communication in the network.

Keywords Wireless body area network · Cryptography · Data freshness · Trust Security

1 Introduction

The aim of wireless body area network (WBAN) is to provide wireless health monitoring of intended persons without reliance of any static infrastructure [1]. A WBAN network consists of wireless nodes that are placed into patient's body, and these nodes form a temporary network where they communicate in multi-hop way [2, 3].

S. Roy
Meghnad Saha Institute of Technology, Kolkata, 700150, India
e-mail: sanjoysss@yahoo.com

S. Biswas (✉)
Maulana Abul Kalam Azad University of Technology, Kolkata 700064, West Bengal, India
e-mail: mailtosuparna@gmail.com

M. Chakraborty et al. (eds.), *Proceedings of International Ethical Hacking Conference 2018*, Advances in Intelligent Systems and Computing 811,
https://doi.org/10.1007/978-981-13-1544-2_19

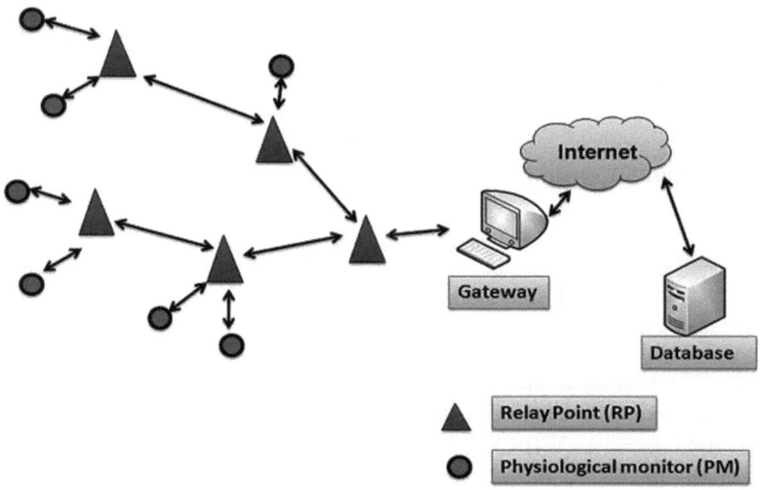

Fig. 1 A sample multi-hop WBAN with relay [2]

In WBAN network, generally three tiers exist. The sensors that are placed into the body of the monitored person are considered as the first tier. The information from this tier are sent to different relay points which constitutes as the second tier, and the data is sent through the gateway to a central database which is considered as the third tier [4]. An illustration of this is given in Fig. 1.

In the context of health monitoring, one of the important factors is data freshness. Through WBAN, the health professionals can continuously keep eye on the patient's health and in any adverse situation, and they can take the prompt decisions to save the person's life [5]. In this regard, the freshness or recentness of data plays a key role. WBANs have the most critical requirement of recent data. The data reached to the health professionals must be as fresh as possible. A delay to send fresh data may result to a delayed detection of the abnormality which may prove fatal to the patient. If any node in the network creates delay in communication or a malicious node is introduced into the WBAN, they can send old data to the destination. To ensure the freshness of data, the need is to identify the node behaving errantly.

The implementations of traditional complex cryptographic approaches to find malicious nodes are not viable due to the stringent requirement of fresh data. In this regard, the trust management scheme can be used to detect the delay contributing nodes and finding alternate fast and reliable route to send data in real time, i.e., within 250 ms, maximum delay permissible in health care.

In WBAN, the nature of the data packets is very sensitive and strong cryptographic functions are needed to maintain the security aspect of the network. But strong cryptographic functions need extensive computation and resources, and this creates a great challenge to implement this mechanism in this resource constrained and energy-starved environment. In one hand, the asymmetric cryptography is very expensive in terms of computational overheads; on the other hand, symmetric cryptography suffers from secret sharing problem. For that reason, an equivalent security solution is needed with reduced energy, memory, and execution time.

To ensure data freshness, several schemes rely on the linear trust evaluation processes to ensure the security of the communication. But the WBAN specified for applications such as health care, military or defense, and security cannot be compromised in any situation. On the other hand, data freshness is to be maintained. The traditional trust evaluation model calculates the trust value considering the successful packet sending or packet drops. Delay factor is not included in the evaluation, which is an important characteristic in the context of data freshness.

In this paper, we present a novel evaluation process, which includes the delay factors in the evaluation of the trust values. The evaluation of this algorithm shows that the delay factors are affecting on the trust components, which have resulted the detection of delay contributing nodes and initiates the need to find an alternate path in the multi-hop WBAN to ensure the recentness of data and maintaining the security of the communication.

In the remaining part of the paper, Sect. 2 presents related work. Section 3 describes the general trust evaluation model. Section 4 describes the novel delay-based trust evaluation algorithm to detect the delay contributing nodes. Section 5 presents the results of algorithm implementation in terms of components of trust value. In Sect. 6, we conclude the paper and present a glimpse of the future work.

2 Related Work

The security approaches applied in wireless sensor networks cannot be directly applied to get good performance in wireless body area networks as it has some unique characteristics and challenges that should be considered while designing security frameworks. The number of sensors on the human body is usually quite limited. And the distributed nature of lightweight sensor nodes has computation and energy constraints. When designing security protocols for WBAN, these characteristics should be taken into account in order to define optimized solutions to satisfy resource limitations of specific environment [6].

Rae Hyun Kim et al. proposed a delay reduced MAC protocol for a health monitoring WBAN. The proposed MAC protocol in this work takes into account the delay parameters to reduce the time delay and packet loss in TDMA-based CSMA/CA environment. It shows that performance of proposed MAC protocol is enhanced in terms of reduced time delay and packet loss over conventional Bio-MAC protocol in WBAN [7].

Feng Li et al. in their work evaluate and quantify trust considering a core dimension that is uncertainty. They specifically use an uncertainty metric to directly reflect a node's confidence in the sufficiency of its past experience and study how the collection of trust information may affect uncertainty in nodes' opinion. They also exploited mobility to efficiently reduce uncertainty and to speed up trust convergence [8].

Jin-Hee Cho et al. in their works discuss on the concepts and properties of trust values and highlighted on some important characteristics of trust. They provided a survey of trust management approaches and highlighted on the accepted metrics [9].

A. Das et al. proposed a dynamic trust model to compute trust considering various aspects of an agent and applied that in a multi-agent scenario. A load-balancing approach is also followed to establish this dynamic trust model [10].

3 Trust Evaluation Model

Trust can be defined as the strong confidence on the competence of an entity to behave reliably in a given context. It represents a WBAN members' anticipation of other nodes' behavior when assessing the risk involved in future interactions [8]. Here the initiator node is called the trusting node, and other nodes are known as trustee. The trust relationship between the nodes is defined by the trust values which is build upon the basis of the past experiences.

Many existing trust evaluation system only considers the trust value as either right or wrong. The problem with that a no of nodes is just deemed as ineligible for just not having the full trust value, and also the past interactions are not reflected properly. The component that is not considered is the uncertainty, which is very important for a mobility-based, real-time network like WBAN. In trust-based model, trust value is considered in terms of belief, disbelief, and uncertainty. And depending on this soft mechanism, the sender can take the decision to send a packet to a particular node.

We use a triplet to represent a node's opinion $(b, d, u) \in [0, 1]^3$: $b+d+u=1$. b, d, and u refer to belief, disbelief, and uncertainty, respectively [8]. In the evaluation process, first the uncertainty component is computed from the normalized variance of the distribution. Belief and disbelief components are calculated on proportional basis.

Trust of a node computed from direct information is the trust that depends on that node's own experience. It is calculated directly from a node's observation. Bayesian inference is statistical inference in which evidence or observations are used to update or to newly infer the probability that a hypothesis may be true. Beta distributions, $Beta(\alpha, \beta)$, are used herein the Bayesian inference, since it only needs two parameters that are continuously updated as observations are made. As we use a triplet to represent the node's opinion, the triplet (b, d, u) is derived from $Beta(\alpha, \beta)$. Therefore, we define uncertainty u as the normalized variance of $Beta(\alpha, \beta)$: $u=(12 \cdot \alpha \cdot \beta)/(\alpha+\beta)^2 \cdot (\alpha+\beta+1)$ [8].

The total certainty is $(1-u)$ which can be divided into b and d according to their proportion of supporting evidence. Since the proportion of supporting evidence for the statement that the transmission between two nodes is reliable is $\alpha/(\alpha+\beta)$, b can be calculated as follows: $b=\alpha/(\alpha+\beta) \cdot (1-u)$. Therefore, $d=(1-u)-b=\beta/(\alpha+\beta) \cdot (1-u)$.

Here the components of the trust depend on the nature of success or failure of the operations. But the critical thing is to detect malicious nodes as well as keep the recentness of data intact.

4 Proposed Novel Algorithm for Evaluating Trust Considering Data Freshness

Data freshness is important characteristics regarding data communication in wireless body area network (WBAN) as healthcare applications need uninterrupted data flow for continuous monitoring. Any delay beyond permissible limit should create a doubt on the malicious nature of that particular node. In trust-based scheme, where the conventional cryptographic methods are not followed, the data freshness or the delay factor is to be considered while trust evaluation to detect the malicious or ineligible nodes and in other words to enhance the security of the network. The delay requirements of IEEE 802.15.6 WBAN are shown in Fig. 2. Here end-to-end delay is considered as the sum of transmission delay, propagation delay, and processing delay.

In a WBAN, each sensor placed in patient's body is considered as a node and they send the data packets to the other nodes which ultimately reach to the medical server. In this network, a malicious node's typical behavior is creating a delay amount which is greater than that of a non-malicious node. Even if the node which is creating a

Application	Bit Rate	Delay
Deep Brain Stimulation	< 320Kbps	< 250ms
Drug Delivery	< 16Kbps	< 250ms
Capsule Endoscope	1Mbps	< 250ms
ECG	192Kbps	< 250ms
EEG	86.4Kbps	< 250ms
EMG	1.536Mbps	< 250ms
Glucose Level Monitor	< 1Kbps	< 250ms
Audio Streaming	1Mbps	< 20ms
Video Streaming	< 10Mbps	< 100ms
Voice	50~100Kbps	< 100ms

Fig. 2 Permissible delay limits for WBAN [7]

Algorithm for trust evaluation considering data freshness factor

```
T_DF_PL : The delay factor that is permissible for a WBAN node
T_DF_L : The mid-point between the permissible limit and the average delay of
the non-malicious nodes in WBAN network
T_DF_C1, T_DF_C2 : The permissible no of delays within T_DF_PL and T_DF_L over a
specified no of communications.
Node_Alpha: Alpha(α) value of a node
Node_Beta: Beta(β) value of a node
C_B1, C_B2: Increment factors for beta values
Algorithm:
For each node to node packet delivery, calculate Delay Factor (DF)
If (Nature of Data transmission is critical)
        {
                If (DF> T_DF_PL) then
                        Update the disbelief component accordingly;
                else if (T_DF_PL > DF> T_DF_L) then
                        {
                        Delay_counter++;
                        If (Delay_counter> T_DF_C1) then
                                Update the disbelief component accordingly;
                        Node_alpha = Node_alpha +1;
                        }
                else
                        Node_alpha = Node_alpha +1;
        }
else
        {
                If (DF> T_DF_PL) then
                        Node_beta = Node_beta + C_b1;
                else if (T_DF_PL > DF> T_DF_L) then
                        {
                        Delay_counter++;
                        If (Delay_counter> T_DF_C2) then
                                Node_beta = Node_beta + C_b2;
                        Node_alpha = Node_alpha +1;
                        }
                else
                        Node_alpha = Node_alpha +1;
        }
                If (belief_value < 0.5)
                        Refer the node as doubtful node;
```

Fig. 3 Algorithm for trust evaluation considering data freshness factor

delay above the threshold value is not non-malicious, that node may be considered ineligible in terms of data freshness.

Here the algorithm is designed for two different scenarios. One is where the data freshness of the sent packet is critical like the ECG data of a heart patient. The other is where the delay is not affecting the overall monitoring of the person like the body temperature of an athlete. Here the alpha(α) and beta(β) values are continuously updated during the data forwarding. When a new forwarding is done, if it is a successful forwarding, then α is incremented, otherwise β is incremented (Fig. 3).

In this algorithm, for every data forwarding the corresponding delay is recorded. In the first case, where the nature of physiological data is critical, any delay beyond the permissible limit (can be defined type or level of illness specific) will make the belief component lying below the threshold belief value. This indicates the maliciousness or ineligibility of the node for being the part of critical data communication. In this case, any alternate path having acceptable trust value or best among the alternate paths will be followed. If for some data forwarding, the delay is nearer permissible limit, then that will be counted and if the counter value is exceeded a predefined permissible counter value for a specified no of operations, then again the belief component will be made dropping below the threshold belief value. In this, the lower threshold is defined as the midpoint between the permissible limit and the average delay factor of other non-malicious nodes of the network.

In the second case, where the nature of data is not critical, then exceeding the permissible limit of delay will not make the node to be considered as malicious or ineligible for this communication. It will penalize this by increasing the beta(β) factor by a constant value, which will change the components of the trust value. Like the critical data packets, in this case the same counter-based detection of nodes is followed where the delay is nearer the permissible limit and if the no of occurrences exceeds the predefined counter threshold value, then the beta(β) will be increased for a constant value. If these updates make the belief value lies below a predefined threshold value (0.5), the corresponding node will be considered as malicious or non-eligible for maintaining recentness of data.

In the following part, the flowchart for the algorithm is provided. The diagram clearly shows two paths for critical and non-critical cases. In critical data cases, if the delay value does not pass through the checking conditions, the disbelief part is updated accordingly. In non-critical data cases, if the delay value does not pass through the checking conditions, the beta value of the node is incremented by the constant value. At the end, the belief value is checked for both the cases which can refer to a node as a doubtful one (Fig. 4).

5 Implementation and Results

In implementation of the algorithm, we aim to test the effectiveness of the algorithm to detect malicious or non-eligible nodes in terms of data freshness by continuous evaluation of trust components, i.e., belief, disbelief, and uncertainty. The algorithm is implemented in MATLAB environment, and the results are achieved for critical data and non-critical data scenario (Fig. 5).

In critical data scenario, the graph presented here shows that if the delay factor is going beyond the permissible limit, then the disbelief component will be updated accordingly and it will result in the belief value dropping below the threshold value. The dotted horizontal line refers to the belief threshold value. Here the last delay value is greater than the permissible limit which makes the belief component drops below the threshold line (Fig. 6).

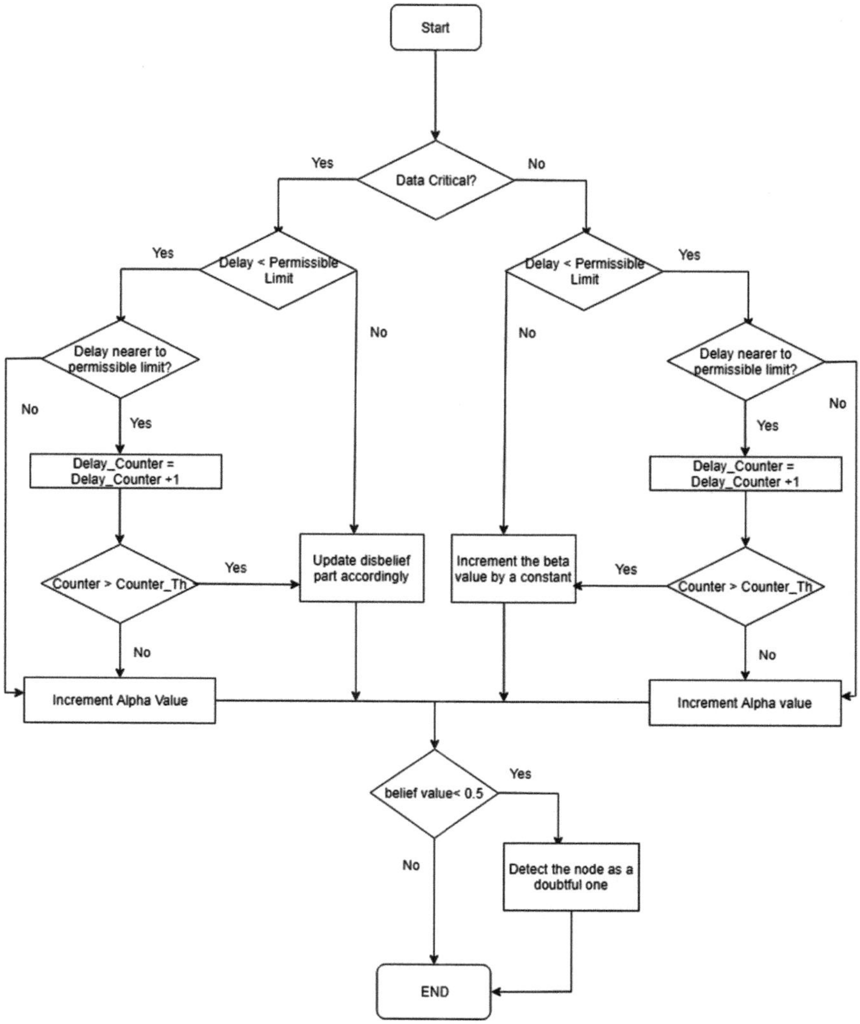

Fig. 4 Flowchart of the algorithm

In non-critical data scenario, the corresponding graph provides the changes in the trust components in terms of recorded delay values. It is shown that any delay over the permissible limit effects on the belief value. If the delay is normal, then the belief value is increasing, otherwise it is decreasing. According to the changes in delay value, the components are varied accordingly. The dotted horizontal line refers to the belief threshold value.

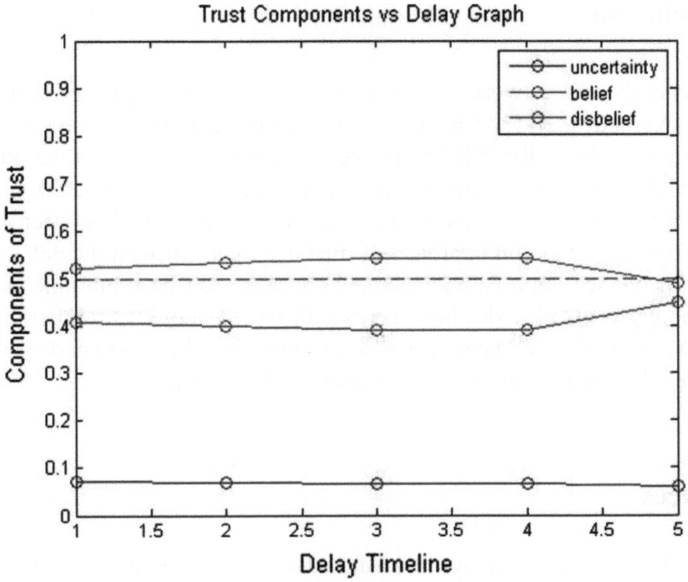

Fig. 5 Trust components versus delay graph for critical data

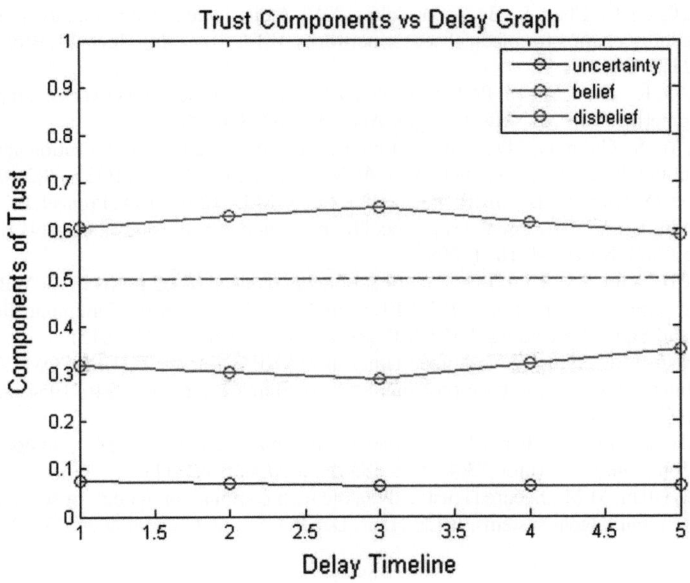

Fig. 6 Trust components versus delay graph for non-critical data

6 Conclusion

In wireless body area network, any security algorithm should give priority to detect the malicious nodes. But the heavyweight cryptographic algorithms are not very suitable for a network like WBAN having low-capacity nodes. On the other hand, recentness of data is to be maintained to carry out the required health monitoring activities at proper time. In this work, a balance is maintained between these two by integrating a lightweight approach of trust management with the delay property. In future to enhance the trust management approach based on multiple assessment factors, more aspects of nodes like energy will be considered. And there is a scope to develop an algorithm to detect whether an abnormally behaving node is malicious or not by exploiting the uncertainty component of the trust value.

References

1. Syed, A.R., Yau, K.L.A.: On cognitive radio-based wireless body area networks for medical applications. In: IEEE Symposium on Computational Intelligence in Healthcare and e-health (CICARE), pp. 51–57. Singapore (2013)
2. Pardeep, K., Jae, L.H.: Security issues in healthcare applications using wireless medical sensor networks: a survey. Sensors **12**(1), 55–91 (2012)
3. He, D., Chen, C., Chan, S., Bu, J., Vasilakos, A.V.: A distributed trust evaluation model and its application scenarios for medical sensor networks. IEEE Trans. Inf. Technol. Biomed. **16**(6), 1164–1175 (2012)
4. Yu, Y., Li, K., Zhou, W., Li, P.: Trust mechanisms in wireless sensor networks: attack analysis and countermeasures. J. Netw. Comput. Appl. **35**, 867–880 (2012)
5. Li, M., Yu, S., Guttman, J.D., Lou, W., Ren, K.: Secure ad hoc trust initialization and keymanagement in wireless body area networks. ACM Trans. Sensor Netw. (TOSN) **9**(2), 18 (2013)
6. Singelee, D., Latre, B., Braem, B., Peeters, M., Soete, M.D., Cleyn, P.D., Preneel, B., Moerman, I., Blondia, C.: A secure low-delay protocol for multi-hop wireless body area networks. Ad-hoc Mobile Wirel. Netw., 94–107 (2008)
7. Kim, R.H., Kim, P.S., Kim, J.G.: An effect of delay reduced MAC protocol for WBAN based medical signal monitoring. In: 2015 IEEE Pacific Rim Conference on Communications, Computers and Signal Processing (PACRIM), pp. 434–437. Victoria, BC (2015)
8. Li, F., Wu, J.: Mobility reduces uncertainty in MANETs. In: IEEE INFOCOM 2007—26th IEEE International Conference on Computer Communications, pp. 1946–1954. Anchorage, AK (2007)
9. Cho, J.H., Swami, A., Chen, I.R.: A survey on trust management for mobile ad hoc networks. IEEE Commun. Surv. Tutor. **13**(4), 562–583 (Fourth Quarter 2011)
10. Das, A., Islam, M.M.: SecuredTrust: a dynamic trust computation model for secured communication in multiagent systems. IEEE Trans. Depend. Secure Comput. **9**(2), 261–274 (2012)

CREnS: A Convolutional Coder-Based Encryption Algorithm for Tiny Embedded Cognitive Radio Sensor Node

S. Roy Chatterjee, S. Mukherjee, J. Chowdhury and M. Chakraborty

Abstract Here an encryption algorithm called cognitive radio encryption standard (CREnS) is proposed for tiny cognitive radio sensor node (CRSN) keeping in mind the limitation of power, memory resource, and computation capability of it. The algorithm uses convolutional coder and pseudorandom sequence which are the integral parts of spread spectrum-based CRSN. The inherent capability of generation of parity bit of convolutional coder is employed to design a dynamic substitution table, and the permutation box is generated from the nonlinear PN sequence. The substitution box satisfies several cryptographic properties like nonlinearity, Walsh spectrum, strict avalanche criterion, bijective, and resiliency. They together in turn provide avalanche effect within acceptable range. The algorithm utilizes simple mathematical operations and does not require any pre-computation before encryption of the plaintext which in turn would overcome the space complexity in hardware realization of the encryption algorithm for embedded tiny CRSN. The architecture has been implemented in Spartan-3E chip XC3s500e-5fg320 using ISE Design Suite 12.1. The result indicates that the algorithm utilizes 22% configurable logic blocks slices and provides 36% avalanche effect.

Keywords Cognitive radio sensor node · Convolutional coder · Encryption
Pseudo random sequence

S. Roy Chatterjee (✉) · S. Mukherjee · J. Chowdhury
Netaji Subhash Engineering College, Kolkata, India
e-mail: rcswagata@gmail.com

S. Mukherjee
e-mail: sumitmukh07@gmail.com

J. Chowdhury
e-mail: jayantachowdhury32@gmail.com

M. Chakraborty
Institute of Engineering & Management, Salt Lake, Kolkata, India
e-mail: mohuyacb@iemcal.com

M. Chakraborty et al. (eds.), *Proceedings of International Ethical Hacking Conference 2018*, Advances in Intelligent Systems and Computing 811,
https://doi.org/10.1007/978-981-13-1544-2_20

1 Introduction

The applications of cognitive radio (CR) are emerging in recent years due to development of spectrum sensing and spectrum sharing techniques. It may be effective to overcome the spectrum scarcity of the traditional emergency wireless communication systems [1, 2]. Emergency communication systems for human safety transmit data in situations when natural calamities or unexpected events occur like leakage of chemicals in factories, presence of explosive in the surroundings. These types of applications need rapid action to safeguard human life. Cognitive radio-based sensor node (CRSN) may overcome the scarcity of the spectrum of the current wireless emergency network and provide rapid transmission of sensory data by sharing the spectrum with the licensed user or utilizing available vacant bands. CRSN uses cognitive radio transceiver unlike the conventional sensor node. The cognitive radio transceiver enables the node to dynamically adapt its communication parameters like carrier frequency, transmission power [3, 4] according to the RF environment. The inherent intelligence of the cognitive radio and the significant growth of the applications of cellular phone encourage us to propose an embedded cellular phone system for public safety from unwanted explosion or any chemical, biological, and radiological attack by utilizing cellular network [5–7]. This system may be effective to safeguard human life form several chemicals and explosives in places like factories, railway stations, shopping malls with appropriate coordination among law enforcement, public safety, and security agencies. This type of emergency communication demands secure data transmission to avoid intentional tampering of sensory information. The data transmitted by CRSN may be sniffed, destroyed, or altered by unauthorized entities. So, acceptable level of security robustness is essential for CRSN. An end-to-end encryption may provide a high level of security as a point-to-point connection with CRSN, and its base station is considered to minimize the delay. But added encryption algorithm causes hardware and computational overhead to the tiny embedded CRSN, whereas the application needs to miniature the CRSN for designing portable embedded cellular phone system. So the major constraint is space and computational complexity. As a consequence, the design of the appropriate cryptographic method for embedded CRSN to utilize in area constrained application is a delicate issue. The main challenge is to design an algorithm that would provide accepted level of security with minimum utilization of space. In [8], the authors discussed an encryption proposal called LWT-PKI based on public key encryption and the authors in [9] proposed public key encryption scheme (PKES) for wireless sensor networks (WSN). However, public key encryptions require high processing power than symmetric key algorithms. The authors in [10] presented an encryption algorithm called scalable encryption algorithm (SEA) based on AES and RC6 algorithms and the authors in [11] discussed AES-based hybrid encryption for secure communication in WSN. In [12], the authors presented an encryption technique combining DES and Blowfish algorithms. It has been proved that Blowfish algorithm is more secure than DES but has similar security capability and less complexity as compared to AES [13, 14]. However, Blowfish may not be suitable for the tiny embedded CRSN as it requires large memory space [15–17].

Here a symmetric key encryption algorithm named cognitive radio encryption standard (CREnS) based on Feistel network is proposed for CRSN. The Feistel scheme is chosen because of its high acceptance in block cipher [13–16]. The CREnS algorithm is designed in Verilog HDL, and performance is evaluated in Spartan-3E FPGA.

After the introduction in Sect. 1, the design methodology of the proposed algorithm is provided in Sect. 2. Section 3 provides the performance evaluation followed by comparative analysis with an allied algorithm. At the end, we discuss future works and conclude the paper in Sect. 4.

2 Design Methodology of CREnS

Several symmetric key algorithms are designed, among which DES and Blowfish are widely accepted symmetric key algorithms using Feistel network. Blowfish algorithm was designed as an alternative to the traditional DES algorithm and to get rid of the limitations present in other encryption models. However, Blowfish requires large memory space, so it may not be suitable for the tiny embedded CRSN.

The CREnS algorithm is designed utilizing basic framework of Blowfish as shown in Fig. 1 with sixteen rounds and 64-bit input size. Unlike Blowfish, CREnS algorithm does not use any predefined P-box array, large S-boxes, and key expansion computation before data encryption. It generates P values from the PN sequence and substitution of bit has been done utilizing convolutional coder. This in turn minimizes the space requirement with acceptable level of security for the embedded tiny CRSN.

The strengths of CREnS are as follows:

i. It provides high security for the code hopping spread spectrum-based CRSN as P values of the P-box dynamically change with the PN sequence.
ii. It does not require the allocation of memory to store the value of S-box and any type of pre-computation for P-box and S-box.
iii. It utilizes PN sequences and convolutional coder which is integral part of any spread spectrum-based communication system. So hardware architecture of reusing the same hardware for encryption further minimizes the size of the CRSN.
iv. The output of the convolutional coder-based function block is entirely decided by the input plaintext bit stream and round key that in turn provides satisfactory level of cryptographic properties.

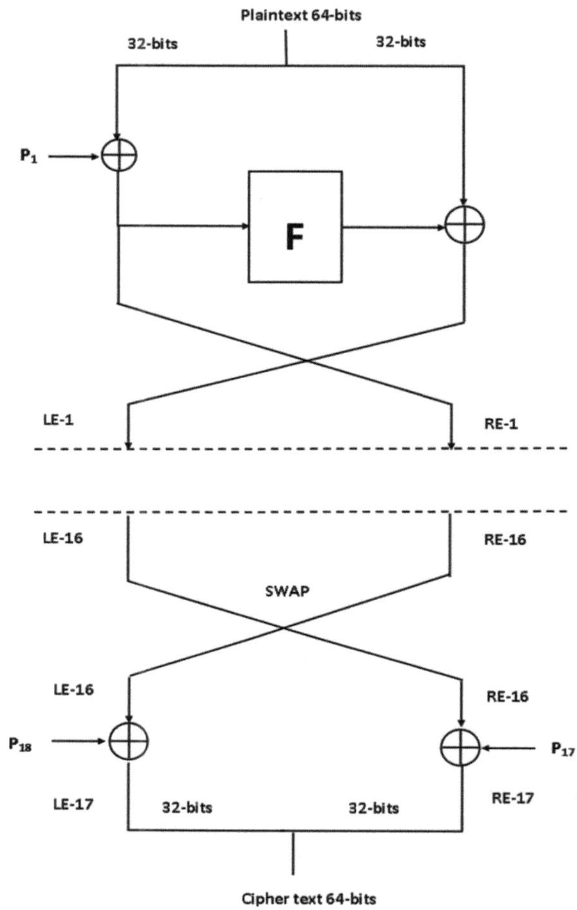

Fig. 1 Basic framework of Blowfish algorithm

2.1 *Design of Function Block*

The function box is designed without any predefined S-box as illustrated with Fig. 2. It employs substitution, splitting, exclusive-OR operation (XOR) and MOD 2^{32} addition for generation of output bit stream. The substitution of input bit stream has done utilizing a serial bit convolutional coder. A simple serial bit (2, 1, 2) convolutional coder is illustrated in Fig. 3. At each time instant, one input bit is fed to the convolutional coder and it produces two output bit as illustrated in Table 1. As a result, the substitution operation produces 64-bit stream from the 32-input bit stream. According to the substitution table, the substituted bit stream (o_n, $o_{n'}$ ... o_{n+32}, $o_{n+32'}$) completely dependent on the input bit stream (i_n ... i_{n+32}) and two initial value b_o, b_1. The output bit stream getting after substitution is split into two 32-bit streams. In the implementation of the algorithm, it is divided into odd and even bit stream. The odd bit stream is XORed with the left 32 bits of the round key, and even bit stream is

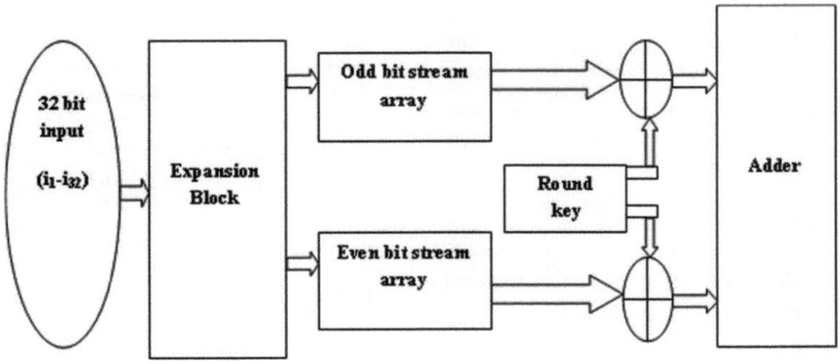

Fig. 2 Block diagram of function block

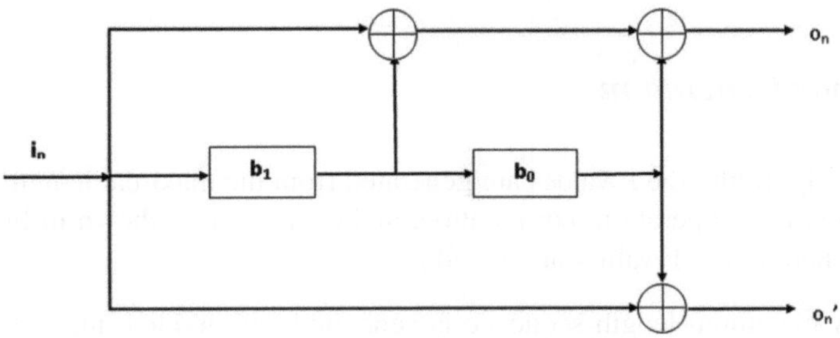

Fig. 3 (2, 1, 2) convolutional coder

Table 1 Parity bit generation of convolutional coder

Input length	Input bit	Expansion equation	Output bit	Output length
32 bit	i_n	$i_n \oplus b1 \oplus b0$	o_n	64 bit
		$i_n \oplus b0$	$o_{n'}$	
	i_{n+1}	$i_{n+1} \oplus i_n \oplus b1$	o_{n+1}	
		$i_{n+1} \oplus b1$	$o_{n+1'}$	
	i_{n+2}	$i_{n+2} \oplus i_{n+1} \oplus i_n$	o_{n+2}	
		$i_{n+2} \oplus i_n$	$o_{n+2'}$	
	⋮	⋮	⋮	
	i_{n+32}	$i_{n+32} \oplus i_{n+31} \oplus i_{n+30}$	o_{n+32}	
		$i_{n+32} \oplus i_{n+30}$	$o_{n+32'}$	

XORed with the right 32 bits of the round key, respectively. After XOR operation, MOD 2^{32} addition is performed to obtain the ultimate output of the function box as shown in Fig. 2 (Fig. 4).

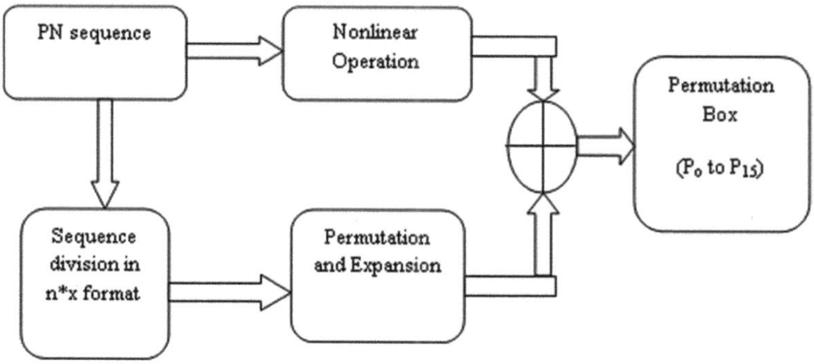

Fig. 4 Generation of P-box

2.2 *P-box Generation*

In CREnS algorithm, the P values are generated from the maximal length sequence utilizing nonlinear operation, permutation, and expansion as shown in Fig. 2. The steps of generation of P values are as follows:

Step 1: A maximum length sequence is generated and divided into $n * x$ format where n equals the number of rounds of the encryption algorithm and x equals the number of binary bits to be permuted in each round. For instance, in the implementation of the algorithm, a maximal length sequence of length $2^7 - 1$, and sixteen number of round have been considered. The maximal length sequence is divided into sixteen subgroup each consisting of eight bits. Bit stream of first subgroup is used to generate first P-box entry (P_0), similarly next subgroup is used to generate P_1.

Step 2: The bit stream of the subgroups is permuted and expanded to generate sixteen subgroups, each consisting of 32-bit streams.
Equations (18) and (19) represent the permutation matrix and expression of the output permuted bit stream for input bit stream $\{a_0, a_1, a_2, a_3, a_4, a_5, a_6, a_7\}$, respectively.

$$[a_0a_1a_2a_3a_4a_5a_6a_7]\begin{bmatrix}00100000\\00000100\\01000000\\00000001\\00000010\\10000000\\00010000\\00001000\end{bmatrix} = [a_5a_2a_0a_6a_7a_1a_4a_3] \tag{18}$$

$$E = a_5x^7 + a_2x^6 + a_0x^5 + a_6x^4 + a_7x^3 + a_1x^2 + a_4x + a_3 \tag{19}$$

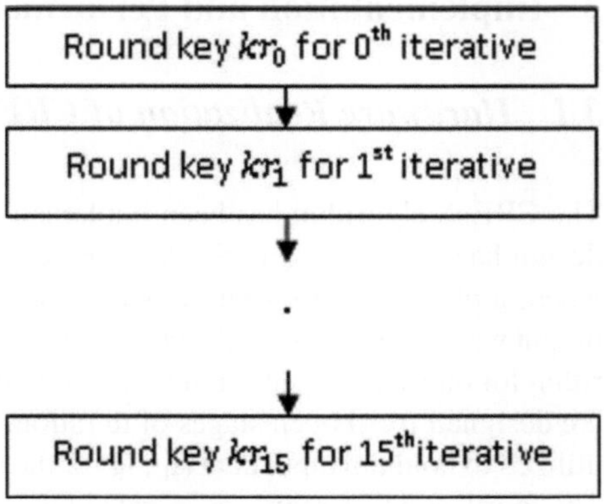

Fig. 5 Round key generation

Equation (20) represents expression of the expansion of each bit in the output permuted stream.

$$E_m = \sum_{i=0}^{i=3} a_{m+i} x^{3-i} \tag{20}$$

where m is the position with respect to the input stream in the permuted stream.

Step 3: A nonlinear operation is applied on the maximal length sequence (p_0, p_1, …) to generate a 32-bit nonlinear PN sequence. The nonlinear operation used in the algorithm is:

$$p_{10} + p_{11}\ AND\ p_8 + p_1\ OR\ p_{12} + p_0 + p_{15} AND\ p_{23} + p_{16}\ OR\ p_{30} + p_{34}\ AND\ p_{41}+$$
$$p_{35}\ OR\ p_{39} + p_{49} + p_{56}\ AND\ p_{61} + p_{64} + p_{72}\ OR\ p_{73} + p_{89}\ AND\ p_{99}\ AND\ p_{101}+$$
$$p_{113}\ OR\ p_{114}\ OR\ p_{12} + p_{22}\ AND\ p_{13}\ AND\ p_{23} + p_{25}\ OR\ p_{33}\ OR\ p_{35} + p_7\ AND$$
$$p_8\ OR\ p_9 + p_0\ AND\ p_{100}\ AND\ p_{110} + p_1\ OR$$
$$p_9\ OR\ p_8 + p_{114}\ OR\ p_{12}\ OR\ p_{22} + p_{89}\ AND\ p_{73}\ AND\ p_{101} + p_{10}\ OR\ p_{15}\ AND\ p_{16}+$$
$$p_{61}\ AND\ p_{34}\ OR\ p_{30} + p_0\ AND\ p_1\ OR\ p_8 + p_9 + p_{99}\ AND\ p_{13}\ \text{OR}\ p_1 + p_{72}\ OR\ p_{25}\ OR$$
$$p_{33} + p_{49}\ \text{OR}\ p_{61}\ OR\ p_{12} + p_{64}\ AND\ p_0\ AND\ p_{23} + p_{113}\ AND\ p_{100}\ AND\ p_8 + p_{34}\ \text{OR}$$
$$p_{72}\ AND\ p_{73} + p_{89}\ AND\ p_{22}\ OR\ p_{56}$$

Step 4: Finally, nonlinear PN sequence is XORed with expanded bit stream to generate P-box entries.

2.3 Round Key Generation

Here a simple circular shift algorithm is used to generate round keys from the initially selected random number as depicted in the flowchart in Fig. 5.

3 Implementation and Performance Evaluation

3.1 Hardware Realization of CREnS

The CREnS algorithm has been implemented using Verilog HDL on ISE 12.1. The design has been routed in Spartan-3E XC3s500e-5fg320. A test bench has written where a plain text of 64 bits was provided as input to the synthesized design. The output was observed using ISim(O.61xd). The hardware architecture of CREnS algorithm for one stage of iteration is shown in Fig. 6. Sixteen similar hardware modules are designed for sixteen stages of iteration. The first, second, and third registers are utilized to hold the input data (I_L, I_R) at the time of fetching the data from the P-box. The fourth and fifth registers operate in pair and hold the data for the time needed for operation of the convolution-based function block and second exclusive-OR operation for the computation of final output data (O_L, O_R) for the subsequent block. The hardware control unit is designed based on finite state machine to generate several control signals as shown in Fig. 6. The control signal named 'Enable1, Enable2' is utilized for selective switching of the registers and shift/load is utilized for proper operation of the convolution coder. CREnS algorithm employs a serial bit convolutional coder. As a consequence, it requires a parallel to serial 32-bit converter to feed the parallel 32-bit stream to the convolutional coder and serial to parallel 32-bit converter for the XOR operation with the right half 32-input bit stream (I_R). The whole hardware module block is initiated with control signal 'Reset.'

3.2 Hardware Utilization

Table 2 illustrates utilization of device hardware components of CREnS, which indicates that it uses only 22% of available CLB slices. Simulation result and comparative performance analysis between CREnS and Blowfish is depicted in Table 3. The results indicate that CREnS requires much less hardware than Blowfish, whereas the achieved frequency is much less in CREnS than Blowfish. But the frequency may be enhanced utilizing parallel bit convolutional coder instead of serial bit convolutional coder and pipelined architecture.

Table 2 Utilization of device hardware

Logic utilization	Used	Available
Number of slices	1039	4656
Number of slice flip flops	1208	9312
Number of 4 input LUTs	1894	9312
Number of bounded IOBs	65	232
Number of GCLKs	1	24

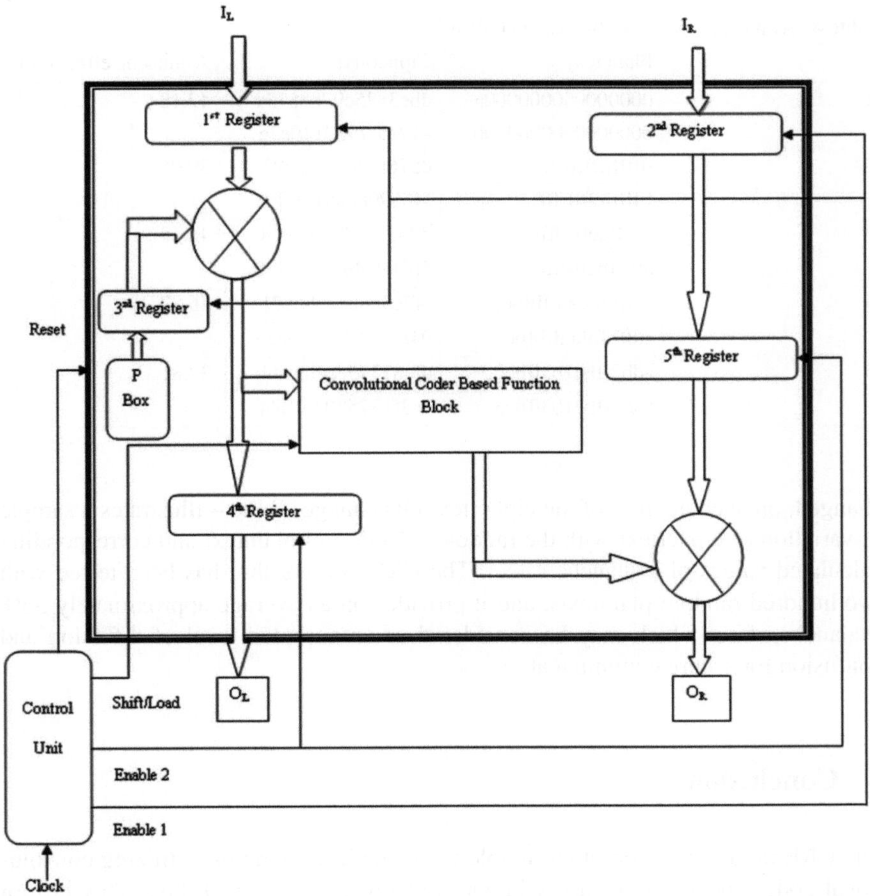

Fig. 6 Hardware module block

Table 3 Comparative performance analysis between CREnS and Blowfish

Parameters	CREnS	Blowfish
Number of slices	1039	2203
Frequency (MHz)	138.203	294.131

3.3 Avalanche Effect

The cornerstone of a block cipher design is to provide confusion and diffusion. The confusion is made to hide the statistical relationship between the ciphertext and the key, while diffusion spreads the influence of individual plaintext symbols over as much of the ciphertext as possible for hiding the statistical features of the plaintext. The effect of confusion and diffusion to the ciphertext is known as avalanche effect [13, 18, 19]. A cipher satisfies the avalanche criterion if a single plaintext bit is

Table 4 Avalanche effect as observed in CREnS

	Plain text	Ciphertext	Avalanche effect (%)
1	000000000000000a	dbe3875c97885739	42.18
	000000000000000b	4723af793211 0e2e	
2	ffffffffffffffff	d576f72cfb30bc92	39.06
	fffffffeffffffff;	5f66998dab190ff0	
3	afffffaeffffffff;	d201a5ed80969590	42.18
	afffffbeffffffff;	5b5b974de33c2470	
4	adffbfbeadffbfbe;	bf7e380cc3b86c71	46.87
	adffbfbfadffbfbe;	948a276512323c25	
5	adb2ffbe1b3ffbe3;	dfe81375ecdbba44	32.81
	adb3ffbe1b3ffbe3;	7a3f13758026ba44	

changed, on average, half of the ciphertext bits change. Table 4 illustrates a sample of variation of ciphertext with the random selection of plaintext and corresponding calculated values of avalanche effect. The CREnS algorithm has been tested with two hundred random plaintexts, and it provides on an average approximately 36% avalanche effect which may be considered as an accepted level of diffusion and confusion for secure communication.

4 Conclusion

The CREnS algorithm employs simple mathematical operations utilizing convolutional coder and PN sequence without predefined S-box and P-box. This in turn reduces the area requirement, minimizes hardware overhead of CRSN. The CREnS algorithm also provides accepted level of avalanche effect. This implementation utilizes a simple round key generation algorithm. Future work on hardware architecture design to optimize both speed and area by utilizing parallel bit convolutional coder and a strong round key generation to further enhance the avalanche effect is on the way.

References

1. Conder, P., Linton, L., Faulkner, M.: Cognitive radio developments for emergency communication systems. In: IEEE International Symposium on Dynamic Spectrum Access Networks (DySPAN), 3–6 May 2011
2. Pawełczak, P., Prasad, R.V., Xia, L., Niemegeers, I.G.M.M.: Cognitive radio emergency networks—requirements and design. In: IEEE International Symposium on New Frontiers in Dynamic Spectrum Access Networks, pp. 601–606, Nov 2005

3. Akan, O.B., Karli, O.B., Ergul, O.: Cognitive radio sensor networks. In: IEEE Network, pp. 34–40, July/Aug 2009
4. Joshi, G.P., Nam, S.Y., Kim, S.W.: Cognitive radio wireless sensor networks: applications, challenges and research trends. In: Sensors, vol. 13, pp. 11196–11228 (2013)
5. Chatterjee, S.R., Chakraborty, M., Chakraborty, J.: Cognitive radio sensor node empowered mobile phone for explosive trace detection. Int. J. Commun. Netw. Syst. Sci. **4**, 33–41 (2011)
6. Chakraborty, M., Chakraborty, J.: Mobile-telephony based secured society: an anti terrorism attempt. In: Proceedings of the International Technical Conference of IEEE Region 10, Hyderabad, pp. 1–6, 18–21 Nov 2008
7. Chakraborty, M., Roy Chatterjee, S., Ray, S.: Performance evaluation of nash bargaining power sharing algorithm for integrated cellular phone system. In: 2016 IEEE Uttar Pradesh Section International Conference on Electrical, Computer and Electronics Engineering (UPCON), pp. 125–131, 9–11 Dec 2016
8. Lokesh, J., Munivef, E.: Design of robust and secure encryption scheme for WSN using PKI (LWT-PKI). In: First International Communication Systems and Networks and Workshops, (COMSNETS '09), pp. 1–2 (2009)
9. Chungen Xu, X., Yanhong, G.: The public key encryption to improve the security on wireless sensor networks. In: Second International Conference on Information and Computing Science (ICIC '09), vol. 1, pp. 11–14, May 2009
10. Cakroglu, M., Bayilmis, C., Ozcerit, A.T., Cetin, O.: Performance evaluation of scalable encryption algorithm for wireless sensor networks. Sci. Res. Essay **5**(9), 856–861 (2010)
11. Ganesh, A.R.: An improved AES-ECC hybrid encryption scheme for secure communication in cooperative diversity based wireless sensor networks. In: International Conference on Recent Trends in Information Technology (ICRTIT '2011), pp. 1209–1214. India (2011)
12. Kumar, G., Rai, M., Lee, G.: Implementation of cipher block chaining in wireless sensor networks for security enhancement. Int. J. Secur. Appl. **6** (2012)
13. Patila, P., Narayankar, P., Narayan, D.G., Meena, S. Md.: A comprehensive evaluation of cryptographic algorithms: DES,3DES, AES, RSA and Blowfish. In: International Conference on Information Security & Privacy (ICISP2015). Nagpur, India, 11–12 Dec 2015
14. Manju Suresh, A., Neema, M.B.: Hardware implementation of blowfish algorithm for the secure data transmission in internet of things. In: Global Colloquium in Recent Advancement and Effectual Researches in Engineering, Science and Technology (RAEREST 2016)
15. Schneier, B.: Description of a new variable-length key, 64-bit block cipher (Blowfish). In: Fast Software Encryption, Cambridge Security Workshop Proceedings, Dec 1993
16. Nadeem: Performance comparison of data encryption algorithms. In: First International Conference on Information and Communication Technology, pp. 84–89, 27–28 Aug 2005
17. Roy Chatterjee, S., Majumder, S., Pramanik, B., Chakraborty, M.: FPGA implementation of pipelined blowfish algorithm. In.: IEEE 2014 Fifth International Symposium on Electronic System Design, 15–17 Dec 2014
18. Heys, H.M., Tavares, S.E.: Avalanche characteristics of substitution-permutation encryption networks. IEEE Trans. Comput. **44**, 1131–1139 (1995)
19. Webster, A.F., Tavares, S.E. (eds.): On the design of S-boxes. In: Advances in Cryptology-CRYPTO '85 Proceedings, pp. 523–534. Springer, Berlin, Heidelberg (1986)

Part VI
Session 2C: Artificial Intelligence

Bilingual Machine Translation: English to Bengali

Sauvik Bal, Supriyo Mahanta, Lopa Mandal and Ranjan Parekh

Abstract The present work proposes a methodology of machine translation system which takes English sentences as input and produces appropriate Bengali sentences as output using natural language processing (NLP) techniques. It first uses a parse tree for syntactic analysis of the sentence structure and then applies semantic analysis for extracting the meaning of the words. An inverse function is then provided to fit that into the Bengali syntax. A dictionary as a separate file is used for mapping between the English words and their Bengali counterparts. The novelty of the present work lies in the fact that it combines both a syntax-based and a meaning-based analysis to arrive at the proper translation. The effectiveness of the algorithm has been demonstrated with examples of different English sentence conversions with several rules, and the results have been compared with that of the Google translator to show the improvements achieved.

Keywords POS tagging · Machine translation · Parse tree · Rule-based system

S. Bal (✉) · S. Mahanta (✉)
University of Engineering & Management, Jaipur, India
e-mail: sauvikbal@gmail.com

S. Mahanta
e-mail: mahantasupriyo@gmail.com

L. Mandal
Institute of Engineering & Management, Kolkata, India
e-mail: mandal.lopa@gmail.com

R. Parekh
Jadavpur University, Kolkata, India
e-mail: ranjan_parekh@yahoo.com

M. Chakraborty et al. (eds.), *Proceedings of International Ethical Hacking Conference 2018*, Advances in Intelligent Systems and Computing 811,
https://doi.org/10.1007/978-981-13-1544-2_21

1 Introduction

Language translation is one of the important applications in the present scenario as today's world is considered to be a global village. If a person has to move from one location to another and is not aware of the regional language of that location, it would be very difficult for him/her to communicate. Not only it is relevant in a global scenario where multiple languages come into consideration, but also in a local setting where two or more neighboring countries might share the same language with similar/dissimilar dialects. For example, in India and Bangladesh, many people use Bengali as their mother tongue though with different dialects. All these make machine translation to be an important area of research. The present work aims to translate a worldwide used language viz. English into a regional language viz. Bengali. The main challenge of language translation is that often a simple mapping between words does not produce expected results. Restructuring of the sentences as well as analysis of the inherent meaning is also necessary for correct outputs. In the existing process, there are so many sentences where the translation does not give meaningful output due to problem of proper analysis of sentences, lack of resources etc. The present work proposed a novel approach where English to Bengali language conversion is done based on some grammatical rules. The proposed work is based on the version of the language used by the Bengali people of West Bengal, India.

2 Literature Survey

A good translator system should contain all words and their corresponding translated words. The main problem of this kind of system is limited available vocabulary. Fuzzy-If-Then-Rule is one of the frequently used methodologies for machine translation [1]. In the process of translation from one language to another, there are some challenges like, lack of resources, different tools, pronunciation dictionary, different language modeling, dialog modeling, content summarization etc. [2]. More research is required to increase the accuracy rate when translation is done in case of low resource languages and in the cases where the volume of target language vocabulary is limited [3]. Another approach of language translation is based on the tense where English sentences can be used as input. This kind of system uses context free grammars for analyzing the syntactical structure of the input which helps to translate the sentence and verify the accuracy of the output [4]. Machine translation may be achieved by deep learning-based neural network (DNN). Memory-augmented neural network is introduced with this mechanism where the memory structure does not consist of any phrase [5]. Another method of machine translation is to retrieve by audio analysis and feature extraction. This kind of process can solve the ambiguity problem in sentence translation to improve the output [6]. Another approach is used for translation where values from the New Testament were used as training values. If the proper resources are not available and the machine is not properly trained, accuracy rate will be decreased [7]. Example-based machine translation is found to be another methodology, used in this case. The problem of this methodology is limited knowledge base. It makes the system inefficient for translation where low-resourced language is used [8]. Machine translation is also important for question–answering sessions. The main problem for this type of system is word ambiguity. By using the matrix factorization, this can be improved. If there are dynamic question–answering sessions, large vocabulary and proper learning would be required for accuracy [9]. For speech to test conversion, machine translation is also important. If the speech is in different language, it is important to have the proper resources for translation. This kind of system extracts the meaning of input sentence. So, proper decision-making algorithm and proper training is needed [10, 11]. Deep learning is one of the important concepts for natural language processing. For language translation, it is important to choose the right decision. Based on the past experience, training can be done and system can take the proper decision by using the concept of deep learning [12]. Sometimes language translation efficiency is reduced when phrase-based translation is required for long sentences. Sequence of the words in inputted language may differ with output language. So, rule-based system is required to improve the translation quality [13]. If there is any complex sentence, tree-based method can be applied for simplification. So, the splitting and decision making should be proper for accurate language translation [14]. If there is any sentence with complicated structure, the parse tree may not be created properly. So, it is very important to generate parse tree, so that the translation can be done efficiently [15]. At the time of machine translation, it is very important to detect sub phrase as well as clause detection. If there is any error in clause detection, the translation may not be done properly [16, 17]. Parsing-based sentence simplification is one of the methods where keywords can be extracted. This process follows dependency-based parsing technique [18].

The study of related works shows that, due to the lack of resources, tools, vocabularies, it is not always possible to translate the English sentence into regional language by using the existing methodologies. If the translation is not properly done, the meaning of the translated statement may not be appropriate. The main reason of this problem is improper analysis of the sentences. Generally, in existing systems, some general rules are applied that fails to do the proper conversion in some cases, e.g., if the first letter of some name is given in small letter, the output of existing system drastically changes. This is one of the major drawbacks of existing system. Parts of Speech (POS) tagging does not work properly in these cases. It shows that priority should be given to make the translation system intelligent enough to analyze of the sentences properly.

3 Methodology

The present work proposes a novel methodology of English to Bengali text translation. Here, an English text or sentence or a paragraph is used as input and the system generates its appropriate Bengali meaning. So, first of all, the English text is taken as input to the system. Then, the sentence is broken into words and then by using the Parts of Speech (POS) Tagger, it retrieves the Parts of Speech of each word. Then, the words are clustered into three groups, i.e.—Subject, Verb, Object, and some other required parts (e.g., WH-words, exclamatory expression etc.). After that, the parse tree is generated for English text and converted into the parse tree of Bengali language by using different Bengali grammatical rules [19]. Here, a separate file is used as database where the English word and the respective Bengali meanings are stored. After judging the syntactical structure of the sentence, the appropriate Bengali words are selected and used. Finally, the output of the system is generated in Bengali language. The proposed system is shown with the help of a bock diagram in Fig. 1.

In the present work, the types of sentences taken as input are shown in Fig. 2.

Here, two examples of assertive and interrogative sentences are taken and demonstrated how they actually work.

Assertive Sentence: First recognize the pattern of the input sentence.

In English, the pattern is: Sub + Verb + Obj

e.g., "I am going to school."

I (sub)	am going (verb)	to school (obj)

Now, as per the Bengali grammar [19], reconstruct as the pattern: Sub + Obj + Verb

I (sub)	school to (obj)	am going (verb)

Fetch corresponding Bengali words.

Interrogative Sentence: First recognize the pattern of the sentence.

e.g., What is the capital of India?

So, the pattern in English is: "wh" word + obj + Sub

What (sub)	is the capital (obj)	of India (Sub)

Now the pattern as per the Bengali language is reconstructed.

So, pattern in Bengali is: sub + obj + "wh" word

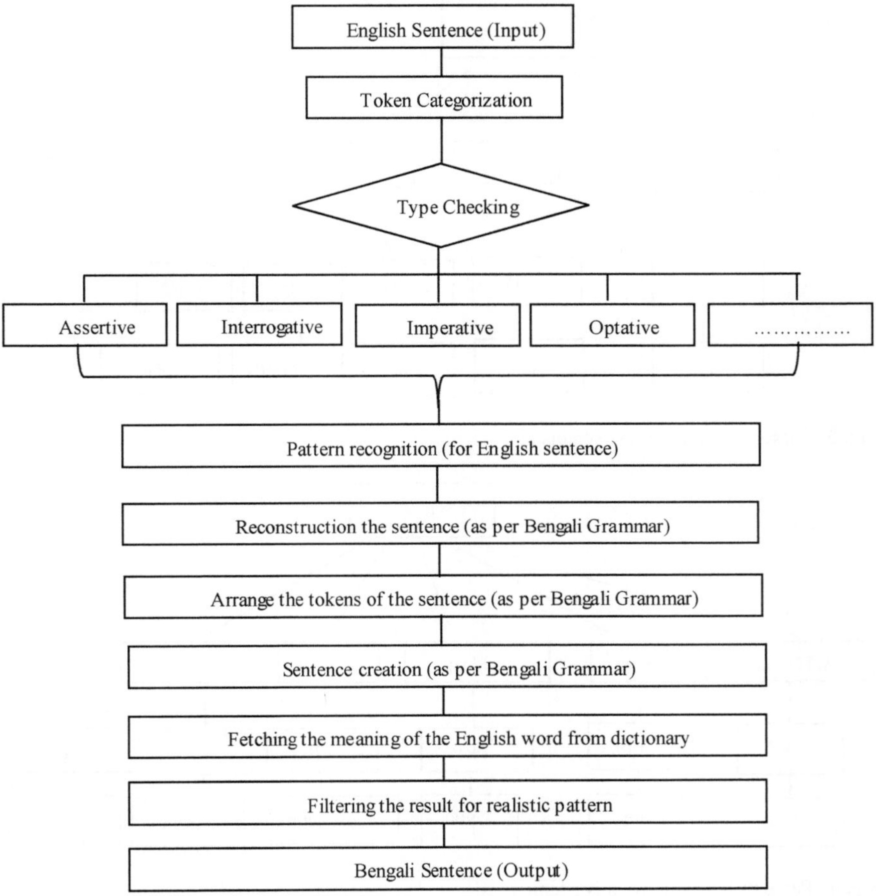

Fig. 1 Block diagram of the proposed system

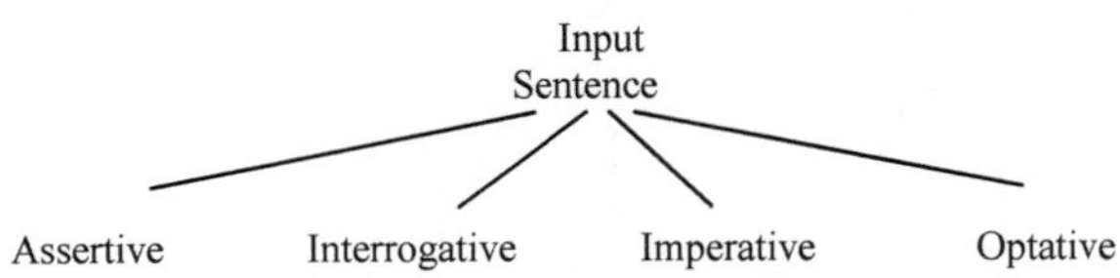

Fig. 2 Different types of sentences considered in the present work

Parse Tree: Here, assertive sentence is taken as an input. Then, analyze the sentence as per the following parse tree (Figs. 3, 4, 5, and 6) [20].

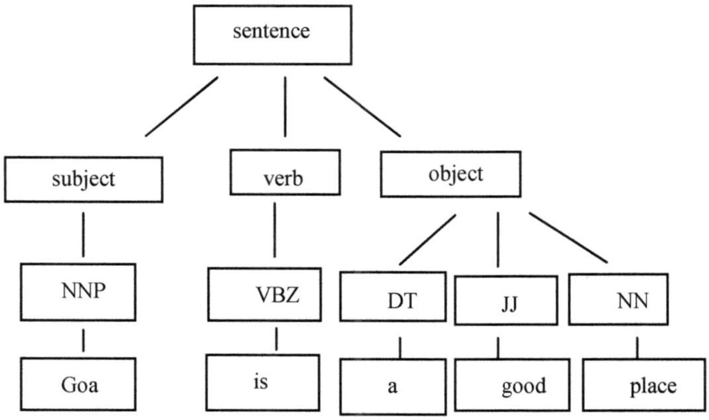

Fig. 3 Parse tree for assertive sentences

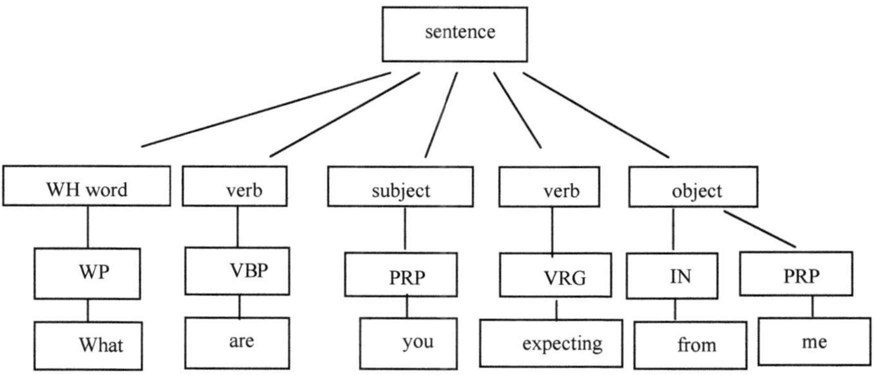

Fig. 4 Parse tree for interrogative sentences

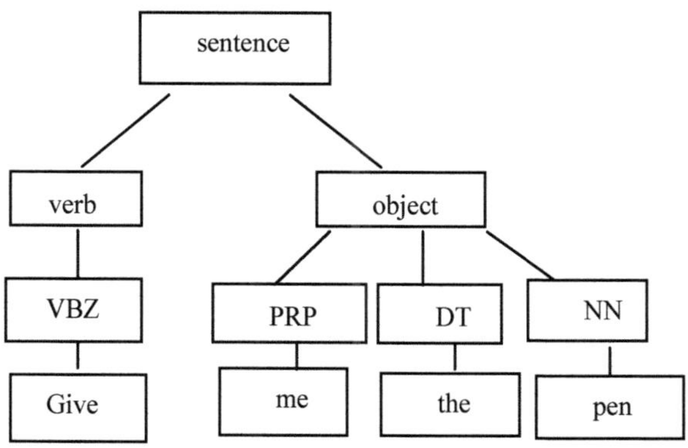

Fig. 5 Parse tree for imperative sentences

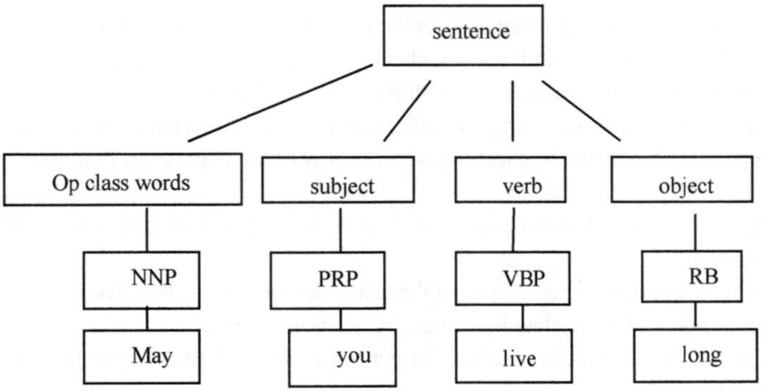

Fig. 6 Parse tree for optative sentences

4 Proposed Algorithm

The proposed algorithm for English to Bengali Translation is based on English Grammar and Bengali Grammar.

Algorithm 1:

Procedure;
Begin:

1. Take the English text as input.
2. Extract the words from the text and put it in a different "Word Class" (i.e: The set of words which are getting from that input text).
3. Extract the Parts of Speech of each word using the POS (Parts of Speech) Tagger.
4. Save them into different "Token Class" (i.e: The set of tokens of the words of that sentence).
5. Create a parse tree based on Subject, Verb and Object and by checking the syntactic structure.
6. If the sentence structure is "Subject + verb + object/ adverb/ complement/adjective", it will mark as Assertive Sentence.
7. If the sentence starting with the help of the verbs (am, are, was, had, were, is, have, has) or modal auxiliaries (shall, should, will, would, can, could, may, might, etc.), it will mark as Interrogative sentence.
8. If the pattern is "Subject (Invisible) + verb + object / where", mark as Imperative Sentence.
9. If the pattern of the sentence is "May + Assertive", It will mark as Optative Sentence
10. Construct the rough structure of the sentence using tokens and words.

11. After creating the Pattern of tokens, set the flag values (i.e. is it a n Auxiliary verb ? If y es, then AU X=True then check whether the prefix verb is in correct form, whic h person does the word signify? if it is in First Person, then First_Per =True, etc.). It triggers different property of the Bengali Grammar and provided classifiers for checking of the syntax of the expression grammat-ically.
12. Analyze the type of the sentence (e.g. Assertive Sentences, Interrogative Sen-tences, etc.).
13. Construct the pattern for each type of the sentences, and check the given sen-tence and put it into that class by using a "C hoice Value".
14. Use the "Dictionary" as da tabase for getting the Bengali meaning of the Eng-lish words .
15. "eng" column contains the English words and "beng" column contains the cor-responding row's Bengali meaning including its type (i.e. verb, adjective, noun , etc), tense (i.e. Present, Past, etc.), Person(1st,2nd ,3rd), etc.
16. After setting the flag values and getting accurate sentence of words and tokens as Bengali meaning, fetch the Bengali meaning as per its English word from the Dictionary.
17. Make the s ynta ctic changes using some rules for better and more realistic re-sults.
18. After fetching the meaning of Bengali words put them into a Bengali sentence as per Token and Word classes.
19. Bengali sentence is the output.

End Procedure;

---

5 Experimentations and Result Analysis

A machine translation system has been implemented as per the algorithm stated in Sect. 4. The system has been tested and compared with that of Google Translate. Here, some test cases are shown where the proposed system results (refer Figs. 8, 10, 12, 14) are better than the respective Google translator results (refer Figs. 7, 9, 11, 13).

It is examined that some of the result of Google translator is not proper and not actually matched with the standard result [20]. The present work compared the output of the Google translator and that of the proposed system on the basis of the Bengali version used by the Bengali people of West Bengal, India. For analysis the sequence of the output, here class "difflib.SequenceMatcher" is used in Python language. There is another class "difflib.Differ", by which the difference between sequences can be identified [21, 22]. Different test cases of input, expected output and generated output of Google translator and that of the proposed system is shown in Fig. 15. For the first example, "after high school, sauvik moves to Jaipur," the Google translator result is 46.51% matched with the expected output, whereas the proposed system matched

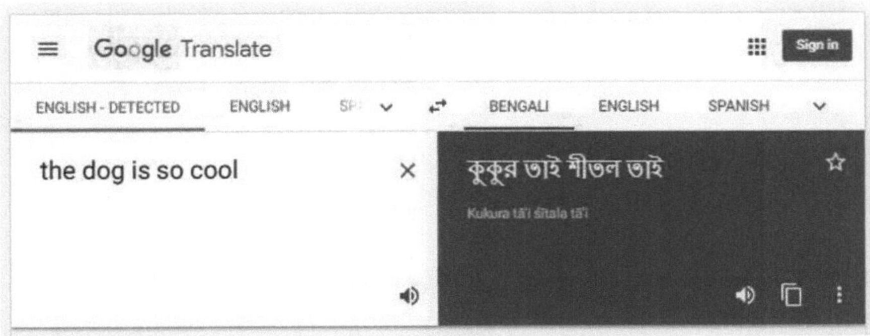

Fig. 7 Sample output of Google translator of the sentence “the dog is so cool”

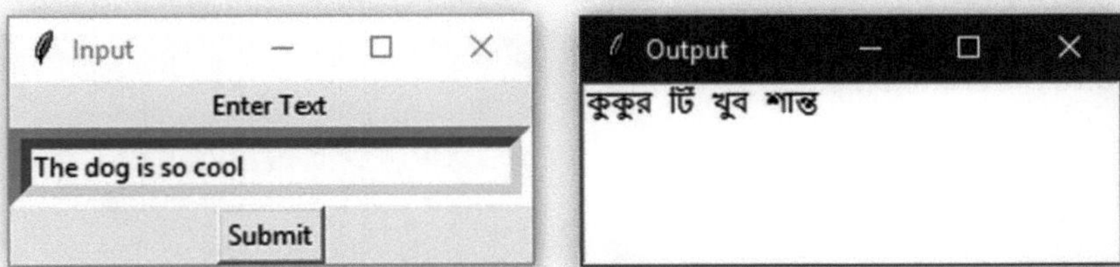

Fig. 8 Sample output of proposed system of the sentence “the dog is so cool”

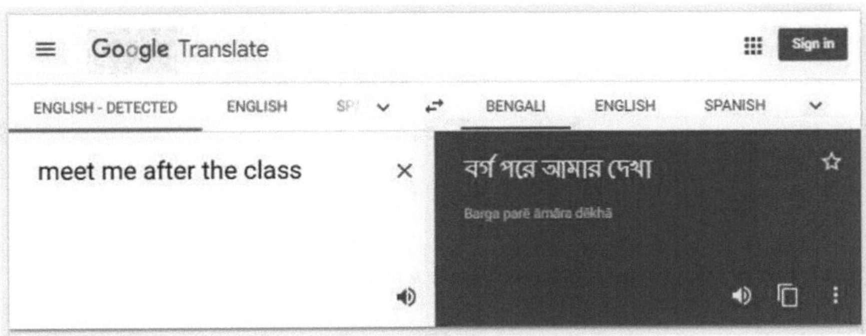

Fig. 9 Sample output of Google translator of the sentence “meet me after the class”

81.31%. This measurement is done on the basis of sequential string matching. In the second example, “I am going to farmland,” Google translator result is 91.89% matching. The proposed system has 95.23%. In the third example, “meet me after the class,” Google translator result is 47.82% and the proposed system has 91.22% matching. In the fourth example, “give me some time,” Google translator result is 76.92% matched and the proposed system has 100% matching. In the fifth example, “the dog is so cool,” Google translator has 50% and the proposed system has 88.88%

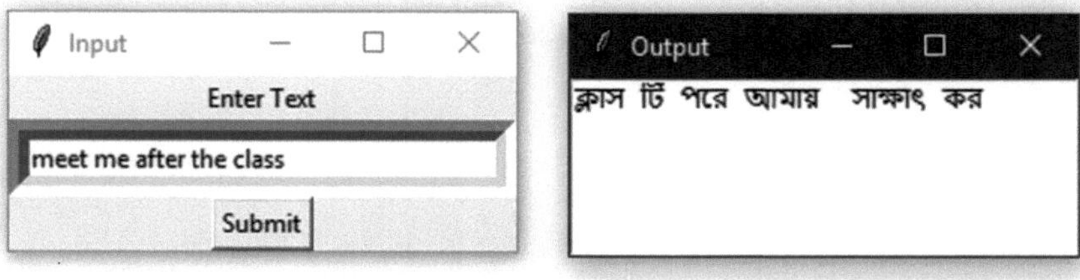

Fig. 10 Sample output of proposed system of the sentence "meet me after the class"

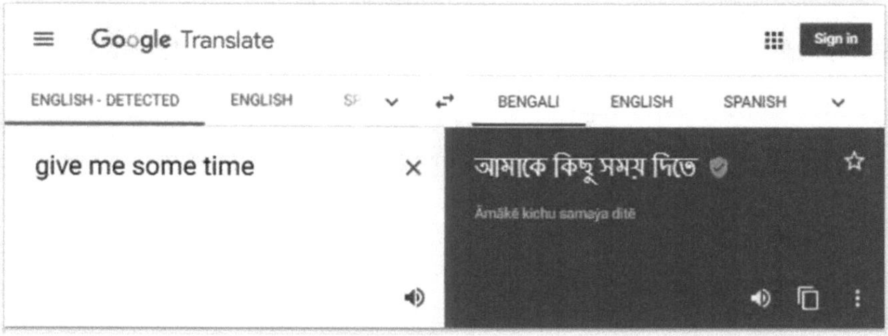

Fig. 11 Sample output of Google translator of the sentence "give me some time"

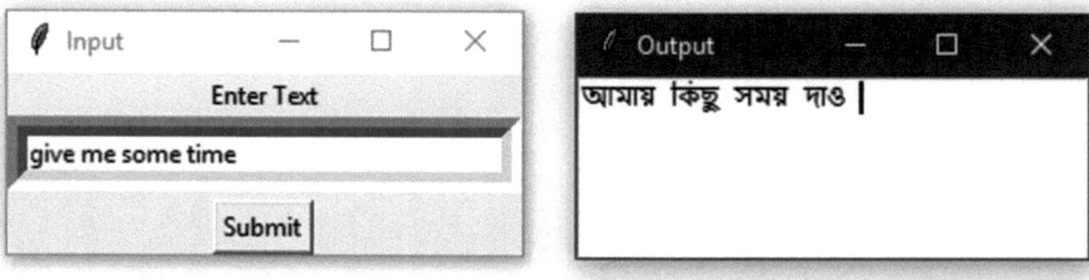

Fig. 12 Sample output of proposed system of the sentence "give me some time"

matching. In the sixth example, "What are you expecting from me," Google translator has 92.59% and the proposed system has 100% matching. In seventh example, "I am going to school," Google translator has 73.68% and the proposed system has 93.33% matching, shown in Fig. 16.

So, based on the above data, the analysis result is shown in Fig. 16.

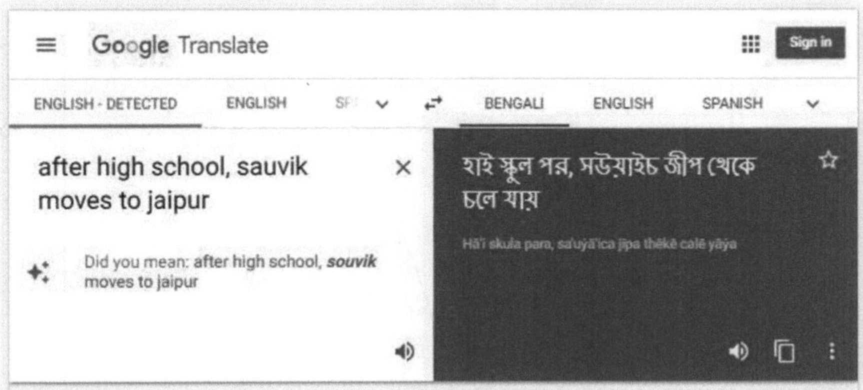

Fig. 13 Sample output of Google translator of the sentence "after high school, sauvik moves to jaipur"

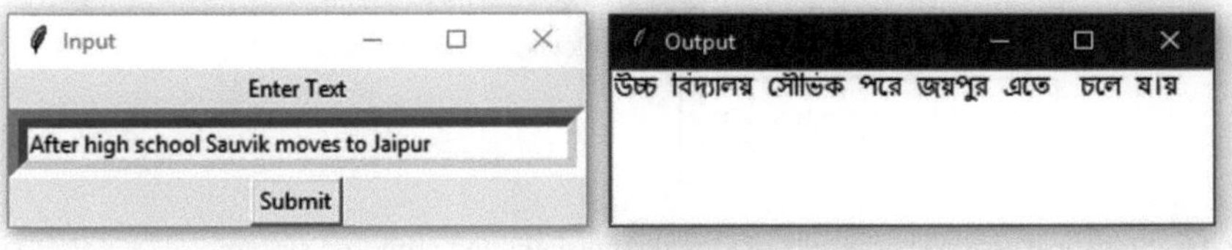

Fig. 14 Sample output of proposed system of the sentence "after high school, sauvik moves to jaipur"

	Input(English)	Standard Statement	Output(Google translator)	Output(Proposed System)
1	the dog is so cool	কুকুরটা খুব শান্ত	কুকুর তাই শীতল তাই	কুকুর টি খুব শান্ত
2	meet me after the class	ক্লাসের পরে আমায় সাক্ষাৎ কর	বর্গ পরে আমার দেখা	ক্লাস টি পরে আমায় সাক্ষাৎ কর
3	give me some time	আমায় কিছু সময় দাও	আমাকে কিছু সময় দিতে	আমায় কিছু সময় দাও
4	after high school, sauvik moves to jaipur	উচ্চ বিদ্যালয় শেষের পরে সৌভিক জয়পুরে চলে যায়	হাই স্কুল পর, সউয়াইচ জীপ থেকে চলে যায়	উচ্চ বিদ্যালয় সৌভিক পরে জয়পুর এতে চলে যায়
5	I am going to farmland	আমি কৃষিজমিতে যাচ্ছি	আমি কৃষিতে যাচ্ছি	আমি কৃষিজমি এতে যাচ্ছি
6	What are you expecting from me	তুমি আমার থেকে কি আশা করছ	তুমি আমার কাছ থেকে কি আশা করছ	তুমি আমার থেকে কি আশা করছ
7	I am going to school	আমি বিদ্যালয়ে যাচ্ছি	আমি স্কুলে যাচ্ছি	আমি বিদ্যালয় এতে যাচ্ছি

Fig. 15 Output comparison with Google translator

6 Conclusions and Future Scopes

The present work proposed a novel approach of translating English sentences into Bengali. This helps people know more about content written in English language. In many cases, the existing machine translation systems fail to translate properly because of improper analysis of sentences. The proposed methodology improves analysis of sentence and translates it in the way how Bengali people of India use it. The present work proposed an algorithm to translate Assertive , Interrogative and

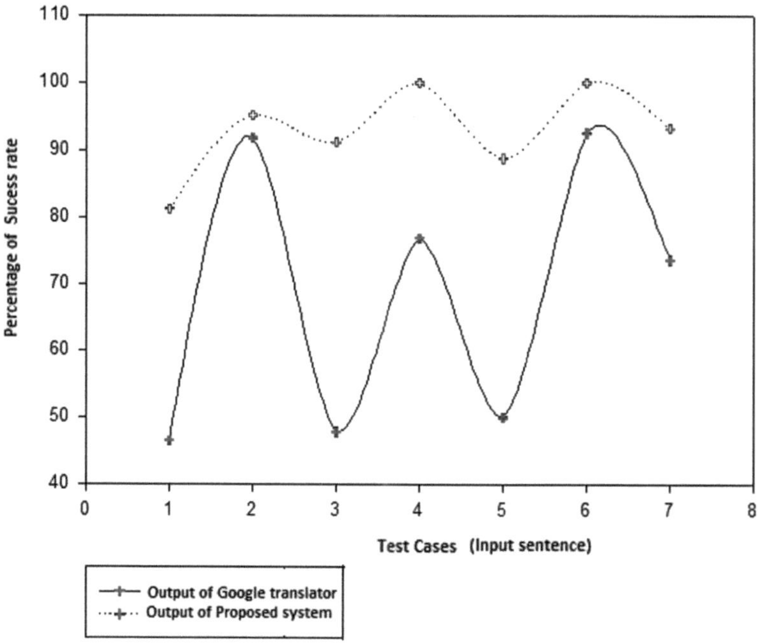

Fig. 16 Statistical analysis of results of Google translator and proposed system

Imperative English sentences by analyzing their sentence pattern. The work will be extended further for the other types of sentences in future. It can be improved to be a bidirectional translator, i.e., Bengali to English and vice versa, as well.

References

1. Adak, C.: An advanced approach for rule based English to Bengali machine translation. J. Nat. Conf. Adv. Comput. Eng. Res. 01–09 (2013)
2. Fung, P., Schultz, T.: Multilingual spoken language processing [Challenges for multilingual systems]. IEEE Signal Process. Mag. (2008)
3. Lee, H.-Y., Lee, L.-S.: Improved semantic retrieval of spoken content by document/query expansion with random walk over acoustic similarity graphs. IEEE/ACM Trans. Audio Speech Lang. Process. **22**(1) (2014)
4. Muntarina, K.Md., Moazzam, G.Md., Bhuiyan, A.-A.: Tense based english to Bangla translation using MT system. Int. J. Eng. Sci. Invent. ISSN (Online): 2319–6734, ISSN (Print): 2319–6726
5. Zhang, S., Mahmut, G., Wang, D., Hamdulla, A.: Memory-augmented Chinese-Uyghur neural machine translation, APSIPA ASC (2017)
6. Deena, S., Ng, R.W.M., Madhyastha, P., Specia, L., Hain, T.: Exploring the use of acoustic embeddings in neural machine translation, (ASRU). IEEE (2017)

7. Macabante, D.G., Tambanillo, J.C., Cruz, A.D., Ellema, N., Octaviano Jr. M., Rodriguez, R., Edita Roxas, R.: Bi-directional English-Hiligaynon statistical machine translation, (TENCON). Malaysia, Nov 5–8 2017
8. Anwarus Salam, K.Md., Khan, M., Nishino, T.: Example based English-Bengali machine translation using WordNet. https://pdfs.semanticscholar.org/3caf/770def3e398b7ca5396e9f79aa6bbab1cc6b.pdf?_ga=2.147250417.779477973.1533393833-37296364.1533231961
9. Zhou, G., Xie, Z., He, T., Zhao, J., Hu, X.T.: Learning the multilingual translation representations for question retrieval in community question answering via non-negative matrix factorization. IEEE Trans. Audio Speech Lang. Process. **24**(7) (2016)
10. Ghadage, Y.H., Shelke, S.D.: Speech to text conversion for multilingual languages. In: International Conference on Communication and Signal Processing. India, 6–8 April 2016
11. Lee, L., Glass, J., Lee, H., Chan, C.: Spoken content retrieval—beyond cascading speech recognition with text retrieval. IEEE/ACM Trans. Audio Speech Lang. Process. **23**(9) (2015)
12. Young, T., Hazarika, D., Poria, S., Cambri, E.: Recent Trends in Deep Learning Based Natural Language Processing. [cs.CL], 16 Aug 2017
13. Hung, B.T., Minh, N.L., Shimazu, A.: Sentence splitting for vietnamese-english machine translation. In: Fourth International Conference on Knowledge and Systems Engineering (2012)
14. Zhu, Z., Bernhard, D., Gurevych, I.: A monolingual tree-based translation model for sentence simplification. In: Proceedings of the 23rd International Conference on Computational Linguistics. Coling (2010)
15. Xiong, H., Xu, W., Mi, H., Liu, Y., Lu, Q.: Sub-sentence division for tree-based machine translation. In: Proceedings of the 47th Annual Meeting of the ACL and the 4th IJCNLP, Short Papers, pp. 137–140. Singapore (2009)
16. Thenmozhi, D., Aravindan, C.: Paraphrase identification by using clause-based similarity features and machine translation metrics. Comput. J. Adv. Access published 5 Oct 2015
17. Kavirajan, B., Anand Kumar, M., Soman, K.P., Rajendran, S., Vaithehi, S.: Improving the rule based machine translation system using sentence simplification (English to Tamil). Int. J. Comput. Appl. (0975–8887) **25**(8) (2011)
18. Tur, G., Hakkani-Tur, D., Heck, L., Parthasarathy, S.: Sentence Simplification for Spoken Language Understanding, Acoustics, Speech and Signal Processing (ICASSP), 2011 IEEE International Conference, 12 July 2011
19. Mitra, M., Mitra, D.: Oxford English-English-Bengali Dictionary, Oxford University Press. ISBN: 9780195689648
20. Basu, S.: English Tutor, Class VI, chhayaprakashanipvt ltd
21. https://docs.python.org/3/library/difflib.html
22. http://pyxr.sourceforge.net/PyXR/c/python24/lib/difflib.py.html#0602

Comparison of Different Classification Techniques Using Different Datasets

Nitesh Kumar, Souvik Mitra, Madhurima Bhattacharjee and Lopa Mandal

Abstract Big data analytics is considered to be the future of information technology in today's world, which incorporates data mining to be one of its most promising tool. The present work illustrates a comparative study to find out which kind of classifiers work better with which kind of datasets. It illustrates comparisons of the efficiency of the different classifiers focusing on numeric and text data. Datasets from IMDb and 20newsgroups have been used for the purpose. Current work mainly focuses on comparing different algorithms such as Decision Stump, Decision Table, K-Star, REPTree and ZeroR in the area of numeric classification, and evaluation of the efficiency of Naive Bayes classifier for text classification. The result in this paper suggests the best and worst of the test parameters, as it widens the scope of their usage on the basis of types and the size of datasets.

Keywords Data mining · Text classification · Numeric data classification Classifier algorithms

1 Introduction

Data mining is the work of gathering knowledge from a dataset. Classification of large datasets is one of the most necessary tasks of big data analytics and data mining. There are different types of datasets according to their data types such as numeric,

N. Kumar (✉)
Tata Consultancy Services Limited, Bengaluru, India
e-mail: niteshkr1126@gmail.com

S. Mitra (✉) · M. Bhattacharjee · L. Mandal
Institute of Engineering & Management, Kolkata, India
e-mail: mitra.souvik97@gmail.com

M. Bhattacharjee
e-mail: madhurimabhattacharya95@gmail.com

L. Mandal
e-mail: mandal.lopa@gmail.com

M. Chakraborty et al. (eds.), *Proceedings of International Ethical Hacking Conference 2018*, Advances in Intelligent Systems and Computing 811,
https://doi.org/10.1007/978-981-13-1544-2_22

string or text type. Classification algorithms such as Naive Bayes, Decision Stump, Decision Table, KStar, REPTree and ZeroR have been used for classification in the present work. For a better performance in different sectors of data mining, evaluation and comparative study on efficiency of different classification algorithms are needed. The present work attempts to accomplish this, by taking dataset of different size and type and applying the relevant algorithms on them. This can provide a wide idea to a data mining work; for example, what kind of algorithm is best for a given dataset (e.g., student performance evaluation for educational institutes uses large numeric datasets, and classifying text documents based on their content uses huge text data).

2 Literature Survey

2.1 Numeric Classification

Different tasks of data mining require different kinds of algorithms. Many researchers have contributed in this field and presented their views in case of usage and efficiency of classification algorithms. In 2007, numeric and string dataset on "Breast Cancer" was used to compare the algorithms J48, Multilayer Perceptron and Bayes Network [1]. The best algorithm from this experiment was reported to be Bayes Network. The proposed future work of this research paper was to evaluate the same on more datasets.

Research work of Deepajothi and Selvarajan in 2012 used numeric- and string-type dataset: "Adult Set." It consists of records of people's income, age, education and other related attributes [2]. The algorithms used were Naïve Bayes, KStar, Random Forest and ZeroR. The best algorithm in this work was found to be Naive Bayes.

Vaithiyanathan et al. in 2013 used Naive Bayes Updatable, J48, Multilayer Perceptron and Bayes-Net on the numeric datasets [3]. From the comparison, the best algorithm was found to be Naive Bayes Updatable, although the time efficiency was not satisfactory according to them.

Another research from 2014 used the algorithms, viz. Naive Bayes, J48, REPTree and Random Tree algorithm [4]. From the experiment of this work, the best algorithm is Random Tree algorithm. The future work of this research paper was proposed to use other classification algorithms and apply the same for different domains.

Another research work done in 2015 used the numeric-type dataset, "Heart risk" having 1865 instances of patient ID and their diseases [5]. The algorithms used are Naive Bayes, J48, REPTree, Random Tree. It has been concluded from the results obtained that Random Tree algorithm is most appropriate for this numeric dataset to classify instances. Random Tree gives 92% accuracy which is relatively higher than the other algorithms.

Raheja et al. in their research on analysis of Linux Kernel Vulnerabilities have explored the last 10 years of Linux kernel Vulnerabilities based on the Common Vulnerabilities Exposures (CVE) [6]. CVE are a dictionary of common names. Here, the algorithms used are Decision Tree-based algorithms, Bayesian classifiers, Instance-based classifiers (K-Nearest Neighbor, KStar). From this experiment, Bayesian, Instance-based and Tree-based ones are found to be giving maximum accuracy for integrity parameter.

An assessment was done using Decision Tree on an educational dataset in 2017 [7]. Dataset from a reputed college was used, consisting of student's name, registration number, marks and grades. Algorithms used are Random Forest, REPTree, Random Tree, Hoeffding, Decision Stump and J48. The Decision Tree algorithms were found to perform well with student's dataset.

The research work by Mettildha et al. [8] used the benchmark dataset of numeric type from Cleveland heart disease containing 14 attributes; for example, the algorithms used are Particle Swarm Optimization (PSO), Bat, Multinomial logistic regression (MLR) and Support Vector Machine (SVM). The optimization models used are feature reduction PSO and Bat. Using the classifiers SVM and MLR, The Bat-SVM model performed best with prediction accuracy of 97%.

2.2 *Text Classification*

A vast number of the algorithms such as Naive Bayes, SVM and Neural Network have been used, and their efficiency can be found out in the works of Ikonomakis et al. [9] and Kotsiantis et al. [10].

In 2011, Ting et al. [11] conducted an experiment to find whether Naïve Bayes is a good classifier for the text-type datasets containing 4000 instances. The experimental result showed that Naive Bayes performed best against algorithms such as Decision Tree, Neural Network and Support Vector Machines.

In 2012, Wahbeh and Al-Kabi [12] used classifiers, viz. SMO, Naive Bayes and J48 on Arabic language works consisting of 1000 text documents from different sectors. Here, Naive Bayes was found to generate the most accurate result.

Purohit A. et al. in their research work in 2015 [13] used 40 abstracts from different research works as dataset for text classification. Association Rule-based Decision Tree and Naive Bayes text classifier were compared to their proposed algorithm. The proposed algorithms used Porter stemmer, Apriori algorithm, Association Rule, Naive Bayes classification which could classify text at 75% accuracy rate.

Rajeswari R. P. et al. in their research work [14] compared Naive Bayes and K-Nearest Neighbor (KNN) classification algorithm. Students' information dataset had been used here containing USN, age, course name, etc. Naive Bayes performed with 66.7% accuracy, whereas KNN had 38.89% of accuracy.

Tilve and Jain K. et al. in 2017 [15] used 20newsgroups and new newsgroup datasets to compare Naive Bayes, Vector Space Model (VSM) and proposed algorithm Stanford tagger. Datasets contained 50 categories of news text data. It was observed that Naive Bayes worked well with both datasets, whereas VSM worked better with new newsgroup dataset as it is small in size. Some researchers also used hybrid classifiers to increase the overall efficiency of the system [16].

From the above survey, it is found that in both types of datasets (numeric and text) mostly used algorithms are Naive Bayes, J48, REPTree, Decision Tree, etc., where Naive Bayes and Decision Tree work well. Along with some of these algorithms, the future work will be to use some other classification algorithms such as REPTree, KStar, Decision Stump and ZeroR to determine how these classification algorithms work in large datasets of numeric or string type and also to evaluate how efficiently Naive Bayes Multinomial algorithm works for text classification for these small or large datasets.

3 Methodology

3.1 Numeric Data Classification

A large numeric dataset from IMDb has been taken. For the purpose of classification, the present work uses Decision Stump, Decision Table, KStar, REPTree and ZeroR algorithms. In each case, tenfold cross-validation has been applied. The results from each classifier are noted and compared with each other to gain a better knowledge of the efficiency of the algorithms.

3.2 Text Classification Process

The steps of text classification start with document collection such as .pdf, .html, .doc. Commonly, the next steps taken are:

Tokenization: A document is treated as a string and then partitioned into a list of tokens.

Removing stop words: Stop words such as "the," "a," "and", frequently occurring, need to be removed.

Stemming word: Stemming algorithm converts different word form into similar canonical form.

4 Experiments and Results

The present work focuses on numeric and text classification. WEKA [17] is used as the experimental tool for both the cases.

For numeric classification: IMDb numeric dataset has been used [18].

For text classification: IMDb and 20newsgroups datasets have been used [19].

4.1 *Numeric Classification*

4.1.1 Experimental Setup

For numeric dataset, the experiment has been done by loading a large numeric dataset taken from IMDb in Weka and applying the algorithms Decision Stump, Decision Table, KStar, REPTree and ZeroR to the same dataset.

The experimental dataset contains average rating and number of votes regarding a movie or show. Here, 10,000 records have been taken into 10 folds of cross-validation for the comparison between Decision Stump, Decision Table, KStar, REPTree and ZeroR. Dataset is converted into ARFF format using WEKA ARFF viewer, and then one by one the algorithms have been applied to the dataset to find comparative results.

4.1.2 Experimental Results

Table 1 shows the test summaries of the algorithms taken in the experiment, and Fig. 1 shows the graphical representation of the data given in Table 1.

Table 2 and Fig. 2 show the comparison of time taken to build model (i.e., learning time) and classification time (excluding the model building time) for each algorithm taken in the experiment. An important observation of the experiment is that Decision Table took exceptionally large length of time to build the model, whereas the model building time of KStar and ZeroR is close to zero seconds. It is also observed that classification time of the KStar algorithm is exceptionally high in comparison with all the other algorithms taken in the experiment.

It has been observed that KStar has the strongest agreement with correlation coefficient of 0.2942 with the least relative absolute error of 93.77%. REPTree and ZeroR failed to perform well with 100% of relative absolute error and negative correlation coefficient. So, accuracy of KStar is comparatively good among other algorithms. Classification time (time taken to classify the dataset after building the model) depends on the size of dataset which has to be classified.

Table 1 Comparison of different algorithms on tenfold cross-validation

	Decision stump	Decision table	KStar	REPTree	ZeroR
Correlation coefficient	0.0986	0.0986	0.2942	−0.0073	−0.0073
Mean absolute error	0.7831	0.7831	0.739	0.7881	0.7881
Root mean squared error	1.0012	1.0012	0.9617	1.0061	1.0061
Relative absolute error (%)	99.37	99.37	93.77	100	100
Root relative squared error (%)	99.51	99.51	95.59	100	100

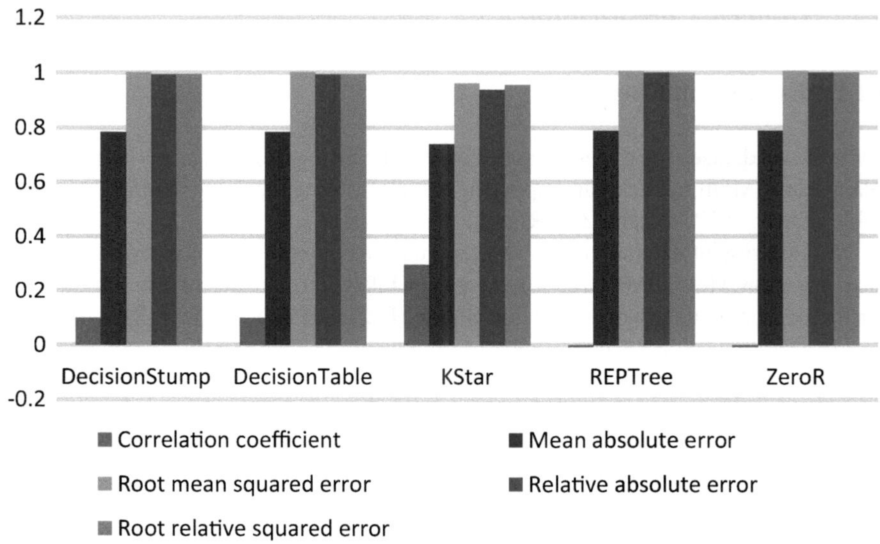

Fig. 1 Comparison of performance for various classifiers

Table 2 Model building time and classification time for different classifiers

Name of the classifier	Time taken to build model	Classification time (excluding model building time)
Decision stump	0.06	0.94
Decision table	11.77	16.23
KStar	0	79
REPTree	0.05	0.95
ZeroR	0	0

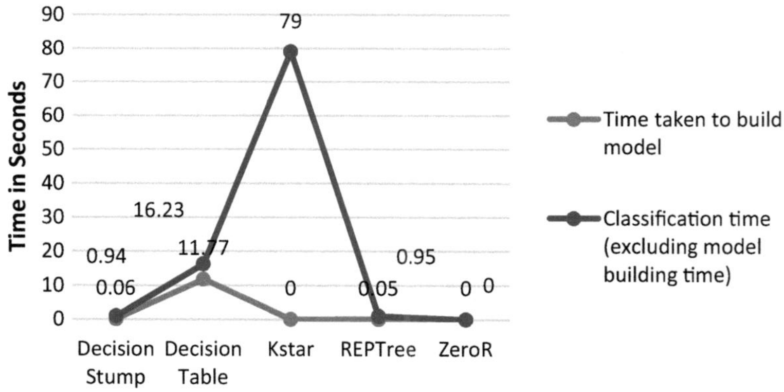

Fig. 2 Comparison on the basis of model building time and classification time

4.2 *Text Classification*

4.2.1 Experimental Setup

For text classification, the experiment has been conducted by loading a large and a small dataset in ARFF format (using simple CLI command) [20] and then text classification on both the cases is performed with desired classification algorithm.

Here, the large dataset taken is from 20newsgroups having 20,000 records that contains a collection of newsgroup documents. Naive Bayes Multinomial Text classification has been applied at consecutively 5, 10, 20 and 50 folds of cross-validation. The results have been compared. The small dataset is taken from IMDb having 1500 records that contains movie reviews. Naive Bayes Multinomial Text classification has been applied at consecutively 10, 20, 100 and 500 folds of cross-validation, and the results have been compared.

4.2.2 Experimental Result

Table 3 shows the test summaries of text classification at different folds of cross-validation for large 20newsgroups dataset. Figures 3 and 4 graphically represent the comparison.

20newsgroups dataset results: The results in Table 3 suggest, at 20-fold cross-validation, Naive Bayes Multinomial Text classification performed the best with 90.03% of correctly classified instances with least absolute error of 10.57%, whereas fivefold cross-validation had the relatively worse performance with 86.69% of cor-

Table 3 Comparison of the text classification process on 20newsgroups dataset with different folds of cross-validation

Test mode	Fivefold cross-validation	Tenfold cross-validation	20-fold cross-validation	50-fold cross-validation
Correctly classified instances (%)	89.69	89.97	90.03	90.02
Incorrectly classified instances (%)	10.31	10.03	9.97	9.98
Root mean squared error	0.098	0.097	0.0966	0.0969
Relative absolute error (%)	10.92	10.66	10.57	10.60
Root relative squared error (%)	44.94	44.50	44.32	44.44
Kappa statistic	0.8915	0.8944	0.895	0.8949
Mean absolute error	0.0104	0.0101	0.01	0.0101
Time taken to build model	34.95	31.06	32.24	34.29
Classification time	158.05	217.94	642.76	1555.71

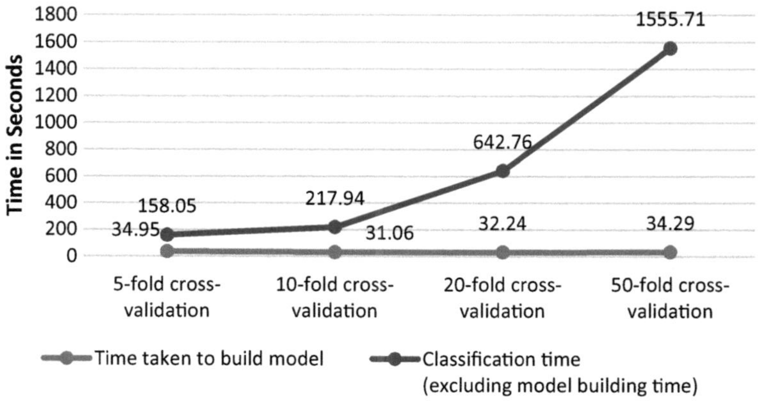

Fig. 3 Comparison of model building time and classification time of Naive Bayes multinomial text classification on 20newsgroup dataset on different folds of cross-validation

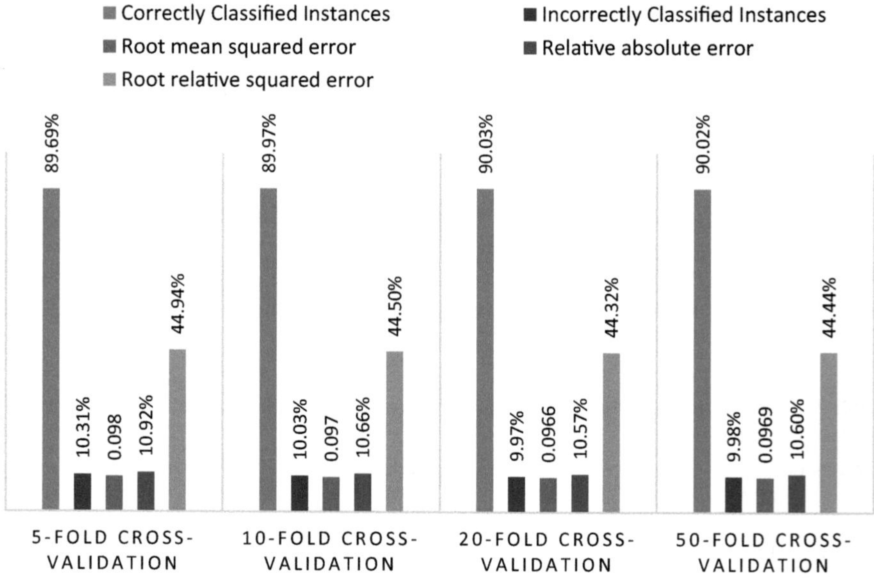

Fig. 4 Naive Bayes multinomial text classification on 20newsgroup dataset on different folds of cross-validation

rectly classified instances. So, the accuracy is comparatively good for 20-fold cross-validation. Also it can be seen from the value of Kappa statistic which is an important measure of classifier performance that 20-fold cross-validation performed best among others.

Model building time is less for 10 folds of cross-validation as compared to other folds of cross-validations, while classification time is less at fivefold cross-validation, then gradually increasing for tenfold cross-validation, 20-fold cross-validation and exceptionally high at 50-fold cross-validation. Here, one important observation can be seen that 20-fold cross-validation is giving higher accuracy and taking almost equal model building time to tenfold cross-validation.

Table 4 shows the test summaries of text classification at different folds of cross-validation for small IMDb dataset. Figures 5 and 6 graphically represent the comparison.

IMDb dataset results: 100-fold cross-validation Naive Bayes Multinomial classification performed the best with 82.29% of correctly classified instances with least absolute error of 36.23%, whereas tenfold cross-validation had the relatively worse performance with 81.29% of correctly classified instances. So, accuracy is best for 100-fold cross-validation. Learning time is relatively high for 100-fold cross-validation than others. Also, it can be seen from the value of Kappa statistic that 100-fold cross-validation performs best compared to others. Classification time depends on the size of dataset which has to be classified using the model which has been built.

Table 4 Comparison of the text classification process on IMDb dataset with different folds of cross-validation

Test mode	Tenfold cross-validation	20-fold cross-validation	100-fold cross-validation	500-fold cross-validation
Correctly classified instances (%)	81.29	81.93	82.29	81.79
Incorrectly classified instances (%)	18.71	18.07	17.71	18.21
Relative absolute error (%)	37.46	36.69	36.23	36.86
Root relative squared error (%)	82.31	81.77	80.99	81.75
Time taken to build model	0.44	0.46	0.49	0.48
Kappa statistic	0.6257	0.6386	0.6457	0.6357
Mean absolute error	0.1873	0.1834	0.1812	0.1843
Root mean squared error	0.4115	0.4088	0.4049	0.4089
Classification time	4.56	6.54	34.51	171.52

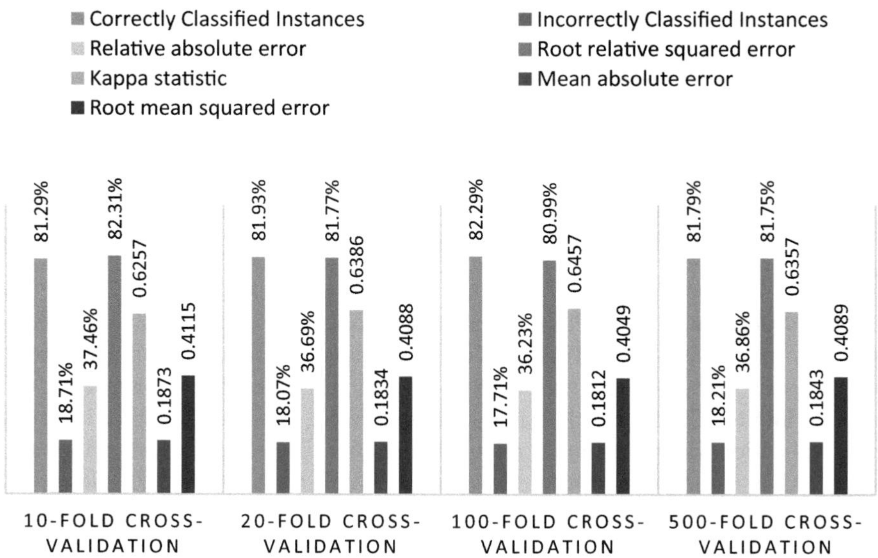

Fig. 5 Comparison of result of Naive Bayes multinomial text classification on IMDb dataset on different folds of cross-validation

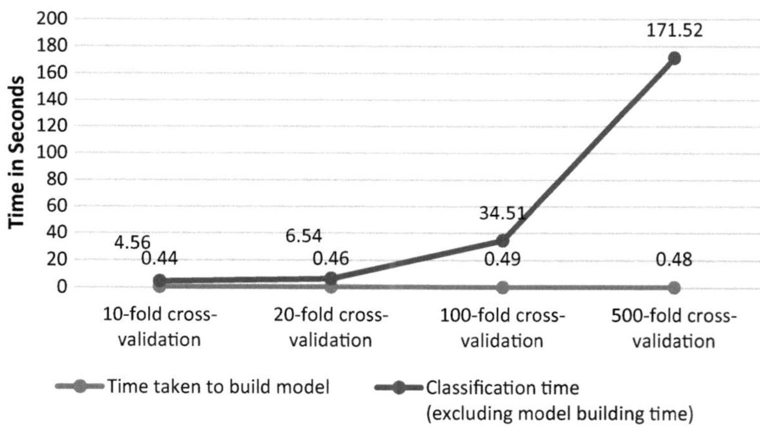

Fig. 6 Comparison of model building time and classification time of result of Naïve Bayes multinomial text classification on IMDb dataset on different folds of cross-validations

5 Conclusion

The present work demonstrates the results of a comparative study of efficiency of different classifier algorithms on numeric datasets as well as evaluating efficiency of Naive Bayes classifier on text classification.

As the results and observations suggest, it can be concluded that for a large numeric dataset, KStar works the best, among the classifiers Decision Stump, Decision Table, KStar, REPTree and ZeroR, whereas Decision Table was slowest to build model for the given dataset and KStar, ZeroR and REPTree were faster.

As for text classification, it has been found out that for a large dataset (e.g., 20newsgroups) Naive Bayes Multinomial Text classification worked best at 20 folds of cross-validation. But for a smaller dataset (e.g., IMDb), the same worked best at 100 folds of cross-validation.

Future goal of the present work will be to use even larger datasets for both numeric and text classification and moreover to apply more algorithms at different folds of cross-validation for large numeric and string datasets to present more detailed results of comparison.

References

1. Bin Othman, M.F., Yau, T.M.S.: Comparison of different classification techniques using WEKA for breast cancer. In: 3rd Kuala Lumpur International Conference on Biomedical Engineering 2006, pp. 520–523. Springer, Berlin, Heidelberg (2007)
2. Deepajothi, S., Selvarajan, S.: A comparative study of classification techniques on adult data set. Int. J. Eng. Res. Technol. (IJERT) **1** (2012)
3. Vaithiyanathan, V., Rajeswari, K., Tajane, K., Pitale, R.: Comparison of different classification techniques using different datasets. Int. J. Adv. Eng. Technol. **6**(2), 764 (2013)
4. Bouali, H., Akaichi, J.: Comparative study of different classification techniques: heart disease use case. In: 2014 13th International Conference on Machine Learning and Applications (ICMLA), pp. 482–486. IEEE (2014)
5. Shinge, S., Zarekar, R., Vaithiyanathan, V., Rajeswari, K.: Comparative analysis of heart risk dataset using different classification techniques. Int. J. Res. Inf. Technol. 489–494 (2015)
6. Raheja, S., Munjal, G.: Analysis of Linux Kernel vulnerabilities. Ind. J. Sci. Technol. **9**(48) (2016)
7. Arundthathi, A., Vijayaselvi, G., Savithri, V.: Assessment of decision tree algorithms on student's recital (2017)
8. Nadu, T.: Cad Diagnosis Using PSO. BAT, MLR and SVM (2017)
9. Ikonomakis, M., Kotsiantis, S., Tampakas, V.: Text classification using machine learning techniques. WSEAS Trans. Comput. **4**(8), 966–974 (2005)
10. Kotsiantis, S.B., Zaharakis, I., Pintelas, P.: Supervised machine learning: a review of classification techniques. Emer. Artif. Intell. Appl. Comput. Eng. **160**, 3–24 (2007)
11. Ting, S.L., Ip, W.H., Tsang, A.H.: Is Naive Bayes a good classifier for document classification. Int. J. Softw. Eng. Appl. **5**(3), 37–46 (2011)
12. Wahbeh, A.H., Al-Kabi, M.: Comparative assessment of the performance of three WEKA text classifiers applied to arabic text. Abhath Al-Yarmouk: Basic Sci. Eng. **21**(1), 15–28 (2012)
13. Purohit, A., Atre, D., Jaswani, P., Asawara, P.: Text classification in data mining. Int. J. Sci. Res. Publ. **5**(6), 1–7 (2015)
14. Rajeswari, R.P., Juliet, K., Aradhana: Text classification for student data set using Naive Bayes classifier and KNN classifier. Int. J. Comput. Trends Technol. (IJCTT) **43**(1) (2017)
15. Tilve, A.K.S., Jain, S.N.: A survey on machine learning techniques for text classification. Int. J. Eng. Sci. Res. Technol. (2017)
16. Mandal, L., Das, R., Bhattacharya, S., Basu, P.N.: Intellimote: a hybrid classifier for classifying learners' emotion in a distributed e-learning environment. Turkish J. Electr. Eng. Comput. Sci. **25**(3), 2084–2095 (2017)

17. https://www.cs.waikato.ac.nz/ml/weka/
18. https://datasets.imdbws.com/
19. http://qwone.com/~jason/20Newsgroups
20. https://weka.wikispaces.com/Text+categorization+with+WEKA

An Algorithmic Approach for Generating Quantum Ternary Superposition Operators and Related Performance Measures

Bipulan Gain, Sudhindu Bikash Mandal, Amlan Chakrabarti and Subhansu Bandyopadhyay

Abstract Quantum computing promises to outperform classical computing in terms of algorithmic speedup for certain classes of computational problems. A quantum algorithm exploits quantum mechanical processes like superposition, interference, and entanglement that works on quantum states of matter providing exponential or super polynomial speedup. In recent times, multi-valued quantum computing is gaining popularity due to its higher state space dimension, and ternary quantum computing is one of the most popular multi-valued quantum computing. In this paper, we propose an algorithmic approach for the generation of quantum ternary superposition operators and evaluating them in terms of trace distance, fidelity, and the entanglement measure. We also propose five new quantum ternary superposition operators, which have larger trace distance and smaller fidelity than the existing ternary superposition operators "Chrestenson gates" and "S-gate" (Sudhindu Bikash, IEEE Comput Soc 2014, [1]). To characterize the amount of entanglement in two given qutrit composite ternary states, we also measure the concurrence of the newly proposed and the existing quantum superposition operators. We have shown that the newly proposed superposition operators generate maximally entangled states with concurrence equal to 1.

B. Gain (✉) · S. Bandyopadhyay
Department of Computer Science and Engineering, Brainwere University,
Kolkata, West Bengal, India
e-mail: bipulan@gmail.com

S. Bandyopadhyay
e-mail: subhansu@ieee.org

S. B. Mandal (✉)
Regent Education and Research Foundation, Kolkata, West Bengal, India
e-mail: sudhindu.mandal@gmail.com

A. Chakrabarti
A.K.Choudhury School of Information Technology, University of Calcutta,
Kolkata, West Bengal, India
e-mail: achakra12@yahoo.com

M. Chakraborty et al. (eds.), *Proceedings of International Ethical Hacking Conference 2018*, Advances in Intelligent Systems and Computing 811,
https://doi.org/10.1007/978-981-13-1544-2_23

Keywords Quantum computing · Ternary quantum computation · Gram–Schmidt decomposition · Quantum ternary superposition operator · Trace distance Fidelity · Concurrence · Entanglement

1 Introduction

Over the limitations of classical computing algorithms, quantum computing was introduced [2]. Initially, it was the concept of qubits, which defines a two-state quantum system [2, 3]. A qubit lives in a two-dimensional complex Hilbert space spanned by two states $|0\rangle$ and $|1\rangle$. A qubit state $|\psi\rangle$ can be either in one of the pure states, i.e., $|\psi\rangle = |0\rangle$ or $|\psi\rangle = |1\rangle$ or in a superposition state represented as follows:

$$|\psi\rangle = a|0\rangle + b|1\rangle \tag{1}$$

where $|a|^2 + |b|^2 = 1$. A qubit state can also be represented as a point on the surface of a Bloch sphere S^2 [4] in the following fashion:

$$|\psi\rangle = sin\frac{\theta}{2}|0\rangle + e^{i\phi}cos\frac{\theta}{2}|1\rangle. \tag{2}$$

θ is related to the proportion of $|0\rangle$ and $|1\rangle$ in the composition of the state, while ϕ is the quantum phase associated with the qubit. It should be noted that diametrically opposite points on the Bloch sphere S^2 correspond to mutually orthogonal vectors in a two-dimensional complex Hilbert space $\mathcal{H}^{(2)}$[4].

The novelty of quantum computing lies in the fact that it shows an exponential or a super polynomial speedup [2, 3, 5] for certain algorithms as compared to their classical counterparts. Ternary quantum systems [4] instead of two-state quantum system provide a higher-dimensional quantum system using lesser number of qubits, which enhances the scope of functional parallelism and degree of entanglement. The unit of information in a ternary quantum system is known as qutrit. A qutrit lives in a three-dimensional complex Hilbert space $\mathcal{H}^{(3)}$ spanned by $|0\rangle$, $|1\rangle$ and $|2\rangle$. With the help of *Poincare* sphere [6] and the conception of Bloch sphere, we can write the pure state of a qutrit [7] as:

$$|\psi\rangle = sin(\frac{\zeta}{2})cos(\frac{\theta}{2})|0\rangle + e^{i\Phi_{01}}sin(\frac{\zeta}{2})sin(\frac{\theta}{2})|1\rangle + e^{i\Phi_{02}}cos(\frac{\zeta}{2})|2\rangle \tag{3}$$

θ and ζ defines the magnitudes of components $|\psi\rangle$ where we can describe Φ_{01} as the phase of $|0\rangle$ relative to $|1\rangle$ and analogously for Φ_{02}. Like binary quantum system, we can also ignore the global phase value in a ternary quantum system [2, 8].

According to *Poincare* sphere, the two unit vectors n and n' representing pure states can be described as follows:

$$0 \leq arccos(n \cdot n') \leq \frac{2\Pi}{3}. \tag{4}$$

So mutually orthogonal vectors in $\mathcal{H}^{(3)}$ do not lead to tipodal or diametrically opposite points on the *Poincare* sphere, but points with a maximum opening angle of $\frac{2\Pi}{3}$.

Key contributions in this paper can be summarized as follows:

- Proposal of a generalized algorithm for finding the possible ternary superposition operators within the ternary operator space based on trace distance and fidelity.
- Proposal of new ternary superposition operators, which are better in terms of trace distance and fidelity as compared to the existing quantum ternary superposition operators ("Chrestenson Gate" [6, 9] and "S-Gate" [1]).
- We also find the concurrences for the new and the existing quantum ternary superposition operations, which characterizes the entanglement in a two qutrit composite ternary quantum system.

In Sect. 2, we discuss the quantum tools used in this work. In Sect. 3, we brief the existing quantum ternary superposition operators. Section 4 describes our proposed algorithm for the generation of ternary superposition operators. In Sect. 5, we demonstrate the evaluation of trace distance, fidelity, and concurrence for ternary quantum operators. Some new quantum ternary superposition operators are presented in Sect. 6. In Sect. 7, we analyze the performance of the existing quantum ternary superposition operators (*CH* and *S*) and that of the newly proposed superposition operators in terms of trace distance, fidelity, and concurrence. results of the analyzed parameters for the four newly proposed and existing quantum ternary superposition operators. Section 8 concludes the paper.

2 Background

2.1 Gram–Schmidt Decomposition

We can produce an orthonormal basis from an arbitrary basis by applying the Gram–Schmidt orthogonalization process [5]. Let $|v_1\rangle, |v_2\rangle, \ldots |v_n\rangle$, be a basis for an inner product space V. With the help of Gram–Schmidt process, we can construct an orthogonal basis $|w_i\rangle$ as follows:

$|w_1\rangle = |v_1\rangle$

$|w_2\rangle = |v_2\rangle - \frac{\langle w_1|v_2\rangle}{\langle w_1|w_1\rangle}|w_1\rangle$

$\vdots$

$$w_n = |v_n\rangle - \frac{\langle w_1|v_n\rangle}{\langle w_1|w_1\rangle}|w_1\rangle - \frac{\langle w_2|v_n\rangle}{\langle w_2|w_2\rangle}|w_2\rangle - \cdots - \frac{\langle w_{n-1}|v_n\rangle}{\langle w_{n-1}|w_{n-1}\rangle}|w_{n-1}\rangle \tag{5}$$

To form an orthonormal set using the Gram–Schmidt procedure, we need to divide each vector by its norm. For example, the normalized vector we can use to construct $|w_2\rangle$ is as follows:

$$|w_2\rangle = \frac{|v_2\rangle - \langle w_1|v_2\rangle|w_1\rangle}{|||v_2\rangle - \langle w_1|v_2\rangle|w_1\rangle||}$$

2.2 Trace Distance

In order to find the the trace distance, the no-cloning theorem of quantum computing do not permit to make an exact copy of an unknown quantum state, but we can make an approximate copy. We can use "trace distance" to determine the similarity between two given quantum states.

Let ρ and σ be two density matrices. The trace distance $\delta(\rho, \sigma)$ is defined as follows:

$$\delta(\rho, \sigma) = \frac{1}{2}Tr|\rho - \sigma| \tag{6}$$

2.3 Fidelity

Fidelity can be defined as the statistical measure of the overlap between two distributions and thus estimates the closeness between the two states. Suppose they are pure states with density operators $\rho = |\psi\rangle\langle\psi|$ and $\sigma = |\phi\rangle\langle\phi|$. Since these are pure states, the trace distance $\delta(\rho, \sigma)$ is defined as $\rho^2 = \rho$, $\sigma^2 = \sigma$, and hence, $\sigma = \sqrt{\sigma}$, $\sigma = \sqrt{\sigma}$. The fidelity F(ρ, σ) is defined as follows:

$$\begin{aligned} F(\rho, \sigma) &= Tr(\sqrt{\sqrt{\rho}\sigma\sqrt{\rho}}) = Tr\sqrt{(\langle\psi|\psi\rangle)(\langle\phi|\phi\rangle)(\langle\psi|\psi\rangle)} \\ &= Tr\sqrt{|(\langle\phi|\psi\rangle)|^2(\langle\psi|\psi\rangle)} = |\langle\phi|\psi\rangle|\sqrt{\langle\psi|\psi\rangle} = |\langle\phi|\psi\rangle| \end{aligned} \tag{7}$$

2.4 Concurrence for Two Qutrits in a Pure State

In quantum computation, there are few composite quantum states, which cannot be factorized and they are commonly known as quantum entangled state [2, 5]. For better entanglement in a quantum ternary system, we consider those states, which have less overlapping character. If the overlapping is higher, then it is easier to separate. Finding the concurrence is a way of characterizing the entanglement.

Now if we extend the concurrence of Wootters [10] by defining the concurrence for two qutrits in pure state of ternary quantum system $|\psi_{3\times3}\rangle = |\psi\rangle$, then the concurrence of the state [5, 10] $|\psi\rangle$ can be expressed as follows :

$$C_3(\psi) = |\langle\psi|\tilde{\psi}\rangle| \tag{8}$$

where $|\tilde{\psi}\rangle = (\mathcal{O} \otimes \mathcal{O})|\psi^*\rangle$ and $|\psi^*\rangle$ is the complex conjugate of the quantu state $|\psi\rangle$ and $\mathcal{O} = \begin{pmatrix} 0 & -i & i \\ i & 0 & -i \\ -i & i & 0 \end{pmatrix}$ [11, 12] where $\mathcal{O}$ is the decomposed representation of Pauli matrix for qutrits.

3 Existing Superposition Operators in Ternary Quantum System

3.1 Chrestenson Gates

We know the Hadamard transform is a special case of the quantum Fourier transform (QFT) in Hilbert space $\mathcal{H}_{(n)}$. The Chrestenson gate [6, 9] in ternary quantum computing is also equivalent to QFT in $\mathcal{H}_{(3)}$. Chrestenson gates are expressed in ternary system as follows:

$$CH_1 = \frac{1}{\sqrt{3}} \begin{pmatrix} 1 & 1 & 1 \\ 1 & w & w^2 \\ 1 & w^2 & w \end{pmatrix}$$
$$CH_2 = \frac{1}{\sqrt{3}} \begin{pmatrix} 1 & 1 & w \\ 1 & w & 1 \\ w & 1 & 1 \end{pmatrix}$$
$$CH_3 = \frac{1}{\sqrt{3}} \begin{pmatrix} 1 & 1 & w^2 \\ 1 & w^2 & 1 \\ w^2 & 1 & 1 \end{pmatrix} \tag{9}$$

These gates can also be referred as quantum ternary superposition operators.

3.2 S-Gate

An S-gate [1] is a ternary superposition operator of the following form:

$$S = \begin{pmatrix} \frac{1}{\sqrt{3}} & \frac{1}{\sqrt{6}} & \frac{1}{\sqrt{2}} \\ \frac{1}{\sqrt{3}} & -\frac{2}{\sqrt{6}} & 0 \\ \frac{1}{\sqrt{3}} & \frac{1}{\sqrt{6}} & -\frac{1}{\sqrt{2}} \end{pmatrix} \tag{10}$$

4 Algorithmic Approach for Generation of Ternary Superposition Operators

In this section, we present the algorithm for generating quantum ternary superposition operators.

Algorithm 1: Generation of Quantum Ternary Superposition Operators

---

1: INPUT :
2: **for** $i \leftarrow 1$ to 3 **do**
3: $\quad \zeta_i, \theta_i = random[0° to\ 120°]$
4: **end for**
5: **for** $i \leftarrow 1$ to 3 **do**
6: $\quad \alpha_i \leftarrow sin(\frac{\zeta_i}{2})cos(\frac{\theta_i}{2})$ where [α_i = *probability amplitudes of the basis state* $|0\rangle$]
7: $\quad \beta_i \leftarrow sin(\frac{\zeta_i}{2})sin(\frac{\theta_i}{2})$ where [β_i = *probability amplitudes of the basis state* $|1\rangle$]
8: $\quad \gamma_i \leftarrow cos(\frac{\zeta_i}{2})$ where [γ_i = *probability amplitudes of the basis state* $|2\rangle$]
9: **end for**
10: **for** $k \leftarrow 1$ to 3 **do**
11: $\quad |\psi_k\rangle \leftarrow \alpha_k|0\rangle + \beta_k|1\rangle + \gamma_k|2\rangle$
12: **end for**
13: **for** $k \leftarrow 1$ to 3 **do**
14: $\quad W_k \leftarrow GRAM - SCHMIDT(|\psi_k\rangle)$
15: **end for**
16: **for** $i \leftarrow 1$ to 3 **do**
17: $\quad$ **for** $j \leftarrow 1$ to 3 **do**
18: $\quad\quad W_i = \begin{pmatrix} W_{ij} \\ W_{ij} \\ W_{ij} \end{pmatrix}$
19: $\quad$ **end for**
20: **end for**
21: $S_B = W_1||W_2||W_3$
22: **if** $S_B.S_B^{\dagger} = I$ **then**
23: $\quad S_B = QuantumTernarySuperpositionOperator$
24: **else**
25: $\quad$ go to step 2
26: **end if**

Further, in the following subsections, we will show the methodology for finding a quantum ternary superposition operator using our proposed algorithm.This will be described with an illustration.

4.1 Normalized and Orthogonal Basis

At first, we try to make the basis normalized and orthogonal. Then, we proceed with Gram–Schmidt orthogonalization. Thus, we produce an orthonormal basis from an arbitrary basis.

4.2 Evaluating the Probability Amplitudes

The three probability amplitudes of the basis states $|0\rangle$, $|1\rangle$ and $|2\rangle$ in a quantum ternary superposition state are called α, β, and γ, where $|\alpha^2| + |\beta^2| + |\gamma^2| = 1$. *for i = 1 to 3.*

$$\alpha_i = sin(\frac{\zeta_i}{2})cos(\frac{\theta_i}{2}) \quad (11)$$

$$\beta_i = sin(\frac{\zeta_i}{2})sin(\frac{\theta_i}{2}) \quad (12)$$

$$\gamma_i = cos(\frac{\zeta_i}{2}) \quad (13)$$

4.3 Applying Gram–Schmidt Orthogonalization

By providing three different sets of ζ and θ in Eqs. (11)–(13), we get three different α, β, and γ. From these values, we can get three different quantum states as follows:

$$|\psi_1\rangle = \alpha_1|0\rangle + \beta_1|1\rangle + \gamma_1|2\rangle \quad (14)$$

$$|\psi_2\rangle = \alpha_2|0\rangle + \beta_2|1\rangle + \gamma_2|2\rangle \quad (15)$$

$$|\psi_3\rangle = \alpha_3|0\rangle + \beta_3|1\rangle + \gamma_3|2\rangle \quad (16)$$

where $|0\rangle = \begin{pmatrix}1\\0\\0\end{pmatrix}, |1\rangle = \begin{pmatrix}0\\1\\0\end{pmatrix}, |2\rangle = \begin{pmatrix}0\\0\\1\end{pmatrix}$.

An orthonormal basis can be produced from an arbitrary basis by application of the Gram–Schmidt orthogonalization process. We can produce the orthonormal basis states as follows: $W_i = \begin{pmatrix}W_{ij}\\W_{ij}\\W_{ij}\end{pmatrix}$ *for i, j = 1 to 3*. i.e.

$$W_1 = \begin{pmatrix} W_{11} \\ W_{12} \\ W_{13} \end{pmatrix}, W_2 = \begin{pmatrix} W_{21} \\ W_{22} \\ W_{23} \end{pmatrix}, W_3 = \begin{pmatrix} W_{31} \\ W_{32} \\ W_{33} \end{pmatrix} \tag{17}$$

by applying "Gram–Schmidt" orthogonalization on Eqs. (14)–(16).

4.4 The Quantum Ternary Superposition Operator

After getting three normalized and orthogonal basis states, i.e., orthonormal basis vectors, we can form the quantum operator by concatenating W_1, W_2, and W_3 from Eq. (17). Let, S_B is the quantum ternary superposition operator, then we have

$$S_B = \begin{pmatrix} W_{11} & W_{12} & W_{13} \\ W_{21} & W_{22} & W_{23} \\ W_{31} & W_{32} & W_{33} \end{pmatrix} \tag{18}$$

We can say S_B is a quantum ternary superposition operator if and only if $S_B \cdot S_B^{\dagger} = I$ [5].

Next, we estimate the entanglement in the ternary quantum system; hence, we have to find those operators whose overlapping is minimum, i.e., larger the trace distance and smaller the fidelity between two quantum states.

4.5 Demonstration with a New Superposition Operator as an Example

We are giving input of three different sets of ζ and θ in Eqs. (11)–(13).

ζ_1	θ_1	ζ_2	θ_2	ζ_3	θ_3
21	36	8	12	85	87

With this input set, we evaluate the values of α_i for $i \leftarrow 1$ to 3:

$$\alpha_1 = 0.1733, \beta_1 = 0.0563, \gamma_1 = 0.9833,$$
$$\alpha_2 = 0.0694, \beta_2 = 0.0073, \gamma_2 = 0.9976,$$
$$\alpha_3 = 0.4901, \beta_3 = 0.4650, \gamma_3 = 0.7373.$$

Using the values of α_i, β_i, and γ_i, $i \leftarrow 1$ to 3 in Eqs. (14)– (16), we can get the following values:

$$|\psi_1\rangle = \begin{pmatrix} 0.1733 \\ 0.0563 \\ 0.9883 \end{pmatrix} \tag{19}$$

$$|\psi_2\rangle = \begin{pmatrix} 0.0694 \\ 0.0073 \\ 0.9976 \end{pmatrix} \tag{20}$$

$$|\psi_3\rangle = \begin{pmatrix} 0.4901 \\ 0.4650 \\ 0.7373 \end{pmatrix} \tag{21}$$

The three vectors thus generated are not orthonormal. Our first target is to make these three vectors orthonormal. By using Gram–Schmidt orthogonalization, we can construct an orthonormal basis set from Eqs. (19)–(21).

$$W_1 = \begin{pmatrix} 0.1733 \\ 0.0563 \\ 0.9833 \end{pmatrix}, W_2 = \begin{pmatrix} -0.8890 \\ -0.4207 \\ 0.1808 \end{pmatrix} and W_3 = \begin{pmatrix} -0.4239 \\ 0.9054 \\ 0.0229 \end{pmatrix}. \tag{22}$$

So we can form a 3×3 matrix named S_{B_1} whose value will look like Eq. (18). We can only say S_{B_1} is a quantum ternary superposition operator if and only if $S_{B_1} \cdot S_{B_1}^{\dagger} = I$.

$$S_{B_1} = \begin{pmatrix} 0.1733 & -0.8890 & -0.4239 \\ 0.0563 & -0.4207 & 0.9054 \\ 0.9833 & 0.1808 & 0.0229 \end{pmatrix} \tag{23}$$

From Eq. (23) we can show $S_{B_1} \cdot S_{B_1}^{\dagger} = I$ [5].

5 Finding the Trace Distance, Fidelity, and Concurrence

After successfully finding the ternary quantum operator, our next aim is to find the "trace distance" and "fidelity" between the quantum states (W_1 and W_2), (W_1 and W_3), and (W_2 and W_3). We inspect the possibility of overlapping between these pair of states to achieve an quantum entanglement, which are not overlapping with each other has higher probability of occurrence.

Three different quantum states in S_{B_1} are W_1, W_2 and W_3. To find the trace distance and fidelity, we need to find the density operator for these three different quantum state of the ternary quantum operator S_{B_1}. Consider the density operators σ, ω, and ρ for the quantum states W_1, W_2 and W_3 as follows:

$$\delta(\sigma, \omega) = \frac{1}{2} Tr|\sigma - \omega| \tag{24}$$

$$\delta(\sigma, \rho) = \frac{1}{2} Tr|\sigma - \rho| \tag{25}$$

$$\delta(\omega, \rho) = \frac{1}{2} Tr|\omega - \rho| \tag{26}$$

where $\sigma = |W_1\rangle\langle W_1|$, $\omega = |W_2\rangle\langle W_2|$ and $\rho = |W_3\rangle\langle W_3|$. Then, the trace distances are as follows:

$$\delta(\sigma, \omega) = 0.9341$$
$$\delta(\sigma, \rho) = 0.9663$$
$$\delta(\omega, \rho) = 0.7071$$

Henceforth, we define the trace distance between (W_1 and W_2) as $\delta(\sigma, \omega)$ that between (W_1 and W_3) as $\delta(\sigma, \rho)$ and between (W_2 and W_3) as $\delta(\omega, \rho)$.

Now we have to find the fidelity between those quantum states

$$F(\sigma, \omega) = 0$$
$$F(\sigma, \rho) = 0$$
$$F(\omega, \rho) = 0$$

Like the tarce distance, fidelity between (W_1 and W_2), (W_1 and W_3), and (W_2 and W_3) can also be expressed in terms of $F(\sigma, \omega)$, $F(\sigma, \rho)$, and $F(\omega, \rho)$, respectively.

It was our aim to find larger trace distance and smaller fidelity to minimize the overlapping, which can give better entanglement.

As $|W_1\rangle$, $|W_2\rangle$, $|W_3\rangle$ are all single qutrit pure state, we need to find the composite state to evaluate the concurrence. By using these three single qutrit pure ternary states, we can compute the composite states as follows:

$$\begin{aligned} W_1W_2 &= W_1 \otimes W_2, \\ W_2W_1 &= W_2 \otimes W_1, \\ W_1W_3 &= W_1 \otimes W_3, \\ W_3W_1 &= W_3 \otimes W_1, \\ W_2W_3 &= W_2 \otimes W_3, \\ W_3W_2 &= W_3 \otimes W_2. \end{aligned}$$

These six composite states are all two qutrit pure states.

The concurrence of these two qutrit composite states can be described with the help of Eq. (8) as follows. The concurrence of the composite quantum states are

$$C(W_1 W_2) = 1$$
$$C(W_2 W_1) = 1$$
$$C(W_1 W_3) = 1$$
$$C(W_3 W_1) = 1$$
$$C(W_2 W_3) = 0$$
$$C(W_3 W_2) = 0$$

6 New Quantum Ternary Superposition Operators

In this section, we propose four new ternary superposition operators, namely S_{B_2}, S_{B_3}, S_{B_4}, and S_{B_5} in addition to S_{B_1}. These operators are generated through random choices of $\zeta_1, \theta_1, \zeta_2, \theta_2, \zeta_3$, and θ_3 based on the algorithm presented in Sect. 4. We can generate a larger list of such operators; however, we present only four new operators, which show remarkably better performance in terms of trace distance, fidelity, and concurrence compared to the existing operators (*CH* and *S*) as presented in Table 1 (Sect. 8).

6.1 S_{B_2}-Operator

ζ_1	θ_1	ζ_2	θ_2	ζ_3	θ_3
3	15	59	66	58	41

$$S_{B_2} = \begin{pmatrix} 0.0260 & 0.8268 & 0.5619 \\ 0.0034 & 0.5620 & -0.8271 \\ 0.9997 & -0.0234 & -0.0118 \end{pmatrix} \tag{27}$$

6.2 S_{B_3}-Operator

ζ_1	θ_1	ζ_2	θ_2	ζ_3	θ_3
26	29	38	46	7	23

$$S_{B_3} = \begin{pmatrix} 0.2178 & 0.7440 & -0.6317 \\ 0.0563 & 0.6365 & 0.7692 \\ 0.9744 & -0.2031 & 0.0976 \end{pmatrix} \tag{28}$$

6.3 S_{B_4}-Operator

ζ_1	θ_1	ζ_2	θ_2	ζ_3	θ_3
21	41	87	49	47	21

$$S_{B_4} = \begin{pmatrix} 0.1707 & 0.8866 & 0.4300 \\ 0.0638 & 0.4255 & -0.9027 \\ 0.9833 & -0.1815 & -0.0161 \end{pmatrix} \tag{29}$$

6.4 S_{B_5}-Operator

ζ_1	θ_1	ζ_2	θ_2	ζ_3	θ_3
3	80	83	72	8	23

$$S_{B_5} = \begin{pmatrix} 0.0201 & 0.8100 & 0.5861 \\ 0.0168 & 0.5858 & -0.8103 \\ 0.9997 & -0.0261 & 0.0019 \end{pmatrix} \tag{30}$$

7 Results and Analysis

In this section, we show the performance measure of the *CH*, *S* and the newly proposed S_{B_1}, S_{B_2}, S_{B_3}, S_{B_4}, and S_{B_5} operators in terms of trace distance, fidelity, and concurrence. We have performed simulation of these operators using a MATLAB program to generate the values for each of these operators for the various performance metrics. In Table 1, we illustrate our experimental results for each of these ternary superposition operators.

In Table 1, we compare the performance of our newly proposed operators with the previously existing superposition operators [1, 6, 9] in terms of trace distance, fidelity, and concurrence. Comparing the results of our newly proposed operators with the existing gates, we find that for the three Chretenson gates all the values of trace distance are 0 and for the S-gate the highest value of trace distance is 0.6667, whereas the trace distance between (W_1 and W_2) and (W_1 and W_3) in our newly proposed operator are nearly equal to 1, and the trace distance between (W_2 and W_3) is low. Now, if we compare fidelity, for CH_1-Gate : the fidelity between (W_1 and W_2) and (W_1 and W_3) is equal to 0 and that between (W_2 and W_3) is 0.3211. For CH_2, all the values are 0, and for CH_3, the fidelity between (W_1) and W_2) and (W_1 and W_3) is 0.5744 and that between (W_2 and W_3) equal to 0. For the S-gate, the fidelity between(W_1 and W_2) is 0.4714, (W_1 and W_3) equal to 0 and the value of the fidelity between (W_2 and W_3) 0.5774, whereas for our newly proposed superposition

Table 1 Comparison of existing superposition operator with the nelwly proposed superposition operator in ternary quantum system

Ternary superposition operators	Trace distance	Fidelity	Concurrence of the two qutrit quantum state
CH_1 Gate	$\delta(\sigma, \omega) = 0$	$F(\sigma - \omega) = 0$	$C(W_1W_2) = 0$; $C(W_2W_1) = 0$
	$\delta(\sigma, \rho) = 0$	$F(\sigma - \rho) = 0$	$C(W_1W_3) = 0$; $C(W_3W_1) = 0$
	$\delta(\omega, \rho) = 0$	$F(\omega - \rho) = 0.3211$	$C(W_2W_3) = 0$; $C(W_3W_2) = 0$
CH_2 Gate	$\delta(\sigma, \omega) = 0$	$F(\omega - \rho) = 0$	$C(W_1W_2) = 0.3333$; $C(W_2W_1) = 0.3333$
	$\delta(\sigma, \rho) = 0$	$F(\sigma - \rho) = 0$	$C(W_1W_3) = 0.3333$; $C(W_3W_1) = 0.3333$
	$\delta(\omega, \rho) = 0$	$F(\omega - \rho) = 0$	$C(W_2W_3) = 0.3333$; $C(W_3W_2) = 0.3333$
CH_3 Gate	$\delta(\sigma, \omega) = 0$	$F(\sigma - \omega) = 0.5744$	$C(W_1W_2) = 0$; $C(W_2W_1) = 0$
	$\delta(\omega, \rho) = 0$	$F(\sigma - \rho) = 0.5744$	$C(W_1W_3) = 0$; $C(W_3W_1) = 0.3333$
	$\delta(\sigma, \rho) = 0$	$F(\omega - \rho) = 0$	$C(W_2W_3) = 0$; $C(W_2W_3) = 0$
S-Gate	$\delta(\sigma, \omega) = 0.3333$	$F(\sigma - \omega) = 0.4714$	$C(W_1W_2) = 1$; $C(W_2W_1) = 1$
	$\delta(\sigma, \rho) = 0.3333$	$F(\sigma - \rho) = 0$	$C(W_1W_3) = 1$; $C(W_3W_1) = 0$
	$\delta(\omega, \rho) = 0.6667$	$F(\omega - \rho) = 0.5774$	$C(W_2W_3) = 1$; $C(W_3W_2) = 0$
S_{B_1}	$\delta(\sigma, \omega) = 0.9341$	$F(\sigma - \omega) = 0$	$C(W_1W_2) = 1$; $C(W_2W_1) = 1$
	$\delta(\sigma, \rho) = 0.9663$	$F(\sigma - \rho) = 0$	$C(W_1W_3) = 1$; $C(W_3W_1) = 0$
	$\delta(\omega, \rho) = 0.6428$	$F(\omega - \rho) = 0$	$C(W_2W_3) = 1$; $C(W_3W_2) = 0$
S_{B_2}	$\delta(\sigma, \omega) = 0.9988$	$F(\sigma - \omega) = 0$	$C(W_1W_2) = 1$; $C(W_2W_1) = 1$
	$\delta(\sigma, \rho) = 0.9992$	$F(\sigma - \rho) = 0$	$C(W_1W_3) = 1$; $C(W_3W_1) = 0$
	$\delta(\omega, \rho) = 0.3682$	$F(\omega - \rho) = 0$	$C(W_2W_3) = 1$; $C(W_3W_2) = 0$
S_{B_3}	$\delta(\sigma, \omega) = 0.9081$	$F(\sigma - \omega) = 0$	$C(W_1W_2) = 1$; $C(W_2W_1) = 1$
	$\delta(\omega, \rho) = 0.9400$	$F(\sigma - \rho) = 0$	$C(W_1W_3) = 1$; $C(W_3W_1) = 0$
	$\delta(\sigma, \rho) = 0.1865$	$F(\omega - \rho) = 0$	$C(W_2W_3) = 1$; $C(W_3W_2) = 0$

(continued)

Table 1 (continued)

Ternary superposition operators	Trace distance	Fidelity	Concurrence of the two qutrit quantum state
S_{B_4}	$\delta(\sigma, \omega) = 0.9338$	$F(\sigma - \omega) = 0$	$C(W_1W_2) = 1$; $C(W_2W_1) = 1$
	$\delta(\sigma, \rho) = 0.9665$	$F(\sigma - \rho) = 0$	$C(W_1W_3) = 1$; $C(W_3W_1) = 0$
	$\delta(\omega, \rho) = 0.6338$	$F(\omega - \rho) = 0$	$C(W_2W_3) = 1$; $C(W_3W_2) = 0$
S_{B_5}	$\delta(\sigma, \omega) = 0.9986$	$F(\sigma - \omega) = 0$	$C(W_1W_2) = 1$; $C(W_2W_1) = 1$
	$\delta(\sigma, \rho) = 0.9993$	$F(\sigma - \rho) = 0$	$C(W_1W_3) = 1$; $C(W_3W_1) = 0$
	$\delta(\omega, \rho) = 0.3133$	$F(\omega - \rho) = 0$	$C(W_2W_3) = 1$; $C(W_3W_2) = 0$

operators, S_{B_1} to S_{B_5} all the values are equal to 0. Our aim was to make the trace distance higher and the fidelity lower in order to generate less overlapping states. The results show that our newly proposed operators are less overlapping in nature as the trace distance is larger and the fidelity is smaller as compared to the existing operators, which proves the goodness of these operators. We can also infer that for our newly proposed operators the maximum value of concurrence equals to 1 thus favoring maximum entanglement.

8 Conclusion

We have proposed a generalized algorithm for churning out quantum ternary superposition operators in the ternary operator space using Gram–Schmidt orthogonalization and based on trace distance and fidelity. We have also proposed five new quantum ternary superposition operators based on our proposed algorithm. The new operators show better performance in comparison with the existing operators in terms of trace distance, fidelity, and concurrence and hence show promise for better entanglement. In future, we will consider implementation of teleportation in ternary quantum system through entanglement, using these newly proposed superposition operators.

References

1. Sudhindu Bikash, M., Amlan, C., Susmita, S.-K.: Synthesis of Ternary Grover's Algorithm, ISMVL '14. IEEE Computer Society (2014). ISBN: 978-1-4799-3535-2
2. Qi-Ping, S., Chui-Ping, Y.: Circuit QED: implementation of the three-qubit refined DeutschJozsa quantum algorithm. J. Quant. Inf. Process. Springer, US (2014). ISSN 1570–0755
3. Michael, A., Nielsen, I., Chuang, L.: Quantum Computation and Quantum Information. Cambridge University Press (2000). ISBN 0-521-63503-9
4. Goyal, S.K., Simon, B.N., Rajeev, S., Sudhavathani, S.: Geometry of the generalized Bloch sphere for qutrit. IOP Publishing, Journal (2004). arXiv:1111.4427
5. McMahon, D.: Quantum Computing Explained (cloth), Wiley Inc., Hoboken, New Jersey. ISBN 978-0-470-09699-4
6. Vamsi, P., Marek, P.: Quantum phase estimation using multivalued logic. In: Proceedings of the 2011 41st IEEE International Symposium on Multiple-Valued Logic. IEEE Computer Society (2011). ISBN: 978-0-7695-4405-2
7. Klimov, A.B., Snchez-Soto, L.L., de Guise, H., Bjrk, G.: Quantum phases of a qutrit. IOP Publishing, J. Phys. A Math. General (2004)
8. Subhash, K., Donald, C., Delaune Elaine T.: On the Realizability of Quantum Computers, journal: Ubiquity, ACM (2006)
9. Moraga, C.: On some basic aspects of ternary reversible and quantum computing. In: Multiple-Valued Logic (ISMVL). ISSN: 0195-623X (2014)
10. Wootters, W.K.: Entanglement of formation and concurrence. J. Quantum Info. Comput., Jan 2001, Rinton Press, Incorporated. ISSN: 1533-7146 (2001)
11. Herreno-Fierro, C., Luthra, J.R.: Generalized concurrence and limits of separability for two qutrits, journal (2005). arXiv:0507223
12. Michel, P., Anne-Cline, B., Metod, S.: Multi-line geometry of qubitqutrit and higher-order pauli operators. J. Int. J. Theoret. Phys. Springer, US (2008). ISSN: 0020-7748

A Survey on Collaborative Filtering: Tasks, Approaches and Applications

H. P. Ambulgekar, Manjiri Kishor Pathak and M. B. Kokare

Abstract Recommendation systems are tools of option used to select the data relevant to a given user from online sources. Collaborative filtering (CF) could be the most thriving approach to build recommendation systems and has been employed in many applications. Collaborative filtering algorithms are the much explored techniques within the field of information mining and data retrieval. In CF, users' past behavior is analyzed so as to determine the connections between the user and their items of interest to suggest an item to the user supported by the opinion of different users with similar liking. CF relies on the actual fact that those customers, who have similar liking within the past, will have similar liking within the future. This paper makes a comprehensive introduction to collaborative filtering, aiming to facilitate readers to grasp this field. First of all, we introduce the background analysis and then try to differentiate between various CF-based social recommendation systems based on the matrix factorization that uses social factors. This paper conjointly describes applications of collaborative filtering in several domains. In our future works, we will utilize user location info (users' check-ins) to advocate additional personalized and real-time items.

Keywords Recommendation systems · Collaborative filtering · Matrix factorization

H. P. Ambulgekar · M. K. Pathak (✉)
Department of Computer Science and Engineering, Shri Guru Gobind Singhji Institute of Engineering and Technology, Nanded 431606, Maharashtra, India
e-mail: 2016mns018@sggs.ac.in; pathakmanjiri25@gmail.com

H. P. Ambulgekar
e-mail: ambulgekar@sggs.ac.in

M. B. Kokare
Department of Electronics & Telecommunication Engineering, Shri Guru Gobind Singhji Institute of Engineering and Technology, Nanded 431606, Maharashtra, India
e-mail: mbkokare@sggs.ac.in

M. Chakraborty et al. (eds.), *Proceedings of International Ethical Hacking Conference 2018*, Advances in Intelligent Systems and Computing 811,
https://doi.org/10.1007/978-981-13-1544-2_24

1 Introduction

The volatile expansion of the accessible digital data and also the range of users created a challenge of data surplus that inhibits suitable access to items of interest on the Internet. Data retrieval system, like Google, has partly resolved this drawback; however, prioritization and personalization of data were absent. This has raised the demand for recommender systems. Recommender systems are very popular for recommending user's favorite things or services from great amount of dynamically generated data. It associates data filtering technique that provides user the knowledge within which they are interested in and additionally manage the matter of data overload [1].

Recently, various approaches have been developed for building recommendation systems such as collaborative, content based or hybrid filtering [2]. Collaborative filtering (CF) is fully grown and therefore the most ordinarily enforced. CF could be a technique that is employed to make customized recommendations supported their past ratings for the product or services. It is a most prominent approach to generating recommendations. The basic hypothesis in CF is that user U and user V's personal interest are matched if each user rates n variety of comparable things [3]. Collaborative filtering needs ratings for associated item to form a prediction. Rating is an association of a user associated an item in terms of a price. Rating is often either implicit or explicit [4]. Explicit rating needs the user to rate associated item in terms of import such as rating from 0 to 5 stars which are commonly found on movie and music recommendation. Implicit rating consists of user's preference from his or her actions. If a user visits a web page, for instance, on Amazon, wherever user purchase things or add them to a listing however if he finishes up shopping for the merchandise, then it over that the user includes robust interest in similar products.

Collaborative recommender systems have been implemented in popular, large, commercial e-commerce Web sites including Amazon, Netflix and Delicious. It contains numerous algorithms to form predictions a few user's interests by collecting preferences from many active users. CF relies on the actual fact that those customers, who have similar liking within the past, will have similar liking within the future. For example, someone who desires to visualize a movie could kindle recommendations from some friends who have similar tastes that are trusty quite recommendations from others. These data are employed to create a choice on that film to visualize.

2 Related Work

In this section, we tend to shortly review the most works within the context. Our main focus is in collaborative filtering which has been with success applied to many real-world problems, like Netflix's movie recommendation.

In 2003, Greg Linden, Brent Smith and Jeremy York presented one industry report on Amazon.com recommendations [5]. They match up to traditional collaborative filtering, clustering models and search-based methods with the item–item collaborative filtering method to solve the recommendation problem. Sarwar et al. [4] proposed item-based CF recommendation algorithms to produce high-quality recommendation for large-scale troubles in 2001. The authors found that users recommend items similar to the users who were previously interested. Bell et al. [6] proposed item-based recommendation model to improve accuracy by using relationships on multiple scales. They found that item-oriented approaches offer higher-quality estimates than user-oriented approaches, whereas additional economical computations are permitted.

There are two winning approaches to CF: First is latent factor models that directly profile each users and merchandise and second is neighborhood models that analyze similarities between merchandise and users. Koren et al. [7] published another paper Multifaceted Collaborative Filtering Model which combines both of these models by building a more precise combined model. Deshpande and Karypis [8] proposed an item-based CF that presents model-based recommendation algorithms that first calculate the similarities by combining condition-based probability similarity and cosine similarity between numerous items and then recognize the set of items to be suggested.

Wang [9] formulated again the memory-based collaborative filtering problem. In this paper, rating is estimated by combining predictions from three sources: item–item similarity, user–user similarity and additional ratings from related users toward related items are employed to flat the predictions. Liu and Zhao [10] built the probabilistic latent preference analysis (pLPA) model to rank predictions. It predicts the rating by using user preferences to a set of items rather than the rating scores on individual items. Harvey et al. [11] presented Bayesian latent variable model for rating prediction that models ratings over each user's latent interests and item's latent topics instead of using user preferences to items. The iExpand [12] model was proposed which makes recommendations by exploiting the information about user's latent interests.

Up to this, we can say collaborative filtering-based recommendation approaches are the first generation of recommender system [13]. Now, we see recommendation models based on social networks. These models are proposed to improve recommendation system performance. There are several basic matrix factorization approaches without considering any social factors BaseMF [14, 15, 16].

Yang and Hill [17] proposed the concept of 'inferred trust circle' based on circles of friends to recommend favorite and useful items to users. Huang et al. [18] proposed an approach to replicating the utility of social recommendation by combining three social factors: receivers' interest, item qualities and interpersonal influences to recommend items based on user–item utility. Ma et al. [19] proposed social recommendation model based on the probabilistic matrix factorization which uses factor analysis approach by using both user's social network information and rating records in 2008. Apart from the interpersonal influence, Jiang et al. [20] proposed social contextual recommendation model (contextMF) and proved that individual preference is also an important factor in social network to make recommendations. All the above algorithms are based on the probabilistic matrix factorization [14]. This method goes beyond traditional item-based collaborative filtering model in [9], influence-based model in [20] and Sorec in [21] by taking interpersonal influence and individual preference into consideration. In 2013, personalized recommendation model [21] shows three social factors: personal interest, interpersonal interest similarity and interpersonal influence, and these factors are combined with a unified personalized recommendation model based on probabilistic matrix factorization. Lyu et al. [22] proposed a matrix factorization framework with social regularization. They combined two social regularization terms to impose constraints between user and their friends. Social regularization that is based on similarity handles different friends in a different fashion. A subset of friends are considered while predicting rating in a specific circle. This type of recommendations is circle-based recommendation. We can have a separate MF for each category by applying circle-based recommendations to the SocialMF [23] model.

Jamali [23] proposed a matrix factorization-based model for recommendation in social rating networks, called SocialMF. They incorporated the propagation of trust into their model to boost the standard of recommendation. It improves the advice accuracy of BaseMF by taking into social trust between users. It forever uses all social links on the market within the data set. In 2016, Qian [24] proposed a user-service rating prediction model based on probabilistic matrix factorization by exploring user's social rating behaviors. In order to predict user-service ratings, they focus on users' rating behaviors. Social users' rating behaviors could be mined from four factors: personal interest, interpersonal interest similarity, interpersonal rating behavior similarity and interpersonal rating behavior diffusion. These factors are fused in a unified probabilistic matrix factorization framework. They proposed this model to directly fuse interpersonal factors together to constrain user's latent features, which can reduce the time complexity of model.

2.1 Preliminary Steps of Collaborative Filtering

Data Gathering. The system must have to collect data in order to make any recommendations. The main goal of gathering info is to urge an inspiration of user preferences, which is able to later be won't to build predictions on user preferences in future [1]. There are two ways to collect the data. The very first method is explicit data collection. During this, user gives rating on a rating scale like rating a movie from one to five stars. It is easy to work with explicit data gathering [25]. The second method is gathering data implicitly. The system collects the user's preferences by observation of various actions of users like the history of purchases, looking history and time spent on some sites, and links followed by the user and content of e-mail will be the factors that may indirectly change opinion by observing its user behavior [13, 26].

Data Filtering. Once information is collected, there are two basic ways in which the info is filtered to form predictions. The foremost basic technique is passive

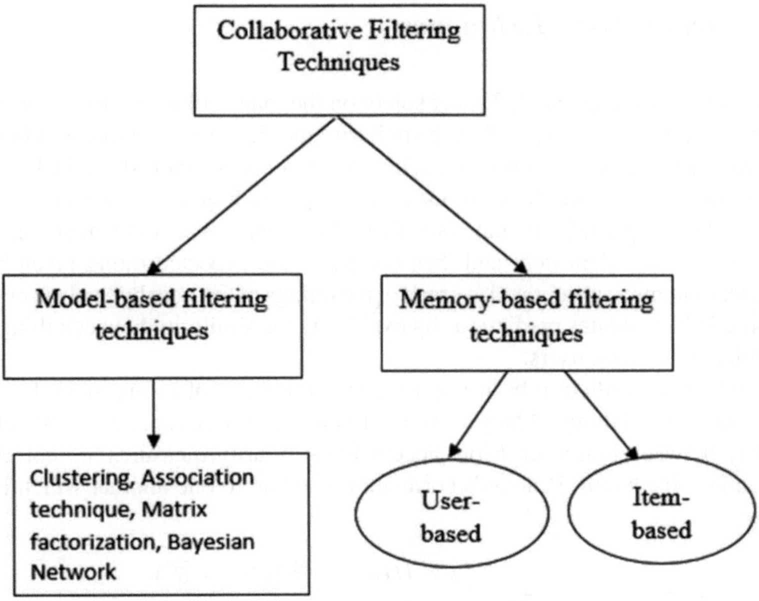

Fig. 1 Collaborative filtering techniques

collaborative filtering that merely aggregates the info to form predictions like the typical rating for associated item. The additional advanced methodology is active collaborative filtering that uses patterns in user history to create predictions [27]. For instance, find similar users to other user and exploit their history to predict a rating.

Prediction. System gives recommendations or predicts things the user could like. This may be created either directly from the information collected in data gathering task that can be memory based mostly or model based or through the system's determined activities of the user (Fig. 1).

3 Collaborative Filtering Techniques

Collaborative filtering may well be a domain-independent prediction technique for content that cannot merely and adequately be delineating by information like movies and music. It works by making information (user-item matrix) of preferences for things by users. Collaborative filtering technique then matches similarities between users' profiles to make recommendations [28]. Such users build a bunch cited as neighborhood. A user gets recommendations to those things that he has not rated before but that was already fully rated by users in his neighborhood. There are two types of techniques identified as memory based and model based [29, 27].

3.1 Memory-Based Techniques

Memory-based strategies [28, 30] act solely on the matrix of user ratings for things. Memory-based methods typically use similarity metrics to get the freedom between two users, or two things, supported each of their ratios. Memory-based CF is often achieved in the following fashion: user-based approach and item-based approach. User-based CF approach calculates similarity between two users by analyzing their ratings on an equivalent item and then computes the expected rating given by the lively user to items as a weighted average of the ratings of the item by lively users [31]. Item-based CF evaluates predictions by exploiting the similarity between things, not the similarity between users.

Systems give prediction by using a weighted average of rating given by active users on the related things. Many sorts of similarity measures are accustomed figure similarity between item/user. More acceptable similarity measures are correlation based and cosine based. Pearson's correlation is relaid to one another which is outlined as:

$$Sim(u, v) = \frac{\sum u \in U\left(R_{u,i} - \overline{R_u}\right)\left(R_{v,i} - \overline{R_v}\right)}{\sqrt{\sum u \in U\left(R_{u,i} - \overline{R_u}\right)^2}\sqrt{\sum u \in U\left(R_{v,i} - \overline{R_v}\right)^2}}, \tag{1}$$

where Sim(u, v) denotes the likeness between two users u and v, $R_{u,i}$ is the rating given to item i by user u and $\overline{R_u}$ is the mean rating given by user u [9]. Also, prediction for associated item is created from the weighted mixture of the chosen neighbor's ratings that are computed as a result of the weighted variation from the neighbors mean. The prediction formula is

$$P(u, i) = \overline{R_u} + \frac{\sum_{i=1}^{n}\left(R_{v,i} - \overline{R_v}\right) \times s(u, v)}{\sum_{i=1}^{n} s(u, v)}. \tag{2}$$

Cosine similarity is completely dissimilar from Pearson's correlation similarity method. It is a vector space model calculating the correspondence among two n-dimensional vectors supported the angle in between them [3]. The similarity between two things u and v is defined as [32].

$$S(\vec{u}, \vec{v}) = \frac{|\vec{u}| * |\vec{v}|}{\vec{u} \cdot \vec{v}} = \frac{\sum i R_{u,i} R_{v,i}}{\sqrt{\sum_i R_{u,i}^2} \times \sqrt{\sum_i R_{v,i}^2}}. \tag{3}$$

Similarity measures are ways accustomed compute the score that specific how related users or things are different. For recommendation generation, scores will be used as it is the basis. Sometimes, similarity metrics can also cited as correlation or distance metrics [13].

3.2 *Model-Based CF Techniques*

The major disadvantage of memory-based CF technique is that it utilizes whole data set associated with user-item data sets since this method is not work as compared to alternative CF system [33]. To beat this issue, model-based CF techniques are proposed by most of the researchers. This technique uses some tiny data sets referred to as model. This model extracts some data from the massive information associated with explicit attribute. This technique uses this model anytime, while not using vast information; as a result, models will increase each speed and quantifiability of advice system. Samples of these techniques embrace dimensionality reduction techniques. There are various decompositions like singular value decomposition (SVD), principle component analysis and matrix factorization techniques such as probabilistic matrix factorization (PMF) and nonnegative matrix factorization (NNMF) which solve the disadvantages of memory-based CF algorithms [1].

Matrix Factorization: Matrix factorization was discovered to increase the performance of CF [34]. In matrix factorization, we are completing the rating matrix with the help of product of two latent feature matrices. Matrix factorization maps both user and items to joint latent factor space of dimensionality f, and user–item interaction is displayed as inner product in that space. Consequently, each item i and each user u are associated with vector $q_i \in R^f$ and vector $p_u \in R^f$, respectively. For an item i, the elements of q_i measure the degree to which the item holds. For the some user u, the elements of p_u measure the degree of interest the user has in items that are high on the equivalent factors positive or negative. The resulting dot product $q_i^T p_u$ captures the interaction between user u and item i. This approximates user u's rating of item i which is denoted by $r_{u,i}$ and calculated as:

$$r_{ui} = q_i^T p_u.$$

To learn the factor vectors p_u and q_i, the system reduces the regularized squared error on the set of known ratings as:

$$\mathrm{L} = min_{q*,p*} \sum_{(u,i)\in k} \left(r_{ui} - q_i^T p_u\right)^2 + \lambda\left(\left\|q_i^2\right\| + \left\|p_u^2\right\|\right). \quad (4)$$

Here, k is the set of the (u, i) pairs for which $r_{u,i}$ is known in training set. Overfitting happens when a model learns noise in the training data to the extent that it negatively impacts the performance of the model on new data. To avoid over fitting to model regularization is used. And this is done by adding parameter λ.

Singular Value Decomposition. SVD is dominant dimensionality reduction technique which reduces dimensionality of rating matrix and identifies latent factors of data. It is a particular realization of the MF approach and also related to PCA [35]. The key issue in an SVD is to find a lowe-dimensional feature space. SVD of an m × n matrix R is of the form.

$$\mathrm{SVD(R)} = \mathrm{P}\sum \mathrm{Q}, \tag{5}$$

where P and Q are m × m and n × n orthogonal matrices, respectively. Σ is the m × n singular orthogonal matrix containing nonnegative elements. An m × m matrix P is named orthogonal if P^TQ is equal to an m × m unit matrix. The diagonal parts in Σ ($\sigma_1, \sigma_2, \sigma_3, \ldots, \sigma n$) are referred to as singular values of matrix R. Generally, singular values are positioned within decreasing direction in Σ. P and Q are the column vectors referred to as left singular vectors and right singular vectors severally.

Nonnegative Matrix Factorization. It is a group of algorithms in multivariate analysis which means it involves observation and analysis of more than one statistical outcome variables at a time and linear algebra where matrix R is factorized into two matrices P and Q with the property that all three matrices have no negative elements [16]. Let the input matrix $X = (x_{1,} x_2, \ldots, x_n)$ contain a collection of n data vectors as columns. Consider factorization of the form:

$$X \approx PQ^T \tag{6}$$

where $X \in R^{N\times M}$, $P \in R^{D\times N}$ and $Q \in R^{D\times M}$. For example, the SVD can be written in this form. In the case of SVD, there is no restriction on the signs of P and Q so the data matrix X is also unconstrained. But in case of NMF, data matrix X is assumed to be nonnegative, as are the factors P and Q. NMF based on the fact that various types of real-world data particularly all images or videos are nonnegative maintains such nonnegativity constraints in matrix factorization.

4 Collaborative Filtering Applications

The CF system GroupLens [36] recommends Usenet news that may be a highly discussed news service on the Internet. It is supported by the client/server architecture. Two major challenges addressed by this method: One is tiny life span of Netnews and second is sparseness of the rating matrices. Users and Netnews are grouped within the system with the present news groups, and then implicit ratings are generated by evaluating the spend time by users for reading Netnews.

The most popular Amazon.com is one of the largest e-commerce websites, and it is also largest recommendation engine. Amazon uses item-to-item CF techniques to suggest online items for various users. The procedure rule scales severally of the number of users and things [5] inside the information. This e-commerce recommendation engine uses a precise data gathering technique to collect user information. User interface is created which is based on the subsequent sections, user browsing details, rating this stuff and getting better recommendations and user profile. User's interests are predicted by the system on the basis of things he/she has rated. Then, users browsing patterns are compared by the system. On the basis of the browsing patterns, system decides the item of user's interest to advocate to the user. Popular

Table 1 Mainstream collaborative filtering systems

Domains	Collaborative filtering systems
E-commerce	Amazon.com, ebay
Music	Music.yahoo.com, Ritigo, CocoA, CDNOW
Video	MovieLens, MovieFinder.com, Youtube, youku.com, Netflix, Rate your music
News	GroupLens, PHOAKS, P-Tango
Books	Douban.com, BookCrossig
Others	Facebook, JesterJokes

feature of Amazon recommenders is that "people who purchased this item additionally bought these items".

Rm and Maes [37] is a user-based CF system that makes recommendations of music albums and artists. In Ringo, once user enters the system, a list of 125 artists has been given to the user to rate consistent with what proportion he likes taking note of them. The list formed from two totally dissimilar sections. The primary session contains most typically rated artists, and active users get chance to rate artists that other users have rated uniformly, so there are similarities between diverse users' profiles. The second session is created upon an arbitrary choice of things from the entire user–item matrix, so as to all artists and albums are able to rate within the initial rating phases at the same point (Table 1).

5 Conclusion

Collaborative filtering is fully grown technique; therefore, it is employed in many domains to make customized recommendations supported their past ratings for the product or services. There are various collaborative filtering methods which made considerable improvement over the last decade. In this paper, we briefly describe numerous collaborative filtering methods which are divided into two categories: memory-based methods and model-based methods. These methods were proposed by many researchers, and several e-commerce websites have taken advantage of this. In this paper, we reviewed some applications of collaborative filtering techniques. We also presented survey on CF-based social recommendation systems and how social network information can be adopted by recommender systems as additional input for improved accuracy.

In this paper, we tried to differentiate between various CF-based social recommendation systems based on the matrix factorization that uses social factors. Due to increasing popularity of social media, new algorithms for recommender systems are needed to extract different meaningful information. Algorithms that we have surveyed are trained and tested offline. Different social components such as individual preference, interpersonal influence, personal interest and interpersonal interest are considered in matrix factorization-based social recommendation systems to improve the prediction and recommendation accuracy. These social factors are important in social network to make recommendation. Some authors also utilized social trust information along with social factors to improve accuracy. But, papers we surveyed here only use different social factors which are related to users and his/her friends. We can also use some additional information like location of users and his/her friends to make personalized recommendation.

6 Future Work

A location of a user is one of the foremost necessary factors while defining user's context. Behavior of user and its preferences will be learned from its locations and location history. The large volume of location-related information generated by users improves the probability that social opinions, e.g., the most favorite dish in an exceedingly restaurant or the foremost widespread activity at some extent of interest, will be accurately assessed by recommender systems. Applications like Foursquare and Yelp encourage individuals to share their current locations, like restaurants or museums that are the foremost widespread kind of location-based social networking services. Existing personalized recommendation model entirely takes user past rating records and social association of social network into thought. In our future works, we are going to take user location info (user's check-ins) to advocate additional personalized and real-time items.

References

1. Konstan, J.A., Riedl, J.: Recommender systems: from algorithms to user experience **22**(1–2), 101–123 (2012)
2. Merve Acilar, A., Arslan, A.: A collaborative filtering method based on artificial immune network **36**(4), 8324–8332 (2009)
3. Zhou, J., Luo, T.: Towards an introduction to collaborative filtering, pp. 576–581 (2009)
4. Sarwar, B., et al.: Item-based collaborative filtering recommendation. In: Proceedings of the 10th International Conference on World Wide Web, pp. 285–295 (2001)
5. Linden, G., Smith, B., York, J.: Amazon. Com. IEEE Internet Comput. **7**(1) (2003)
6. Bell, R.M., Koren, Y., Volinsky, C., Ave, P., Park, F.: Modeling relationships at multiple scales to improve accuracy of large recommender systems, pp. 95–104 (2007)
7. Koren, Y., Ave, P., Park, F.H.D.: Management, D. applications: factorization meets the neighborhood: a multifaceted collaborative filtering model, pp. 426–434 (2008)
8. Deshpande, M., Karypis, G.: Item-Based Top-N Recomm. Alg. **22**(1), 143–177 (2004)
9. Wang, J., De Vries, A.P., Reinders, T.: Unifying user-based and item-based collaborative filtering approaches by similarity. In: 29th Annual International ACM SIGIR Conference, Jan 2006

10. Liu, N.N., Zhao, M.: Probabilistic Latent Preference Analysis for Collaborative Filtering, pp. 759–766 (2009)
11. Harvey, M., Carman, M.J., Ruthven, I., Crestani, F.: Bayesian Latent Variable Models for Collaborative Item Rating Prediction Categories and Subject Descriptors, pp. 699–708 (2011)
12. Liu, Q., et al.: Enhancing collaborative filtering by user interest expansion via personalized ranking **42**(1), 218–233 (2012)
13. Adomavicius, G., Tuzhilin, A.: Toward the next generation of recommender systems: a survey of the state-of-the-art and possible extensions **17**(6), 734–749 (2005)
14. Salakhutdinov, R., Mnih, A.: Probabilistic Matrix Factorization, pp. 1–8 (2007)
15. Takács, G., et al.: Investigation of various matrix factorization methods for large recommender systems. In: 2nd KDD Working Large Scale Recommended Systems Netflix Prize Competition, vol. 1, pp. 553–562 (2008)
16. Aghdam, M.H., Analoui, M., Kabiri, P.: Application of nonnegative matrix factorization in recommender systems. In: 6th International Symposium Telecommunication, IST 2012, pp. 873–876 (2012)
17. Yang, X., Hill, M.: Circle-based recommendation in online social networks. In: 18th ACM SIGKDD International Conference on Knowledge Discovery and Data Mining, pp. 1267–1275 (2012)
18. Huang, J., Cheng, X., Guo, J., Shen, H., Yang, K.: Social Recommendation with Interpersonal Influence. IOS Press Amsterdam (2010)
19. Ma, H., Yang, H., Lyu, M.R., King. I.: SoRec: Social Recommendation Using Probabilistic Matrix Factorization, pp. 0–9 (2008)
20. Jiang, M., Cui, P., Liu, R., Yang, Q., Wang, F.: Social Contextual Recommendation, pp. 45–54 (2012)
21. Mei, T., Qian, X., Feng, H., Zhao, G., Mei, T., Member, S.: Personalized recommendation combining user interest and social circle. IEEE Trans. Knowl. Data Eng. **26**(7) (2014)
22. Lyu, M.R., Ave, P., Park, F.: Recommender Systems with Social Regularization, pp. 287–296 (2011)
23. Jamali, M., Ester, M.: A matrix factorization technique with trust propagation for recommendation. In: Social Networks Proceedings of the 2010 ACM Conference on Recommender Systems (RecSys) (2010)
24. Qian, X., Zhao, G.: User-service rating prediction by exploring social users rating behaviors. IEEE Trans. Multimedia **18**(3) (2016)
25. Peska, L.: Multimodal implicit feedback for recommender systems **1885**, 240–245 (2017)
26. Hu, Y., Park, F., Koren, Y., Volinsky, C., Park, F.: Collaborative filtering for implicit feedback datasets. Data Mining ICDM '08 (2008)
27. Breese, J.S., Heckerman, D., Kadie, C.: Empirical analysis of predictive algorithms for collaborative filtering. In: 14th Annual Conference Uncertainty Artificial Intelligence, pp. 43–52 (1998)
28. Candillier, L., Meyer, F., Boull, M.: Comparing state-of-the-art collaborative filtering systems. In: MLDM '07 5th International Conference Machine Learning and Data Mining in Pattern Recognition (2007)
29. Bobadilla, J., Ortega, F., Hernando, A.: Knowledge-based systems recommender systems survey. Knowl. Syst. **46**, 109–132 (2013)
30. Symeonidis, P., Nanopoulos, A., Manolopoulos, Y.: MoviExplain: a recommender system with explanations. In: Third ACM Conference Recommended Systems—RecSys '09, pp. 317–320 (2009)
31. Isinkaye, F.O.: Recommendation systems: principles, methods and evaluation, pp. 261–273 (2015)
32. Herlocker, J.L., Konstan, J.A., Terveen, L.G., Riedl, J.T.: Eval. Collab. Filter. Recomm. Syst. **22**(1), 5–53 (2004)
33. Thomas, P.A.: Survey on recommendation system methods. no. Icecs, pp. 1603–1608 (2015)
34. Li, C.: The research based on the matrix factorization recommendation algorithms, pp. 691–698 (2017)

35. He, B., Lublin, M.: Matrix factorization: objective and ALS algorithm on a single machine. CME **323**, 1–4 (2015)
36. Resnick P., Iacovou N., Suchak M., Bergstrom P., Riedl J.: GroupLens: an open architecture for collaborative filtering of netnews. In: Proceedings of the 1994 ACM Conference on Computer Supported Cooperative Work (1994)
37. Rm, A.S., Maes, P.: Social information filtering: algorithms for automating 'Word of Mouth', pp. 210–217 (1995)

Part VII
Session 3A: Artificial Intelligence

Feature Subset Selection of Semi-supervised Data: An Intuitionistic Fuzzy-Rough Set-Based Concept

Shivam Shreevastava, Anoop Tiwari and Tanmoy Som

Abstract We are surrounded by a spate of data generating from various sources. To extract some relevant information from these data sets, many pre-processing techniques have been proposed, in which feature selection technique is widely used. However, most of the feature selection approaches focus on supervised learning, which operates on labelled data only. In real-world applications, such as medical diagnosis, forensic science, both labelled and unlabelled data instances are available. Semi-supervised learning handles these types of situations. Some of the researchers have presented rough set as well as fuzzy-rough set-based methods for feature selection of semi-supervised data sets but these approaches have their own limitations. Intuitionistic fuzzy sets maintain a stronger potency of exhibiting information and better drawing and representing intricate ambiguities of the uncertain character of the objective world when compared with fuzzy sets, as it considers positive, the negative and hesitancy degree simultaneously. In this paper, we have proposed a novel feature selection technique for partially labelled data set based on intuitionistic fuzzy-rough set theory. Moreover, we have presented supporting theorems and proposed a novel algorithm to compute reduct based on our method. Finally, we have presented supremacy of our approach over fuzzy-rough technique by considering a partially labelled information system.

Keywords Feature selection · Semi-supervised learning · Rough set · Fuzzy set
Intuitionistic fuzzy set

1 Introduction

Feature selection is one of the fundamental and most widely used dimensionality reduction techniques that is frequently used in the area of data mining, statistics, signal processing, machine learning, pattern recognition and bioinformatics [1–8].

S. Shreevastava · A. Tiwari (✉) · T. Som
Varanasi, India
e-mail: anoop.phd2014@gmail.com

M. Chakraborty et al. (eds.), *Proceedings of International Ethical Hacking Conference 2018*, Advances in Intelligent Systems and Computing 811,
https://doi.org/10.1007/978-981-13-1544-2_25

Feature selection is the process of selecting a subset of features, by eliminating irrelevant and insignificant or highly correlated features from the data set by retaining optimum relevant characteristics required for the process of pattern recognition. Feature selection reduces computational cost and removes noise available in the information system, which increases learning efficiency and enhances prediction accuracies. Due to generation of huge amount of data from various information technology sources, many of the decision class labels are found to be missing in the high-dimensional data sets such as gene expression microarray data [9, 10].

On the basis of decision class labels, feature selection techniques can be divided into three categories, viz. supervised, semi-supervised and unsupervised [11–14]. Supervised feature selection techniques [15] are applied to the data sets with all the decision class labels available, and unsupervised feature selection concepts [16, 17] are practiced to the data sets with no decision class labels while semi-supervised attribute reduction methods are applied to the data sets with a combination of available decision class labels and missing decision class labels. Supervised learning approaches learn underlying functional relationship available in the data, while unsupervised learning concepts use some inherent structure available in data and find some groups in the data such that objects in the same group are different from the objects of the other group on the basis of some criteria. Traditional learning methods are unable to exploit unlabelled data for pattern recognition and knowledge discovery. Therefore, semi-supervised learning approaches can play vital role in order to deal with the information system containing both labelled and unlabelled data.

Rough set theory (proposed by Pawlak) [18–20] has been effectively implemented to perform feature selection, but struggles while dealing with real-valued data set as it requires discretization and which may lead to information loss. Dubois and Prade [21, 22] combined the concept of fuzzy set (proposed by Lotfi Zadeh [23, 24]) and rough set and proposed the concept of fuzzy-rough set. Fuzzy-rough set theory has been successfully applied to deal with real-valued data sets in order to reduce the dimension of the data set, especially for feature selection. Very few researches have been presented semi-supervised feature selection techniques based on fuzzy-rough set theory. Atanassov [25–27] proposed the concept of intuitionistic fuzzy set which is logically supposed as an extension of Zadeh's fuzzy set. A fuzzy set presents only the degree to which an object belongs to a set, while intuitionistic fuzzy set provides both membership grade and non-membership grade of an object. The membership and non-membership grades define an indeterminacy index that models the hesitancy in determining the degree to which an object meets a particular property. So, intuitionistic fuzzy set can tackle uncertainty in a much better way than fuzzy set.

In the recent years, some of the intuitionistic fuzzy-rough set models [28–30] were proposed and successfully implemented for pattern recognition and decision-making [31, 32].

In this study, we present a novel intuitionistic fuzzy-rough set assisted feature selection which can easily deal with the data set having both labelled and unlabelled data. Furthermore, we give theorems supporting our concept and prove the validity of the theorems. Moreover, an algorithm based on our proposed method has been presented. Finally, we apply this approach to an information system containing both labelled and unlabelled data and show that it performs better than semi-supervised fuzzy-rough feature selection approach proposed by Jensen et al. [14].

2 Preliminaries

In this section, we discuss some basic definitions regarding intuitionistic fuzzy set and intuitionistic fuzzy information system.

Definition 2.1 [26] Let $\langle p, q\rangle$ be an ordered pair, then $\langle p, q\rangle$ is said to be an intuitionistic fuzzy value if $0 \leq p, q \leq 1$ and $0 \leq p+q \leq 1$.

Definition 2.2 [27] Let X be non-empty universe of discourse of objects. An intuitionistic fuzzy set G in X is collection of objects having the form $G = \{\langle x, m_G(x), n_G(x)\rangle | x \in X\}$, where $m_G : X \rightarrow [0, 1]$ and $n_G : X \rightarrow [0, 1]$ satisfy $0 \leq m_G + n_G \leq 1, \forall x \in X$ and $m_G(x)$ and $n_G(x)$ are positive and negative membership grades of the element $x \in X$ to G, respectively, $\pi_G(x) = 1 - m_G(x) - n_G(x)$ represents hesitancy degree of x in G. It is evident that $0 \leq \pi_G(x) \leq 1,\ \forall x \in X$.

A fuzzy set $G = \{\langle x, m_G(x)\rangle | x \in X\}$ can be identified as intuitionistic fuzzy set in the form $\{\langle x, m_G(x),\ 1 - m_G(x)\rangle | x \in X\}$. Thereby, an intuitionistic fuzzy set is said to be an extension of fuzzy set.

The cardinality of an intuitionistic fuzzy set G can be defined as follows [33]:

$$|\mathrm{G}| = \sum_{x \in X} \frac{1 + m_G(x) - n_G(x)}{2} \tag{1}$$

Definition 2.3 An intuitionistic fuzzy information system (IFIS) is represented as quadruple (X, P, V_{IF}, IF), where X is a non-empty set of finite objects, P is a non-empty set of finite attributes, V_{IF} and IF are sets of all intuitionistic fuzzy values and an information function, respectively, where IF is defined as $IF : X \times P \rightarrow V_{IF}$ such that $IF(x, b) = \langle m_b(x), n_b(x)\rangle \in V_{IF}, \forall b \in P$. When $P = C \cup D$ and $C \cap D = \phi$, where C and D are sets of conditional and decision attributes, respectively, then (X, P, V_{IF}, IF) is recognized as intuitionistic fuzzy decision system.

3 Intuitionistic Fuzzy-Rough Feature Selection

Jensen et al. [14] proposed fuzzy-rough set-based feature selection technique to handle semi-supervised data. However, in the literature, none of the feature selection technique has considered semi-supervised intuitionistic fuzzy information system till date.

Now, in this section and in the next section, we extend concept of feature selection based on fuzzy-rough set for semi-supervised data as follows:

A subset B of set of conditional attributes C can be defined using the intuitionistic fuzzy similarity relation as follows:

$$\left\langle \mu_{R_B}(x, y), \vartheta_{R_B}(x, y) \right\rangle = T\left(\left\langle \mu_{R_a}(x, y), \vartheta_{R_a}(x, y) \right\rangle \right), \forall a \in B \tag{2}$$

where $x = \langle x_1, x_2 \rangle$ and $y = \langle y_1, y_2 \rangle$ are two intuitionistic fuzzy values and T is an intuitionistic fuzzy triangular norm or t-norm. Now, lower and upper approximations of an intuitionistic fuzzy set A in X (universe of discourse) based on the intuitionistic fuzzy similarity relation R is defined as follows [29]:

$$\begin{aligned} R_B \downarrow_I A(x) &= \inf_{y \in X} I(R_B(x, y), A(y)) \\ R_B \uparrow_T A(x) &= \sup_{y \in X} T(R_B(x, y), A(y)), \forall x, \mathrm{y} \in X. \end{aligned} \tag{3}$$

where T and I are intuitionistic fuzzy triangular norm and intuitionistic fuzzy implicator, respectively. Now, on the basis of above defined lower approximation, we can define intuitionistic fuzzy positive region by:

$$pos_B(x) = (R_B \downarrow_I [x]_d)(x). \tag{4}$$

where $[x]_d$ contains all objects having same decision value as x. Now, we consider following intuitionistic fuzzy triangular norm T_w and intuitionistic fuzzy implicator I_w as mentioned in [29]:

$$\begin{aligned} T_w(x, y) &= \langle \max(0, x_1 + y_1 - 1), \min(1, x_2 + y_2) \rangle \\ I_w(x, y) &= \langle \min(1, 1 + y_1 - x_1, 1 + x_2 - y_2), \max(0, y_2 - x_2) \rangle \end{aligned}$$

For every data instance, we can redefine intuitionistic fuzzy positive region as follows:

$$
\begin{aligned}
pos_B(x) &= (R_B \downarrow_I [x]_d)(x) = \inf_{y\in X} I(R_B(x,y), [x]_d(y)) \\
&= \min\left\{ \inf_{y\in[x]_d} I(R_B(x,y), [x]_d(y)), \inf_{y\notin[x]_d} I(R_B(x,y), [x]_d(y)) \right\} \\
&= \min\left\{ \inf_{y\in[x]_d} I\left(\left\langle \mu_{R_B}(x,y), \vartheta_{R_B}(x,y)\right\rangle, \langle 1,0\rangle\right), \right. \\
&\left. \inf_{y\notin[x]_d} I\left(\left\langle \mu_{R_B}(x,y), \vartheta_{R_B}(x,y)\right\rangle, \langle 0,1\rangle\right) \right\} \\
&= \min\left\{ \inf_{y\in[x]_d} \left\langle \min\left(1, 1+1-\mu_{R_B}(x,y), 1+\vartheta_{R_B}(x,y)-0\right), \max\left(0, 0-\vartheta_{R_B}(x,y)\right)\right\rangle, \right. \\
&\left. \inf_{y\notin[x]_d} \left\langle \min\left(1, 1+0-\mu_{R_B}(x,y), 1+\vartheta_{R_B}(x,y)-1\right), \max\left(0, 1-\vartheta_{R_B}(x,y)\right)\right\rangle \right\} \\
&= \min\left\{ \inf_{y\in[x]_d} \langle 1,0\rangle, \inf_{y\notin[x]_d} \left\langle \vartheta_{R_B}(x,y), 1-\vartheta_{R_B}(x,y)\right\rangle \right\} \\
&= \min\left\{ \langle 1,0\rangle, \inf_{y\notin[x]_d} \left\langle \vartheta_{R_B}(x,y), 1-\vartheta_{R_B}(x,y)\right\rangle \right\} \\
&= \inf_{y\notin[x]_d} \left\langle \vartheta_{R_B}(x,y), 1-\vartheta_{R_B}(x,y)\right\rangle
\end{aligned} \tag{5}
$$

So, we define the degree of dependency of decision feature on a subset of conditional features as follows:

$$\Gamma_B = \frac{|pos_B|}{|X|} \tag{6}$$

where X is the universe of discourse and $|.|$ in the numerator is the cardinality of an intuitionistic fuzzy set as defined in Sect. 2 and in denominator, it denotes cardinality of a crisp set.

4 Semi-supervised Intuitionistic Fuzzy-Rough Feature Selection

It is very expensive and time-consuming for data experts to deal large amount of labelled data, and this motivates us for some better technique, viz. semi-supervised techniques in order to learn about small amounts of labelled data and larger amounts of unlabelled data. For handling both labelled and unlabelled data, some modification is required in the definition of positive region as defined in Sect. 3 as follows:

Theorem 4.1 *Let L and U be the sets of labelled and unlabelled objects, respectively, and $\{L, U\}$ is a partition of X (universe of discourse), i.e. $L \cap U = \phi$ and $L \cup U = X$, then positive region in the system can be defined by:*

$$pos_B^{ssl}(x) = \begin{cases} \inf\limits_{y \neq x} \langle \vartheta_{R_B}(x, y), 1 - \vartheta_{R_B}(x, y) \rangle & if x \in U \\ \inf\limits_{y \in (U \cup co([x]_d^L))} \langle \vartheta_{R_B}(x, y), 1 - \vartheta_{R_B}(x, y) \rangle & if x \in L \end{cases} \tag{7}$$

where $[x]_d^L$ *represents the set of labelled objects having same decision value as x and* $co(\cdot)$ *is the complement operator.*

Proof Let $x \in U$. Its decision class contains only x. Now Eq. (5) instantaneously simplifies to

$$pos_B^{ssl}(x) = \inf_{y \neq x} \langle \vartheta_{R_B}(x, y), 1 - \vartheta_{R_B}(x, y) \rangle.$$

If $x \in L$, then decision class of x consists of all labelled instances y satisfying $d(x) = d(y)$. All unlabelled objects are not element of it, as all of them belong to their own individual classes. Therefore, infimum is taken over $U \cup co[x]_d^L$ and it results in

$$pos_B^{ssl}(x) = \inf_{y \in (U \cup co[x]_d^L)} \langle \vartheta_{R_B}(x, y), 1 - \vartheta_{R_B}(x, y) \rangle$$

Hence, we get the required result.

Now for unlabelled data and $L \neq \phi$, new degree of dependency can be redefined as:

$$\Gamma_B^{ssl} = \frac{|pos_B^{ssl}|}{|X|} \tag{8}$$

Theorem 4.2 *For every* $B \subseteq C$, $\Gamma_B^{ssl} \leq \Gamma_B$.

Proof For any given function f and sets R and S along with condition $R \subseteq S$, it is obvious that

$$\inf_{x \in S} f(x) \leq \inf_{x \in R} f(x) \tag{9}$$

If $x \in X$, then, for any semi-supervised model, either $x \in U$ or $x \in L$. Let $x \in U$, then according to Eq. (5), we can conclude that:

$$\begin{aligned} pos_B^{ssl}(x) &= \inf_{y \neq x} \langle \vartheta_{R_B}(x, y), 1 - \vartheta_{R_B}(x, y) \rangle \\ &= \inf_{y \in (X \setminus \{x\})} \langle \vartheta_{R_B}(x, y), 1 - \vartheta_{R_B}(x, y) \rangle \\ &\leq \inf_{y \in co([x]_d)} \langle \vartheta_{R_B}(x, y), 1 - \vartheta_{R_B}(x, y) \rangle \\ &\text{(using Eq. (9) along with } co([x]_d) \subseteq (X \setminus \{x\})) \end{aligned}$$

$$= \inf_{y \notin [x]_d} \langle \vartheta_{R_B}(x, y),\ 1 - \vartheta_{R_B}(x, y) \rangle$$

where $[x]_d$ is established within the completely labelled system. Now, let $x \in L$, then

$$\begin{aligned} pos_B^{ssl}(x) &= \inf_{y \in \left(U \cup co[x]_d^L\right)} \langle \vartheta_{R_B}(x, y), 1 - \vartheta_{R_B}(x, y) \rangle \\ &\leq \inf_{y \in co([x]_d)} \langle \vartheta_{R_B}(x, y), 1 - \vartheta_{R_B}(x, y) \rangle \\ &\quad \left(\text{using Eq. (9) along with } co([x]_d) \subseteq \left(U \cup co\left([x]_d^L\right)\right)\right) \\ &= \inf_{y \notin [x]_d} \langle \vartheta_{R_B}(x, y), 1 - \vartheta_{R_B}(x, y) \rangle = pos_B(x). \end{aligned}$$

Now, in semi-supervised model either objects are labelled or have no label. So, we can conclude that $\forall x \in X, pos_B^{ssl}(x) \leq pos_B(x)$, which implies that $\frac{|pos_B^{ssl}(x)|}{|X|} \leq \frac{|pos_B(x)|}{|X|}$ and hence, $\Gamma_B^{ssl} \leq \Gamma_B$.

5 Algorithm: Semi-supervised Intuitionistic Fuzzy-Rough Feature Selection

In this section, we give the algorithm for feature selection using semi-supervised intuitionistic fuzzy-rough set using the degree of dependency concept, i.e. Γ_B^{ssl}. The algorithm starts with a null set and adds those attributes one by one, which provide a maximum increase in the degree of dependency of decision attribute over a subset of conditional attributes until it achieves highest potential value for any semi-supervised intuitionistic fuzzy information system (it will be 1 in case of a consistent system). This algorithm gives a close-to-minimal reduct of a decision system without exhaustively checking all possible subsets of conditional attributes, which is the key advantage of our proposed algorithm. The resulting algorithm can be given as follows:

Input : *C*, *Collection of all conditional attributes*;
Output : *B*, *the reduct set*;
1. Z ← {}; $\Gamma_{\text{best}}^{ssl} = 0$;
2. do
3. L ← Z
4. foreach $p \in$ (C\B)
5. if $\Gamma_{Z \cup \{p\}}^{ssl} > \Gamma_L^{ssl}$
6. L ← Z ∪ {p}
7. $\Gamma_{\text{best}}^{ssl} = \Gamma_L^{ssl}$

8. $Z \leftarrow L$

9. until $\Gamma^{ssl}_{\text{best}} \neq \Gamma^{ssl}_{C}$

10. return Z

6 Worked Example

An arbitrary example of semi-supervised fuzzy information system inspired from Jensen et al. [34] is given in Table 1 with universe of discourse $X = \{x_1, x_2, x_3, x_4, x_5, x_6\}$, set of conditional attributes $C = \{a, b, c, d, e, f\}$ and one decision attribute q. In this information system, class labels of the objects x_1 and x_5 are missing while class labels for other objects are available.

Now, similarity degree of two objects can be calculated using following similarity relation [35]:

$$R_a(x, y) = 1 - \frac{|\mu_a(x) - \mu_a(y)|}{\left|\mu_{a_{\max}} - \mu_{a_{\min}}\right|} \tag{10}$$

where $\mu_a(x)$, $\mu_a(y)$ are membership grades of objects x, y respectively, and $\mu_{a_{\max}}$, $\mu_{a_{\min}}$ are maximum and minimum membership grades for an attribute a, respectively.

Now, we can calculate the degree of dependency of decision feature q over conditional feature $\{a\}$ using [35] as follows:

$$\gamma^{\text{ssl}}_{\{a\}} = \frac{0.25 + 0 + 0.25 + 0.25 + 0.25 + 0}{6} = \frac{1}{6}$$

Similarly, dependencies functions over $\{b\}$, $\{c\}$, $\{d\}$, $\{e\}$ and $\{f\}$ are:

$$\gamma^{\text{ssl}}_{\{b\}} = \frac{0 + 0 + 0 + 0.33 + 0 + 0.33}{6} = \frac{0.66}{6}$$

Table 1 Fuzzy information system

Objects	Attributes						
	a	b	c	d	e	f	q
x_1	0.4	0.4	1.0	0.8	0.4	0.2	–
x_2	0.6	1.0	0.6	0.8	0.2	1.0	0
x_3	0.8	0.4	0.4	0.6	1.0	0.2	1
x_4	1.0	0.6	0.2	1.0	0.6	0.4	0
x_5	0.2	1.0	0.8	0.4	0.4	0.6	–
x_6	0.6	0.6	0.8	0.2	0.8	0.8	1

$$\gamma^{ssl}_{\{c\}} = \frac{0.25 + 0.25 + 0.25 + 0.25 + 0 + 0}{6} = \frac{1}{6}$$
$$\gamma^{ssl}_{\{d\}} = \frac{0 + 0 + 0.25 + 0.25 + 0.25 + 0.25}{6} = \frac{1}{6}$$
$$\gamma^{ssl}_{\{e\}} = \frac{0 + 0.25 + 0.50 + 0.25 + 0 + 0.25}{6} = \frac{1.25}{6}$$
$$\gamma^{ssl}_{\{f\}} = \frac{0 + 0 + 0.50 + 0.25 + 0.25 + 0.25}{6} = \frac{1.25}{6}$$

Since $\{e\}$ and $\{f\}$ have same degree of dependencies, so we can take any one of them as reduct candidate. Taking $\{e\}$ as reduct candidate, we add other attributes one by one and find degree of dependencies as follows:

$$\gamma^{ssl}_{\{a,\, e\}} = \frac{2.25}{6},\ \gamma^{ssl}_{\{b,\, e\}} = \frac{2.33}{6},\ \gamma^{ssl}_{\{c,\, e\}} = \frac{2.50}{6},\ \gamma^{ssl}_{\{d,\, e\}} = \frac{2.25}{6},\ \gamma^{ssl}_{\{e, f\}} = \frac{2.25}{6}.$$

Now, we insert other attributes to the next reduct candidate, i.e. $\{c, e\}$ and get degree of dependencies as:

$$\gamma^{ssl}_{\{a,\, c, e\}} = \frac{2.50}{6},\ \gamma^{ssl}_{\{b,\, c,\, e\}} = \frac{2.84}{6},\ \gamma^{ssl}_{\{c,\, d,\, e\}} = \frac{3.0}{6},\ \gamma^{ssl}_{\{c,\, e, f\}} = \frac{3.50}{6}.$$

Since $\{c, e, f\}$ provides maximum value of degree of dependency. Hence, other attributes are added to the potential reduct set $\{c, e, f\}$ and corresponding degree of dependencies are:

$$\gamma^{ssl}_{\{a,\, c, e, f\}} = \frac{3.50}{6},\ \gamma^{ssl}_{\{b,\, c,\, e, f\}} = \frac{3.92}{6},\ \gamma^{ssl}_{\{c,\, d,\, e, f\}} = \frac{3.50}{6}.$$

So, we get $\{b, c, e, f\}$ as next potential reduct set and after adding rest of the attributes to this set, we obtain degree of dependencies as follows:

$$\gamma^{ssl}_{\{a,\, \mathrm{b}, c, e, f\}} = \frac{3.92}{6},\ \gamma^{ssl}_{\{b,\, c,\, d,\, e, f\}} = \frac{3.92}{6}.$$

On adding other attributes to the potential reduct set $\{b, c, e, f\}$, we get no increment in degree of dependency. Hence, the final reduct is $\{b, c, e, f\}$.

Now we convert the above semi-supervised fuzzy information system into semi-supervised intuitionistic fuzzy information system using Jurio et al. [36] concept with hesitancy degree 0.2. The reduced information system is given in Table 2.

We define an intuitionistic fuzzy tolerance relation using [37] as follows:

Let $\alpha = 1 - \frac{|\mu_a(x) - \mu_a(y)|}{|\mu_{a_{\max}} - \mu_{a_{\min}}|}$, $\beta = \frac{|\nu_a(x) - \nu_a(y)|}{|\nu_{a_{\max}} - \nu_{a_{\min}}|}$

$$\left\langle \mu_{R_a}(x, y), \nu_{R_a}(x, y) \right\rangle = \begin{cases} \langle \alpha, \beta \rangle & \text{if}\, \alpha + \beta \le 1 \\ \langle 1, 0 \rangle & \text{if}\, \alpha + \beta > 1 \end{cases} \tag{11}$$

where μ_{R_a} and ν_{R_a} are membership and non-membership grades of intuitionistic fuzzy tolerance relation. If R_P is the intuitionistic fuzzy tolerance relation induced by the subset of features P, *then*

$$\langle \mu_{R_P}(x, y), \upsilon_{R_P}(x, y)\rangle = \inf_{a\epsilon P}\langle \mu_{R_a}(x, y), \upsilon_{R_a}(x, y)\rangle$$

Now, we calculate the reduct set of intuitionistic fuzzy information system as given in Table 2 by using Sect. 4 as follows:

$\upsilon_{R_a}(x, y)$ can be calculated by using Eq. (11) and recorded in Table 3.

Now, positive region for object x_1 over attribute $\{a\}$ can be given as:

$$\begin{aligned} pos^{ssl}_{\{a\}}(x_1) &= \inf_{y\neq x_1} \langle \upsilon_{R_a}(x_1, y), 1 - \upsilon_{R_a}(x_1, y)\rangle \\ &= \inf(\langle 0.25, 0.75\rangle, \langle 0.50, 0.50\rangle, \langle 0.75, 0.25\rangle, \\ &\quad \langle 0.25, 0.75\rangle, \langle 0.25, 0.75\rangle) = \langle 0.25, 0.75\rangle \end{aligned}$$

Similarly, positive region for other objects are:

$$\begin{aligned} &pos^{ssl}_{\{a\}}(x_2) = \langle 0, 1\rangle, \; pos^{ssl}_{\{a\}}(x_3) = \langle 0.25, 0.75\rangle, \\ &pos^{ssl}_{\{a\}}(x_4) = \langle 0.25, 0.75\rangle, \; pos^{ssl}_{\{a\}}(x_5) = \langle 0.25, 0.75\rangle, \\ &pos^{ssl}_{\{a\}}(x_6) = \langle 0, 1\rangle. \end{aligned}$$

Table 2 Intuitionistic fuzzy information system

Objects	Attributes						
	a	b	c	d	e	f	q
x_1	⟨0.32, 0.48⟩	⟨0.32, 0.48⟩	⟨0.80, 0.00⟩	⟨0.64, 0.16⟩	⟨0.32, 0.48⟩	⟨0.16, 0.64⟩	–
x_2	⟨0.48, 0.32⟩	⟨0.80, 0.00⟩	⟨0.48, 0.32⟩	⟨0.64, 0.16⟩	⟨0.16, 0.64⟩	⟨0.80, 0.00⟩	0
x_3	⟨0.64, 0.16⟩	⟨0.32, 0.48⟩	⟨0.32, 0.48⟩	⟨0.48, 0.32⟩	⟨0.80, 0.00⟩	⟨0.16, 0.64⟩	1
x_4	⟨0.80, 0.00⟩	⟨0.48, 0.32⟩	⟨0.16, 0.64⟩	⟨0.80, 0.00⟩	⟨0.48, 0.32⟩	⟨0.32, 0.48⟩	0
x_5	⟨0.16, 0.64⟩	⟨0.80, 0.00⟩	⟨0.64, 0.16⟩	⟨0.32, 0.48⟩	⟨0.32, 0.48⟩	⟨0.48, 0.32⟩	–
x_6	⟨0.48, 0.32⟩	⟨0.48, 0.32⟩	⟨0.64, 0.16⟩	⟨0.16, 0.64⟩	⟨0.64, 0.16⟩	⟨0.64, 0.16⟩	1

Table 3 Intuitionistic fuzzy tolerance relation

Objects	x_1	x_2	x_3	x_4	x_5	x_6
x_1	0	0.25	0.50	0.75	0.25	0.25
x_2	0.25	0	0.25	0.50	0.50	0
x_3	0.50	0.25	0	0.25	0.75	0.25
x_4	0.75	0.50	0.25	0	1	0.50
x_5	0.25	0.50	0.75	1	0	0.50
x_6	0.25	0	0.25	0.50	0.50	0

So, degree of dependency can be calculated as mentioned in Sect. 4 and given by:

$$\Gamma^{ssl}_{\{a\}} = \frac{0.25 + 0 + 0.25 + 0.25 + 0.25 + 0}{6} = \frac{1}{6}.$$

Similarly, degree of dependencies over $\{b\}$, $\{c\}$, $\{d\}$, $\{e\}$ and $\{f\}$ are:

$$\Gamma^{ssl}_{\{b\}} = \frac{0.66}{6}, \Gamma^{ssl}_{\{c\}} = \frac{1}{6}, \Gamma^{ssl}_{\{d\}} = \frac{1}{6}, \Gamma^{ssl}_{\{e\}} = \frac{1.25}{6}, \Gamma^{ssl}_{\{f\}} = \frac{1.25}{6}.$$

Since, degree of dependencies for $\{e\}$ and $\{f\}$ are same and the highest. Hence, we can take any one of them as potential reduct candidate. Taking $\{e\}$ as reduct candidate and adding other attributes to this set, we obtain degree of dependencies as follows:

$$\Gamma^{ssl}_{\{a,e\}} = \frac{2.25}{6}, \Gamma^{ssl}_{\{b,e\}} = \frac{2.33}{6}, \Gamma^{ssl}_{\{c,e\}} = \frac{2.25}{6}, \Gamma^{ssl}_{\{d,e\}} = \frac{2.25}{6}, \Gamma^{ssl}_{\{e,f\}} = \frac{2.25}{6}.$$

Since the value of degree of dependency is maximum for $\{b, e\}$, we choose $\{b, e\}$ as a new reduct candidate and adding other attributes to this set, we get degree of dependencies as follows:

$$\Gamma^{ssl}_{\{a,b,e\}} = \frac{3.17}{6}, \Gamma^{ssl}_{\{b,c,e\}} = \frac{2.83}{6}, \Gamma^{ssl}_{\{b,d,e\}} = \frac{3.68}{6}, \Gamma^{ssl}_{\{b,e,f\}} = \frac{3.08}{6}.$$

Since degree of dependency is the largest for attribute set $\{b, d, e\}$, we chose $\{b, d, e\}$ as the next potential reduct candidate and add other attributes to this set and get degree of dependencies as follows:

$$\Gamma^{ssl}_{\{a,b,d,e\}} = \frac{3.17}{6}, \Gamma^{ssl}_{\{b,c,d,e\}} = \frac{3.08}{6}, \Gamma^{ssl}_{\{b,d,e,f\}} = \frac{3.68}{6}$$

Now, we get no increment in degree of dependency. So, process exits and we obtain the reduct set as $\{b, d, e\}$.

We observe that, in case of semi-supervised fuzzy-rough set-based approach the obtained reduct is $\{b, c, e, f\}$ and after applying the proposed technique, we get the reduct set as $\{b, d, e\}$. It is obvious from the given example that our approach provides the smallest reduct set when compared to already existing approach for semi-supervised data sets. Moreover, we can adjust different types of intuitionistic fuzzy t-norms and implicators to handle uncertainty and noise available in the decision system in a better way.

7 Conclusion

Semi-supervised approaches are essential to deal with the abundance of unlabelled data available in high-dimensional data set as it is often costly and heavily time-consuming for domain experts to find decision class labels. This study has proposed a novel concept to feature selection for data set with labelled and unlabelled data. In this paper, we presented an intuitionistic fuzzy-rough set model and generalized it for attribute selection for semi-supervised data. Furthermore, we proposed supporting theorems and an algorithm has been presented in order to demonstrate our approach. Finally, the proposed algorithm applied to a data set and observed that our proposed approach is performing better than previously reported semi-supervised fuzzy-rough feature selection method in terms of selected attributes.

There is further scope of the above approach for feature selection based on discernibility matrix and interval intuitionistic fuzzy-rough set models in order to tackle uncertainty in a much better way. Moreover, a generalized conversion approach of fuzzy information system into intuitionistic fuzzy information system can be presented so that our proposed approach could be applied to real-valued data sets in a wider sense.

References

1. Webb, A.R.: Statistical pattern recognition. Wiley (2003)
2. Jain, A.K., Duin, R.P.W., Mao, J.: Statistical pattern recognition: a review. IEEE Trans. Pattern Anal. Mach. Intell. **22**(1), 4–37 (2000)
3. Kwak, N., Choi, C.H.: Input feature selection by mutual information based on Parzen window. IEEE Trans. Pattern Anal. Mach. Intell. **24**(12), 1667–1671 (2002)
4. Langley, P.: Selection of relevant features in machine learning. In: Proceedings of the AAAI Fall Symposium on Relevance, vol. 184, pp. 245–271, Nov, 1994
5. Kohavi, R., John, G.H.: Wrappers for feature subset selection. Artif. Intell. **97**(1–2), 273–324 (1997)
6. Iannarilli, F.J., Rubin, P.A.: Feature selection for multiclass discrimination via mixed-integer linear programming. IEEE Trans. Pattern Anal. Mach. Intell. **25**(6), 779–783 (2003)
7. Jäger, J., Sengupta, R., Ruzzo, W.L.: Improved gene selection for classification of microarrays. In: Pacific Symposium on Biocomputing, vol. 8, pp. 53–64, Dec 2002
8. Xiong, M., Fang, X., Zhao, J.: Biomarker identification by feature wrappers. Genome Res. **11**(11), 1878–1887 (2001)
9. Xing, E.P., Jordan, M.I., Karp, R.M.: Feature selection for high-dimensional genomic microarray data. In: ICML, vol. 1, pp. 601–608, June 2001
10. Ding, C., Peng, H.: Minimum redundancy feature selection from microarray gene expression data. J. Bioinf. Comput. Biol. **3**(02), 185–205 (2005)
11. Zhao, Z., Liu, H.: Semi-supervised feature selection via spectral analysis. In: Proceedings of the 2007 SIAM International Conference on Data Mining, pp. 641–646. Society for Industrial and Applied Mathematics, April 2007
12. Xu, Z., King, I., Lyu, M.R.T., Jin, R.: Discriminative semi-supervised feature selection via manifold regularization. IEEE Trans. Neural Netw. **21**(7), 1033–1047 (2010)
13. Zhao, J., Lu, K., He, X.: Locality sensitive semi-supervised feature selection. Neurocomputing **71**(10), 1842–1849 (2008)
14. Jensen, R., Vluymans, S., Mac Parthaláin, N., Cornelis, C., Saeys, Y.: Semi-supervised fuzzy-rough feature selection. In: Rough Sets, Fuzzy Sets, Data Mining, and Granular Computing, pp. 185–195. Springer International Publishing (2015)
15. Guyon, I., Elisseeff, A.: An introduction to variable and feature selection. J. Mach. Learn. Res. **3**(Mar), 1157–1182 (2003)

16. Dy, J.G.: Unsupervised feature selection. In: Computational Methods of Feature Selection, pp. 19–39 (2008)
17. Mitra, P., Murthy, C.A., Pal, S.K.: Unsupervised feature selection using feature similarity. IEEE Trans. Pattern Anal. Mach. Intell. **24**(3), 301–312 (2002)
18. Pawlak, Z.: Rough sets. Int. J. Parallel Program. **11**(5), 341–356 (1982)
19. Pawlak, Z., Grzymala-Busse, J., Slowinski, R., Ziarko, W.: Rough sets. Commun. ACM **38**(11), 88–95 (1995)
20. Pawlak, Z.: Rough sets: theoretical aspects of reasoning about data, vol. 9. Springer Science & Business Media (2012)
21. Dubois, D., Prade, H.: Putting rough sets and fuzzy sets together. In: Intelligent Decision Support, pp. 203–232. Springer Netherlands (1992)
22. Dubois, D., Prade, H.: Rough fuzzy sets and fuzzy rough sets. Int. J. General Syst. **17**(2–3), 191–209 (1990)
23. Zadeh, L.A.: Fuzzy sets. information and control **8**(3), 338–353 (1965)
24. Klir, G., Yuan, B.: Fuzzy sets and fuzzy logic, vol. 4. Prentice Hall, New Jersey (1995)
25. Atanassov, K.T.: More on intuitionistic fuzzy sets. Fuzzy Sets Syst. **33**(1), 37–45 (1989)
26. Atanassov, K.T.: Intuitionistic fuzzy sets. Fuzzy Sets Syst. **20**(1), 87–96 (1986)
27. Atanassov, K.T.: Intuitionistic Fuzzy Sets: Theory and Applications. Physica-Verlag (1999)
28. Chakrabarty, K., Gedeon, T., Koczy, L.: Intuitionistic fuzzy rough set. In: Proceedings of 4th Joint Conference on Information Sciences (JCIS), Durham, NC, pp. 211–214 (1998)
29. Cornelis, C., De Cock, M., Kerre, E.E.: Intuitionistic fuzzy rough sets: at the crossroads of imperfect knowledge. Expert Syst. **20**(5), 260–270 (2003)
30. Zhang, X., Zhou, B., Li, P.: A general frame for intuitionistic fuzzy rough sets. Inf. Sci. **216**, 34–49 (2012)
31. Huang, B., Zhuang, Y.L., Li, H.X., Wei, D.K.: A dominance intuitionistic fuzzy-rough set approach and its applications. Appl. Math. Model. **37**(12), 7128–7141 (2013)
32. Zhang, Z.: Attributes reduction based on intuitionistic fuzzy rough sets. J. Intell. Fuzzy Syst. **30**(2), 1127–1137 (2016)
33. Iancu, I.: Intuitionistic fuzzy similarity measures based on Frank t-norms family. Pattern Recogn. Lett. **42**, 128–136 (2014)
34. Jensen, R., Shen, Q.: Fuzzy–rough attribute reduction with application to web categorization. Fuzzy Sets Syst. **141**(3), 469–485 (2004)
35. Jensen, R., Shen, Q.: Computational Intelligence and Feature Selection: Rough and Fuzzy Approaches, vol. 8. Wiley (2008)
36. Jurio, A., Paternain, D., Bustince, H., Guerra, C., Beliakov, G.: A construction method of Atanassov's intuitionistic fuzzy sets for image processing. In: 2010 5th IEEE International Conference on Intelligent Systems (IS), pp. 337–342. IEEE, July 2010
37. De, S.K., Biswas, R., Ranjan Roy, A.: Intuitionistic fuzzy database. In: Second International Conference on IFS, NIFS, vol. 4. no. 2 (1998)

An Efficient Indoor Occupancy Detection System Using Artificial Neural Network

Suseta Datta and Sankhadeep Chatterjee

Abstract Accurate occupancy information in a room helps to provide different valuable applications like security, dynamic seat allocation, energy management etc. This paper represents the detection of human in a room on the basis of some identical features which has been done by using the artificial neural network with three data sets of training and testing with the help of a suitable algorithm from which 97% accuracy for detecting occupancy is being calculated.

Keywords Occupancy detection · Artificial neural network · Security

1 Introduction

Presently, security has become a major issue in order to prevent crimes or discriminatory situations. Occupancy sensors are broadly used to provide security in such cases. Occupancy detection means motion detection in a room or in an office building by which security purpose can be ensured properly. Occupancy detection can be done for both like commercial and noncommercial purposes. Detecting the presence of a human in a room can possibly be done based on some features such as varying the volume of CO_2, light, increasing temperature in the room, and humidity measurement. Cali [1] introduces an algorithm by which the detection of the occupants in a room presented , validated, and evaluated on the basis of concentration of CO_2

S. Datta
Department of Computer Application, University of Engineering and Management, Kolkata, India
e-mail: susetadatta999@gmail.com

S. Chatterjee (✉)
Department of Computer Science and Engineering, University of Engineering and Management, Kolkata, India
e-mail: chatterjeesankhadeep.cu@gmail.com

M. Chakraborty et al. (eds.), *Proceedings of International Ethical Hacking Conference 2018*, Advances in Intelligent Systems and Computing 811,
https://doi.org/10.1007/978-981-13-1544-2_26

most accurately and also can identify the number of occupants in that room. Pedersen [2] initiated the occupancy detection based on the rules applied to the trajectory of sensor data using plug-and-play occupancy detection method where the most reliable performance was found when PIR was used to detect when the building became unoccupied to occupied, and volatile organic compound (VOC) or CO_2 sensor data were used to determine when the building was occupied by human. Candanedo [3] stated that the occupancy detection in an office room using all the data of temperature, light, humidity, and CO_2 may decrease the accuracy of the system. Here three data sets were used such as one for training and two for testing the models on the basis of opening and closing the office door. The result using random forest (RF), linear discriminant analysis (LDA), and classification and regression trees (CART) reflects that the proper selection of the features with a suitable classification model can have a major impact on the accuracy prediction. In this paper, sensing the presence of human in a room based on some identical features such as $CO_{2,}$ humidity, temperature using artificial neural network (ANN) has been done applying the algorithm named scaled conjugate gradient (SCG) to get much accurate result. This appraisal was conducted based on three sets of data where the first purpose being training that consumes one of the data sets and the other which is the most important one, the testing of models that consumes two of the other data sets considering the door of the room has been opened and closed multiple times during the period of occupancy. The current work mainly focuses on ANN-based model only. A rigorous parameter tuning has been performed to find the optimal ANN setup. The performance of the proposed method has been judged in terms of accuracy, precision, recall, and specificity [4, 5].

2 Proposed Method

2.1 *Artificial Neural Network*

Artificial neural network [6–9] is a basic engineering model based on the biological neural structure of a human brain. The neurons are connected by links, and they interact with each other. The nodes are used to take input, and the inputs are passed to the hidden layer where the processing is being done, and the processed data are passed to the output layer. An ANN initially endures a training phase where the recognition of data is being done.

2.2 *Back-Propagation Algorithm*

Back-propagation algorithm [10–15] is one of the most popular training algorithm for ANNs. Back-propagation neural network is basically the pictorial representation by which the error function can be calculated for the uncertain domain. The activity learning of artificial neural network is BPNN. In the current study, ANN has been employed to detect human presence in a room. Scaled conjugate gradient descent algorithm has been used to train the ANN model.

3 Results

A confusion matrix is a specific table layout that is used to describe the performance of a classification algorithm on a set of test data for which the true values are known. When any classification model is confused, the confusion matrix solves the problem by making predictions. It shows not only the errors made by a classifier but also the type of errors that has been made. So it is also called error matrix. Positive (P) condition predicts the number of real positive cases in the data. Negative (N) condition predicts the number of real negative cases in the data. True positive is the proportion of positive cases that were correctly identified. True negative (TN) is the proportion of negative cases that were correctly identified. False positive (FP) is the proportion of negative cases that were incorrectly classified as positive. False negative (FN) is the proportion of positive cases that were incorrectly classified as negative.

$$\mathbf{Accuracy} = \frac{TP + TN}{TP + TN + FN + FP} = \frac{TP + TN}{P + N}$$

$$\mathbf{Precision} = \frac{TP}{TP + FP}$$

$$\mathbf{Specificity} = \frac{TN}{TN + FP} = \frac{TN}{N}$$

$$\mathbf{Sensitivity} = \frac{TN}{TN + FP} = \frac{TN}{N}$$

Figure 1a represents confusion matrix for training, validation, and test. It depicts 95.2% correct predictions. Figure 1b represents ROC curve when neuron number is 10. Figure 1c represents validation performance graph where the best validation performance is 1.8939e−07. Figure 1d represents the error histogram of 20 bins

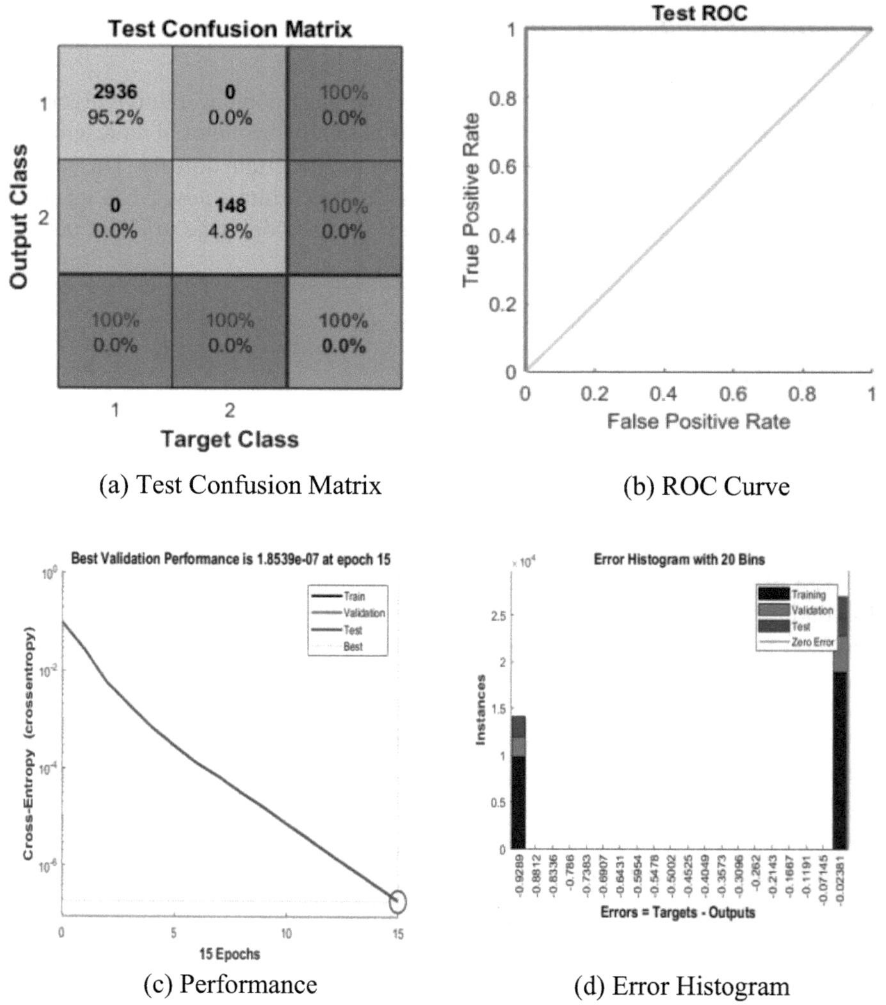

(a) Test Confusion Matrix (b) ROC Curve

(c) Performance (d) Error Histogram

Fig. 1 When considering the number of neuron is 10

for training, validation, and test. Figure 2a represents confusion matrix for training, validation, and test. It depicts 31.3% correct predictions. Figure 2b represents ROC curve when neuron number is 12. Figure 2c represents validation performance graph where the best validation performance is 2.1077e−07. Figure 2d represents the error histogram of 20 bins for training, validation, and test.

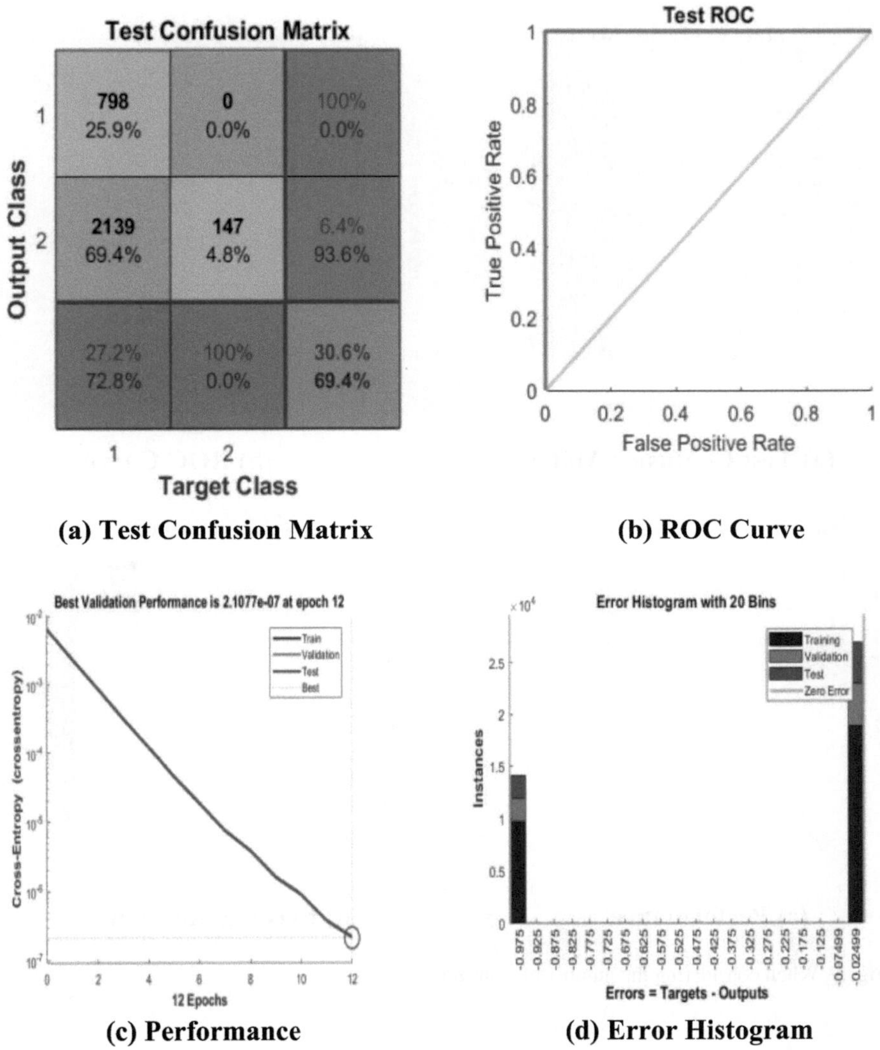

(a) Test Confusion Matrix (b) ROC Curve

(c) Performance (d) Error Histogram

Fig. 2 When considering the number of neuron is 12

Figure 3a represents confusion matrix for training, validation, and test. It depicts 100% correct predictions. Figure 3b represents ROC curve when neuron number is 14. Figure 3c represents validation performance graph where the best validation performance is 1.4098e−07. Figure 3d represents the error histogram of 20 bins for training, validation, and test. Figure 4a represents confusion matrix for training, validation, and test. It depicts 31.3% correct predictions. Figure 4b represents ROC

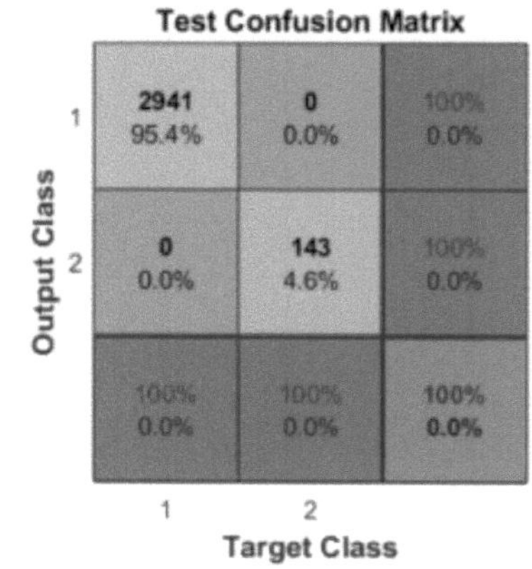

(a) Test Confusion Matrix

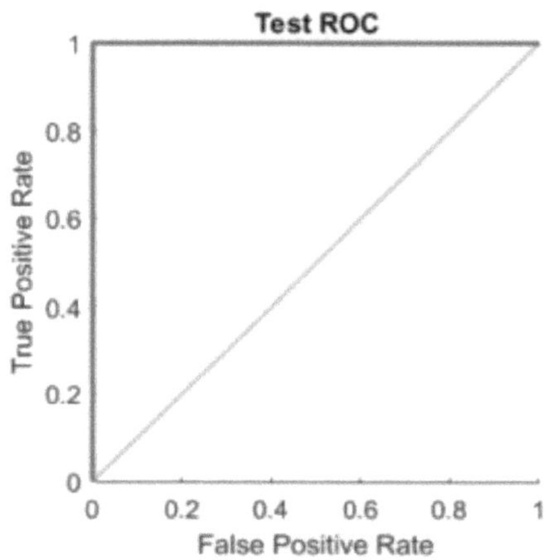

(b) ROC Curve

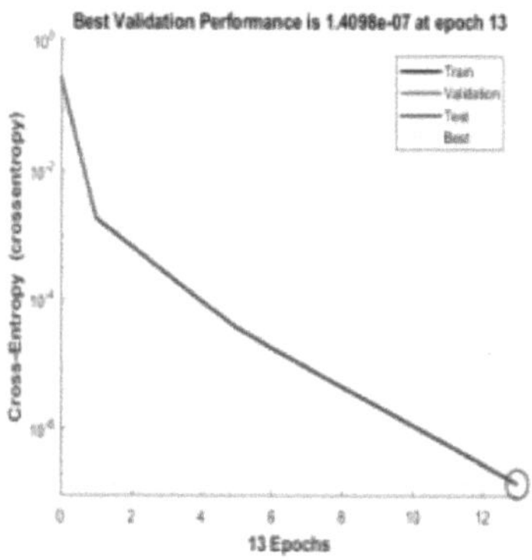

(c) Performance

(d) Error Histogram

Fig. 3 When considering the number of neuron is 14

curve when neuron number is 16. Figure 4c represents validation performance graph where the best validation performance is 1.8416e−07. Figure 4d represents the error histogram of 20 bins for training, validation, and test. Figure 5a represents confusion matrix for training, validation, and test. It depicts 31.3% correct predictions. Figure 5b represents ROC curve when neuron number is 18. Figure 5c represents validation performance graph where the best validation performance is 1.6906e−07. Figure 5d represents the error histogram of 20 bins for training, validation, and test. Figure 6a represents confusion matrix for training, validation, and test. It depicts 31.3% correct predictions. Figure 6b represents ROC curve when neuron number is 20. Figure 6c represents validation performance graph where the best validation performance is 1.6491e−07. Figure 6d represents the error histogram of 20 bins for training, validation, and test.

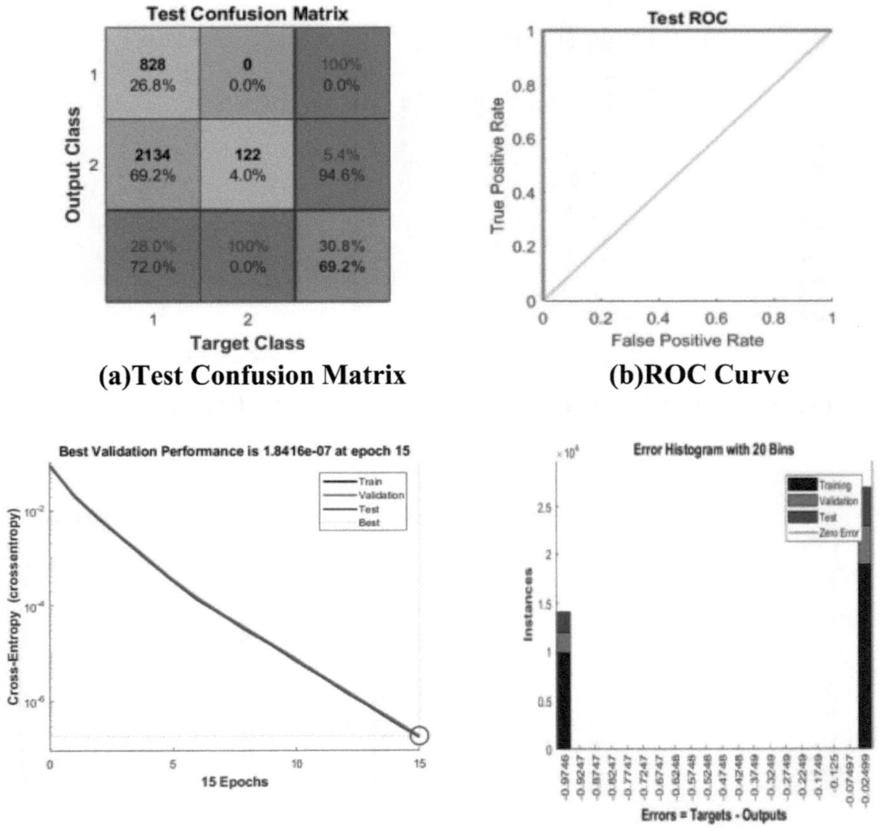

(a)Test Confusion Matrix (b)ROC Curve

(c)Performance (d)Error Histogram

Fig. 4 When considering the number of neuron is 16

Figure 7a represents confusion matrix for training, validation, and test. It depicts 98.5% correct predictions. Figure 7b represents ROC curve when neuron number is 22. Figure 7c represents validation performance graph where the best validation performance is 0.013098. Figure 7d represents the error histogram of 20 bins for training, validation, and test. Figure 8a represents confusion matrix for training, validation, and test. It depicts 99.6% correct predictions. Figure 8b represents ROC curve when neuron number is 24. Figure 8c represents validation performance graph where the best validation performance is 0.0049658. Figure 8d represents the error histogram of 20 bins for training, validation, and test.

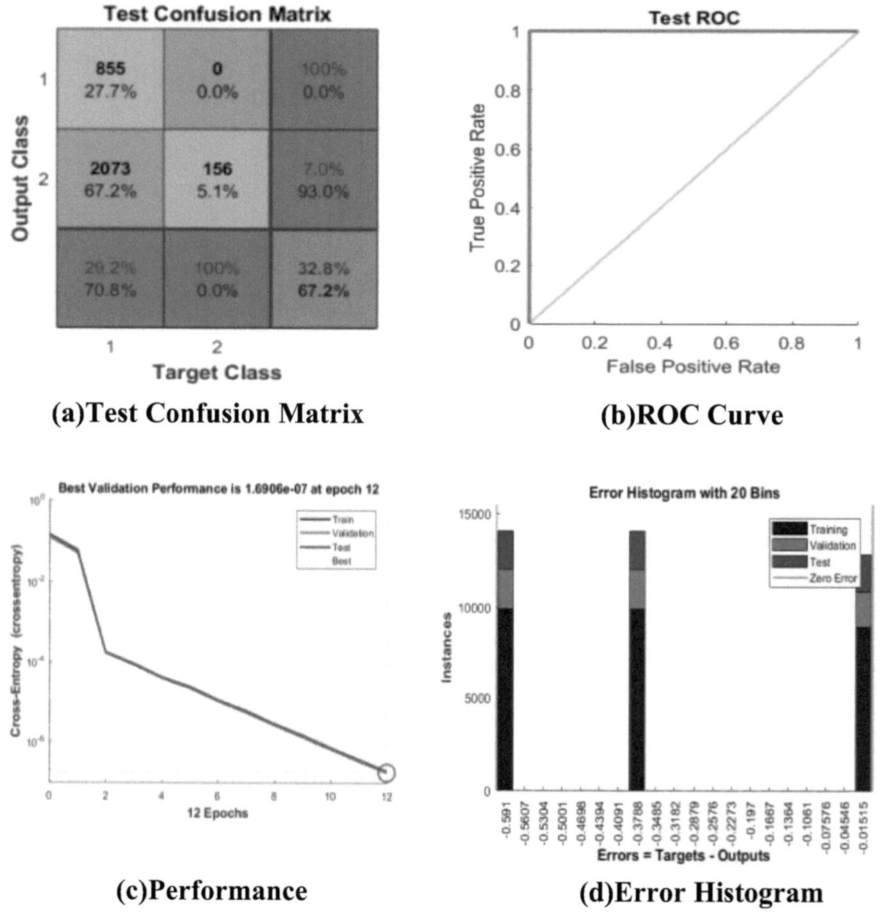

Fig. 5 When considering the number of neuron is 18

Figure 9 depicts the plot of accuracy for different ANN setups. Accuracy (ACC) is calculated as the number of all correct predictions divided by the total number of the data set. Figures 10, 11, and 12 represent the same for precision, recall, and specificity.

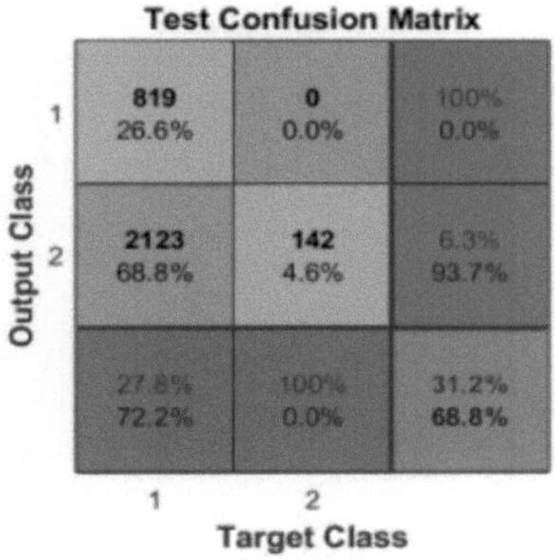

(a)Test Confusion Matrix

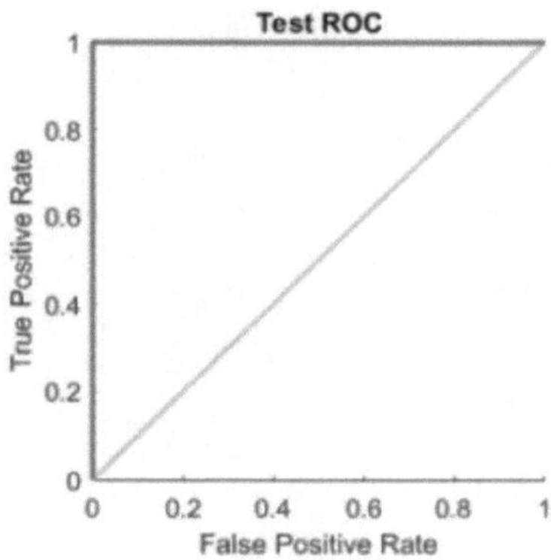

(b)ROC Curve

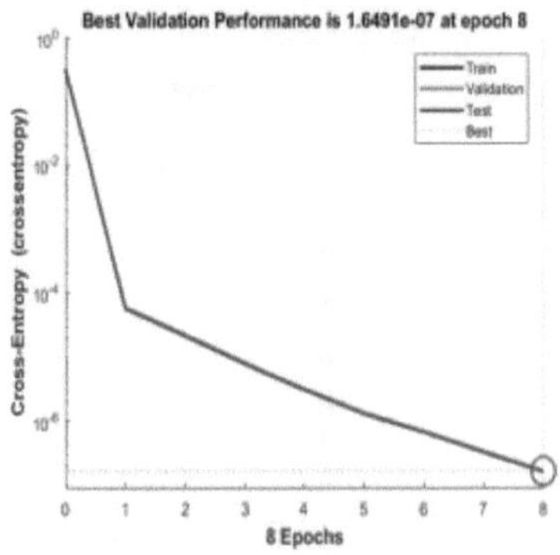

(c)Performance

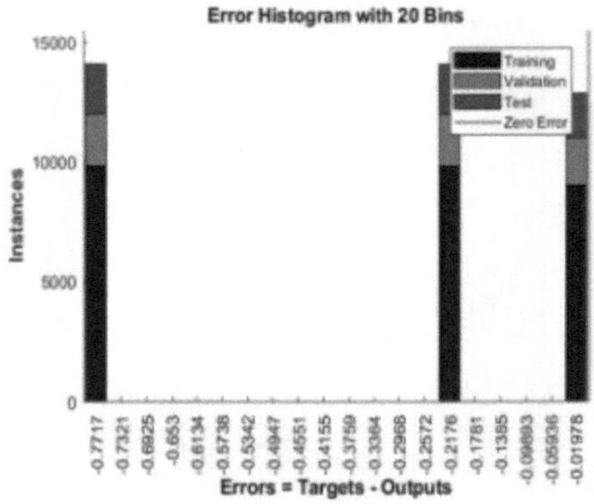

(d)Error Histogram

Fig. 6 When considering the number of neuron is 20

4 Conclusion

The occupancy detection in a room can be done at accuracy level of 95–97% with the artificial neural network using scaled conjugate gradient algorithm. The number of neuron should not be 22 or above as from this point the ROC curve is getting degraded again. We are getting the best result when the number of neuron is 10. This project can be extended in the future using the concept of deep neural network where the extraction of the features can be done without manual operation, so that we may get more authentic result by using it.

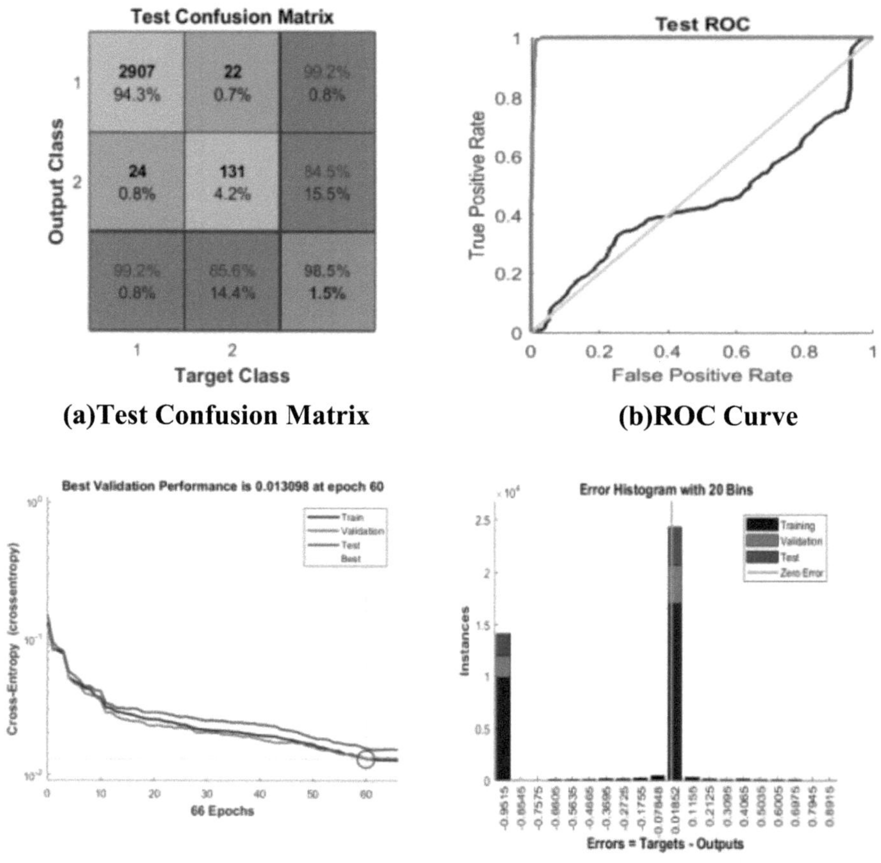

(a)Test Confusion Matrix
(b)ROC Curve
(c)Performance
(d)Error Histogram

Fig. 7 When considering the number of neuron is 22

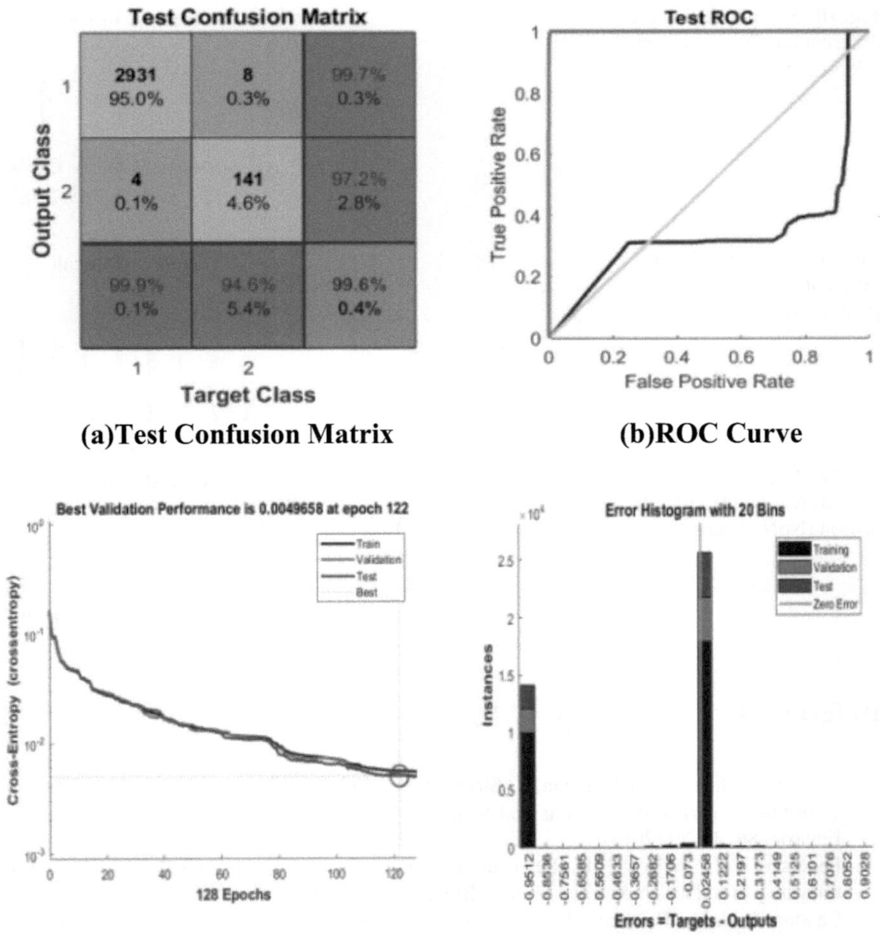

(a)Test Confusion Matrix (b)ROC Curve

(c)Performance (d)Error Histogram

Fig. 8 When considering the number of neuron is 24

Fig. 9 Plot of accuracy for different number of neuron (NoN) setups

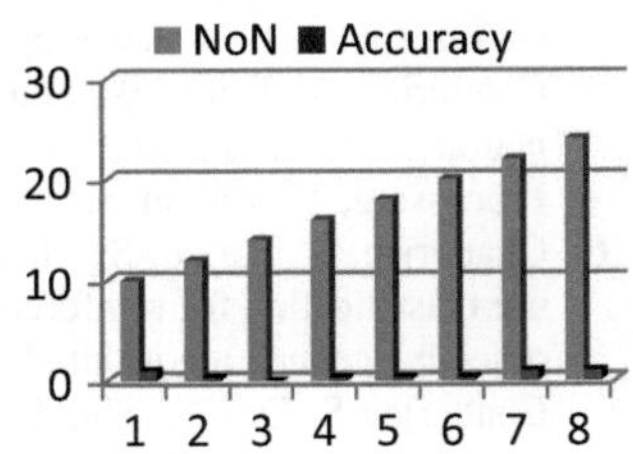

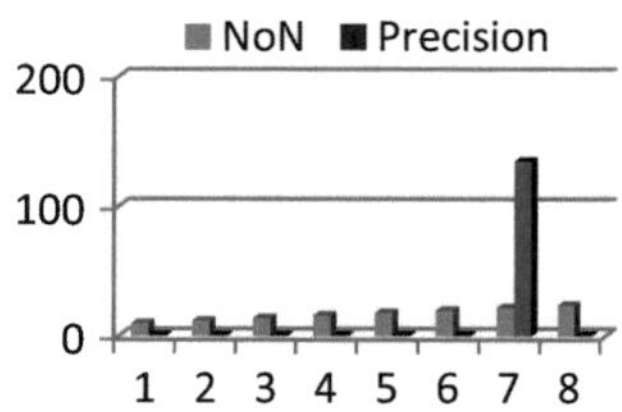

Fig. 10 Plot of precision for different number of neuron (NoN) setups

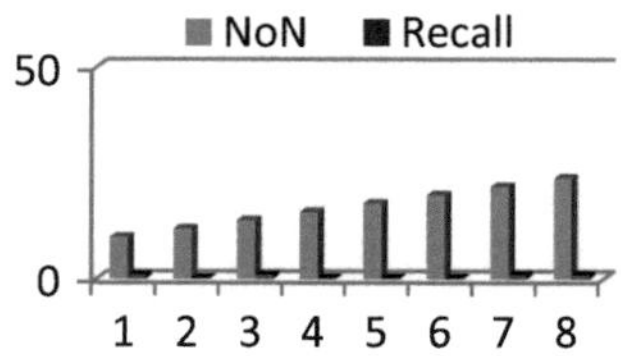

Fig. 11 Plot of recall for different number of neuron (NoN) setups

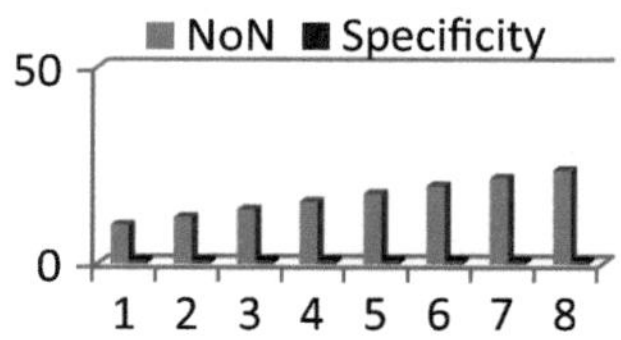

Fig. 12 Plot of specificity for different number of neuron (NoN) setups

References

1. Cali, D., Matthes, P., Huchtemann, K., Streblow, R., Müller, D.: CO_2 based occupancy detection algorithm: experimental analysis and validation for office and residential buildings. Build. Environ. **86**, 39–49 (2015)
2. Pedersen, T.H., Nielsen, K.U., Petersen, S.: Method for room occupancy detection based on trajectory of indoor climate sensor data. Build. Environ. **115**, 147–156 (2017)
3. Candanedo, L.M., Feldheim, V.: Accurate occupancy detection of an office room from light, temperature, humidity and CO_2 measurements using statistical learning models. Energy Build. **112**, 28–39 (2016)
4. Chatterjee, S., Hore, S., Dey, N., Chakraborty, S., Ashour, A.S.: Dengue fever classification using gene expression data: a PSO based artificial neural network approach. In: Proceedings of the 5th International Conference on Frontiers in Intelligent Computing: Theory and Applications (pp. 331–341). Springer, Singapore (2017)
5. Chatterjee, S., Dutta, B., Sen, S., Dey, N., Debnath, N.C.: Rainfall prediction using hybrid neural network approach. In: 2nd International Conference on Recent Advances in Signal Processing, Telecommunications & Computing (SigTelCom)—2018, Vietnam (In press)
6. Chatterjee, S., Sarkar, S., Hore, S., Dey, N., Ashour, A.S., Shi, F., Le, D.N.: Structural failure classification for reinforced concrete buildings using trained neural network based multi-objective genetic algorithm. Struct. Eng. Mech. **63**(4), 429–438 (2017)
7. Chatterjee, S., Dey, N., Shi, F., Ashour, A.S., Fong, S.J., Sen, S.: Clinical application of modified bag-of-features coupled with hybrid neural-based classifier in dengue fever classification using gene expression data. Med. Biol. Eng. Comput. 1–12 (2017)
8. Chatterjee, S., Sarkar, S., Dey, N., Ashour, A.S., Sen, S., Hassanien, A.E.: Application of cuckoo search in water quality prediction using artificial neural network. Int. J. Comput. Intell. Stud. **6**(2–3), 229–244 (2017)

9. Chatterjee, S., Banerjee, S., Mazumdar, K.G., Bose, S., Sen, S.: Non-dominated sorting genetic algorithm—II supported neural network in classifying forest types. In: 2017 1st International Conference on Electronics, Materials Engineering and Nano-Technology (IEMENTech) (pp. 1–6). IEEE, April 2017
10. Chatterjee, S., Banerjee, S., Basu, P., Debnath, M., Sen, S.: Cuckoo search coupled artificial neural network in detection of chronic kidney disease. In: 2017 1st International Conference on Electronics, Materials Engineering and Nano-Technology (IEMENTech) (pp. 1–4). IEEE, April 2017
11. Chatterjee, S., Dey, N., Ashour, A.S., Drugarin, C.V.A.: Electrical energy output prediction using cuckoo search based artificial neural network. In: Smart Trends in Systems, Security and Sustainabilty (pp. 277–285). Springer, Singapore (2018)
12. Chakraborty, S., Dey, N., Chatterjee, S., Ashour, A.S.: Gradient Approximation in Retinal Blood Vessel Segmentation
13. Chatterjee, S., Sarkar, S., Dey, N., Ashour, A.S., Sen, S.: Hybrid non-dominated sorting genetic algorithm: II-neural network approach. Adv. Appl. Metaheuristic Comput. **264** (2017)
14. Chatterjee, S., Sarkar, S., Hore, S., Dey, N., Ashour, A.S., Balas, V.E.: Particle swarm optimization trained neural network for structural failure prediction of multistoried RC buildings. Neural Comput. Appl. **28**(8), 2005–2016 (2017)
15. Chatterjee, S., Ghosh, S., Dawn, S., Hore, S., Dey, N.: Forest type classification: a hybrid NN-GA model based approach. In: Information Systems Design and Intelligent Applications (pp. 227–236). Springer, New Delhi (2016)

Real-Time Facial Recognition Using Deep Learning and Local Binary Patterns

B. Venkata Kranthi and Borra Surekha

Abstract Today, surveillance is everywhere where the operators continuously observe the video captured by the camera to identify the human/object for public safety. Automated systems are being developed for real-time facial recognition as it is highly difficult for the operators to track and identify in highly crowded areas. The feature selection process is generally used to represent faces, and a machine learning-based approach is used to classify the faces in face recognition. A variety of poses, expressions and illumination conditions make the manual feature selection process error-prone and computationally complex. This paper proposes a less computationally complex real-time face recognition algorithm and system based on local binary patterns and convolutional neural networks (CNNs). A modified version of LENET is used instead for face recognition. The recognition accuracy of the proposed method is tested on two publicly available datasets. A new database covering most of the challenges like illumination and oriental variations, facial expressions, facial details (goggles, beard and turban) and age factor is also developed. The proposed architecture proved accurate up to 97.5% in offline mode and an average accuracy of 96% in the real-time recognition process. In the real-time process, frame reading and frame processing are done in two separate threads to improve the frame rate from 28 to 38 FPS.

Keywords Face recognition · Deep learning · Real-time system · Face detection LBP · Computer vision

B. Venkata Kranthi (✉)
ECE Department, GITAM University-Bengaluru Campus, Bengaluru, India
e-mail: bvkranthi1@gmail.com

B. Surekha
ECE Department, K. S. Institute of Technology, Kanakapura Road, Bengaluru, India
e-mail: borrasurekha@gmail.com

M. Chakraborty et al. (eds.), *Proceedings of International Ethical Hacking Conference 2018*, Advances in Intelligent Systems and Computing 811,
https://doi.org/10.1007/978-981-13-1544-2_27

1 Introduction

Person identification is a highly recommended application for real-time systems such as biometrics, surveillance systems, video conferencing, interactive games and device authorization. Recognizing a person is achieved in many ways: fingerprint verification, iris detection and by recognition techniques where information is to be provided by the user. In a surveillance system, the person had to be recognized from a distance without her/his involvement personally. The face can be found and tracked using a regular surveillance camera or even with a webcam or mobile cam. The face recognition process includes detection of a face in a frame and identification of him/her. Sometimes, the application may require tracking the face continuously in real time.

The face recognition algorithms from literature can be categorized into statistical-based methods such as principal component analysis (PCA), independent component analysis (ICA) and linear discriminant analysis (LDA), graphical-based active appearance model (AAM) and elastic bunch graph matching (EBGM), machine learning-based support vector machine (SVM) and neural network-based methods which further rely on the backpropagation and combinations of them [1]. The statistical approaches are less effective because of the singularity problem and cannot recognize the faces when images are acquired at different scales. Further, they are computationally complex to make it work in real time. Graph-based algorithms are mathematically complex. Further, the selection of number of fiducial points on the face is challenging which leads to imperfect eye localization and reduction in recognition accuracy. Machine learning and neural network-based approaches are latest and are able to classify faces with good accuracy but the selection of these features for a variety of facial variations is a challenging task. The recently developed CNN-based object detection approaches can be used for face recognition as they do not require feature selection process and are leading in object recognition tasks. CNN takes a raw image and finds the necessary features by itself for object recognition which can be converted to face recognition.

Face detection is the crucial stage in identifying/tracking the person. Zou and Kamata [2] proposed human skin colour segmentation for face localization. The skin colour model was generated using Gaussian mixture model (GMM) and the expectation–maximization (EM) algorithm to assign the pixels to either face or background [3, 4]. These methods fail when dealing with different skin tones. Viola and Jones [5] developed face recognition based on Haar features which are further passed through boosted cascade classifiers [4–7]. Haar cascades require more amount of data and is also computationally expensive for training compared to LBP. To reduce the computational complexity, Bilaniuk et al. [8] proposed a local binary pattern (LBP)-based method for detecting faces at a rate of 5 FPS on a low-power embedded system. The literature on LBP confirms its effectiveness in terms of speed and computational complexity on the Raspberry Pi (RP) [9]. Benzaoui et al. [10] proposed a new one-dimensional local binary pattern (1DLBP) to produce a binary code that acts as a feature vector for representing the face.

Ge et al. [11] proposed LBP feature-based facial recognition in which LBP features are found at each pixel location followed by construction of histograms. The histograms are combined sequentially for different sizes of cells to form a feature vector of the face. The feature vectors are compared and classified using Carle square dissimilarity and principal component analysis (PCA). Aizan et al. [12] proposed nearest neighbour classifier along with chi-square function as a dissimilarity measure for classification of feature vectors. This feature vector is compressed using wavelets without loss of information and is recognized using the classical SVM to obtain good recognition rate. Zhang and Xiao [13] proposed local binary pattern calculated at multiple scales, and the histogram is combined for each cell at multiple scales to consider the scale factor. Dahmouni et al. [14] represented faces using confidence interval concept along with local binary probabilistic pattern (LBPP) on each pixel, and compression of the feature vector is done by using global 2D-DCT frequency method. The LBP feature itself is not enough to represent the faces in machine learning-based techniques like SVM, as there are various challenges associated with expression, scale factor, illumination conditions, and orientation and pose changes.

To handle the expressions and pose variations, Huang et al. [15] proposed multi-scale LBP features which are calculated only on the landmarks instead of the complete image at multiple scales. Later, the features are combined with the Gabor features at kernel level and are then used for face recognition. Wei et al. [16] proposed a remote monitoring system using cloud storage. The LBP along with Gabor features provides rotational invariance in face detection and recognition process. Krishna Kishore and Varma [17] proposed a hybrid technique that combines global and local descriptors. Majeed [18] fused global features calculated using PCA and local features computed using LBP using sym4 wavelet method. Radial basis function (RBF) finds the face or non-face class for the feature vector. Tyagi et al. [19] proposed LBP and Zernike moments (ZMs) for global features. While the local ternary pattern (LTP) with genetic algorithm (GA) finds the feature vector, the support vector machine (SVM) defines the class based on feature vector.

All the above-mentioned methods associated LBP with other statistical-based methods such as PCA transforms (DWT, DCT), genetic algorithm and RBF for detecting the faces. The common drawback of all these techniques is the computational load associated with them when implementing these algorithms in real-time applications. Yan et al. [20] proposed CNN with three convolution layers, two pooling layers, two full-connected layers and one Softmax regression layer along with dropout layers for handling overfitting applied on the ORL and AR face database. Ding and Tao [21] proposed multiple CNNs that provide a high-dimensional facial feature vector and stacked auto-encoder (SAE) to compress the feature vector dimensions. Moon et al. [22] proposed a method for face recognition from the images taken from a distance ranging from 1 to 9 m away and trained them using CNN. All these methods are computationally expensive. Liu et al. [23] proposed VIPLFaceNet architecture which is better than AlexNet on the LFW database using the only single network and released open-source SDK under BSD licence. VIPLFaceNet can recognize a face in 150 ms on the i7 desktop. Jain and Patel [24] proposed GPU-based implementation of robust face detection using Viola–Jones face detection algorithm and NVIDIAs compute unified device architecture (CUDA). Xil et al. [25] proposed LBPNet similar to the convolutional neural network where the kernels are replaced by the LBP descriptor by which high accuracy is obtained without a costly model learning and extensive data. The literature on CNN proves its accuracy and computational efficiency compared to LBP-based methods and can be computed over a GPU.

It is observed that the existing models based on CNN for face recognition require more computational resources. This paper proposes a simple, fast and yet accurate face recognition model that makes use of LBP cascades for face detection and the recently developing convolutional neural networks (CNNs) for face recognition with a modified version of LENET architecture. The newly developed architecture is also compared with the existing LENET architecture on two databases (ORL and YALE-B) and also on a database that is newly proposed in this paper for evaluating real-time recognition.

The paper is structured as below: Sect. 2 covers the methodology used for detection of faces by LBP cascades. Section 3 elaborates the recognition process using CNN. Section 4 presents the databases used. Section 5 discusses the comparison of results. Section 6 concludes and presents the future scope of the paper.

2 CNN and LENET

The manual selection of features in case of statistical and machine learning techniques for face recognition makes it suboptimal in a real application. These are also time-consuming which makes the recognition process hard in real time. The possible solution for this is to directly apply the neural networks to learn the features from the raw input image itself by adjusting the weights. Such an implementation is already available as CNN in the recent years and is a present pioneer in object recognition tasks.

CNN is a combination of convolutional layers, subsampling layers and fully connected layers connected in a sequential manner. LeCun et al. [26] proposed first CNN shown in Fig. 1 which is a significant breakthrough in neural network research. This model is trained by backpropagation and does not consider the dropout layers which play a key role in the overfitting problem.

Khalajzadeh et al. [27] showed that CNN can perform both representation and recognition of objects in a single network architecture directly from the raw input image with no or minimal pre-processing. Later Tivive and Bouzerdoum [28] showed that it is also robust to some small amount of illumination variations, geometric distortions and occlusions which are perfect for face recognition application. The training process of CNN is computationally very expensive, which makes it not useful for running on small, low-end computers. The solution for this is to run the training process on a high-performance computer and store the weights for testing

Fig. 1 LENET architecture

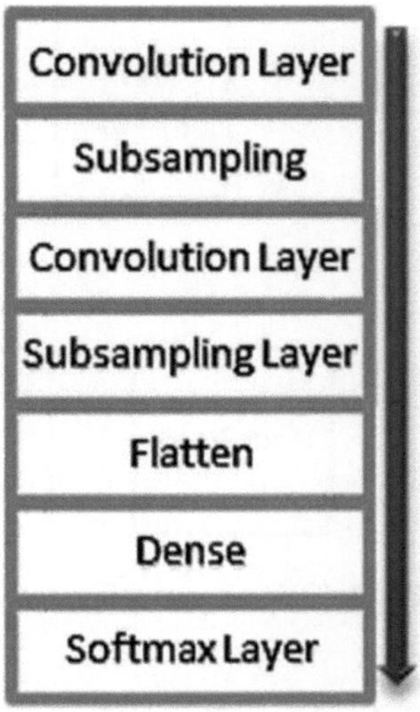

on a low-performance computer. The benefit of using CNN is that there is no need for storing or transferring the database information for the screening process.

3 Methodology

Face recognition in done in two stages: training and testing phases as shown in Fig. 2. In training phase, the image database is divided into 75% for training and 25% for validation. These face images are extracted from daily life gallery of images using LBP face detection scheme. The extracted faces are then cropped with 10% extra size in all directions after the face is detected from the original image. These images are fed into the proposed CNN architecture for training and validation process for 1000 epochs. In each epoch, the weights of each neuron are updated by adjusting according to the backpropagation gradients with optimizer adadelta having a standard learning rate of 1.0. The weights, number of subjects referred as classes and the CNN model are stored after 1000 epochs for next phase.

In testing phase, video is captured using Raspicam with every frame of size (240, 360). Each frame is passed to the LBP detector for face extraction with a scale factor of 1.2 and minimum face size of (30, 30) that can be detected. The faces in the frame are resized to a size of (32, 32) as this is the size that the proposed CNN uses at its first stage. The face image is then passed to the CNN model with learned weights. The output of this block is the name of the subject/class. The class and the face detector output as bounding box are overlaid on the streaming video in real time.

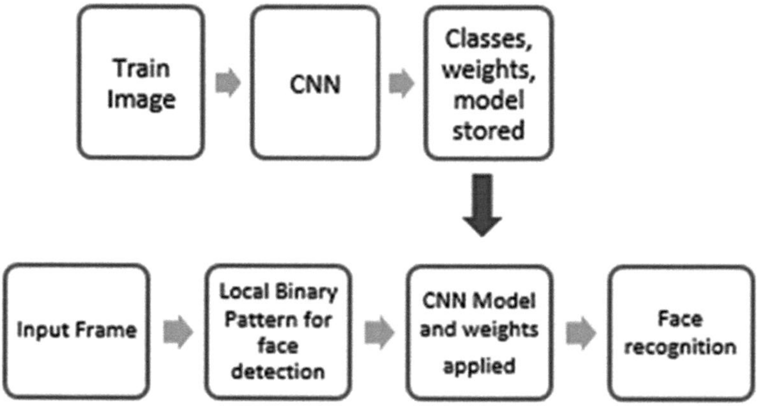

Fig. 2 Training and testing stages of face recognition

3.1 Face Detection

Face detection plays a crucial role in person identification and/or tracking applications. Detection of the face from a cluttered background in a given frame is a challenging task. A well-known algorithm, LBP cascades is used in this work, for its speed compared to Haar cascades with a little compromise on the accuracy. Both are similar except Haar calculates rectangular facial features whereas LBP calculates features related to textures. Real-time applications require low computational complex algorithms, so the LBP cascades is used to detect the face.

LBP cascades use the LBP feature shown in Fig. 3 which is calculated directly from the face image. The detailed explanation of LBP can be found in Pietikainen et al. [29] and Liao et al. [30]. The complete image is divided into small parts of size 3×3 pixels usually, and LBP feature is calculated as below:

1. A square of a defined size or circular window of the specified radius is chosen from the image. The centre pixel value is compared with the neighbouring pixels; if greater than the neighbouring pixels it assigns 1 to the neighbouring pixels, otherwise it assigns 0 bit.
2. According to the chosen window sizes, the weighted window of the same size is chosen and will multiply point by point. The weights of the weighted window are chosen in proportion to the distance from the centre pixel.
3. The value of the processing pixel in the output is calculated by applying binary-to-decimal conversion in this paper. This process is repeated for the whole image.
4. A cell size is chosen, and LBP values are calculated for all the pixel locations in the image. A histogram is formed using these LBP values inside the cell. This process is repeated for all cells in the image.
5. These histograms are combined sequentially for all the cells to form a feature of the image.

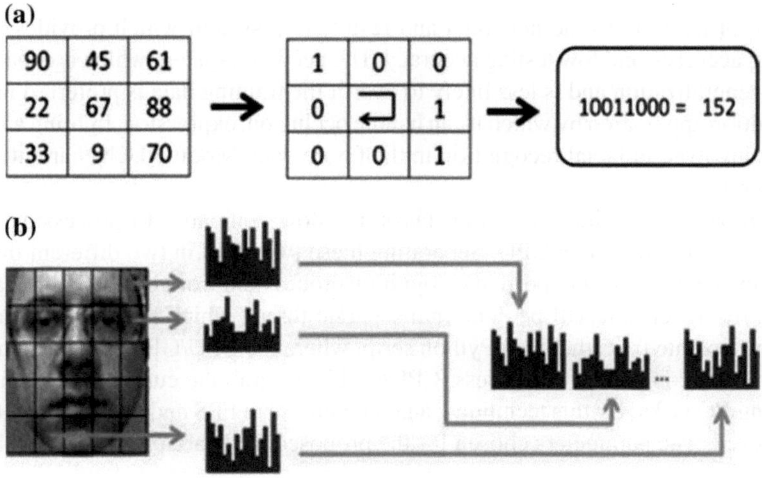

Fig. 3 LBP feature extraction

6. The histogram features are used to train the system to detect a face from a given frame with high probability and are then sorted using boosting cascade classifier.
7. These features that satisfy most of the real faces in the training phase are formed as an .XML file which will be used in the testing process.

3.2 *Face Recognition*

A lot of facial images of different persons are provided to CNN architecture in the initial stage. These images go through the training procedure, all through different layers in CNN, and the weights of the proposed CNN are updated in every loop until good amount of accuracy is achieved. In this stage, cross-validation on a subset of facial images is provided for training to avoid any overfitting. The final weights along with the model architecture are stored. In the second phase, the proposed CNN is tested on offline images and on live video frames as input. The faces are extracted with LBP with the scale parameter of 20%, min face size as (30, 30). The extracted faces are resized to (32, 32) and are provided to CNN for recognition. The weights and model from the training phase are provided as inputs to the testing phase where recognition is done.

LENET does not fit for the face database because of overfitting. So, a modified LENET-based CNN architecture is presented in this paper. Some selected neurons are dropped out according to the percentage provided in training the network but not in validation. The weights of selected neurons are normally adjusted for particular features related to specific criteria. So, the nearby neurons slowly depend on them and can result in overfitting overtime. Because of this technique, the network is not capable of learning to the new data and results in a system which provides better training accuracy and low testing accuracy. The need of a system which is capable of better generalization and is less likely to overfit the training data is preferred in face recognition application by which it can handle occlusion, expression, lighting and age factors involved in facial recognition in the future data. Modified CNN architecture is shown in Fig. 4.

Input/output (reading the frame) tasks are slow compared to processing tasks (processing the frame) on CPU. Separating these two tasks in two different threads will help in improving the speed of recognition process and work in real time. Reading frames from webcam will be done in a separate thread which is continuous and is entirely separate from the main Python script where the CPU/GPU is processing the frame. After completion of process, CPU/GPU will grab the current frame without blocking I/O tasks. By this technique, a great increase in FPS and decrease in latency is achieved. The parameters chosen for the proposed CNN are presented in Table 1.

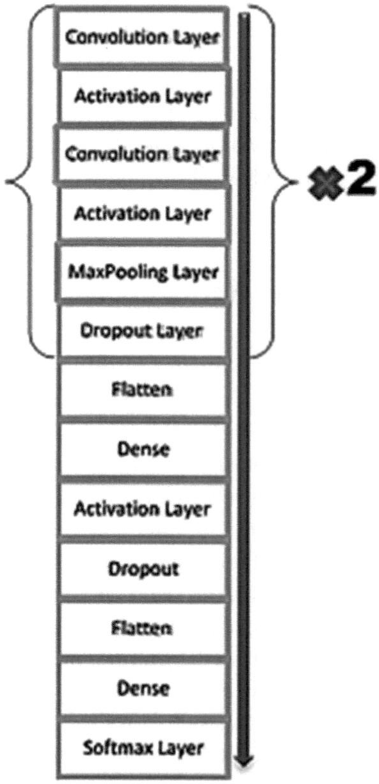

Fig. 4 Proposed CNN architecture

Table 1 Parameters of the proposed CNN

Layers	Parameters
Input image	Size= [32, 32, 3]
Convolution layer	Number of filters = 32 Convolution Kernel = [3 × 3] Stride = 1
Activation layer	Adadelta
Pooling layer	Maximum value, sliding window size = [2, 2], Stride = 2
Dropout layer	25%

4 Results and Discussion

4.1 *Database*

The proposed CNN is evaluated on the existing famous face datasets ORL (now called as AT&T) and Cropped Yale-B databases for verifying the accuracy of the algorithm so that the same can be applied to the proposed face database. The ORL database helps in learning about orientation and detail changes, and the Yale database helps to understand illumination variations which perfectly suits and is recommended for the proposed database needs.

The ORL database (Fig. 5a) contains 40 persons and 10 images of each person taken in a span of 2 years. It covers various challenges faced in face recognition like orientational changes with a small side movement, occlusions like wearing glasses, turban and various expressions on the face taken in front of a constant background. Each subject will be standing upright and looking front to the camera.

Cropped Yale-B database (Fig. 5b) consists of 2432 images of 40 subjects taken by varying the illumination in 64 different ways with a single light source.

The proposed database (Fig. 5c) contains 6 subjects with 400 images for each taken from the personal gallery over a span of a decade with various cameras. These facial images cover illumination variations, orientation changes, facial details (beard, glasses and turban), facial expressions and even the age factor. LBP cascades are applied to the images in the gallery to extract the faces.

4.2 *Offline Results*

In the evaluation process, LENET architecture is applied to ORL, Cropped Yale-B and the proposed databases. A large gap between the training and validation accuracies can be observed in Fig. 6 for each database which is caused by overfitting.

The new CNN architecture proposed is then applied to ORL and Cropped Yale-B databases for verifying the behaviour of learning the orientational and illumina-

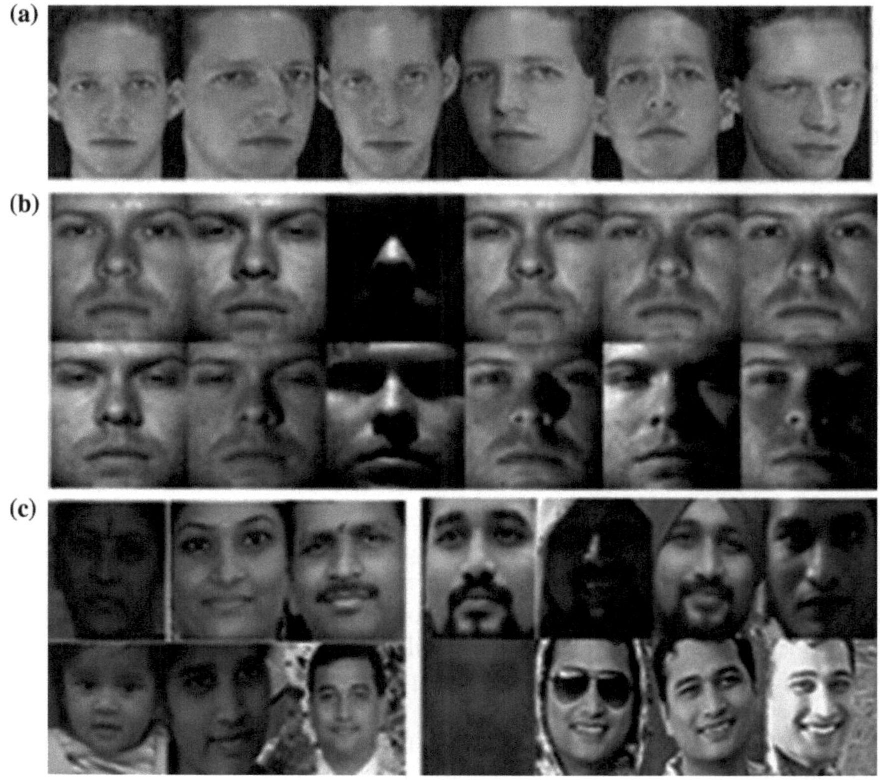

Fig. 5 a ORL (AT&T), b Cropped Yale-B and c proposed face database

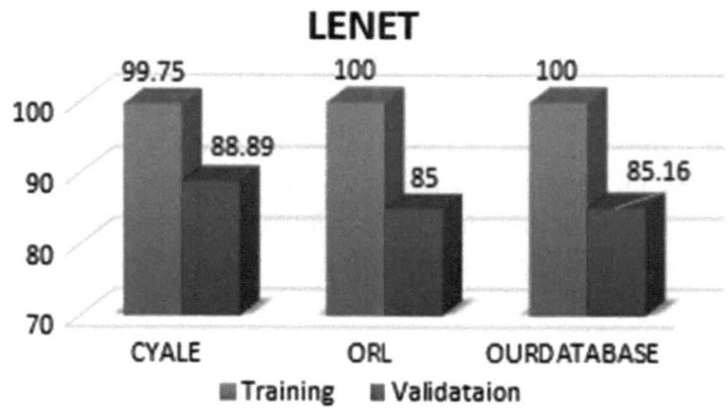

Fig. 6 Accuracy of LENET on databases

tion changes of faces in databases for recognition. In this, the number of stacked layers convolution-convolution-maxpooling-dropout (CACAMD) in the network is changed along with the dropout percentage (25 and 50%) where 25% or 50% neurons are dropped out for achieving the best accuracy on all three databases.

Table 2 Accuracy comparison of the proposed CNN models on all databases

Model no.	Database	Training	Validation	|Sum/Diff|	AVG on model
1	CYALE	99.23	99.17	3306.6	1148.0
	ORL	98.67	95	52.77	
	Proposed database	99.83	97.5	84.69	
2	CYALE	97.31	99.5	89.87	81.94
	ORL	99.33	93	30.38	
	Proposed database	98.72	97.16	125.56	
3	CYALE	99.83	98.68	172.62	94.66
	ORL	100	97	65.67	
	Proposed database	99.94	95.66	45.70	
4	CYALE	99.45	99.17	709.36	273.34
	ORL	100	92	24.00	
	Proposed database	99.94	97.66	86.67	

CNNs always need a big database to learn properly while training. CYALE-B has a good number of face images for learning. The accuracy for CYALE-B for all the cases is more than 99% within 500 loops/epochs as shown in Table 2. The best model for this database is model-3 (CACAMD-D25) and resulted in training accuracy of 99.83% and the validation accuracy of 98.68%. But the model-1 (CACAMD2-D25) is chosen as it results in less overfitting of the data even though the number of epochs or loops required by the model-1 is more. Further, the testing process using the same model will also be time-consuming compared to model-3.

ORL database has very few images per subject and is not sufficient for learning the orientation variations of faces even for 1000 epochs shown in Table 2. This paper is not considering the data augmentation at this stage which will improve the accuracy. The best model for ORL database is model-3 with less overfitting and higher accuracy of 100 and 97% for training and validation. Model-1 is chosen for maintaining uniformity across all the databases. The model-2, CACAMD2-D50, and model-4, CACAMD-D50, are less performing compared to others.

The proposed database is intermediate compared to both the databases discussed in the previous paragraphs. It covers orientation changes, illumination variations, face expressions and age factor as well. For these reasons, the best model based on the generalized model and optimum model-1 from the above discussions is chosen and applied to the proposed database and got training accuracy of 99.83% and validation accuracy of 97.5%. For comparison, all the CNN models are applied to the proposed database and results are shown in Table 2. The sum over the difference value on training and validation accuracies is used for calculating the best model by which the overfitting problem can be eliminated.

The results of validation accuracies on all the databases with the LENET and modified CNN architecture are also compared in this paper. Training accuracy in both the cases for all the databases is more or less the same. In all the databases, the proposed CNN performs well compared to LENET. The results are shown in Fig. 7.

The results are also compared with the existing algorithms that used LBP as features and are presented in Table 3. The method that used LBP-GA-SVM combination produced best results in the case of ORL database in which the facial attributes of extended Yale-B are estimated via local ternary pattern, which is genetically reformed. The resulting feature vectors are classified using SVM and led to recognition rates of 99.37 and 98.75%. The LBPP-DCT-Euclidean-based method is found best for Cropped Yale-B database as compared to the proposed method where local feature extraction method is used to represent a face, and the global DCT frequency method is used for compressing the feature vector. The recognition rates obtained are 95.5 and 100%.

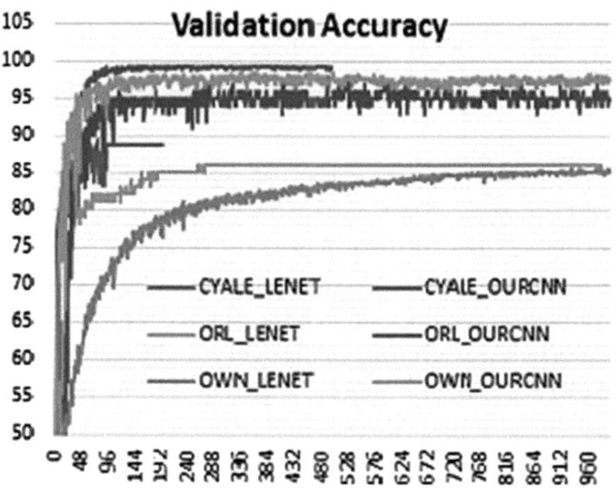

Fig. 7 Accuracy comparison of LENET and proposed CNN on all three databases (OWN refers to the proposed database)

Table 3 Comparison of the proposed method with LBP-based methods

Method	ORL (%)	Cropped Yale-B (%)
LBPP-DCT-Euclidean [14]	95.5	100
LBP-GA-SVM [19]	99.37	98.75
LBP	92	96
HSLGBP	93	97
MTLBP	93	97
MW-LBP [13]	95	98
LBP-CNN (proposed method)	97	99.50

4.3 Real-Time Results

Real-time testing is done indoor with moderate lighting, slight orientation, expression changes and occlusion with a laptop camera. The average accuracy achieved is 97% for detection of faces at a distance of less than a metre and when standing still in front of the camera. With slight changes in orientation, lighting and occlusions, and with a scale factor of 20% for face detector, the accuracy decreases according to the increase in scale factor. The recognition accuracy is more than 90% always, with an average accuracy of 96%. The proposed algorithm runs at a frame rate of 28 FPS without threading and 38 FPS with threading frames.

The proposed method is also evaluated for its computational time taken to detect and recognize a face from the frame. The results are given in Table 4.

The platform used is i7 processor and NVIDIA GPU that supports CUDA libraries. The recognition using the proposed CNN architecture and threading of frames enabled took the maximum of 16 ms on CPU and 4 ms on CPU + GPU, whereas both detection and recognition took 140 ms on CPU and 74 ms on CPU + GPU. The computational time analysis shows that the detection process takes more amount of time indicating scope for improvement. The proposed method is able to perform detection and recognition of one face in a frame in real time and is better compared to [21] in which 150 ms required. In the proposed method, the CUDA-enabled NVIDIA GPU and threading of frames enable face detection and recognition process possible in real time.

Some of the frames on the live video are presented in Fig. 8 for the purpose of understanding the detection and recognition accuracies. The frames are taken at random where maximum acceptable accuracy is achieved while handling challenges involved in facial recognition. Figure 8a shows the orientation and expression change. Figure 8b shows occlusion factor. Figure 8c, d is captured from a short and long distance with varying illumination conditions. In all the frames, the person is properly identified to the correct class.

This paper presented a model based on LBP-CNN that can behave uniformly even at new situations and to achieve better results for a variety of databases in both offline mode as well as online mode. The best accuracy obtained in four CNN models for an individual database is chosen to compare with other LBP-based methods existing in the literature. The proposed CNN architecture comparatively gave best result in YALE database and moderate in ORL database.

Table 4 Computational time of the proposed algorithm for detection and recognition of the face

Platform	Only recognition		Detection and recognition	
	Min (ms)	Max (ms)	Min (ms)	Max (ms)
Only CPU(i7)	6	16	110	140
CPU(i7) + GPU(CUDA)	1.2	4	61	74 (150 ms in [27])

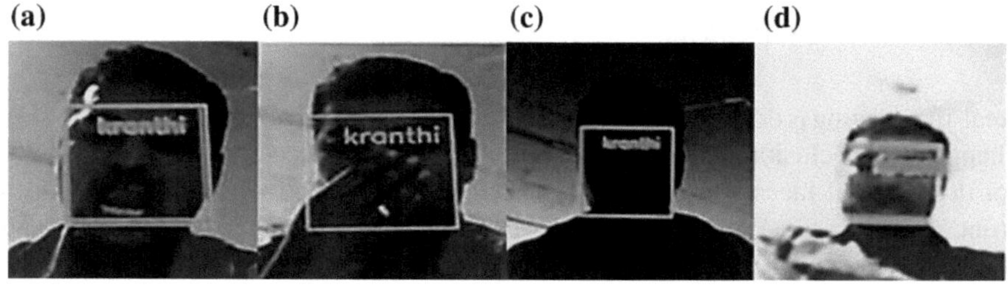

Fig. 8 Examples of real-time recognition

5 Conclusion and Future Scope

A new LENET-based CNN architecture in combination with LBP for real-time face detection and face recognition is presented in this paper. The proposed CNN is better than the LENET in face recognition without overfitting. Two separate threads for reading the images and processing the images are presented which improved frame rate from 28 to 38 FPS. The CUDA-enabled GPU helps in performing detection and recognition of a face in only 74 ms. The proposed CNN provides better training and validation accuracy of 99.83 and 97.5% by regularizing data using dropout layers on a database which covers most of the challenges in the face recognition process. The proposed method is also compared with the results of existing methods which are based on LBP and demonstrated that the proposed CNN method is better than the others and maintains uniformity across different databases.

In this paper, it is observed that the time required for the detection process is more compared to recognition as it has to search throughout the image in every location at every possible scale. One future improvement can be reducing the computational complexity of detecting faces from the image. The online results of using the proposed CNN on the live video are good and thus can be tested on a single board computer like Raspberry Pi or NVIDIA Jetson TX2. Separating the process of frame reading and frame processing (threading) is also supported by Raspberry Pi and so the increase in frame rate is also achievable. One stage of CACAMD can be used in low-power devices to reduce the time with a small compromise on accuracy.

6 Compliance with Ethical Standards

6.1 Research Involving Human Participants

The proposed database shown in Fig. 5c consists of facial images of the author's family, and the list is written in Table 5 for convenience according to the order. The large database requirement for deep neural networks forces the author to make use of author's family pictures as they were available for the past many years in evaluating

Table 5 Human participants' list with the relationship to the author

Name of person	Relationship with author
Budigi Ramadevi	Mother
Budigi Swapna	Sister
Chittipineni Nagarjuna	Brother-in-Law
Budigi Shreyan	Son
Budigi Gouthami	Wife
Venkata Kranthi Budigi	Author

the proposed algorithm. The face image used in Fig. 8 is the author's face himself. On behalf of all of these members, author Venkata Kranthi B will be held responsible for using the face images of all his family members in his study, and the pictures were taken and used in the study with all their prior concern. I as an author give my concern to publish the following human participants' images in the chapter.

6.2 *Other Open-Sourced Databases Used*

The other databases used in the study are AT&T's Database of Faces formerly called ORL database from AT&T Laboratories Cambridge and Cropped Yale-B face database from Athinodoros Georghiades et al. [31]. Both databases are free to use and were properly acknowledged by author.

References

1. Chellappa, R., Wilson, C.L., Sirohey, S.: Human and machine recognition of faces: a survey. Proc. IEEE **83**(5), 705–741 (1995)
2. Zou, L., Kamata, S.I.: Face detection in color images based on skin color models. In: Proceedings of IEEE Conference TENCON 2010, pp. 681–686 (2010)
3. Redner, R.A., Walker, H.F.: Mixture densities, maximum likelihood and the EM algorithm. SIAM Rev. **26**(2), 195–239 (1984)
4. Zhou, H., Sadka, A.H.: Combining perceptual features with diffusion distance for face recognition. IEEE Trans. Syst. Man. Cybern. Part C (Appl. Rev.) **41**(5), 577–588 (2011)
5. Viola, P., Jones, M.: Rapid object detection using a boosted cascade of simple features. In: Proceedings of IEEE Computer Society Conference on Computer Vision and Pattern Recognition, pp. 1–511 (2001)
6. Krishna, M.G., Srinivasulu, A.: Face detection system on AdaBoost algorithm using Haar classifiers. Int. J. Mod. Eng. Res. **2**(5), 3556–3560 (2012)
7. Surekha, B., Nazare,K.J., Raju, S.V., et al.: Attendance recording system using partial face recognition algorithm. In: Intelligent Techniques in Signal Processing for Multimedia Security, pp. 293–319 (2017)
8. Bilaniuk, O., Fazl-Ersi, E., Laganiere, R., et al.: Fast LBP face detection on low-power SIMD architectures. In: Proceedings of Conference on Computer Vision and Pattern Recognition Workshops (CVPRW), pp. 630–636 (2014)

9. Fernandes, S., Bala, J.: Low power affordable and efficient face detection in the presence of various noises and blurring effects on a single-board computer. In: Proceedings of the 49th Annual Convention of the Computer Society of India (CSI), pp. 119–127 (2015)
10. Benzaoui, A., Boukrouche, A., Doghmane, H., et al.: Face recognition using 1DLBP, DWT and SVM. In: Proceedings of International Conference on Control, Engineering & Information Technology, pp. 1–6 (2015)
11. Ge, W., Quan, W., Han, C.: Face description and identification using histogram sequence of local binary pattern. In: Proceedings of International Conference on Advanced Computational Intelligence, pp. 415–420 (2015)
12. Aizan, J., Ezin, E.C., Motamed, C.: A face recognition approach based on nearest neighbor interpolation and local binary pattern. In: Proceedings of International Conference on Signal-Image Technology & Internet-Based Systems, pp. 76–81 (2016)
13. Zhang, J., Xiao, X.: Face recognition algorithm based on multi-layer weighted LBP. In: Proceedings of International Symposium on Computational Intelligence and Design, pp. 196–199 (2016)
14. Dahmouni, A., Aharrane, N., Satori, K., et al.: Face recognition using local binary probabilistic pattern (LBPP) and 2D-DCT frequency decomposition. In: Proceedings of International Conference on Computer Graphics, Imaging and Visualization, pp. 73–77 (2016)
15. Huang, K.K., Dai, D.Q., Ren, C.X., et al.: Fusing landmark-based features at kernel level for face recognition. Pattern Recogn. **63**, 406–415 (2017)
16. Li, C., Wei, W., Li, J., et al.: A cloud-based monitoring system via face recognition using Gabor and CS-LBP features. J. Supercomput. **73**(4), 1532–1546 (2017)
17. Krishna Kishore, K.V., Varma, G.P.S.: Hybrid framework for face recognition with expression & illumination variations. In: Proceedings of International Conference on Green Computing Communication and Electrical Engineering, pp. 1–6 (2014)
18. Majeed, S.: Face recognition using fusion of local binary pattern and zernike moments. In: Proceedings of International Conference on Power Electronics. Intelligent Control and Energy Systems, pp. 1–5 (2016)
19. Tyagi, D., Verma, A., Sharma, S.: An improved method for face recognition using local ternary pattern with GA and SVM classifier. In: Proceeedings of International Conference on Contemporary Computing and Informatics, pp. 421–426 (2016)
20. Yan, K., Huang, S., Song, Y., et al.: Face recognition based on convolution neural network. In: 2017 36th Chinese Control Conference (CCC), pp. 4077–408 (2017)
21. Ding, C., Tao, D.: Robust face recognition via multimodal deep face representation. IEEE Trans. Multimed. **17**(11), 2049–2058 (2015)
22. Moon, H.M., Seo, C.H., Pan, S.B.: A face recognition system based on convolution neural network using multiple distance face. Soft. Comput. **21**(17), 4995–5002 (2017)
23. Liu, X., Kan, M., Wu, W., et al.: VIPLFaceNet: an open source deep face recognition SDK. Front. Comput. Sci. **11**(2), 208–218 (2017)
24. Jain, V., Patel, D.: A GPU based implementation of robust face detection system. Proc. Comput. Sci. **87**, 156–163 (2016)
25. Xi1, M., Chen1, L., Polajnar1, D., et al.: Local binary pattern network: a deep learning approach for face recognition. In: Proceedings of International Conference on Image Processing, pp. 3224–3228 (2016)
26. LeCun, Y., Bottou, L., Bengio, Y., et al.: Gradient-based learning applied to document recognition. Proc. IEEE **86**(11), 2278–2324 (1998)
27. Khalajzadeh, H., Mansouri, M., Teshnehlab, M.: Face recognition using convolutional neural network and simple logistic classifier. Stud. Comput. Intell. **223**, 197–207 (2013)
28. Tivive, F.H.C., Bouzerdoum, A.: A gender recognition system using shunting inhibitory convolutional neural networks. In: International Joint Conference on Neural Networks, pp. 5336–5341 (2006)
29. Pietikainen, M., Hadid, A., Zhao, G., et al.: Local binary patterns for still images. Computer vision using local binary patterns. Comput. Imaging Vis. **40**, 13–47 (2011)

30. Liao, S., Zhu, X., Lei, Z., et al.: Learning multi-scale block local binary patterns for face recognition. In: International Conference on Biometrics, pp. 828–837 (2007)
31. Georghiades, A.S., Belhumeur, P.N., Kriegman, D.J.: From few to many: illumination cone models for face recognition under variable lighting and pose. IEEE Trans. Pattern Anal. Mach. Intell. **23**(6), 643–660 (2001)

Hepatocellular Carcinoma Survival Prediction Using Deep Neural Network

Chayan Kumar Kayal, Sougato Bagchi, Debraj Dhar, Tirtha Maitra and Sankhadeep Chatterjee

Abstract Hepatocellular carcinoma is one of the most common types of liver cancer in adults. In patients having this disease, prediction of survival is very strenuous. Through this eminent experiment, the authors have proposed a new improved classification approach using DNN (deep neural network) for predicting survival of patients with hepatocellular carcinoma. The dataset was obtained at a University Hospital in Portugal and contains several demographic, risk factors, laboratory and overall survival features of 165 real patients diagnosed with HCC. Authors have selected 15 risk factors out of 49 risk factors which are significantly responsible for HCC in this proposed method. The outcome of this experiment has proved to be of significant increase in accuracy of the prediction of survival over the conventional methods like multivariable Cox model or unsupervised classification.

Keywords Hepatocellular carcinoma · Classification · Deep neural network Survival

C. K. Kayal · S. Bagchi · D. Dhar · T. Maitra · S. Chatterjee (✉)
Department of Computer Science & Engineering, University of Engineering and Management, Kolkata, India
e-mail: chatterjeesankhadeep.cu@gmail.com

C. K. Kayal
e-mail: chayankayal32@gmail.com

S. Bagchi
e-mail: sougato97@gmail.com

D. Dhar
e-mail: debrajdhar100@gmail.com

T. Maitra
e-mail: tirthamaitra0@gmail.com

M. Chakraborty et al. (eds.), *Proceedings of International Ethical Hacking Conference 2018*, Advances in Intelligent Systems and Computing 811,
https://doi.org/10.1007/978-981-13-1544-2_28

1 Introduction

Hepatocellular carcinoma (HCC) is one of the primary liver cancers found in adults, and it is the most common cause of death in people with cirrhosis, accounting for an estimated 5 lakhs deaths annually [1]. HCC happens as a result of liver inflammation and is most closely linked to chronic viral hepatitis infection (hepatitis B or C) or exposure to various toxins such as alcohol or aflatoxin. The majority of HCC is prevalent in the southern part of the world, mainly Southeast Asia and some sub-urban areas of Africa. The incidence of HCC has doubled in the USA over the past two and a half decades, and incidence and mortality rates are likely to double over the next two decades [2]. Today data-propelled statistical research has become an attractive complement for analytical research. Survival prediction is one of the most challenging tasks among these medical researchers. These statistical predictions help the analysis of the physician and also save huge resources. These researches are done with the help of some computational techniques/methods such as ANN, SVM and KNN as mentioned in [3]. These techniques are able to model unknown/complex relationships which are nonlinear or noisy and are difficult to analyze.

Recent researches have focused on application of machine learning in cancer prediction. Llovet et al. [4] proposed a method of classification using a new staging system, known as the Barcelona Clinic Liver Cancer (BCLC) staging classification, which selects the best candidate/patient for the most optimal therapy currently available using four stages of operation. Another method for prediction of HCC has been reported by Chevret et al. [5] using a new prognostic classification, which selected five major prognostic parameters. In their proposed work, they analyzed the data of 761 patients with HCC. The splitting was done randomly based on which they established a new classification system. Lee et al. [6] published their work on classification and prediction of survival of HCC using gene expression profiling in 2004. The primary aim of the researchers was to derive the molecular characteristics of any tumor and the secondary aim to test their prognostic value based on their expression profile. In 2015, Santos et al. [7] presented their work on the HCC using k-means clustering and SMOTE algorithm to build a representative dataset and use it as training example for different machine learning procedures.

2 Proposed Method

Artificial neural network (ANN) [8–13] can be considered as the mere replica of the biological neurons present in the human brain. The human brain is the most complex and powerful functional unit. It is capable of handling complex relations and taking important decisions in less than a fraction of second and also capable of modeling complex/unknown functional relationships with interconnected processing units (artificial neurons) [14–24]. That is the reason for the interest of replicating this enormous powerful model of computing, and this gave birth to ANN. ANN is different for traditional hardcoded algorithms. These learn the functional relationships from a given dataset during its learning (training) stage. ANN mainly consists of an input layer, an output layer and one hidden layer in between input and output layers. The hidden layers are responsible for all the computations in a neural network model. In the proposed method, authors have used the DNN (deep neural network) model to perform the HCC classification task.

DNN is basically an extension of the ANN, where the number of hidden layers is more than one. To evaluate the effectiveness of the DNN model, authors have compared it with other models such as SVM (support vector machine) and KNN (k-nearest neighbor).

In SVM, the variables are mapped on a 3d/2d plane using some mapping functions and an optimal hyperplane is drawn to classify the variables. This hyperplane is drawn by considering the worst type of variables of different kinds. Therefore, this model is more robust as this takes into account the worst conditions. The variables are classified on the basis of their location inside/outside the kernel, or on the side of the hyperplane.

KNN is a nonparametric method used for classification and regression problems in machine learning. Here, a new variable is categorized by comparing the distance (mainly Euclidian distance) between that point and other points of different categories. The category having the maximum number of neighboring variables (or maximum number of variables with least distance) with that new variable is termed as the type of that variable.

In the current study, the DNN model has been constructed using the well-known "keras" library. The model consists of four hidden layers in between the input and output layers. Each layer containing, respectively, 1024, 512, 256 and 128 neurons which were chosen randomly based on trial-and-error approach by the authors. The weight variables have been uniformly assigned in each layer for better optimization. Activation function used in the hidden layers is the commonly used "Linear Rectifier unit" (Relu), and at the output layer, the sigmoid activation function has been used to retrieve predicted values in a probabilistic way. The optimizer used in the DNN architecture is one of the variants of the gradient descent optimizer, known as "Adam" optimizer, which was found to have better optimization result in the current study [1].

Figure 1 depicts the flow of the experiments in the current study. The basic flow of the experiment conducted by the authors is as follows:

1. Preprocessing: The following preprocessing is done in the dataset before the classification.

    (i) Feature extraction—This step involves extraction of significant features which are most important in classification. In the presented work, the feature extraction had been performed by finding out the correlation between the feature and the target. During this phase, out of 49 features, 14 unique features were extracted out of the dataset which were found to have the most significant effect on the class prediction result, resulting in lesser distortion and higher accuracy [3].

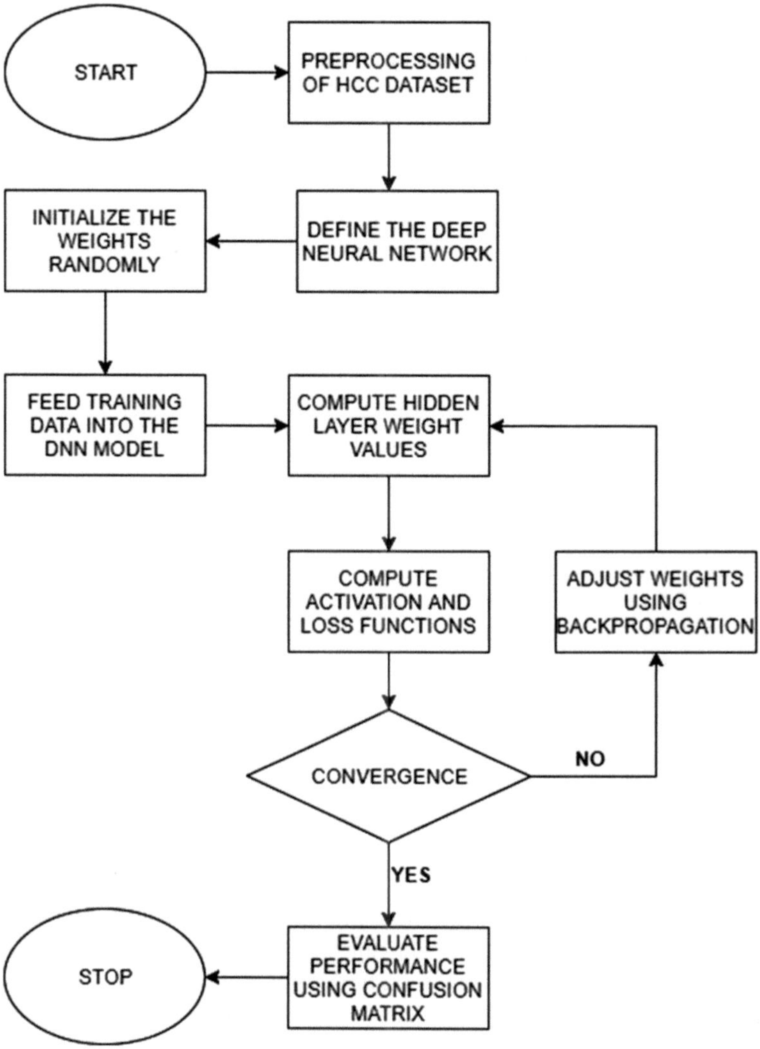

Fig. 1 Flowchart of the DNN training

(ii) Data cleaning—The dataset might contain missing or inconsistent values. To deal with such issues, statistical methods are used in this step of pre-processing.

(iii) Data normalization—The normalization of the dataset is carried out to reduce the distance between attribute values. It is generally achieved by keeping the value range in between -1 and $+1$.

2. Then, the whole dataset was divided into training and testing sets in a ratio of 7:3. Both the training and testing set data have been shuffled to get the optimal result at the end of the experiment.
3. In the training phase, the training dataset was supplied to the DNN classifier model to train. The optimizer used to reduce the generated error is "Adam" optimizer with a learning rate of 0.03, and the loss function used was "binary cross-entropy" function. The model was trained with a batch size of 20 and number of epochs being 100.
4. During the evaluation phase, the test dataset was supplied to the model, to predict the outcomes with a probabilistic value ranging from 0 to 1. The threshold value used to determine the output was 0.5.

3 Dataset Description

The dataset has been obtained from UCI Machine Learning Repository, which contains data of 165 patients who were detected with HCC. The dataset contains a total of 49 features which were selected by a European Organization, EASL-EORTC, who are specialized in research and treatment of cancer [25]. The dataset is a divergent one, with a total of 23 quantitative variables and 26 qualitative variables. The number of missing values contained in the dataset is over 10%, while only eight instances of the patient record have complete information in all the attributes, with a percentage value of 4.85% among the whole datasets. The target variables are the survival at 1 year, and it was encoded as a binary variable: 0 (dies) and 1 (lives).

In the dataset, each feature has a missing value percentage between 0 and 48.48%. The conventional approaches for dealing with such missing values are deletion of a record or imputation of a value. One of the most commonly used strategies when dealing with missing values is the elimination of that particular instance, but this approach has been ruled out since the beginning due to the large number of incomplete data. Thus, the authors used the imputation-based approach to handle the records using the mean and most frequent strategy.

Table 1 depicts a detailed description of the HCC dataset showing each feature's type/scale and the range of values of each feature along with the missing values contained in each of the feature columns [2, 4].

4 Results and Discussion

After the execution of both the training and testing phases, the proposed DNN model performance was evaluated based on some matrices. Theses metrics are: (i) the accuracy, which is defined as a ratio of sum of the specimens classified correctly to the total number of specimens, (ii) precision, which is known as the ratio of correctly classified data in positive class to the total number of data classified as to be in positive

Table 1 Characterization of HCC data according to medical reports

Prognostic factors	Type/scale	Range	Missing (%)
Gender	Qualitative/dichotomous	0/1	0
Symptoms	Qualitative/dichotomous	0/1	10.91
Alcohol	Qualitative/dichotomous	0/1	0
Hepatitis B surface antigen	Qualitative/dichotomous	0/1	10.3
Hepatitis B e antigen	Qualitative/dichotomous	0/1	23.64
Hepatitis B core antibody	Qualitative/dichotomous	0/1	14.55
Hepatitis C virus antibody	Qualitative/dichotomous	0/1	5.45
Cirrhosis	Qualitative/dichotomous	0/1	0
Endemic countries	Qualitative/dichotomous	0/1	23.64
Smoking	Qualitative/dichotomous	0/1	24.85
Diabetes	Qualitative/dichotomous	0/1	1.82
Obesity	Qualitative/dichotomous	0/1	6.06
Hemochromatosis	Qualitative/dichotomous	0/1	13.94
Arterial hypertension	Qualitative/dichotomous	0/1	1.82
Chronic renal insufficiency	Qualitative/dichotomous	0/1	1.21
Human immunodeficiency virus	Qualitative/dichotomous	0/1	8.48
Nonalcoholic steatohepatitis	Qualitative/dichotomous	0/1	13.33
Esophageal varices	Qualitative/dichotomous	0/1	31.52
Splenomegaly	Qualitative/dichotomous	0/1	9.09
Portal hypertension	Qualitative/dichotomous	0/1	6.67
Portal vein thrombosis	Qualitative/dichotomous	0/1	1.82
Liver metastasis	Qualitative/dichotomous	0/1	2.42
Radiological hallmark	Qualitative/dichotomous	0/1	1.21
Age at diagnosis	Quantitative/ratio	20–93	0
Grams/day	Quantitative/ratio	0–500	29.09
Packs/year	Quantitative/ratio	0–510	32.12
Performance status	Qualitative/ordinal	0, 1, 2, 3, 4	0
Encephalopathy	Qualitative/ordinal	1, 2, 3	0.61
Ascites degree	Qualitative/ordinal	1, 2, 3	1.21
International normalized ratio	Quantitative/ratio	0.84–4.82	2.42
Alpha fetoprotein (ng/mL)	Quantitative/ratio	1.2–1,810,346	4.85
Hemoglobin (g/dL)	Quantitative/ratio	5–18.7	1.82
Mean corpuscular volume (fl)	Quantitative/ratio	69.5–119.6	1.82
Leukocytes (G/L)	Quantitative/ratio	2.2–13,000	1.82
Platelets (G/L)	Quantitative/ratio	1.71–459,000	1.82
Albumin (mg/dL)	Quantitative/ratio	1.9–4.9	3.64
Total bilirubin (mg/dL)	Quantitative/ratio	0.3–40.5	3.03
Alanine transaminase (U/L)	Quantitative/ratio	11–420	2.42

(continued)

Table 1 (continued)

Prognostic factors	Type/scale	Range	Missing (%)
Aspartate transaminase (U/L)	Quantitative/ratio	17–553	1.82
Gamma glutamyl transferase (U/L)	Quantitative/ratio	23–1575	1.82
Alkaline phosphatase (U/L)	Quantitative/ratio	1.28–980	1.82
Total proteins (g/dL)	Quantitative/ratio	3.9–102	6.67
Creatinine (mg/dL)	Quantitative/ratio	0.2–7.6	4.24
Number of nodules	Quantitative/ratio	0–5	1.21
Major dimension of nodule (cm)	Quantitative/ratio	1.5–22	12.12
Direct bilirubin (mg/dL)	Quantitative/ratio	0.1–29.3	26.67
Iron (mcg/dL)	Quantitative/ratio	0–224	47.88
Oxygen saturation (%)	Quantitative/ratio	0–126	48.48
Ferritin (ng/mL)	Quantitative/ratio	0–2230	48.48

Table 2 Confusion matrix of the DNN model

Actual class	Predicted class	
	Predicted: 0	Predicted: 1
True: 0	22	6
True: 1	8	27

class, (iii) recall (true positive rate), which is defined as the ratio of true positive to the total number of instances classified under positive class. Table 2 reveals the confusion matrix for DNN model where class "0" indicates negative survivability and "1" denotes positive survivability.

Table 3 reveals that the accuracy of the DNN model is 78%, which is certainly better compared with the accuracy of KNN model which is 64% and it is also better than support vector machine (SVM) model which is having an accuracy of 58%. On the other hand, the precision of the DNN model is 83.58%, whereas KNN model is having a precision of 63.41% and SVM model is having a precision of 58%. Recall value percentage of the DNN is at 81.25% in contrast to KNN being 89.65% and SVM having 100%. Another one of the important evaluation parameters F-measure percentage of DNN is 80% which is greater than the 74.28% and 73.41% of the KNN and SVM models, respectively. The SVM established the worst result among the other two models, whereas DNN model has been established as the most efficient model in the testing phase of the dataset.

The observations obtained from Fig. 2 show which errors are there in the predicted values of the test dataset. Corresponding red line denotes the wrong predicted values, and the blue line denotes the original target values.

Table 3 Performance comparison between different models

	DNN (%)	KNN (%)	SVM (%)
Accuracy	78	64	58
Precision	83.58	63.41	58
Recall	81.25	89.65	100
F-Measure	80	74.28	73.41

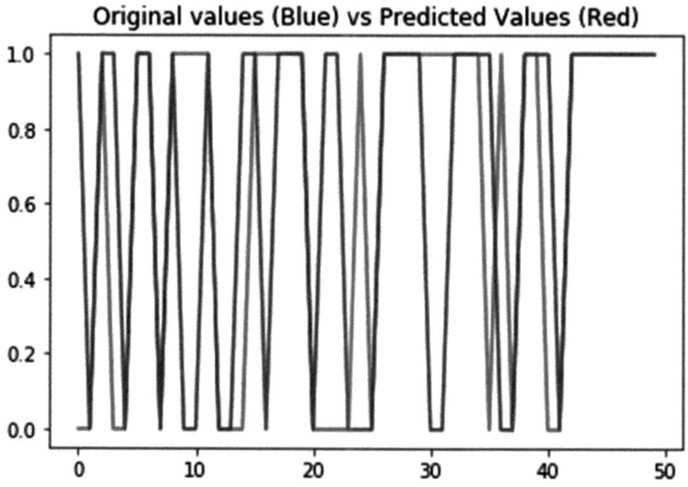

Fig. 2 Graph between original values and predicted values of DNN model

5 Conclusion

The current study proposed a deep neural network-based prediction of survival of patients suffering of hepatocellular carcinoma. Traditional methods such as SVM- and KNN-based methods have found to be not suitable for such predictions. Experimental results have revealed that the performance of DNN is superior to other classifiers. Nevertheless, future studies could be focused on appropriate feature selection to build more efficient and trustworthy classifier to predict survivability.

References

1. Parkin, D.M., Bray, F., Ferlay, J., Pisani, P.: Estimating the world cancer burden: Globocan 2000. Int. J. Cancer **94**, 153–156 (2001)
2. El Serag, H.B., Mason, A.C.: Rising incidence of hepatocellular carcinoma in the United States. N. Engl. J. Med. **340**, 745–750 (1999)
3. Thorgeirsson, S.S., Grisham, J.W.: Molecular pathogenesis of human hepatocellular carcinoma. Nat. Genet. **31**, 339–346 (2002)

4. Llovet, J.M., Brú, C., Bruix, J.: Prognosis of hepatocellular carcinoma: the BCLC staging classification. Semin. Liver Dis. **19**(3): 329–338 (1999)
5. Chevret, S., Trinchet, J.-C., Mathieu, D., AbouRached, A., Beaugrand, M., Chastang, C.: A new prognostic classification for predicting survival in patients with hepatocellular carcinoma. J. Hepatol. **31**(1), 133–141 (1999)
6. Lee, J.-S., Chu, I.-S., Heo, J., Calvisi, D.F., Sun, Z., Roskams, T., Durnez, A., Demetris, A.J., Thorgeirsson, S.S.: Classification and prediction of survival in hepatocellular carcinoma by gene expression profiling. **40**(3), 667–676 (2004)
7. Santos, M.S., Abreu, P.H., Garcia-Laencina, P.J., AdeliaSimao, A.C.: A new cluster-based oversampling method for improving survival prediction of hepatocellular carcinoma patients. J. Biomed. Inform. **58**, 49–59 (2015)
8. Chatterjee, S., Sarkar, S., Hore, S., Dey, N., Ashour, A.S., Shi, F., Le, D.N.: Structural failure classification for reinforced concrete buildings using trained neural network based multi-objective genetic algorithm. Struct. Eng. Mech. **63**(4), 429–438 (2017)
9. Chatterjee, S., Dey, N., Shi, F., Ashour, A.S., Fong, S.J., Sen, S.: Clinical application of modified bag-of-features coupled with hybrid neural-based classifier in dengue fever classification using gene expression data. Med. Biol. Eng. Comput. 1–12 (2017)
10. Chatterjee, S., Sarkar, S., Dey, N., Ashour, A.S., Sen, S., Hassanien, A.E.: Application of cuckoo search in water quality prediction using artificial neural network. Int. J. Comput. Intell. Stud. **6**(2–3), 229–244 (2017)
11. Chatterjee, S., Banerjee, S., Mazumdar, K.G., Bose, S., Sen, S: Non-dominated sorting genetic algorithm—II supported neural network in classifying forest types. In: 2017 1st International Conference on Electronics, Materials Engineering and Nano-Technology (IEMENTech) (pp. 1–6). IEEE, April 2017
12. Chatterjee, S., Banerjee, S., Basu, P., Debnath, M., Sen, S.: Cuckoo search coupled artificial neural network in detection of chronic kidney disease. In: 2017 1st International Conference on Electronics, Materials Engineering and Nano-Technology (IEMENTech) (pp. 1–4). IEEE, April 2017
13. Chatterjee, S., Dey, N., Ashour, A.S., Drugarin, C.V.A.: Electrical energy output prediction using cuckoo search based artificial neural network. In: Smart Trends in Systems, Security and Sustainability (pp. 277–285). Springer, Singapore (2018)
14. Chakraborty, S., Dey, N., Chatterjee, S., Ashour, A.S.: Gradient Approximation in Retinal Blood Vessel Segmentation
15. Chatterjee, S., Sarkar, S., Dey, N., Ashour, A.S., Sen, S.: Hybrid non-dominated sorting genetic algorithm: II-neural network approach. Adv. Appl. Metaheuristic Comput. **264** (2017)
16. Chatterjee, S., Sarkar, S., Hore, S., Dey, N., Ashour, A.S., Balas, V.E.: Particle swarm optimization trained neural network for structural failure prediction of multistoried RC buildings. Neural Comput. Appl. **28**(8), 2005–2016 (2017)
17. Chatterjee, S., Ghosh, S., Dawn, S., Hore, S., Dey, N.: Forest type classification: a hybrid NN-GA model based approach. In: Information Systems Design and Intelligent Applications (pp. 227–236). Springer, New Delhi (2016)
18. Chatterjee, S., Hore, S., Dey, N., Chakraborty, S., Ashour, A.S.: Dengue fever classification using gene expression data: a PSO based artificial neural network approach. In: Proceedings of the 5th International Conference on Frontiers in Intelligent Computing: Theory and Applications (pp. 331–341). Springer, Singapore (2017)
19. Hore, S., Chatterjee, S., Sarkar, S., Dey, N., Ashour, A.S., Balas-Timar, D., Balas, V.E.: Neural-based prediction of structural failure of multistoried RC buildings. Struct. Eng. Mech. **58**(3), 459–473 (2016)
20. Chatterjee, S., Raktim C., Dey, N., Hore, S.: A quality prediction method for weight lifting activity 95–98 (2015)
21. Hore, S., Chakraborty, S., Chatterjee, S., Dey, N., Ashour, A.S., Van Chung, L., Le, D.N.: An integrated interactive technique for image segmentation using stack based seeded region growing and thresholding. Int. J. Electr. Comput. Eng. **6**(6), 2773 (2016)

22. Chatterjee, S., Paladhi, S., Hore, S., Dey, N.: Counting all possible simple paths using artificial cell division mechanism for directed acyclic graphs. In: 2015 2nd International Conference on Computing for Sustainable Global Development (INDIACom) (pp. 1874–1879). IEEE, March 2015
23. Hore, S., Chatterjee, S., Chakraborty, S., Shaw, R.K.: Analysis of different feature description algorithm in object recognition. In: Feature Detectors and Motion Detection in Video Processing, p. 66 (2016)
24. Chakraborty, S., Chatterjee, S., Dey, N., Ashour, A.S., Ashour, A.S., Shi, F., Mali, K.: Modified cuckoo search algorithm in microscopic image segmentation of hippocampus. Microsc. Res. Tech. **80**(10), 1051–1072 (2017)
25. Lichman, M.: UCI machine learning repository (http://archive.ics.uci.edu/ml). Irvine, CA: University of California, School of Information and Computer Science (2013)

Detection and Retrieval of Colored Object from a Live Video Stream with Mutual Information

Debayan Chatterjee and Subhabrata Sengupta

Abstract Detection and segmentation refer to a process of segregating a part of any body from its whole. In the case of image segmentation, it refers to the method of obtaining a particular portion of that image. As we know, a video or any mp4 file is an orderly orientation of frames passing the screen per second; hence, video detection schemes can also use this segmentation process. With the advancements of AI and machine learning in our daily life, video segmentation has become a necessity for much technological advancement. Some of the important applications of video segmentation could be a face recognition system, color detection schemes, an object tracking system, etc. If we want to track anyone or some object, video surveillance becomes the primary key to such an application, and hence, a lot of importance is given to it. To find the best results on object detection techniques, different algorithms put forward. Here we would cover all the technological advancements and researches, works done on object detection and live stream of videos. With today's computing powers, the vision to tracking objects as well as moving objects has advanced a lot. Various needs like video surveillance and facial recognition systems in many security checks use color detection and object recognition schemes. Although there are many challenges like high resolution of the videos, machines capable of analyzing higher frame rates in the videos and many more.

Keywords Image segmentation · Video segmentation · Skin detection · Edge and corner detection · Color detection and tracking

D. Chatterjee · S. Sengupta (✉)
Department of Information Technology, IEM, Salt Lake, Kolkata, India
e-mail: ssg365@gmail.com

D. Chatterjee
e-mail: debayanc100@gmail.com

M. Chakraborty et al. (eds.), *Proceedings of International Ethical Hacking Conference 2018*, Advances in Intelligent Systems and Computing 811,
https://doi.org/10.1007/978-981-13-1544-2_29

1 Introduction

1.1 The RGB Color Model

With the onset and advancements of artificial intelligence, video segmentation and tracking have taken the center stage in many technological fields and have promised a better future with these advancements. Many researchers have already been done to reach the required segmentation from any live stream and many more are also in the budding stage. Color discrimination from a given image or a video is another crucial part of object segmentation domain. In this method, we can obtain the colors from an image or set of frames in the video. Representation of RGB color model is displayed in Fig. 1.

The RGB color model comprises the three primary colors—red, green, and blue. This model mainly states that combining these three primary colors in various proportions, various other colors could be obtained. Being a hardware model, RGB color is used in the process of device rendering, processor of image, and capturing images.

1.2 Hue, Saturation, and Value

Hue, saturation, and value are most fundamental aspects of HSV model. A. Smith proposed this color model for different shades of image frames and objects. User's tint, shade, and tone are used in HSV color model, and it is being defined in a hexagonal cone. Figure 2 represents the illustration of this model. Studying the colors of an object or a video enriches the information carried by it. We will discuss the advancements in this domain and how this is leading to the gate of success in the

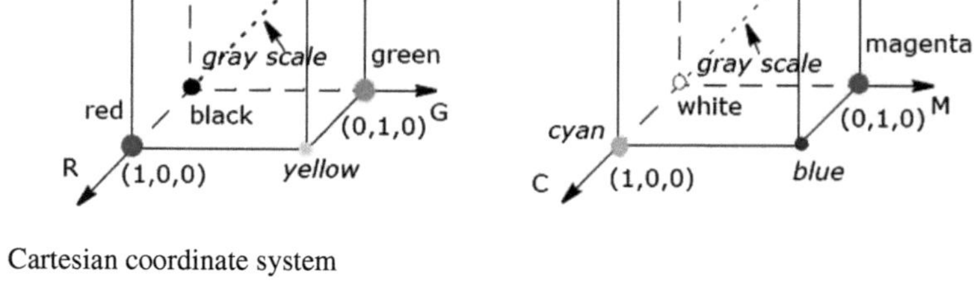

Fig. 1 Cartesian coordinate system

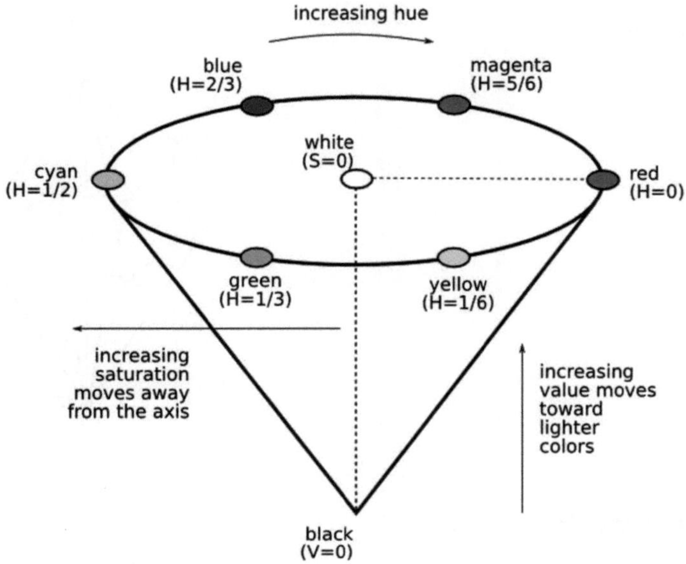

Fig. 2 Hue, saturation, and value framework

future. Various facial recognition policies as well as object detection use this scheme for successful recognition applications.

1.3 Framework of Hue, Lightness, and Saturation

Basic attributes like hue, lightness, and saturation are most integral parts of HLS model. This name of the color model was given by W. Ostwald. The definition of hue remains as it is in the HSV model. In the HSL color model, black occupies the bottom portion of hex cone, while white being the extremity of black lies at the tip at the top of the cone. This is actually the measure of lightness. So, we can say that the value of lightness is 1 for white, whereas it is 1 for black. By having some deformities, HLS is very much similar. This HLS model serves us in obtaining the details of the image, and hence, only after receiving the details, we can move forward for color detection and segmentation schemes. Figure 3 depicts the internal architecture which will be looking like a double hex.

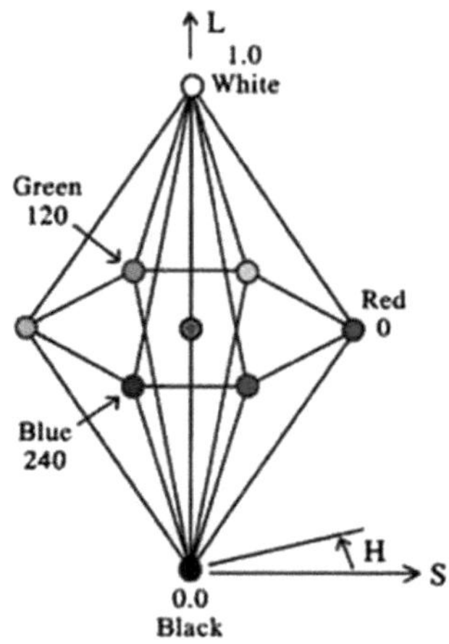

Fig. 3 Hue, lightness, and saturation

2 Related Work

Kalisa Wilson et al. showed the path to track multiple color objects in real time in the publication "Real-Time Tracking for Multiple Objects Based on Implementation of RGB Color Space in Video" [1]. By using the method of thresholding, system architecture is developed. Main attributes of the system are preprocessing and identification of colored object with segmented shots.

Ssu-Wei Chen et al. proposed a novel idea of tracking objects in the publication "Moving Object Tracking Based on Background Subtraction Combined Temporal Difference" [2]. Entire concepts have been described with the content of continuous image subtraction process.

Vandana S. Bhatai et al. developed an idea on unique colored object feature in "Face detection system using HSV color model and morphing operations" [3]. Main objective of this publication is to recognize of face values with the help of skin color. By transformation of image frames in different color models, object will be recognized.

Ravikant Gupta et al. illustrated the concept of an algorithm on objects' feature in "Human Face Detection by YCbCrHs Technique" [4]. His work was mainly to detect the feature points from the image implementation of spatial filters have been introduced.

Prasad Kalane et al. developed a novel idea to detect any target by using Kalman Filter, and he described it in his paper "Target Tracking Using Kalman Filter" [5]. A target can be any and every object, maybe a human or an umbrella or anything. He further stated that there are two main steps for detecting an object:

1. Detecting the moving objects which are subject to our experiment.
2. Keeping the track of the concerned objects from frame to frame.

Isaac Cohen et al. in his paper on "Detecting and Tracking Moving Objects for Video Surveillance" proposed to choose a complete model which gives a well-explained outline about the current estimation, which is also quite simple in perspective of numerical entanglement [6].

Andres Alarcon Ramirez et al. developed an algorithm to find out the moving object in his paper "A New Algorithm for Tracking Objects in Videos of Cluttered Scenes" [7]. By this algorithm, object position can be predicted in the next consecutive frames.

Chris Harris et al. in his paper on "A Combined Corner and Edge Detector" proposed a technique to detect extremities and corners in an image, intersections would then be considered as points where various edges meet [8]. To continue in the footsteps of this approach, he started from the Moravec's corner detector.

Katja Nummiaro et al. proposed a colored object tracking in more than one camera setup [9]. In this object was placed in a system having multiple cameras. To support multiple cameras, they used more than one histogram for a concerned object. The cameras were placed at various angles in order to obtain the features of the object from numerous angles.

Marcus Thaler et al. proposed a real-time human or objects detection technique in a 360° full angle view [10]. This is really helpful in the frontier of any sports like cricket, tennis, and football where close-up views are needed for vital decision making. Hence, this technique can be proficiently used in these spheres where the full view comes of great use.

Wen-cheng Wang et al. gave a novel idea where face recognition is used for colored images [11]. He concluded with four kinds of color space: RGB, HIS, YUV, and YCbCr. Main agenda of this idea is to cluster the skin color data in different color spaces and to represent a suitable color mixing model.

Hani K. Mohair et al. proposed an algorithm to predict the skin color of human beings given a certain image in "Human skin color detection: A review on neural network perspective" [12]. Here, the raw image was first processed, and further erosion and dilation were applied to find out the skin color ignoring the various sizes and features of a human face.

Monika Deswal et al. proposed an algorithm to recognize the character and composition of an image in her paper on "A Fast HSV Image Color and Texture Detection and Image Conversion Algorithm" [13]. With the filtration of noise, the RGB image was converted to HSV format, and a corresponding set of HSV values was constructed.

Douglas Chai et al. gave an interesting procedure of face detection and partitioning using the skin color map in "Face Segmentation Using Skin-Color Map in Videophone Applications" [14]. The algorithm works as follows: First the head and shoulder image was taken as input and then color detection followed by density regularization, geometric rectification, and shape extraction was done, and finally, segmented facial portion was collected.

S. Chitra et al. proposed a comparative analysis on two different color spaces HSCBCr and YCbCr in extraction of skin color in "Comparative Study for Two Color Spaces HSCbCr and YCbCr in Skin Color Detection" [15]. The main objective of this article was to highlight the process of conversion of image from HSV to YCbCr model according to the threshold values.

Intaek Kim et al. proposed a multi-object detection using color schemes. Author proposed an algorithm that tracks multiple objects in a video applying extended Kalman filter and color information [16].

Akriti Kaushik et al. proposed a RGB color sensing technique [17]. They used various sensors to catch hold of the RGB color. In the field of medical science, color sensing has great impact. The main trick behind this approach was that when some colored object is kept before the sensor, then the color gets detected and the LED of the same color glows; it uses eight-color LEDs that include three primary colors—red, green, blue and along with magenta, yellow, and cyan as well as black and white.

Sagar Pandey et al. represented a novel idea on detection of red color channel in Color Detection and Tracking from Live Stream—An Extensive Survey [18]. In this paper, authors have demonstrated the idea of video segmentation with color detection technique. Contextual preprocessing technique has been applied to detect the RGB color model.

3 Proposed Architecture

The following flow chart shows how the color and object detection and tracking algorithm is processed (Fig. 4).

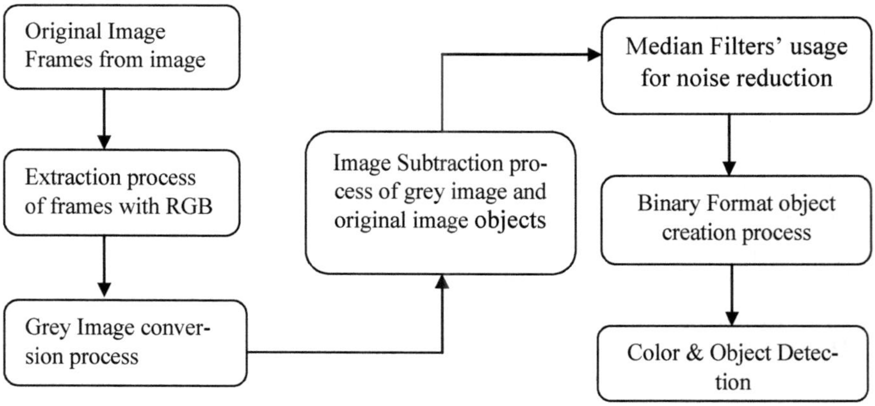

Fig. 4 Proposed architecture view

3.1 Algorithmic Steps

Step 1: From the base of RGB framework, red color objects are extracted from a live stream.
Step 2: Red-colored object will be transformed to gray color object, and image subtraction process will be carried on between red color object and gray color object.
Step 3: By the use of different median filters, different types of noises are removed.
Step 4: Difference factor value of two different frames will be converted to binary image with respect to its actual threshold.
Step 5: Retrieval process of different frames will be done by the implementation of precision and recall.
Step 6: Identification of red-colored object is justified from a live stream video.

4 Result and Interpretation

Using video segmentation techniques, the following were applied on a few sample videos, and the results were tested as per performance measures mentioned above (Tables 1, 2, 3 and 4).

A: Total number of frames.
B: Total number of cuts present.
C: Total number of correct cuts detected.
D: Total number of cuts went undetected.
E: Total number of false cuts detected.
R = Recall = C/(C + D). P = Precision = C/(C + E).

Table 1 Mutual information

Video	A	B	C	D	E	R	P
Sports	412	68	45	23	12	66.1	78.9
Cartoon	600	147	65	82	10	44.2	86.6
AD	599	90	51	39	12	56.6	80.9
News	720	219	26	193	16	11.8	61.9

Table 2 Chi-square

Video	A	B	C	D	E	R	P
Sports	412	68	27	41	7	39.7	79.4
Cartoon	600	147	8	139	0	5.4	100
AD	599	90	25	65	5	27.7	83.3
News	720	219	219	0	501	100	30.4

Table 3 Histogram equalization

Video	A	B	C	D	E	R	P
Sports	412	68	40	28	10	58.8	80
Cartoon	600	147	79	68	14	53.7	84.9
AD	599	90	50	40	8	55.5	86.2
News	720	219	125	94	12	57.0	91.2

Table 4 Edge detection

Video	A	B	C	D	E	R	P
Sports	412	68	56	12	56	82.3	50.0
Cartoon	600	147	106	41	16	72.1	86.8
AD	599	90	67	23	12	74.4	84.8
News	720	219	151	68	21	68.9	87.7

5 Simulation Results

See Figs. 5, 6, 7, and 8.

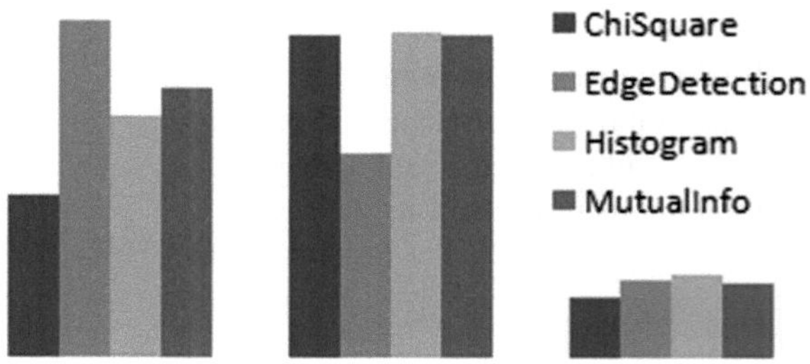

Fig. 5 Video category—sports

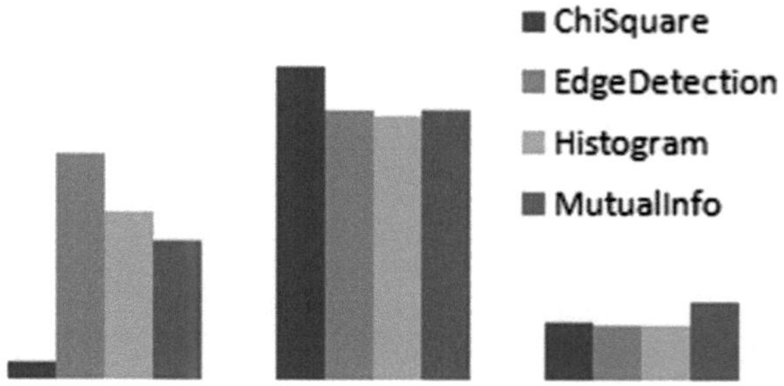

Fig. 6 Video category—cartoon

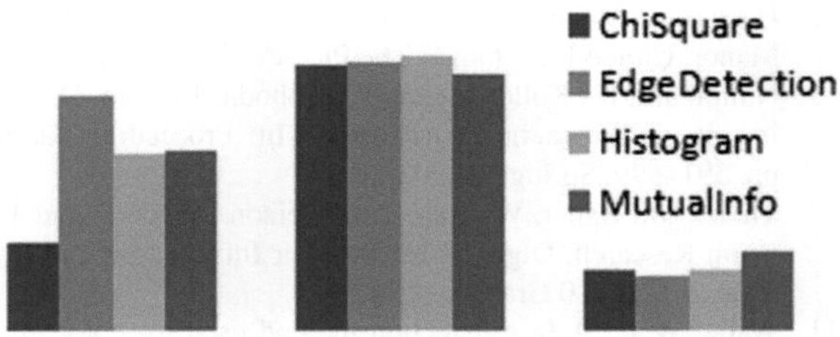

Fig. 7 Video category—advertisement

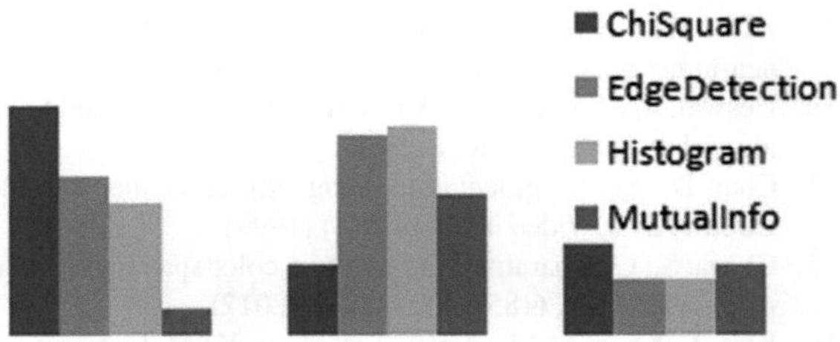

Fig. 8 Video category—news

6 Conclusion and Future Work

MATLAB and many image processing tools can easily implement our working algorithm that is given above and gives successful results with high percentage of precision as related to the given image or video file as input. In all these researches, the main objective still remains the same—to make the accuracy of detection closer and closer to 100%. Various algorithms are proposed by numerable researchers where they have shown the path of better precision works related to this topic. Furthermore, we still believe there is still a lot of more work to do to make a smarter world to live in.

References

1. Wilson, K.: Real-time tracking for multiple objects based on implementation of RGB color space in video. Int. J. Signal Process. Image Process. Pattern Recognit. **9**(4), 331–338 (2016)
2. Chen, S.-W., Wang, L.K., Lan, J.-H.: Moving object tracking based on background subtraction combined temporal difference. In: International Conference on Emerging Trends in Computer and Image Processing (ICETCIP'2011), Bangkok, December 2011
3. Bhata, V.S., Pujaria, J.D.: Face detection system using HSV color model and morphing operations. Int. J. Curr. Eng. Technol.
4. Gupta, R., Pandey, S., Tayal, Y., Pandey, P.K., Singh, D.V.B.: Human face detection ByYCbCrHs technique. Int. J. Emerg. Technol. Comput. Appl. Sci. (IJETCAS)
5. Kalane, P.: Target tracking using Kalman filter. Int. J. Sci. Technol. **2**(2) (2012)
6. Cohen, I., Medioni, G.: Detecting and tracking moving objects for video surveillance. In: IEEE Proceedings of the Computer Vision and Pattern Recognition, 23–25 June 1999, Fort Collins CO (1999)
7. Ramirez, A.A., Chouikha, M.: A new algorithm for tracking objects in videos of cluttered scenes. Int. J. Inf. Technol. Model. Comput. (IJITMC) **1**(2) (2013)

8. Harris, C., Stephens, M.: A Combined Corner and Edge Detector. Plessey Research Roke Manor, United Kingdom © The Plessey Company pic. (1988)
9. Nummiaro, K., Koller-Meier, E., Svoboda, T., Roth, D., Gool, L.V.: Color-based object tracking in multi-camera environments. In: Proceedings of the DAGM'03. LNCS, vol. 2781, pp. 591–599. Springer (2003)
10. Thaler, M., Bailer, W.: Real-Time Person Detection and Tracking in Panoramic Video. Joanneum Research, Digital—Institute for Information and Communication Technologies Steyrergasse 17, 8010 Graz, Austria
11. Wang, W.-c.: A face detection method used for color images. Int. J. Signal Process. Image Process. Pattern Recognit. **8**(2), 257–266 (2015)
12. Al-Mohair, H.K., Mohamad-Saleh, J., Suandi, S.A.: Human skin color detection: a review on neural network perspective. Int. J. Innov. Comput. Inf. Control **8**(12) (2012)
13. Deswal, M., Sharma, N.: A fast HSV image color and texture detection and image conversion algorithm. Int. J. Sci. Res. (IJSR)
14. Chai, D.: Face segmentation using skin-color map in videophone applications. IEEE Trans. Circuits Syst. Video Technol. **9**(4) (1999)
15. Chitra, S.: Comparative study for two color spaces HSCbCr and YCbCr in skin color detection. Appl. Math. Sci. **6**(85), 4229–4238 (2012)
16. Kim, I., Khan, M.M., Awan, T.W., Soh, Y.: Multi-target tracking using color information. Int. J. Comput. Commun. Eng. **3**(1) (2014)
17. Kaushik, A., Sharama, A.: RGB color sensing technique. Int. J. Adv. Res. Sci. Eng.
18. Pandey, S., Sengupta, S.: Color detection and tracking from live stream—an extensive survey. Int. J. Comput. Appl. **168**(3) (2017)

A Machine Learning Framework for Recognizing Handwritten Digits Using Convexity-Based Feature Vector Encoding

Sourav Saha, Sudipta Saha, Suhrid Krishna Chatterjee and Priya Ranjan Sinha Mahapatra

Abstract Handwritten digit recognition has always been an active topic in OCR applications as it stems out of pattern recognition research. In our day-to-day life, character image recognition is required while processing postal mail, bank cheque , handwritten application form, license plate image, and other document images. In recent years, handwritten digit recognition has been playing a key role even for user authentication applications. In this proposed work, we develop a gradient descent ANN model using novel and unique geometric feature extraction technique for handwritten digit recognition system which can be further extended to identify any alphanumeric character images. We have extracted geometric features of handwritten digit based on computational geometric method and applied artificial neural network (ANN) technique for classification of handwritten digits through machine learning approach. The characteristics of extracted feature for a digit class are found to be distinct despite wide variations within the class and thereby lead to reasonably good recognition rate even with small trainee samples.

Keywords Handwritten digit recognition · Machine learning · Artificial neural network · Computer vision · Shape analysis · Pattern recognition

S. Saha (✉) · S. Saha · S. K. Chatterjee
Institute of Engineering & Management, Kolkata, India
e-mail: souravsaha1977@gmail.com

S. Saha
e-mail: subho040995@gmail.com

S. K. Chatterjee
e-mail: c.suhrid@gmail.com

P. R. S. Mahapatra
Department of Computer Science & Engineering, University of Kalyani, Kalyani, India
e-mail: priya_cskly@yahoo.co.in

M. Chakraborty et al. (eds.), *Proceedings of International Ethical Hacking Conference 2018*, Advances in Intelligent Systems and Computing 811,
https://doi.org/10.1007/978-981-13-1544-2_30

1 Introduction

With the growing demand for advanced smart robotics technology, automated identification of textual images comprising alphanumeric characters has been vastly explored in machine learning community. The task of character image recognition is dealt in processing postal mail, bank check, manually entered application form data, license plate image, and other document images. Particularly, automated handwritten character recognition has always been a challenging problem in computer vision-based document processing task. Handwritten document processing offers many interesting pattern classification problems like handwritten character recognition, writer authentication, signature verification, and script recognition. Handwriting is subject to large variations in scaling, rotational, and sheering effects. Various approaches have been proposed by the researchers for the recognition of isolated handwritten numerals/characters [1–4]. Even though recently high accuracy is reported for deep learning-based models, it may be noted that such deep learning-based approaches require intense computation as well as very large amount of samples for training the model due to lack of computational intelligence in domain-specific feature space exploration. In this proposed work, we develop a gradient descent ANN model which uses novel and unique geometric feature extraction technique for forming discriminative feature vectors to recognize handwritten digits. Being generic by nature, the proposed framework can be further extended to identify any isolated alphanumeric character images. We have extracted geometric features of handwritten digit based on computational geometric method and applied Artificial Neural Network (ANN) technique for classification of handwritten digits through machine learning approach. The characteristics of extracted feature for a digit class are found to be unique even though there are appearance-based large variations among the digit figures representing same digit class, and the proposed model achieves reasonably good recognition rate even with small training sample space.

1.1 Related Work

The performance of automated handwritten character recognition mostly depends on the feature extraction approach and the learning scheme. Over the years, various state-of-the-art approaches [1] have been explored for feature extraction of character recognition. Many researchers have reported that the stroke direction feature [2, 3] is one of the most distinguishable features for handwritten character recognition [4]. This feature can be represented in terms of histograms of local direction elements. In order to improve the discrimination ability, blurring operation is applied to local direction elements [5]. Addition of curvature feature [6] or local structure features [7–9] may lead to further feature enhancement in terms of uniqueness. The statistical features and local structural features are represented in vector form. Each feature vector corresponds to a point in the feature space. An effective classifier partitions the feature space into regions such that a partitioned region enclosing some feature points can mostly correspond to a specific class. Neural networks [10, 11] and statistical techniques [12–14] are generally used for the classification due to the implementation efficacy. A comparative study of various feature extraction and classification schemes for handwritten digit recognition has been reported by Liu et al. [8]. According to their observation, the chain code feature, the gradient feature, and the gradient, stroke, and concavity (GSC) feature offer high accuracies as compared to others. While comparing performances of various classification schemes for handwritten digit recognition, they found that the k-NN rule is a better performer than MLP and a binomial classifier. The advantage of the structural approach for forming features is that it provides a good symbolic description of the image as illustrated in [9]. Structural features describe a pattern in terms of its topology and geometry by giving its global and local properties. From the survey of the existing solutions on characters recognition, it appears that there is a necessity to explore more on feature enhancement to remove the confusion between similar shaped characters for their recognition. Most of the deep learning-based approaches [15] for OCR applications require very large amount of samples to train the model due to poor feature space exploration. In this proposed work, we develop a gradient descent ANN model which uses unique geometric features for handwritten digit recognition system.

2 Proposed Framework

This section discusses our proposed framework in detail. The proposed scheme for handwritten digit recognition assumes digit image as vascular chord generated using thick digital curve. The model extracts feature vectors based on convexity of significant chord segments and their relative orientation. Fig. 1 intuitively demonstrates the overall flow of the proposed strategy. The description of each step of the overall flow is detailed next.

2.1 Polygonal Approximation

The approximation of arbitrary two-dimensional curves by polygonal figures is an imperative technique in digital image processing. A digital curve can be effectively simplified by polyline without loss of its visual property. Techniques for polyline approximation of digital curve have been driving interest among the researchers for decades [16]. The number of line segments used in the process of creation of a polygon determines the accuracy of the approximation algorithm. For an algorithm to be effective and accurate, it must not exceed the minimum number of required sides necessary to preserve the actual shape of the curve. A polygon thus created with only the minimum requisite number of line segments is often named as a minimum-perimeter polygon. A higher number of edges in an approximated polygonal figure add to the

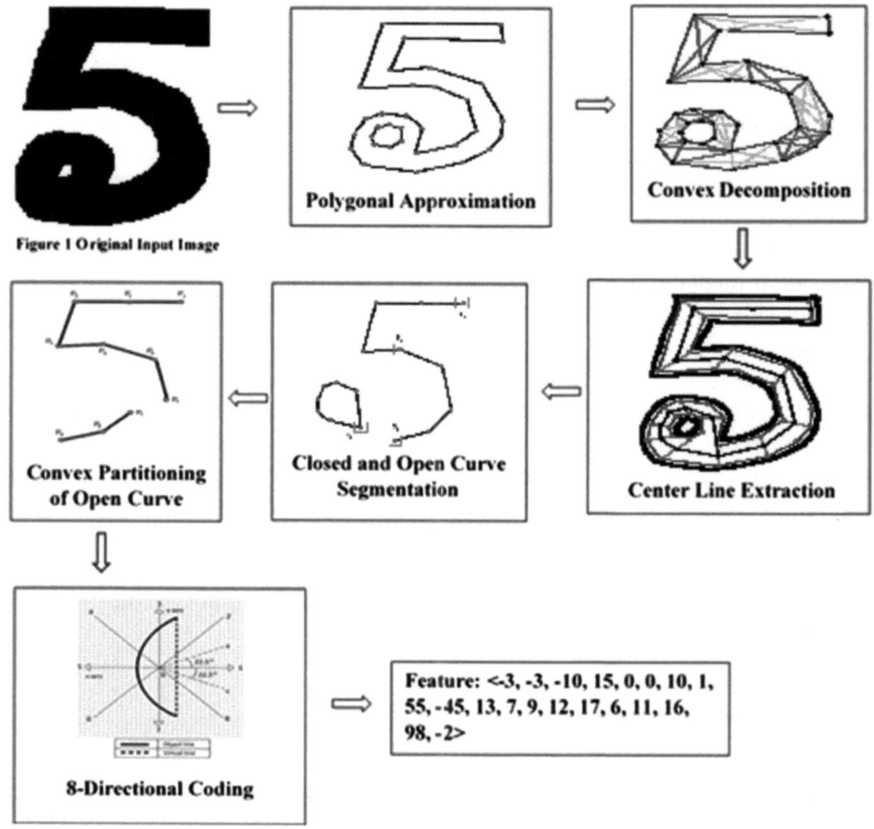

Fig. 1 Overall flow of the proposed scheme

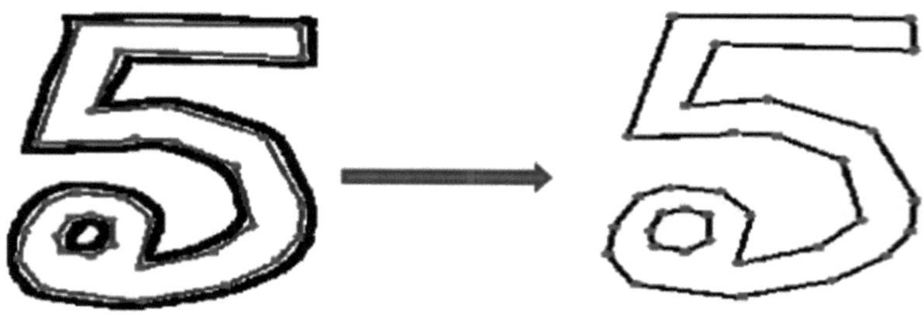

Fig. 2 Polygonal approximation

source of noise to the model. Polygon approximation removes the nonessential curve points and provides us with a discrete figure that is bounded by line segments. The resultant image consists of a simple polygonal region in the plane bounded by a non-self-intersecting, closed, polygonal path as shown in Fig. 2.

2.2 *Convex Decomposition/Partitioning of Simple Polygonal Shape*

Human vision organizes object shapes in terms of separable parts and their spatial relationships. A shape decomposition model helps to dissociate a separable part from the shape based on its spatial relationships with other parts, and thereby, it provides a natural way to represent a complex polygonal shape in terms of simple graph. The proposed work develops a graph-theoretic framework based on the principle presented in [16] to decompose the shape of an image object into substantial parts. The proposed framework primarily explores polygonal approximation to generate corresponding vertex visibility graph. Further, an iterative clique extraction strategy is employed to decompose the shape into convex pieces based on its vertex visibility graph (Fig. 3).

A clique is a complete subgraph of a graph. A maximal complete subgraph is called a maximal clique if cannot be extended by including any more adjacent vertices. An object can be visualized as a composition of convex pieces. One of the interesting observations with respect to such composition would be that a convex piece of a shape corresponds to a maximal clique in many cases. Such cliques are enclosed by sides of the polygonal shape and boundary cuts. A boundary cut is an interface between two clique representative convex parts. On the basis of this idea, we have developed a heuristic strategy which iteratively extracts maximal cliques from the visibility graph of a polygonal shape. The most popular technique to find maximal cliques of a given undirected graph was presented by Bron and Kerbosch [17]. The proposed algorithm which is presented as Algorithm: GetShapePartition (Fig. 5) adopts their principle for generating maximal cliques and uses heuristics to find the most suitable clique corresponding a convex piece of shape.

Illustration of the Proposed Shape Partitioning Algorithm:

The main objective of Algorithm: GetShapePartition (Fig. 5) is to discover a suitable clique partitioning of polygonal shape based on a heuristic. At every step, the objective of the heuristic is to obtain a maximal clique from the residual graph which covers maximal polygonal area with minimal cut length. The iterative procedure is illustrated in Fig. 4 where a red region indicates an extracted clique as convex part.

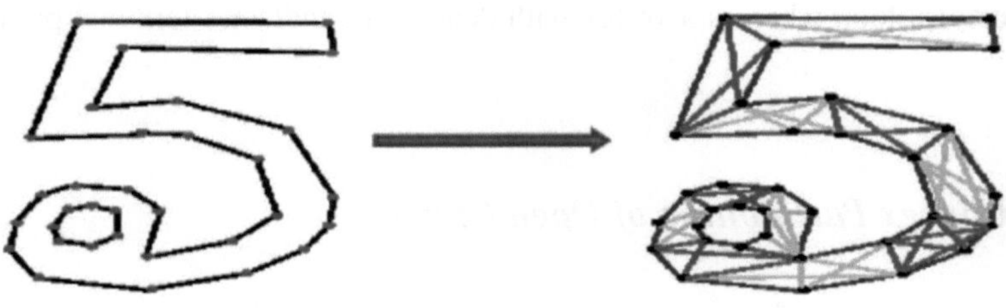

Fig. 3 Convex partitioning of digit polygonal approximation

Illustration of the Proposed Shape Partitioning Algorithm:

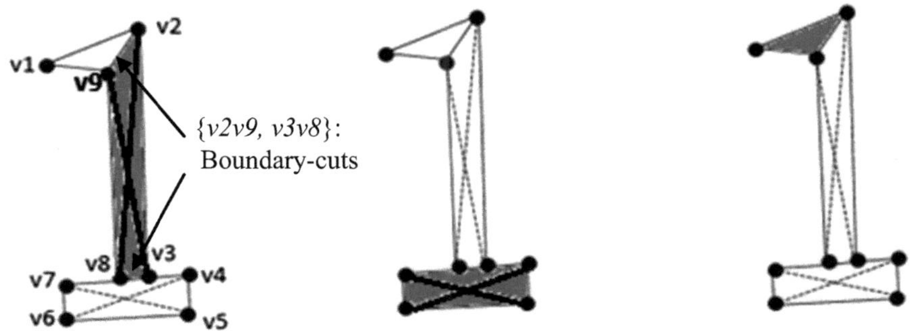

Fig. 4 GetShapePartition: **a** Step 1, **b** Step 2, and **c** Step 3

2.3 *Centerline Extraction*

Centerline of a figure simplifies the structure to a great extent. Traditional skeleton extraction methods fail to extract proper centerline of a digit image due to irregular thickness of the curve. Our proposed convex partitioning strategy helps in meaningful determination of the centerline of thick digit shape. As shown in Fig. 6, the centerline of the polygonal shape is obtained through joining midpoints of cut faces of adjacent convex pieces. These midpoints act as pivotal points of the centerline. A tailpiece is a convex part with one cut face, and its centerline is extracted based on the diameter of the convex polygon encompassing the piece.

2.4 *Open–Closed Curve Segmentation*

An open curve has two terminal ends, whereas a closed curve forms a loop. The centerline is treated in our framework as a partitioning graph where pivotal points along the centerline act as vertices. We identify loop/closed curve by exploring the degree of a vertex. As shown in Fig. 7, a vertex with degree three indicates the presence of a loop whereas a vertex with degree one indicates terminal point of a curve.

2.5 *Convex Partitioning of Open Curve*

Convex partitioning of an open curve is an important step for decomposing a non-convex curve into distinct convex segments. The partitioning strategy is illustrated using Fig. 8. We assume that the pivotal points of the curve are numbered as P_1, P_2, P_3, P_4, P_5, P_6, P_7, P_8, and P_9 in clockwise manner. We consider $\Delta P_1P_2P_3$ as

Algorithm 2: GetShapePartition

```
Input: G : A graph in terms of adjacency matrix; R : A set of vertices, Initially it is
       set as empty; P: VertexSet(G); X: A set of vertices, Initially it is set as
       empty;
Output: Global ParitioningList: A list partitioning of G
Function GetShapePartition (G, R, P, X)
    if G is null-graph then
        /* terminate since no more partition is required */
        return;
    end
    if P is empty AND X is empty then
        /* R is a maximal clique. Apply heuristic to obtain
           current most suitable maximal clique. */
        heuristicCost ← GetHeuristicCost(R);
        if currentMinHeuristicCost > heuristicCost then
            suitableMaximalClique ← R;
            currentMinHeuristicCost ← heuristicCost;
        end
        S ← S ∪ R;
        if S = VertexSet(G) then
            /* all possible maximal cliques are generated for
               current G */
            Identify and make a list(neighborMaximalCliqueList) of all
             disconnected cliques in the neighborhood of suitableMaximalClique;
             identifiedCliqueList ←
             suitableMaximalClique ∪ neighborMaximalCliqueList;
            Output most suitable partition described by identifiedCliqueList ;
            Add identifiedCliqueList to ParitioningList;
            MaximalCliqueAdjMat ←
             GetAdjacencyMatrix(identifiedCliqueList);
            for each edge described by pair (i, j) in G do
                /* Update G: Remove an edge from G if it
                   belongs the sub-complete graph stored as
                   current identifiedCliqueList such that it is not
                   a boundary cut interfacing residual graph */
                if edge(i, j) ≠ boundary-cut then
                    G(i, j) ← {G(i, j) − MaximalCliqueAdjMat(i, j)};
                end
            end
        end
    end
    u ← a pivot vertex ∈ P ∪ X;
    for each vertex v ∈ P \ NeighboSet(u) do
        GetShapePartition(G, R ∪ {v}, P ∩ NeighborSet(v), X ∩
         NeighborSet(v));
        P ← P \ {v};
        X ← X ∪ {v};
    end
```

Fig. 5 Proposed algorithm for shape partitioning

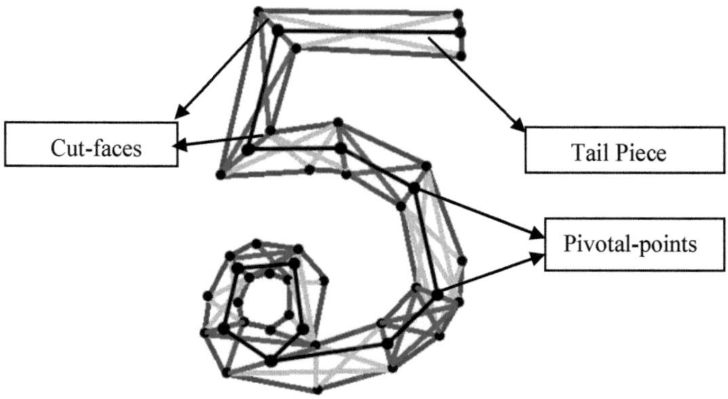

Fig. 6 Centerline extraction

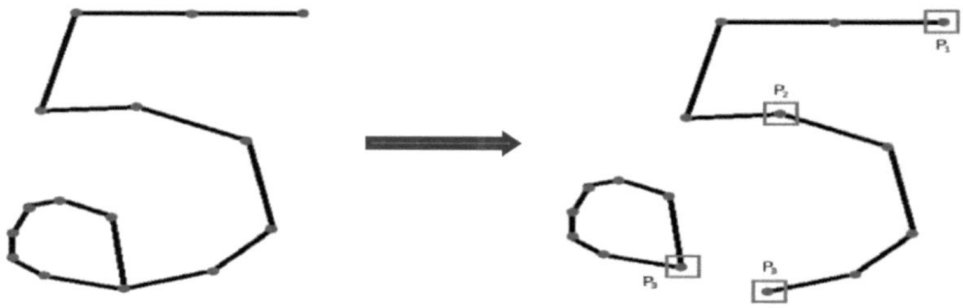

Fig. 7 Curve segmentation

initial convex polygon and form a convex sequence C. We proceed by checking the convexity of the polygon $P_1P_2P_3P_4$ after joining a candidate point P_4 with $\Delta P_1P_2P_3$. If its inclusion results in a convex polygon, we append the vertex in convex sequence C. In Fig. 9, we can grow the initial convex sequence until P_8 as its inclusion results in non-convex polygon. We treat convex sequence $\{P_1, P_2, P_3, P_4, P_5, P_6, \text{and } P_7\}$ as a convex segment. Further, a new convex sequence is formed using subsequent three vertices, namely P_7, P_8, and P_9. Thereafter, the same process as stated above is repeated to extract another convex segment.

2.6 *Eight-Directional Feature Encoding of Convex Segments*

Each convex part undergoes special eight-directional coding to form a feature vector component as discussed below. Here, we assume that the endpoints of the convex chord are joined by a virtual line resulting in generation of a convex closed curve. The closed curve is partitioned into eight equal angular divisions with each division representing a directional code. For each division, we compute a value by subtracting virtual chord length from real chord length residing inside the division. These values

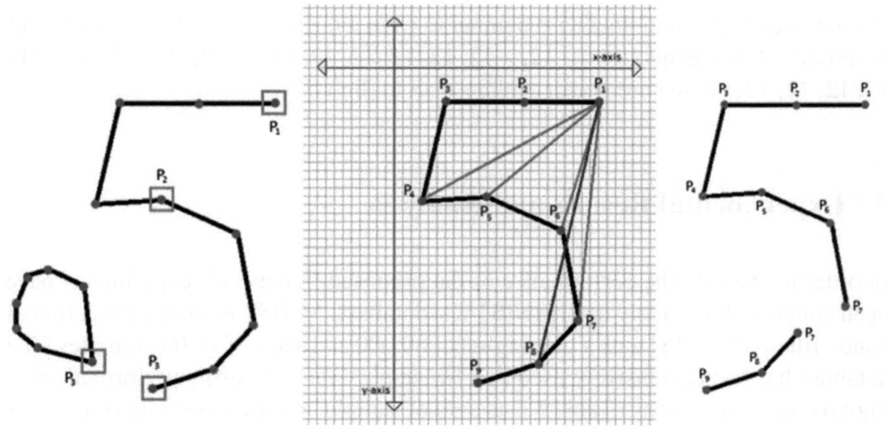

Fig. 8 Convex partitioning of open curve

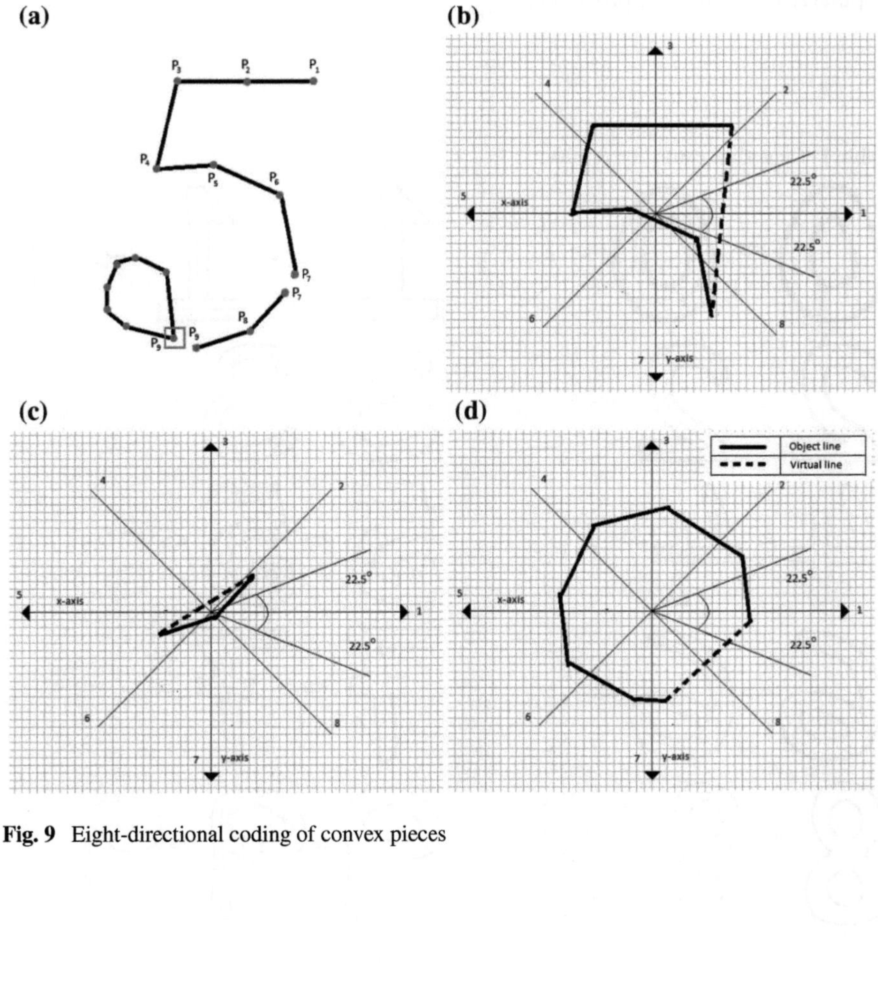

Fig. 9 Eight-directional coding of convex pieces

are concatenated in anticlockwise manner to form a feature vector. For example, our proposed scheme generates ([−12, −12, 12, 12, 12, 12, 12, −12], [12, 12, 12, −10, 12, 12, 12, 12]) as feature vector with respect to digit 6.

3 Experimental Results and Analysis

In order to evaluate the performance of the proposed framework, experiments have been conducted on widely used MNIST test database. MNIST is a large collection of handwritten digits [18] with a training set of 60,000 and a test of 10,000 samples. The database has been extensively used as a benchmark for comparing performances of digit recognition systems. Table 1 lists a set of sample images along with centerlines extracted by our algorithm. The centerlines obtained through our strategy appear to be better in terms of qualitative skeleton representation ability in comparison with popularly used thinning procedures.

Table 1 Centerline extraction results

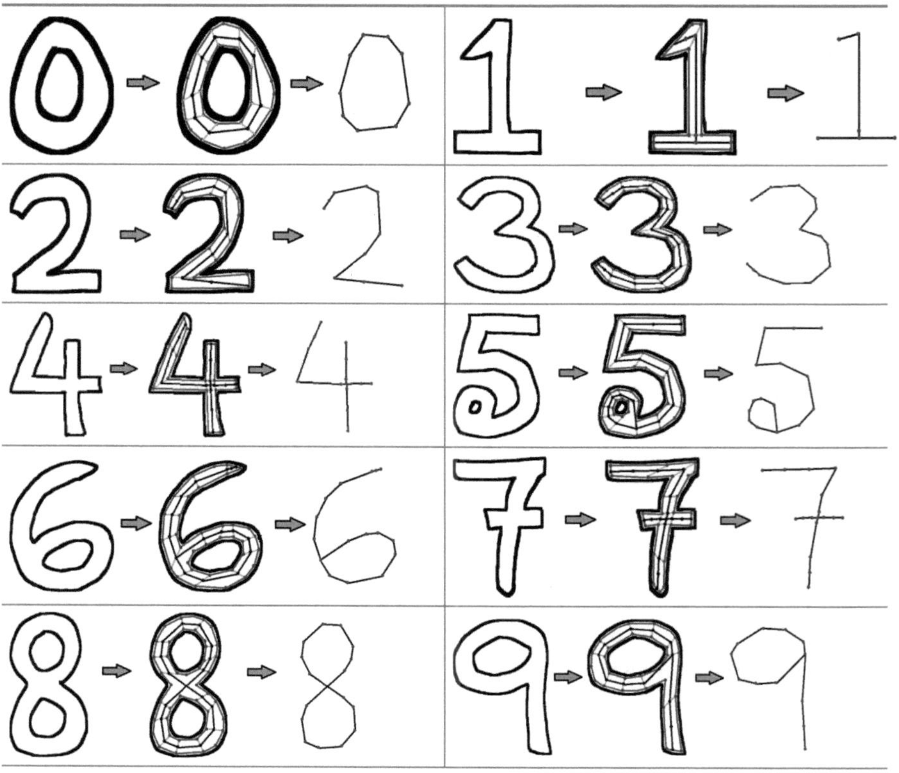

Table 2 Confusion matrix

	Predicted digit (%)										
Actual digit image		0	1	2	3	4	5	6	7	8	9
	0	99	0	0	0	0	0	0	0	0	1
	1	0	99	0	0	0	0	0	1	0	0
	2	0	0	98	0	0	0	0	0	0	2
	3	0	0	0	98	0	1	0	0	0	2
	4	0	0	0	0	97	0	0	1	0	2
	5	0	0	0	0	0	100	0	0	0	0
	6	0	0	0	0	0	1	99	0	0	0
	7	1	0	0	0	0	0	0	98	0	1
	8	0	0	0	2	0	0	0	0	97	1
	9	0	0	0	1	0	0	0	2	0	97

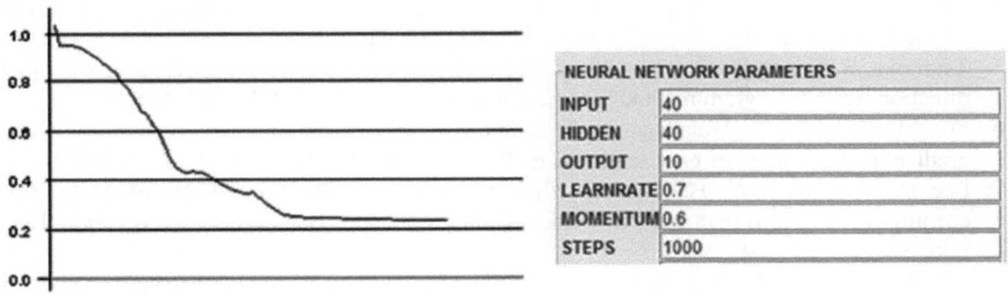

Fig. 10 ANN training profile

Performance Evaluation Metric:

Evaluation of performance is a difficult for such framework, mainly due to the subjectivity of the human vision-based judgment. The merit of such a classification scheme is usually presented in terms of confusion matrix (Table 2), whereas the accuracy is measured based on true positive (TP), true negative (TN), false positive (FP), and false negative (FN) values [Accuracy = $(TP + TN)/(TP + TN + FP + FN)$]. As per our observation, the proposed framework seems to perform reasonably well with overall accuracy nearly 99%. Figure 10 shows the progression of error graph while training our ANN model along with its various parameters.

4 Conclusion

This paper presents a relatively new direction for identification of handwritten characters. We employ a gradient descent ANN model to explore the potential of a unique geometric feature extraction technique for handwritten digit recognition system which can be further extended to identify any alphanumeric character images. One of the interesting qualities of the proposed framework seems to be its ability to perform reasonably well even with small trainee set. The proposed method also demonstrates robustness to affine as well as a few non-affine transformations of numeric character images.

References

1. Trier, O.D., Jain, A.K., Taxt, T.: Feature extraction methods for character recognition—a survey. Pattern Recognit. **29**(4), 641–662 (1996)
2. Yasuda, M., Fujisawa, H.: An improvement of correlation method for character recognition. Trans. IEICE Jpn. **J62-D**(3), 217–224 (1979)
3. Yamashita, Y., Higuchi, K., Yamada, Y., Haga, Y.: Classification of hand printed Kanji characters by the structured segment matching method. Pattern Recognit. Lett. **1**, 475–479 (1983)
4. Kimura, F., et al.: Evaluation and synthesis of feature vectors for handwritten numeral recognition. IEICE Trans. Inf. Syst. **E79-D**(5), 436–442 (1996)
5. Tsuruoka, S., et al.: Handwritten Kanji and Hiragana character recognition using weighted direction index histogram method. Trans. IEICE Jpn. **J70-D**(7), 1390–1397 (1987)
6. Shi, M., Fujisawa, Y., Wakabayashi, T., Kimura, F.: Handwritten numeral recognition using gradient and curvature of gray scale image. Pattern Recognit. **35**(10), 2051–2059 (2002)
7. Lee, D.-S., Srihari, S.N.: Hand-printed digit recognition: a comparison of algorithms. In: Proceedings of the Third International Workshop on Frontiers of Handwriting Recognition, Buffalo, NY, pp. 153–164 (1993)
8. Liu, C.-L., Liu, J.-Y., Dai, R.W.: Preprocessing and statistical/structural feature extraction for handwritten numeral recognition. In: Downton, A.C., Impedovo, S. (eds.) Progress of Handwriting Recognition, pp. 161–168. World Scientific, Singapore (1997)
9. Heutte, L., Paquet, T., Moreau, J.V., Lecourtier, Y., Olivier, C.: A structural/statistical feature based vector for handwritten character recognition. Pattern Recognit. Lett. **19**(7), 629–641 (1998)
10. Bishop, C.M.: Neural Networks for Pattern Recognition. Clarendon Press, Oxford (1995)
11. Holmstroom, L., et al.: Neural and statistical classifiers—taxonomy and two case studies. IEEE Trans. Neural Netw. **8**(1), 5–17 (1997)
12. Fukunaga, K.: Introduction to Statistical Pattern Recognition, 2nd edn. Academic Press, New York (1990)
13. Duda, R.O., Hart, P.E., Stork, D.G.: Pattern Classification, 2nd edn. Wiley Inter Science, New York (2000)
14. Jain, A.K., Duin, R.P.W., Mao, J.: Statistical pattern recognition: a review. IEEE Trans. Pattern Anal. Mach. Intell. **22**(1), 4–37 (2000)
15. Lauer, F., Suen, C.Y., Bloch, G.: A trainable feature extractor for handwritten digit recognition. Pattern Recognit. **40**(6),1816–1824 (2007)
16. Saha, S., et al.: A computer vision framework for partitioning of image-object through graph theoretical heuristic approach. In: International Conference on Computational Intelligence, Communications, and Business Analytics (CICBA-2017) (2017)
17. Bron, C., Kerbosch, J.: Algorithm 457: finding all cliques of an undirected graph. ACM **16**(9), 575–577 (1973)
18. LeCun, Y., Cortes, C.: MNIST handwritten digit database (2010). http://yann.lecun.com/exdb/mnist/
19. Liu, Cheng-Lin, et al.: Handwritten digit recognition: investigation of normalization and feature extraction techniques. Pattern Recognit. **37**(2), 265–279 (2004)

Part VIII
Session 3B: Internet of Things

A Secure Framework for IoT-Based Healthcare System

Arup Kumar Chattopadhyay, Amitava Nag, Debalina Ghosh and Koustav Chanda

Abstract The usage of IoT at healthcare domain emerged as Internet-of-Medical-Things (IoMT). It brings convenience to the patients and physicians by providing services like real-time health monitoring, patient information management, disease and epidemic outbreak tracking, diagnostic and treatment support, digital medicine etc. Wearable IoT devices or tailored bio-sensors enable continuous monitoring of different physiological parameters of the patients. A wearable ubiquitous health care monitoring system is nothing but interconnected body sensors forming body sensor network (BSN). The BSN is collection of tiny-powered and lightweight wireless sensor nodes with very limited computation and storage capabilities. Simultaneously, without secure communication in BSN the privacy of the patient is vulnerable. The objective of the paper to propose a secure framework for IoT based healthcare system which provides confidentiality, integrity and authentication within public IoT-based communication network. We utilize cryptosystem to ensure secure communication and entity authentication between the smart sensors, local processing units (LPU) and gateway. Moreover, to develop the proposed framework, we have designed an IoT-based test-bed that is connected and programmed with a Raspberry Pi series platform.

Keywords IoT · IoMT · BSN · Secure communication · Local processing unit Raspberry Pi

A. K. Chattopadhyay (✉) · D. Ghosh
Institute of Engineering and Management, Salt Lake City, Kolkata, West Bengal, India
e-mail: ardent.arup@gmail.com

D. Ghosh
e-mail: debalinag1986@gmail.com

A. Nag
Central Institute of Technology, Kokrajhar, Kokrajhar, India

K. Chanda
Academy of Technology, Hooghly, India

M. Chakraborty et al. (eds.), *Proceedings of International Ethical Hacking Conference 2018*, Advances in Intelligent Systems and Computing 811,
https://doi.org/10.1007/978-981-13-1544-2_31

1 Introduction

The term "Internet of Things" (IoT) was originally coined by British technology pioneer Kevin Ashton in the year of 1999. He was first to describe a system in which objects in the physical world could be connected to the Internet. Nowadays, IoT has become a very popular term, and it describes the scenario where network connectivity and computing capability are extended to integrate a variety of "smart things" of our day to day life. "Smart things" are smart objects embedded with sensors, actuators, software and hardware components and enabled with Internet connectivity. As predicted by experts, IoT is the future of Internet and it will comprise billions of personal and professional smart devices connected to Internet.

Authors in [1] discussed the basic building blocks of IoT, which are smart objects/things that can sense, unand react to their surrounding environment. They presented an architecture of IoT which is loosely coupled and decentralized system of smart things. The smart objects/things are autonomous physical objects augmented with sensing, processing and network capabilities. The smart objects are categorized into three types—(a) activity-aware smart objects: These are the objects which are able to record information about its own use and its real-time activities; (b) policy-aware smart objects: These are the activity-aware objects that are able to interpret events and activities with respect to predefined organizational policies; and (c) process-aware smart objects: These process-aware smart objects represent the most accomplished of the three object types and play a fundamental role in the automation of industrial work management and operation.

The scope of IoT is not just limited to connect the smart things; it is extended to create an environment where the smart things can communicate with each other, exchange data, perform filtering, processing, categorization, condensing and contextualization to extract information from the data, such that meaningful applications can be executed to achieve users' goal. In IoT, every physical objects in the virtual world are locatable, addressable and reachable. IoT includes not only computers and laptops, rather other physical devices like home appliances and vehicles connected with wireless sensor-based distributed communication architecture. The use of IoT is spread over a verity of domains (as shown in Fig. 1) which includes energy, healthcare, M2M applications, home automation and home security, manufacturing, asset management, retailing, smart display, tracking, cash payment, smart inventory and replenishment management, cargo handling, smart medicine etc. But along with different opportunities of IoT, a few new security threats also are raised. For example, "Hello Barbie", for children it is a novel IoT-based commercial product. But it helps the intruder to find out all important details in the house.

IoT extends the concept of Internet and makes it more pervasive. It allows seamless interactions among different type of devices such as medical sensor, monitoring cameras, home appliances (as discussed by Gope and Hwang in [2, 3]), and for these reasons, IoT has become more productive in healthcare sector. In healthcare domain, IoT includes different kinds of cheap sensors. Sometimes these sensors are wearable, sometimes implanted. These sensors help patients to enjoy modern medical

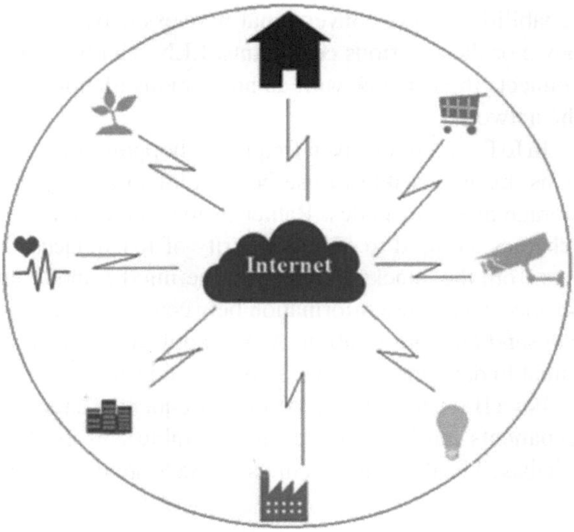

Fig. 1 An overview of IoT applications

healthcare services anywhere, anytime. It can also result in improvement in the quality of life for the aged people. The different body sensors are used for the purpose of collecting data like body temperature, heart rate, oxygen saturation level(SPo2), blood pressure, blood sugar, electrocardiogram (ECG), electroencephalogram (EEG). The body sensor network (BSN) [4] is the most powerful technology used by IoT for the improvisation of modern healthcare system. This BSN is a collection of tiny power and lightweight sensor nodes. To monitor the human body functions and surrounding environment, these sensors are used. BSN nodes collect very sensitive (life-critical) information, so they need strict security mechanisms to prevent malicious interactions with the system.

The rest of paper id arranged as follows: in Sect. 2, we discuss the security requirements of IoT based healthcare systems; in Sect. 3, we focus on some of the current IoT securities related works; we proposed our security framework for IoT in Sect. 4; and Sect. 5 concludes the paper.

2 Security Requirements in IoT-Based Healthcare Systems

Security is one of the most important requirements in any communication medium. The communication in the applications of sensor networks is mostly wireless in nature. Because of that, a number of security threats may arise. Information and network security are very much important in wireless networks because wireless networks have open and shared characteristics. But there exists a number of significant differences between conventional wireless network and IoT network in terms of privacy and security [5]. The deployment of IoT is based on low-power and lossy network (LLN). LLNs are limited by their dynamism, computational and storage capabilities where conventional wireless networks have extremely dynamic topology. For these various constraints, LLNs can be easily attacked by an attacker who connects the network with an anonymous identity and acts as an authentic node in the network.

In IoT perception layer, frequency-hopping communication and public key encryptions are impossible to use because of low computation power and low memory storage at sensor nodes. Rather in IoT perception layer, lightweight cryptographic schemes are used to ensure security of IoT devices. The network layer of IoT suffers from the attacks like man-in-the-middle and counterfeit attacks. Both of these attacks can capture information between communicating nodes and at the same time can send fake information. At IoT application layer, data sharing can raise security threat to data privacy, access control and can cause disclosure of information.

For a BSN, the major problem is frequent and rapid change in topology of network as patients can freely move with wearable sensors. The key security requirements in IoT-based healthcare system using BSN are described in [6] are as follows:

2.1 Data Privacy

In BSN, data privacy is the most important issue. These sensors are carried by the patient body. All important details regarding the patient are sensed by these sensors. So, patient's vital information should not be leaked by the BSN to the external world. In IoT-based healthcare application, the sensor nodes collect the sensitive data of the patient and forward it to a coordinator node. If any intruder overhears the vital information and uses the data for any illegal purpose, then it may cause severe damage to the patient.

2.2 Data Integrity

Only maintaining the confidentiality is not enough. Data should be protected from external modifications also. In any case, if the confidential and vital data of a patient have been manipulated, then this altered data will be forwarded to the coordinator. This lack of integrity results in severe damage for life-critical patients. In bad communication environment, data loss can also occur.

2.3 Data Freshness

Sometimes the attacker may capture data from communication channel and replay them later using some previously used key to confuse the coordinator node. So data freshness is very important because fresh data imply no replaying of old messages.

2.4 Authentication

In BSN-based healthcare system, all the sensor nodes send their data to a coordinator. The coordinator sends periodic updates of the patient to a server. So authentication is very much important in any IoT-based healthcare system. Authentication is needed to confirm the identity of both the coordinator and the server.

2.5 Anonymity

This a very important property which guarantees that the adversary never can identify the patient and the originating conversation. Anonymity thus hides the source of a packet or the sensor data during wireless communication.

2.6 Secure Localization

The accurate estimation of the patient location is needed in most of the BSN applications. So smart tracking mechanisms are required else incorrect reports about the patient location will be delivered.

To ensure a secure IoT-based healthcare system, all the aforesaid security requirements are essential to resist various security threats and attacks like data modification, impersonation, eavesdropping, replaying. Yeh [7] pointed out a number of security requirements by implementing a strong cryptosystem that includes: (1) a session key between communicating parties, (2) removal of inappropriate use of the bitwise XOR operations as it can only resist against "ciphertext-only" attacks, (3) GPS information need to be protected from spoofing attack, (4) prevent man-in-the-middle attack, (5) using multiple security and privacy properties simultaneously.

3 Related Study

The development of IoT applications and related security measures got considerable attention by industry and academia in recent years. In 2013, a lightweight multicast authentication scheme was presented by Yao et al. [8] targeting small-scale IoT applications. In that paper, the properties of fast accumulator proposed by Nyberg [9] were utilized. Among those proposed properties, we found the absorbency property and quasi-communicative property; those can be used to fabricate a lightweight multicast authentication scheme. When the scheme has been tested, the authors experimented with all seven major criteria required by a multicast authentications for resource-constrained applications. The authors claimed that the proposed scheme was more efficient than other similar systems.

Keoh et al. [10] presented an overview of the security solutions for IoT ecosystems proposed by the Internet Engineering Task Force (IETF). The authors also investigated CoAP and Datagram Transport Layer Security (DTLS), and based on the performance, these authors developed a refined and lightweight DTLS capable of providing robust security for IoT objects. They have also pointed out some unresolved issues like device bootstrapping, key management, authorization, privacy and message fragmentation issues in IoT.

Later, Bello and Zeadally [11] explored the possibility to device decentralize control for self-collaborated device-to-device communications. They identified the main challenges as (1) the computation cost of smart objects and (2) network heterogeneity. The authors also analysed the different communication mechanisms in licensed and unlicensed spectra and routing techniques which are able to support intelligent inter-device communications. But in their analysis, a few unresolved issues were identified as follows: (1) maximizing the usage of available network resources; (2) optimization in routing; (3) load balancing using inter-device-based cooperation; and (4) resistance to new types of attacks.

In 2015, Kawamoto et al. [12] proposed an effective data collection scheme in IoT networks for location-based authentication. They improved the accuracy in authentication process by dynamically adjusting the parameters related to network control based on the real-time requirements from the system and its surrounding network environment. Also, an investigation for optimization of authentication accuracy was presented in the paper. There was a suggestion regarding the future work that is on controlling the data distribution from inhomogeneous IoT devices.

Ning et al. [13] proposed an aggregated proof-based hierarchical authentication scheme for layered U2IoT architecture to provide security protection among ubiquitous smart objects. There are some security properties like entity anonymity, mutual authentication and hierarchical access control which are achieved by the techniques of user authorization, aggregated proof-based verifications, homomorphism functions and Chebyshev chaotic maps.

In 2015, Gope and Hwang [2] proposed an authentication protocol for distributed wireless sensor networks. Their proposal is compatible with client–server–server (i.e. the sensor–gateway–server) architecture. It also satisfies important security properties such as mutual authentication, sensor anonymity and untraceability, system scalability, resistance against impersonation attack, replay attack and cloning attack. So authors claimed the proposed scheme is secure and efficient.

In 2016, Gope and Hwang [6] further proposed an authentication scheme for a distributed IoT-based healthcare system where the proposed protocol was based on body sensor networks (BSNs). BSN consists of lightweight and healthcare-oriented smart objects. Lightweight cryptography functions, such as a one-way hash function, random number generator and bitwise XOR operation, are adopted to simultaneously achieve system efficiency and security robustness. The authors also investigated the security density and protocol efficiency by using BAN logics analysis and computation cost comparison [14]. In 2015, Wortmann and Flüchter [14] put forward that the innovations in IoT can be characterized by the combination of the physical and digital components, such that new products can be created and novel business models can be generated. They have suggested that the function of one product can be enhanced if it can be connected to other related products. Thus one product can be a part of a large product system.

4 Proposed Scheme

In this section, we first describe the proposed framework for IoT-based healthcare system and then introduce the secure communication processes for the proposed IoT-based healthcare system. Figure 2 shows the deployment of the proposed IoT-based healthcare system which could be used in hospitals or healthcare organizations.

As shown in the Fig. 2, the proposed model has three different kinds of communication channels: the biometric sensor nodes (the edge sensors) to the internal processing unit (IPU), internal processing unit to gateway (router) and gateway to the cloud. As these channels are open (public), the data transmissions on these channels are not secure. Malicious users may attack these insecure channels. Thus, security in the IoT-based healthcare system is one of the main concerns.

In this paper, we consider the following four security issues:

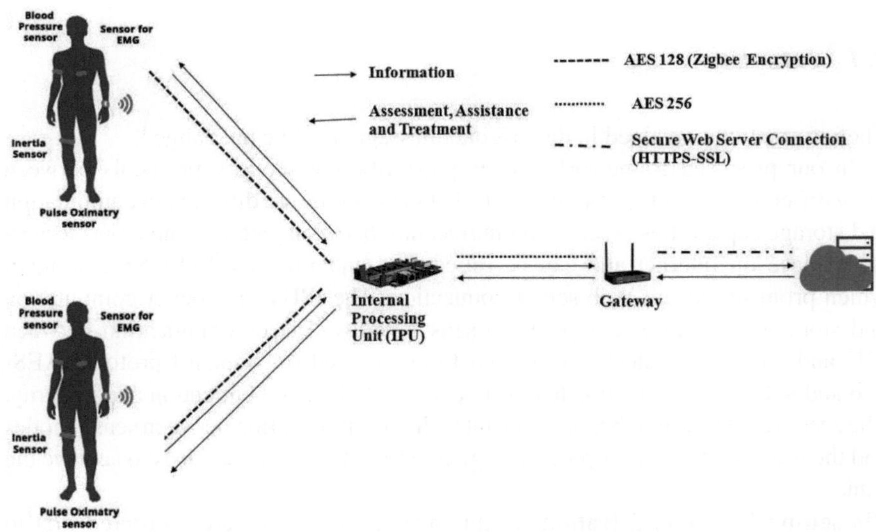

Fig. 2 The proposed framework for secure IoT-based healthcare system

4.1 Access Control

The biometric sensors that are attached to the user or patient, the IPUs and the gateway are the only valid members of the healthcare system. The security system must prevent any other external device to interface or to break into this system.

4.2 Authenticity

Biometric sensors that are attached to the user or patient are the real senders, and on other side, doctors and patients are the authorized stakeholders.

4.3 Confidentiality

The data sent by the biometric sensors that are attached to the user or patient are only readable by the authorized stakeholders.

4.4 Integrity

The information contained in the original message must be unchanged.

In our proposed framework, we propose different security protocols between three different kinds of communication channels as they are different in computation and storage capabilities. Secure communication between gateway and cloud servers takes place on Internet and can be relied on standardized HTTPS-SSL protocol which promises secure Web server connection. The IPUs have better computation and storage capabilities compared to sensor nodes. The communication between IPU and gateway (router) is proposed to be secured by standard protocol AES-256 and SHA-3 hash function to ensure confidentiality, authentication and integrity. The access control is much more relevant to the communication between sensor nodes and the nearest IPU. We propose the registration of bio-sensor devices to achieve the same.

Bio-sensor Device registration: In this phase, bio-sensors are registered first to join the healthcare system. For the registration process, a bio-sensor device (say jth sensor) first sends its ID_j to the healthcare system as follows (different parameters used in registration process are shown in Table 1):

1. The ith IPU uses RSA-1024 to generate its key pair and shares the public key to the nearby bio-sensors.

Table 1 Parameters used in sensor node registration process

Variable	Description
ID_j	Identity of jth sensor
CID_j	encrypted ID_j
SK_j	Secret key of sensor node j
IPU_i	Internal process unit i
H(.)	Hash function

Table 2 The solutions of the four security issues

Security issue	Description	Solution
Access control	To avoid external bio-sensors joining the healthcare system	Use AES 128 with common, pre-shared key
Authenticity	To validate biometric sensors are the real senders and clinicians, doctors, patients are the authorized stakeholders	RSA-1024 (public/private keys)
Confidentiality	To prevent access of information that are transmitted between the sensors and IPU	Point-to-point encryption scheme, AES 128
Integrity	To validate the originality of data from the sensor to the IPU	Message authentication code (MAC) created using AES-128

2. The nearby jth sensor uses RSA-1024 public key of the ith IPU to encrypt (the RSA-1024 will be only used in resignation of sensor node) its symmetric key SK_j and sends it to IPU_i.
3. IPU_i decrypts and gets the SK_j. Now, IPU_i and sensor node j share the same secret key SK_j.
4. The jth sensor computes $CID_j = Encrypt(ID_j, SK_j)$ and sends it to IPU_i.
5. The nearby IPU_i receives the request. Then, the IPU_i processes the request as follows: $ID_j = Decrypt(CID_j, SK_j)$
6. The IPU_i stores $H(ID_j)$ (we propose SHA-3 to be used as hash function on IPUs) in its database for future reference.
7. When an authorized bio-sensor device wants to send data to the system, it send data along with ID_j. Before accepting the data from jth sensor, the IPU_i extracts and downloads ID_j first and computes $H(ID_j)$ and then verifies it with entry for jth sensor. Once verified, the rest of the information will be accessed. This ensures that only registered sensors can participate in communication with IPUs.

The solutions for the security issues are described in Table 2.

For practical implementation of the proposed healthcare system, we have designed an IoT-based test bed that is connected and programmed with a Raspberry Pi series platform. As body sensors, we have used heartbeat and pressure sensors with ZigBee

Table 3 A summary of basic implementation environment

Environment	Description
Raspberry Pi 3 B	Broadcom BCM 2837, 4XARM Cortex-A53 Architecture, 1.2 GHz, 1 GB LPDDR2, 16 GB SD card
Operating system	Ubuntu core
Programming language	Python
Programming IDE	Spyder
Crypto API	PyCrypto

modules (for data transmission). Table 3 illustrates summary of basic implementation environment.

5 Conclusion

In this paper, we have presented a secure framework for IoT-based healthcare system. The proposed model satisfies almost all major security demands for IoT-based healthcare system. To develop the proposed framework, we have programmed the bio-sensors using the Raspberry Pi II and uploaded the collected data from sensors into the cloud storage securely. The captured data from the proposed framework can be used for data analytics.

References

1. Kortuem, G., Kawsar, F., Sundramoorthy, V., Fitton, D.: Smart objects as building blocks for the Internet of Things. IEEE Internet Comput. **14**(1), 44–51 (2010). https://doi.org/10.1109/MIC.2009.143
2. Gope, P., Hwang, T.: Untraceable sensor movement in distributed IoT infrastructure. IEEE Sens. J. **15**(9), 5340–5348 (2015)
3. Gope, P., Hwang, T.: A realistic lightweight authentication protocol preserving strong anonymity for securing RFID system. Comput. Secur. **55**, 271–280 (2015)
4. Kumar, P., Lee, H.-J.: Security issues in healthcare applications using wireless medical sensor networks: a survey. Sensors **12**(1), 55–91 (2011)
5. Alaba, F.A., Othman, M., Hashem, I.A.T., Alotaibi, F.: Internet of Things security: a survey. J. Netw. Comput. Appl. **88**, 10–28 (2017)
6. Gope, P., Hwang, T.: BSN-care: a secure iot-based modern healthcare system using body sensor network. IEEE Sens. J. **16**(5), 1368–1376 (2016)
7. Yeh, K.-H.: A secure IOT-based healthcare system with body sensor networks. IEEE Access **4**, 10288–10299 (2016)
8. Yao, X., Han, X., Du, X., Zhou, X.: Alotaibi, F.: A lightweight multicast authentication mechanism for small scale IoT Applications. IEEE Sens. J. **13**(10), 3696–3701 (2013)

9. Nyberg, K.: Fast accumulated hashing. In: International Workshop on Fast Software Encryption, pp. 83-87. Springer, Cambridge, UK (1996)
10. Keoh, S.L., Kumar, S.S., Tschofenig, H.: Securing the Internet of Things: a standardization perspective. IEEE Internet Things J. **1**(3), 265–275 (2014)
11. Bello, O., Zeadally, S.: Intelligent device-to-device communication in the Internet of Things. IEEE Syst. J. **10**(3), 1172–1182 (2016)
12. Kawamoto, Y., Nishiyama, H., Kato, N., Shimizu, Y., Takahara, A., Jiang, T.: Effectively Collecting Data for the Location-Based Authentication in Internet of Things. IEEE Syst. J. **11**(3), 1403–1411 (2017)
13. Ning, H., Liu, H., Yang, L.T.: Aggregated-proof based hierarchical authentication scheme for the Internet of Things. IEEE Trans. Parallel Distrib. Syst. **26**(3), 1045–9219 (2014)
14. Wortmann, F., Flüchter, K.: Internet of Things—technology and value added. Bus. Inf. Syst. Eng. **57**(3), 221–224 (2015)

Smart Irrigation: IOT-Based Irrigation Monitoring System

Ajanta Dasgupta, Ayush Daruka, Abhiti Pandey, Avijit Bose, Subham Mukherjee and Sumalya Saha

Abstract The project aims at autonomous monitoring of irrigation system in both large- and small-scale plantation estates with a view to eradicating the manual system which involves personal liability concerns and the ignorance of the field workers. Even sometimes the experienced people cannot assure how much fertilizers or water must be used for the maximum yield. Hence, our system will monitor the temperature, humidity, moisture content of the soil and other physical factors like presence of major pollutants in air like PM2.5, PM10, CO, NOx. The factors and the crop yield are compared with dataset of past surveys and will try to predict whether irrigation is necessary or not. With the help of this information, the rate of releasing water from pumps is decided and fed to a microcontroller system which supervises and controls the whole irrigation system. Besides, there is also provision to monitor plant growth in both longitudinally and horizontally.

Keywords IOT · Irrigation · Smart irrigation

1 Introduction

Irrigation is the application of controlled amounts of water to plants at needed intervals. Irrigation aids in developing agricultural crops, upholding landscapes and revegetating disturbed soils in dry areas and during periods of less than average rainfall.

Irrigation has various other applications too, for example, safeguarding plants from frost, curbing the growth of weed in crop fields and preventing soil consol-

A. Dasgupta (✉) · A. Daruka · A. Pandey · A. Bose
Department of Information Technology, Institute of Engineering & Management, Kolkata, India
e-mail: ajanta.dasgupta10@gmail.com

S. Mukherjee · S. Saha
Department of Electronics & Communication Engineering, Institute of Engineering & Management, Kolkata, India

M. Chakraborty et al. (eds.), *Proceedings of International Ethical Hacking Conference 2018*, Advances in Intelligent Systems and Computing 811,
https://doi.org/10.1007/978-981-13-1544-2_32

idation. On the other hand, agriculture that counts only on unmediated rainfall is referred to as rain-fed or dryland farming.

Irrigation systems are also used for control of dust and sewage disposal and in mining. Study of irrigation is mostly accompanied by the study of drainage, which is the natural or artificial expulsion of surface and subsurface water from a given region.

Irrigation has been succoring agriculture for years and is the commodity of many cultures. Historically, it was the foundation for economy, from Asia to the Southwestern United States.

Smart irrigation is sustainably managed, accountable, responsible and trusted irrigation. Smart irrigation aims to diminish their environmental footprint through well-planned water applications and to guarantee a successful business. This enables them to reinvest in new and improved technologies which ensure sustainable irrigation over time.

New irrigation technologies and support tools are consistently being innovated in New Zealand and globally. Water use efficiency and energy use efficiency are the main focuses of these innovations. Over the last two decades, there has been a shift from manual flood irrigation to remotely controlled spray irrigation using techniques like center pivots, dripline and micro-sprinklers.

2 Need of the Device

India is an agro-based country, and now-a-days the small fields and farms are being merged with the large plantation farms. Due to the increase of 8% of foreign direct investment (FDI) in agricultural sphere, more and more farms are globalized. The multi-national companies cannot bear the loss due to the farmers (who are employed as field laborers) by means of excessive use of fertilizers and pesticides. The system will assist to implement optimal usage of man power and endeavor to reduce the burgeoning expenditure. Since the whole system will be integrated with a central server and will have mobile and web-app-based user interfaces, the corporate supervisors can control the system from their own work desk in offices. There will be just one-time investment for the purchase and installation of the system in the farm leading to a long-term benefit. The increase in yield will also benefit the consumers as the price of basic food materials will decelerate with the supply hike as a consequence of which the inflation in field of daily commodities may decrease.

3 Related Work

Technology is improving every minute. Even though irrigation ensures maximum overall crop yield, it might cause wastage of water resources. Let us introduce some of the systems proposed to improve irrigation processes and their advantages and disadvantages.

In [1], focus is on optimization of water usage and shows that this technique requires 10% of the water needed. The system turns on the water supply as soon as it detects the soil moisture values below a certain value or when the soil temperature exceeds the set threshold. This system provides irrigation for a fixed duration when switching on manually at a particular date and time exactly through an application running on another device as mobile or a laptop. Date and time details are then stored in the end nodes sensing unit, whereas data from other sensors and other irrigation results are uploaded to an application using the GPRS. Irrigation also depends on a lot of other parameters. That is one of the disadvantages of this system.

In [2], the focus is on the automated valve and a manual valve which is controlled by using sensors and wireless networks. All the end nodes send the soil moisture values to the base station after a fixed interval that is neither too long nor too small. Based on the moisture sensor value, commands are sent to the node containing valve actuator to either open or close the water supply valve. The node that contains the valve actuator is equipped with a boost regulator for relay operation. All these operations are executed via the application interface. Through the web interface, the user gets valve opened and closed. The advantage of this system is that by means of application interface the user can view the irrigation details and/or manually setup time-based irrigation and/or schedule-based irrigation irrespective of user's location. The major issues with this system are: not considering air humidity, sunshine brightness intensity and wind speed values which can have a great impact on the irrigation efficiency.

This process [3] is focused on automatic irrigation based in the greenhouses using actuator networks and wireless sensors. On the basis of knowledge of the environmental parameters and plant growth, a decision will be made on the regulation of water supply. The system uses machine learning process to enhance the diagnosis process for plants. Machine learning totally depends on logging data, and it is used to set rules and parameters for irrigation threshold. To derive these parameters and rules, the system uses a rule editor tool. This rule editor tool provides the visualization of evaluated states and measured constraints too. Quality indicators are used for handling the uncertainty of data. The advantages of this system are the integration of plant-based method with soil moisture sensor method and the fast configuration of sensor nodes with OS (Tiny OS) used which improves the accuracy of irrigation. The less coverage area of about 120 m because of XBee devices is the major contention with this system.

In [4], the focus is on closed loop distant observing of precise irrigation by means of Citect configuration software. All the end nodes transmit the data of soil temperature, humidity from DHT11 and soil moisture values to the sink node. As the sink node receives the data, the values received re-compared with the set threshold values as per the process mentioned above. Based on that, sink node sends a command to open as well as to close the valve. Information obtained from sensors like soil temperature, moisture and humidity and valve status at various time intervals is transmitted to the web server using the GPRS module. The end user has the freedom to monitor the process remotely via web interfaces. The advantages of the developed system include the conservation of water up to 25% when compared to normal irrigation systems and also real-time collection and transmission of data. The major disadvantage is tapered irrigation efficiency by reason of not considering the wind speed values and sunshine intensity and duration as parameters for reference evapotranspiration.

Process [5] is based on the use of wireless sensor networks for dynamic automatic irrigation and to avoid the use of pesticides. The on stream camera compares the measured value with reference values as soon as the wireless sensor nodes measure the soil moisture and soil fertility. When the soil is wet and there are no pesticides found, the valve remains closed. It gets open when the soil is dry and pesticides are found. The microcontroller is put in sleep mode when there is no need of irrigation, and when needed, the microcontroller will go into active mode for power consumption. The advantages of this system are improved energy efficiency using power saving modes, dynamic irrigation and pesticide avoidance. Reduced irrigation efficiency because of not considering bright sunshine duration, air temperature and wind speed values for reference evapotranspiration are considered as the main issues with this system.

In [6], focus is on irrigation system using wireless sensor network and fuzzy logic to preserve the water resource and to improve the soil fertility. Soil humidity sensors are equipped to all the end node areas. The coordinator receives the measured soil moisture value and different crop growth information during different periods by the end node. All the data from the coordinator node are then transmitted to the monitoring station using RS232. The inputs to the fuzzy logic controller are deviation of soil moisture value and the time at which the deviation occurs. From that, opening and closing of the irrigation valve will be computed. The main problem with this system includes inconsistency of fuzzy logic and lesser bandwidth coverage for as much as Xbee is confined to 120 m.

In [7], focus is on the automated irrigation system to ensure low-cost and high power efficiency. The wireless sensing unit (WSU) is built with soil temperature sensor and humidity sensor. Once the soil temperature and humidity are read by the WSU, it forwards those values to wireless interface unit (WIU). Then, WIU actuates the solenoid valve for the irrigation process on the basis of threshold-based algorithm. All the irrigation details will be intimated via short message service (SMS) and also forwarded as an email to the farmer using general packet radio service (GPRS) module. The main issues with this system are the signal quality of the wireless sensing unit that differs time to time due to soil moisture dynamics and other environmental parameters like sunshine duration, wind speed which might have not been used for the irrigation decision which significantly affects the irrigation efficiency.

4 Proposed System

The system can be operated in two modes—(i) manual and (ii) autonomous. The rate of irrigation, physical factors, etc. are continuously uploaded in the server. The manual mode gives option to select the rate of releasing water by pumps, duration of irrigation, etc. In the first phase, the autonomous mode decides the rate of irrigation according to the present physical parameters by the analysis of previous standard surveys uploaded initially in server. The next phase of automation will recognize the ideal rate of irrigation by using machine learning where the physical factors, rate of irrigation and the rate of growth in the first phase are used as training data. The pumps can also be controlled from a distant place via web-based apps or mobile apps. The product will be like a thin slab having all the sensors embedded on it. The product possesses a vertical track along which an ultrasonic sensor traverses to and fro to measure the longitudinal plant growth. Another sonar moves in a horizontal track to map the distance between the crops (in a particular plot or area) and the product itself to monitor the secondary growth.

When the implementation of fertilizers and pesticides is executed, the system administrator will have the option to switch on to a special mode where the whole system becomes dedicated in supervising the change in moisture content, acidity of the soil and the rate of photosynthesis and transpiration in a more precise way for studying how the plants react immediately to the fertilizers. It also observes how the Air Quality Index (AQI) is changing for the application of fertilizers.

The project has four fields in technical sphere:

Sensors:

A. Temperature and humidity sensor DHT11 will measure the ambient atmospheric temperature and humidity. There is a control unit in DHT11. The output of the control unit is used to control the irrigation system by switching it on and off depending on the soil moisture contents. If the moisture value obtained is less than the preset value, then the motor will be automatically turned ON. The change in moisture is proportional to the amount of current flowing through the soil.

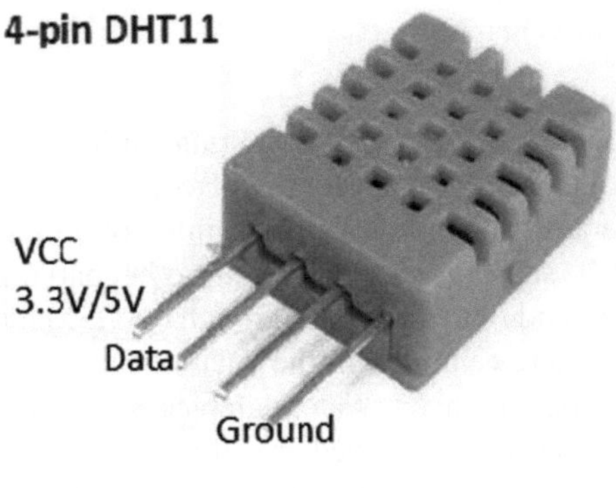

1. DHT11 temperature humidity sensor

B. Hygrometer sensor measures the soil moisture content

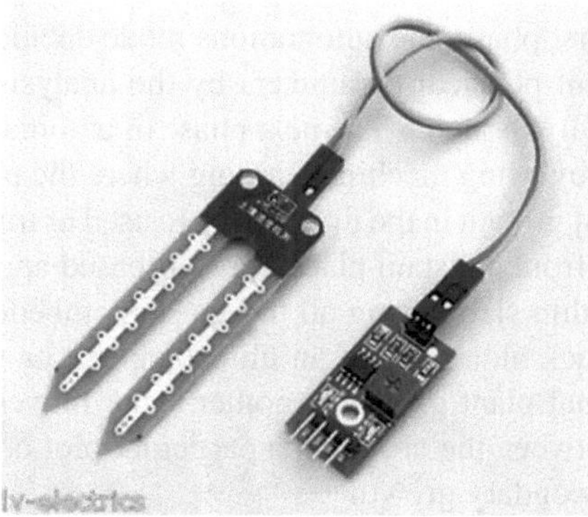

2. Hygrometer sensor
C. PH meter sensor calculates the acidity of the soil
D. MQ135, MQ131, MQ2, MQ 9 sensors are used to measure the pollutants in air to evaluate AQI.
E. Ultrasonic sensors are used for pest control and also to monitor the plant growth.
F. Water level indicators are used to fill the field with water up to the required level.

MCUs and wireless communication modules:

A. MCU plays the vital role in making judgments and taking vital decision and is the main apparatus for interfacing the sensors and connecting to network.
B. Wi-Fi module is used to upload the sensor data to web-cloud.
C. GSM module is used to control the pump.
3. Apps and dedicated web server and APIs: These will be required to analyze the data and develop various GUIs.
4. Miscellaneous: DC geared motors will be used to control the movement of ultrasonic sensors. Stepper motors are used to move the water level indicator sensor to the required height.

DC Geared Motor

The device is divided into two parts: one is the transmitter and other is the receiver. The transmitter part attached with sensors is placed in the field to detect various parameters like temperature, humidity. The transmitter portion senses the parameters from the field through its sensors and sends it to the other part that is the receiver. The receiver portion in turn sends it to the server through the GSM module that is attached with it. The server is where centrally all the data related to the various parameters that is sensed from the field is saved. Water pumps are placed at various portions of the field that supply water in a concerned area if required so as a result of analysis on the various data of various parameters that is saved in the central server at various time from the field conditions. Water pumps operate through the transmitter portion that sends commands for its operation. The data that are saved in the server are taken into consideration to analyze the field condition and predict whether irrigation is necessary or not (Figs. 1 and 2).

Dataset Collected from the Central Server

Fig. 1 Picture of the receiver module of the smart device

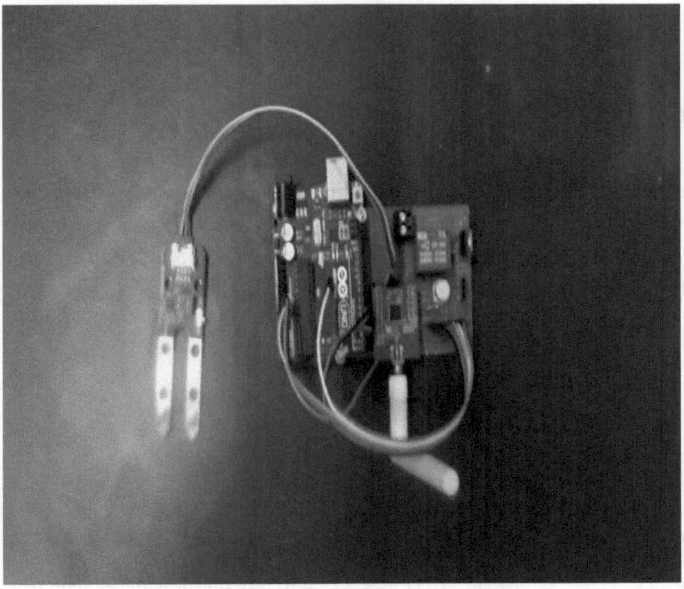

Fig. 2 Picture of the transmitter module of the smart device

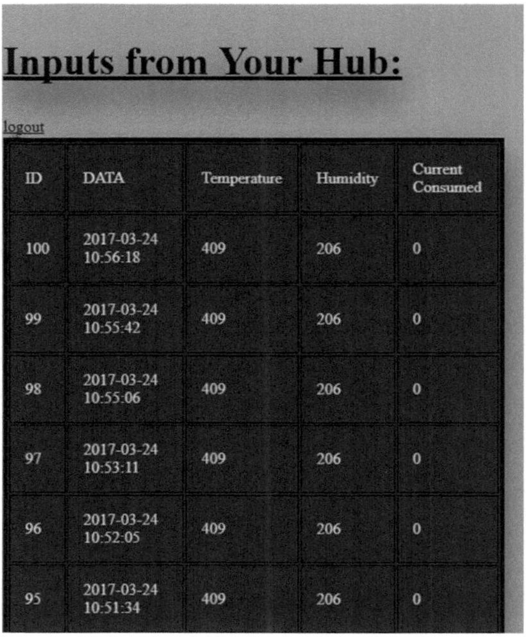

Inputs from Your Hub:

logout

ID	DATA	Temperature	Humidity	Current Consumed
100	2017-03-24 10:56:18	409	206	0
99	2017-03-24 10:55:42	409	206	0
98	2017-03-24 10:55:06	409	206	0
97	2017-03-24 10:53:11	409	206	0
96	2017-03-24 10:52:05	409	206	0
95	2017-03-24 10:51:34	409	206	0

5 Conclusion

Agriculture is a field that still lacks the mass innovation and applications based on modern techniques. Our proposal of smart irrigation will make optimized use of resources and solve the problem of water shortage. The data are stored in the server. Based on the conditions, data would be retrieved, so that the system can adjust itself according to that.

References

1. Gutierrez, J., Villa-Medina, J., Nieto-Garibay, A., Porta-Gandara, M.: Automated irrigation system using a wireless sensor network and GPRS module. IEEE Trans. Instrum. Meas. **63**(1), 166–176 (2014)
2. Coates, R., Delwiche, M., Broad, A., Holler, M.: Wireless sensor network with irrigation valve control. Comput. Electron. Agric. **96**, 13–22 (2013)
3. Goumopoulos, C., O'Flynn, B., Kameas, A.: Automated zone-specific irrigation with wireless sensor/actuator network and adaptable decision support. Comput. Electron. Agric. **105**, 20–33 (2014)
4. Yu, X., Han, W., Zhang, Z.: Remote monitoring system for intelligent irrigation in hybrid wireless sensor networks. Int. J. Control Autom. **8**(3), 185–196 (2015)
5. Merlin Suba, G., Jagadeesh, Y.M., Karthik, S., Sampath, E.R.: Smart irrigation system through wireless sensor networks. ARPN J. Eng. Appl. Sci. **10**(17), 7452–7455 (2015)

6. Gao, L., Zhang, M., Chen, G.: An intelligent irrigation system based on wireless sensor network and fuzzy control. J. Netw. **8**(5), 1080–1087 (2013)
7. Nallani, S., Berlin Hency, V.: Low power cost effective automatic irrigation system. Indian J. Sci. Technol. **8**(23), 1–6 (2015); Mamun, A.A., Ahmed, N., Ahamed, N.U., Matiur Rahman, S.A.M., Ahmad, B., Sundaraj, K.: Use of wireless sensor and microcontroller to develop water-level monitoring system. Indian J. Sci. Technol. **7**(9), 1321–1326 (2014)

Secure Data Transmission Beyond Tier 1 of Medical Body Sensor Network

Sohail Saif and Suparna Biswas

Abstract Medical body sensor network (MBSN), a three-tier architectural network, has been in wide use on demand for remote health monitoring and support in both urban and rural areas. Primary concern of such system is security of sensitive health data along with low end-to-end delay and energy consumption among others. This paper implements secure patient data transmission between tier 2 and tier 3 by ensuring confidentiality and integrity. Man-in-middle attack and distributed denial of service attack can be detected based on end-to-end delay in data transmission. Hash-based secret key is used for encryption which is generated using extracted biological information of user at coordinator PDA of MBSN. Using shared extracted biological information, secret key is regenerated at cloud-based medical server for decryption of data. Experimental results show using different symmetric key encryption techniques, maximum end-to-end delay is only 11.82% of 250 ms which is the maximum permissible delay limit for healthcare application.

Keywords Medical body sensor network · Confidentiality · Integrity · Hash Secret key · End-to-end delay · Man-in-middle attack · Denial of service attack

1 Introduction

With the rapid advancement in semiconductor technology and communication networks, wireless sensor network has been realized for solving problems in various domains including defence, health care, gaming and entertainment. Sensors are wearable or implanted in human body for regular sensing of vital physiological parameters for continuous monitoring and support purposes. The specialized wireless sensor net-

S. Saif · S. Biswas (✉)
Department of Computer Science & Engineering, Maulana Abul Kalam Azad University of Technology, Bidhan Nagar, West Bengal, India
e-mail: mailtosuparna@gmail.com

S. Saif
e-mail: sohailsaif7@gmail.com

M. Chakraborty et al. (eds.), *Proceedings of International Ethical Hacking Conference 2018*, Advances in Intelligent Systems and Computing 811,
https://doi.org/10.1007/978-981-13-1544-2_33

work applied for remote health monitoring and support may be termed as wireless body area network (WBAN) or wireless body sensor network (WBSN) or medical body sensor network (MBSN), and people who may be in demand of such system are from both rural and urban areas [1]. Moreover, the purpose of using such system has both the perspectives: fitness regime and continuous health monitoring remotely without being manned and without any mobility restriction for elderly people living in rural areas. This system is advantageous over traditional doctor–patient system for mainly: (i) precision in data measurement, (ii) timeliness or proper interval of physiological data measurement and (iii) physically seeing a doctor for measuring blood pressure, sugar, temperature, heart rate [2], etc., which are irrelevant. But complete success of such remote healthcare system depends completely on how accurately actual measured data are being received at the diagnosis end within permissible delay limit of 10–250 ms [3]. Vital signals sent wirelessly may face threats in the form of (i) modified information—attacks towards integrity, (ii) leakage to sensitive patient information—attacks towards confidentiality, (iii) data access by illegitimate user—attacks towards authentication and (iv) data lost or delayed in transit—may cause havoc on patient's life. A lot of works have been done to address these issues [4]. In tier 1, sensor nodes send signal to the coordinator node over short range (distance around 2 m). Coordinator node then forwards the signals to the tier 2 through access point (AP). Here, various intra-BSNs, cellular networks and Internet connectivity take place. In tier 2 communication, APs and Internet are connected with cloud-based medical server from where concerned caregiver can access the vital signals. Signal being transmitted beyond tier 1 is vulnerable to all types of security attacks in transit through insecure wireless channel.

Figure 1 depicts security measures taken at tier 2 and tier 3 to ensure confidentiality and integrity here.

Rest of the paper is organized as follows: Sect. 2 describes related works, proposed work and algorithm are illustrated in Sect. 3, Sect. 4 elaborates experimental set-up followed by detailed experimental results in Sect. 5, and finally, the whole work is concluded in Sect. 6.

2 Related Work

Physiological data of patients transmitted through MBSN from patient to doctor are vulnerable for different attacks. Intruders can easily alter this vital health information which can be life threatening. Therefore, strong security techniques are needed to achieve proper confidentiality, authenticity and integrity of data. Researchers proposed key generation techniques from biological information [5], and some researches show the authentication [6, 7] techniques using biometric information of human body such as ECG, finger print and heart rate. Table 1 shows the recent works on various security aspects for BSN applications.

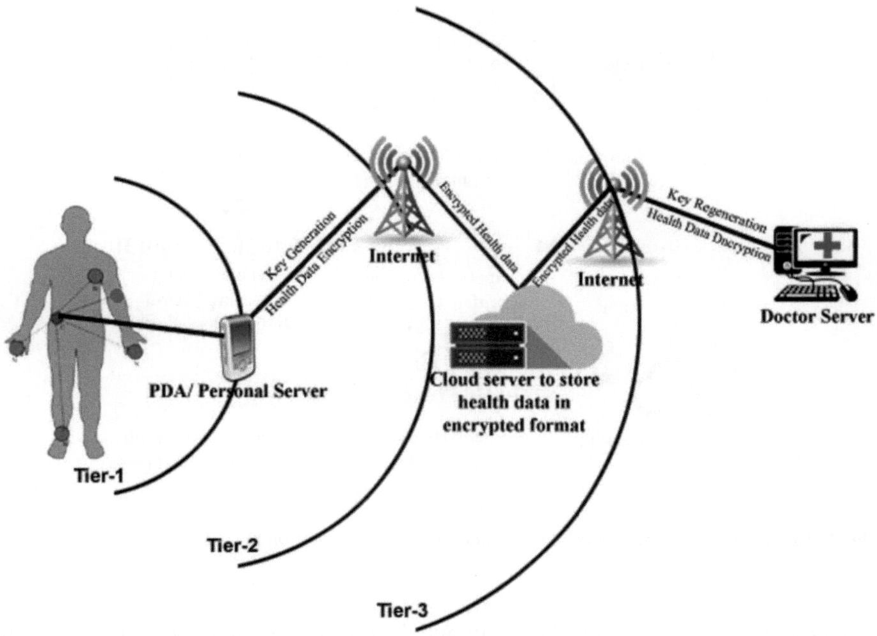

Fig. 1 Three-tier architecture of MBSN

Table 1 Comparative literature survey

Authors, year	Confidentiality	Authentication	Integrity	Implementation	Delay
Liu et al. [8], (2016)	✓	✓	×	✓	✓
Debiaoetal. [9], (2017)	×	✓	×	✓	×
Zhengetal. [10], (2016)	✓	✓	×	×	×
Ramlietal. [7], (2013)	×	✓	×	×	×
Razaetal. [4] (2016)	✓	✓	×	✓	✓

3 Proposed Work

Our proposed security system is implemented in cloud which will secure the data transmission between tier 2 and tier 3. Figure 2 shows the key generation process from fingerprint of patient, key sharing process, encryption and decryption.

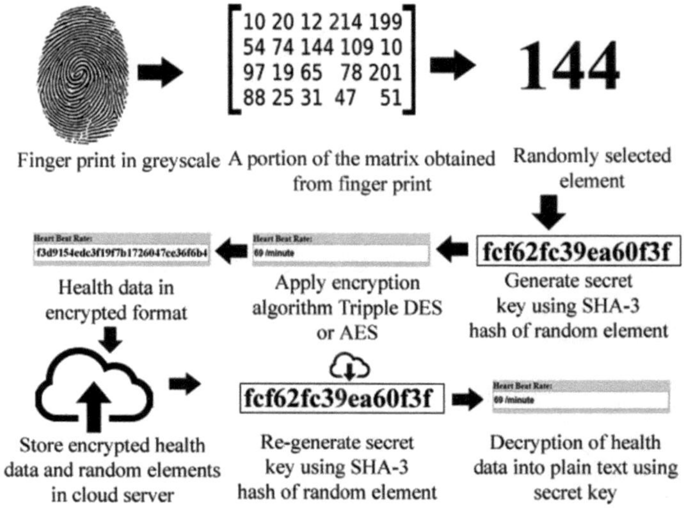

Fig. 2 Key generation, encryption, key regeneration and decryption process

Firstly, fingerprint images of patients and doctors are captured, and it is converted to its equivalent matrix, and then the matrix is stored to cloud server along with doctor and patient id, respectively. For the authentication of doctor's identity, doctor's fingerprint information is used, and for the key generation purpose, patient's fingerprint information is used. A random element is selected from the fingerprint matrix. We have used two different encryption techniques: Algorithm 1 shows the steps using triple DES [11], while Algorithms 2 shows the steps using AES-128 [6, 12].

For AES, 128-bit hash of randomly selected element is computed using SHAKE-128(SHA-3) algorithm, and then, this hash is used as the encryption key. After encryption, encrypted health data and the random element are stored in the cloud database. During decryption, random element and encrypted health data are fetched, and decryption key is generated by computing the hash of random element again.

For 3DES, three numbers of 64-bit key are generated; for that, three random elements are chosen and 64-bit hash is calculated for all these elements using SHAKE-128(SHA-3) algorithm. Here, 3DES is followed by encryption–decryption–encryption for encryption, where in each step random element is stored in cloud along with key id (K1, K2, K3). For decryption, reverse process of encryption is followed which is decryption–encryption–decryption; here, for each step keys are fetched from cloud. Then, the hash of K1, K2, K3 is again computed which are used as a decryption key.

Algorithm 1: Implementation using Triple DES

```
BEGIN:
Procedure: Thumb image to matrix conversion
For i = 1 to n do
scan thumb and get image
Img[i]= scanned image
grayscale[i]= img[i]
M[i]:=imread('grayscale[i]')
generate matrix M
thumb_info:=M
store thumb_info in cloud server
end for
Procedure: Authentication, key generation, encryption and transmission of health data
if doctor_request = true then
if doc_thumb_stored = doc_thumb_sent then
for i = 1 to n do
a[i] = vital signals of the patient's measured
for i = 1 to 3 do
fetch thumb_info from cloud server
rand_ele[i]= choose random element from thumb_info
pkey[i]= SHA-3(rand_ele[i]) // compute 64 bit hash of element

end for
end for
enc[i] = DES encryption of the a[i] using pkey[i]
dec[i] = DES decryption of enc[i] using pkey[i]
enc[i] = DES encryption of dec[i] using pkey[i]
Store enc[i] and rand_ele[i] in cloud server
Else
Request decline
end if
end if
Procedure: key re-generation and decryption of health data
for i = 1 to ndo
Fetch enc[i] and rand_ele[i] from cloud server
for i = 1 to 3 do
dkey[i]= SHA-3(rand_ele[i]) // compute 64 bit hash of random element
end for
```

```
dec[i] = DES decryption of enc[i] using dkey[i]
enc[i] = DES encryption of dec[i] using dkey[i]
a[i]  = DES decryption of enc[i] using dkey[i]
end for
END
```

Algorithm 2: Implementation using AES-128

```
BEGIN:
Procedure: Authentication, key generation, encryption and transmission of health data
if doctor_request = true then
ifdoct.biosignal_pat = doct.biosignal_sentthen
for i = 1 to ndo
a[i] = vital signals of the patient's measured
        Fetch thumb_info from cloud server
rand_ele[i]= choose random element from thumb_info
pkey[i]= SHA-3(rand_ele[i]) // compute 128 bit hash of element
enc[i] = AES encryption of the a[i] using pkey[i]
end for
Store enc[i] and rand_ele[i] in cloud server
 Else
Request decline
end if
end if
Procedure: key re-generation and decryption of health data
for i = 1 to n do
Fetch enc[i] and rand_ele[i] from cloud server
dkey[i]= SHA-3(rand_ele[i]) // compute 128 bit hash of random element
dec[i] = AES decryption of enc[i] using dkey[i]
end for
END
```

Notations used in the above algorithms are as follows:

1. doctor_request: Doctor sending request to patient for health data
2. doc_thumb_sent: Doctor sending his thumb signal with request of health data.
3. doc_thumb_stored: Thumb information of doctor stored in cloud.
4. thumb_info: Thumb information of patient and doctor in matrix form.
5. n: Number of vital signals.

Table 2 Simulation environment of cloud server

Description	Value/Name
Server location	Burlington, Massachusetts, USA
Processor	2.30 GHZ dual core
RAM	2 GB
Storage	5 GB HDD in RAID
Bandwidth	10 GB/month
Apache version	2.4.27
PHP version	7.1.9
MYSQL version	5.7.19
Architecture	X84_64
OS	Cent OS 6.5

Table 3 Simulation environment of PDA

Description	Value/name
System model	HP G-62 Notebook
Processor	Intel(R) Core (TM) i3 CPU M380 @ 2.53 GHz
RAM	4 GB
Storage	500 GB
Architecture	X84_64
OS	Windows 7.1

4 Experimental Set-Up

For our experiments, a cloud-based database server is used to store physiological signal of patients in encrypted format. A cloud server is also known as virtual private server, but its hardware components are physically located in different places. The main reasons behind using the cloud server are remote access of data, flexibility, cost-effectiveness, etc. Back end of our implemented system is controlled by PHP and AJAX, and front end is designed using HTML 5 and CSS. We have considered a notebook PC here as a PDA device which forwards the physiological signal of patients to the cloud. Table 2 and Table 3 show the simulation environment of PDA device and cloud-based server, respectively. Tables 4 and 5 show the hardware resource utilization during the secure transmission of patient's health data from patient to doctor via cloud server.

We can see from the above tables that resource utilization is very low, so our implementation can be easily adopted in a low-end system also.

Table 4 Resource utilization of cloud server

Description	Usage	Limit	Faults
CPU usage	16.7%	100%	0
I/O usage (Kbps)	1410	8192	0
IOPS	16	1024	0
No. of processes	13	100	0
RAM usage	321 MB	2048 MB	0

Table 5 Resource utilization of PDA

Description	Usage	Limit
CPU usage (%)	29	100
RAM usage (GB)	2.04	4.00
No. of processes	42	100
Network usage	9%	4 MBPS

5 Results and Discussion

We have performed the experimental test in five phases with five runs of 10 min each. The physiological information of patient such as body temperature, heart rate and blood pressure is transmitted in real time from patient to doctor upon request of doctor. In the first phase, physiological data of patients are sent to doctor via cloud server in plain text format without any security with an interval of 1 min. Secondly, physiological data are sent to doctor during attack on transmission. Here, in our scenario we have considered application layer-based http flood attack (DDOS) and man-in-middle (MITM) attack. For the DDOS attack simulation, we have used a Python-based attacking tool named Golden Eye. To simulate the environment, an attack is targeted to apache-based PDA/personal server's IP address from an another PDA device; during simulation, we first sent 20 GET method request to the targeted IP address per second using http protocol for 12 min (with an interval of 1 min) and recorded the end-to-end delay. Next, we sent 50, 110 and 150 GET method requests and recorded the end-to-end delay, respectively, but for the 150 request the targeted system is crashed (system turned-off) because of memory exhaust. We also recorded the CPU utilization of the targeted PDA device during the attack. For the man-in-middle attack, it is periodically and randomly targeted where an attacker reads the health information during transmission and again re-transmits it to doctor. Comparison of average end-to-end delay and resource utilization between DDOS attack and non-attacked period are shown in Figs. 3 and 4, respectively. Comparison of average transmission delay between man-in-middle attack, dos attack and normal traffic is shown in Fig. 5.

Figure 3 shows that end-to-end delay for transmission of physiological signal is increasing during the attack. It is under the permissible delay when requests are 6

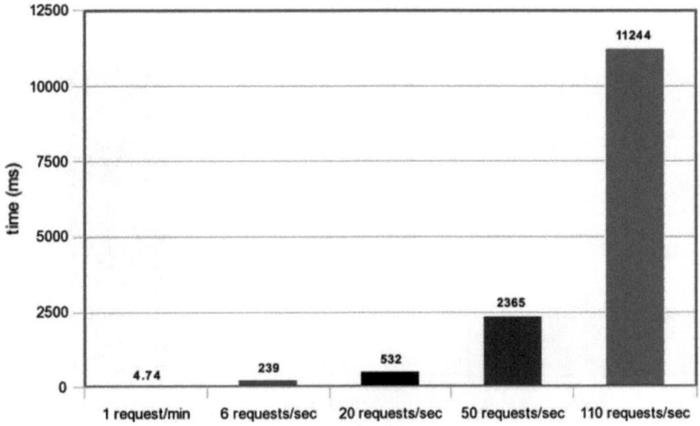

Fig. 3 End-to-end delay of physiological data transmission in plain text during DDOS attack

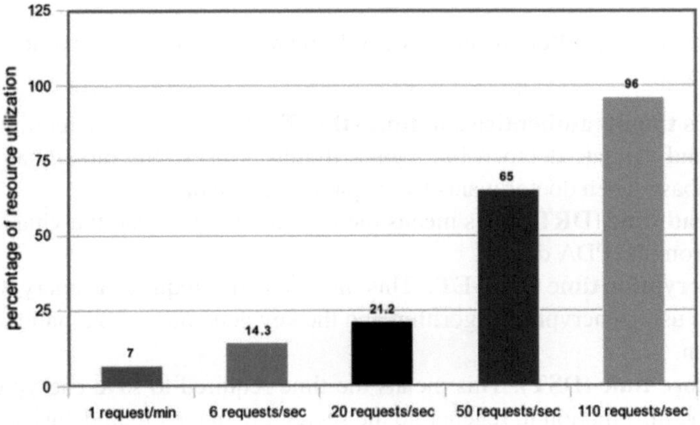

Fig. 4 Resource utilization of PDA device during transmission of physiological data in plain text during DDOS attack

per second but after that it crossed the limit, so here we can detect the attack because of delay. Figure 5 shows that during man-in-the-middle attack, end-to-end delay is more than normal transmission, so if there is any extra delay then we could say that there is an attack.

In third stage, physiological data of patients are transmitted from PDA to doctor via cloud server with confidentiality and authenticity. Here, secure key distribution is done in order to prevent man-in-middle attack on secret key. For our implementation, we have used doctor's thumb information to achieve authentication, for confidentiality light weight faster symmetric encryption algorithm AES and triple DES is used. To generate secret key, patient's thumb information has been used. Average data size of the physiological signal of patients is 3.21 KB. Here, we have provided

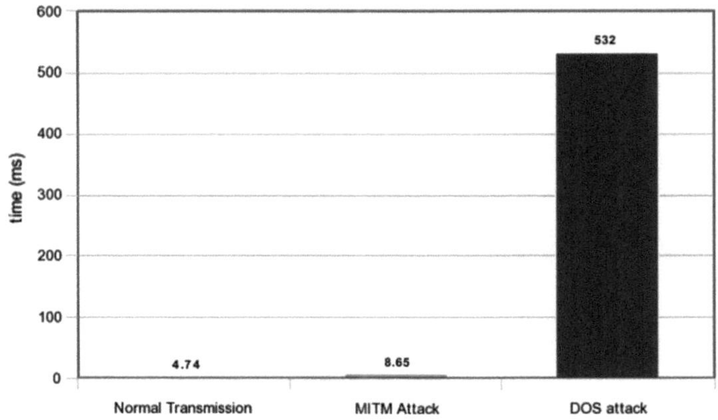

Fig. 5 Comparison of end-to-end delay during normal transmission, MITM and DDOS attack

the averages of five different sets of data. Explanations of various parameters used are as follows:

1. Doctor's thumb authentication time (DTAT): This reflects the required time to compare and validate doctor's bio-signal (thumb print) to the bio-signal stored in cloud database when doctor wants to see patient health data.
2. Data read time (DRT): This means the required time to read the vital signals of patient's from the PDA device.
3. AES encryption time (AES-ET): This shows the time required to encrypt patient's health data using encryption algorithm and the key generation from patient's thumb information.
4. Data store time (DST): This means the time required to save encrypted health data and the information to regenerate the secret key in the cloud database server.
5. Data fetch time (DFT): This shows the time required to fetch encrypted health data and the secret key information.
6. AES decryption time (AES-DT): This reflects the time required to regenerate the secret key from the fetched information and to decrypt the health data into readable format.
7. DES encryption–decryption–encryption time (DES-EDET): This shows the time required for encryption process and the generation of 3 no's of 64-bit key from patient's thumb information.
8. DES decryption–encryption–decryption time (DES-DEDT): This shows the time required for decryption process and the regeneration of 3 no's of 64-bit key from patient's thumb information.

For detailed view and better understanding, total end-to-end delay and its different components of secure patient health data transmission using AES and triple DES encryption with authenticity and confidentiality are plotted graphically in Fig. 6 and Fig. 8, respectively. Results of Fig. 7 show that DTAT lies between 2.23 and 2.67 ms, DRT lies between 0.29 and 0.41 ms, AES-ET lies between 5.32 and 6.01 ms, DST

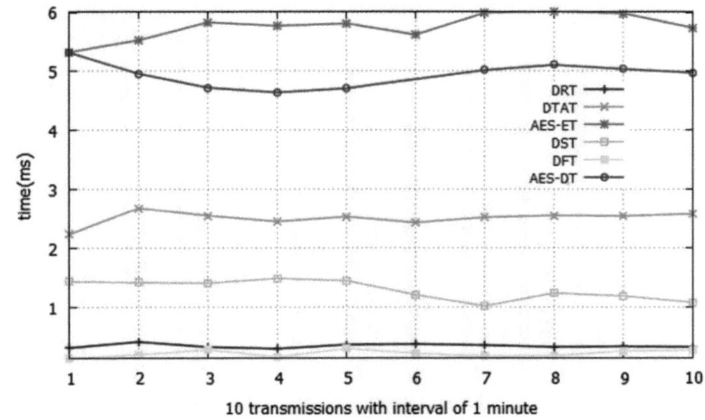

Fig. 6 End-to-end delay of physiological data transmission using AES

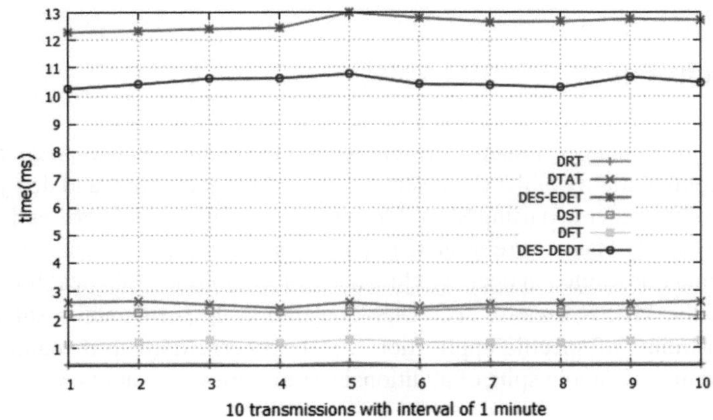

Fig. 7 End-to-end delay of physiological data transmission using triple DES

lies between 1.02 and 1.49 ms, DFT lies between 0.12 and 0.31 ms and AES-DT lies between 4.71 and 5.32 ms. Similarly from Fig. 8, we can understand that DES-EDET and DES-DEDT are increased slightly because of three layers of DES encryption and decryption process and generation of 3 no's of 64-bit key. Average total time required for both AES and triple DES using 128- and 192-bit keys is, respectively, 15.05 ms and 29.54 ms. Hence, end-to-end delay is only 6.02 and 11.82% of 250 ms which is the permissible delay limit for healthcare application.

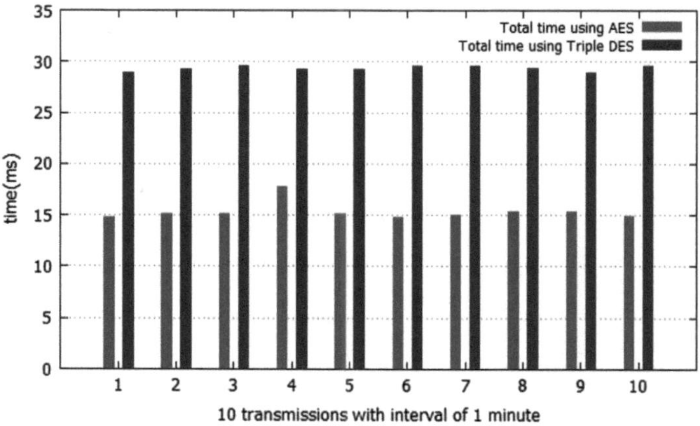

Fig. 8 Comparison between end-to-end delay of physiological data transmission using AES and triple DES

6 Conclusion

This work implements symmetric key encryption techniques such as AES and triple DES for confidentiality of data, and secret key used for encryption and decryption is hash-based key generated using SHA-3 to ensure integrity beyond tier 2 and between tier 2 and tier 3. This secure system implements a complete security to the vital signals being transmitted in open wireless network exploiting biological information extract of the MBSN user for secret key generation. Experimental results satisfy delay constraints of specific application, e.g. healthcare domain demanding real-time streaming of data in spite of additional security measures applied.

References

1. Li, H.-B., Takahashi, T., Toyoda, M., Katayama, N., Mori, Y., Kohno, R.: An experimental system enabling WBAN data delivery via satellite communication links. In: 2008 IEEE International Symposium on Wireless Communication Systems, Reykjavik, pp. 354–358 (2008)
2. Mukhopadhyay, S.C.: Wearable sensors for human activity monitoring: a review. IEEE Sens. J. **15**, 1321–1329 (2015)
3. Movassaghi, S., Abolhasan, M., Lipman, J., Smith, D., Jamalipour, A.: Wireless body area networks: a survey. IEEE Commun. Surv. Tutor. 1–29 (2013)
4. Raza, S.F., Naveen, C., Satpute, V.R., Keskar, A.G.: A proficient chaos based security algorithm for emergency response in WBAN system. In: 2016 IEEE Students' Technology Symposium (TechSym), Kharagpur, pp. 18–23 (2016)
5. Karmakar, K., Saif, S., Biswas, S., Neogy, S.: WBAN Security: study and implementation of a biological key based framework. In: International Conference on Emerging Applications of Information Technology, 12–13 Jan 2018

6. Saif, S., Gupta, R., Biswas, S.: Implementation of cloud assisted secure data transmission in WBAN for healthcare monitoring. In: International Conference on Advanced Computational and Communication Paradigms, 8–10 Sept 2017
7. Ramli, S.N., Ahmad, R., Abdollah, M.F., Dutkiewicz, E.: A biometric-based security for data authentication in wireless body area network (WBAN). In: ICACT 2013, 27–30 Jan 2013, pp. 998–100 (2013)
8. Liu, J., Zhang, Z., Chen, X., Kwak, K.S.: Certificateless remote anonymous authentication schemes for wireless body area networks. IEEE Trans. Parallel Distrib. Syst. **25**(2), 332–342 (2014)
9. He, D., Zeadally, S., Kuma, N., Hyouk Lee, J.: Anonymous authentication for wireless body area networks with provable security. IEEE Syst. J. 1–12 (2017)
10. Zheng, G., Fang, G., Shankaran, R., Orgun, M., Zhou, J., Qiao, L., Saleem, K.: Multiple ECG fiducial points-based random binary sequence generation for securing wireless body area networks. IEEE J. Biomed. Health Inf. **21**(3), 655–663 (2017)
11. He, D., Chan, S., Zhang, Y., Yang, H.: Lightweight and confidential data discovery and dissemination for wireless body area networks. IEEE J. Biomed. Health Inform. **18**(2), 440–448 (2014)
12. Saleem, S., Ullah, S., Yoo, H.S.: On the security issues in wireless body area networks. **3**(3), 178–184 (2009)

Multi-target-Based Cursor Movement in Brain-Computer Interface Using CLIQUE Clustering

Shubham Saurav, Debashis Das Chakladar, Pragnyaa Shaw, Sanjay Chakraborty and Animesh Kairi

Abstract Brain-computer interfacing (BCI) is a bridging technology between a human brain and a device that enables signals from the brain to direct some external activity, such as control of a cursor or a prosthetic limb. In practice, brain signals are captured by the popular EEG technique and then the scalp voltage level is transferred into corresponding cursor movements. In multi-target based BCI, the set of targets are assigned to the different clusters initially and then the cursor is mapped to the nearest cluster using clustering technique. Finally, the cursor hits all the targets sequentially inside its own cluster. In this work, the famous CLIQUE clustering technique is chosen to assign the cursor into a proper cluster and if the cursor movement will be optimum in time, then the disabled persons can communicate efficiently. CLIQUE clustering is an integration of density based and grid based Clustering methods which is used to measure the cursor movement as bit transfer rate in a cell within the grid. This technique will lead us to improve the performance of the BCI system in terms of multi-targets search.

Keywords Electroencephalography (EEG) · Brain-computer interface (BCI) Grid clustering · Density-based clustering · CLIQUE clustering · Multi-targets

S. Saurav (✉) · D. D. Chakladar · P. Shaw
Computer Science and Engineering Department,
Institute of Engineering and Management, Kolkata, India
e-mail: shubhamsaurav001@gmail.com

D. D. Chakladar
e-mail: ddaschakladar@gmail.com

P. Shaw
e-mail: pragnyaashaw@gmail.com

S. Chakraborty
Department of Information Technology, Techno India, Kolkata, India
e-mail: schakraborty770@gmail.com

A. Kairi
Department of Information Technology,
Institute of Engineering and Management, Salt Lake City, Kolkata, India
e-mail: animesh.kairi@iemcal.com

M. Chakraborty et al. (eds.), *Proceedings of International Ethical Hacking Conference 2018*, Advances in Intelligent Systems and Computing 811,
https://doi.org/10.1007/978-981-13-1544-2_34

1 Introduction

A Brain-computer interface sometimes referred as a neural-control interface (NCI) is a straight communication way of an enhanced brain and an outside device. BCIs have a comprehensive range of utilizations starting from a robotic arm controller to cursor movement in various dimensions. One major role of BCIs is to help physically disabled people to interact with the outside world through BCI spellers [1, 2]. Besides that, BCIs also help victims to monitor external devices such as wheelchairs and prosthetic devices or to control their environment using a cursor [3–5]. In the case of cursor control, for instance, the signal is dispatched directly from the brain to a process controlling the cursor, rather than following the general route through the body's neuromuscular system from the brain to the finger on a mouse [6]. One-dimensional (1D) cursor movements can be determined by analyzing the magnitude of a single spectral band at a single location on the scalp. The user can handle this amplitude through continuous learning. In our proposed work, the user moves the cursor through digitized data collected by EEG signals in order to select more than one target on a screen. A cursor associated with the brain will allow the person with motor disabilities to interact with their environment which definitely has a vast impact on the advancement of medical technologies [7]. A typical human Brain-computer interaction system architecture is demonstrated in Fig. 1. Figure 1 depicts a typical BCI system in which a person supervises a computer system and its various activities through a series of functional units. The user keeps tracking the phase of the computer system to decide the result of his/her control efforts.

BCI can empower a paralyzed person to write something or control a motorized wheelchair or prosthetic limb through his or her perception alone. This can be achieved by reading signals from an array of neurons and using various computer chips and software programs to convert the signals into real-life actions. BCI can be broadly classified into three main components, such as signal acquisition, signal processing, and effector device [8]. The signal acquisition is further divided into two categories: noninvasive and invasive. Invasive BCI is directly installed into the gray matter of the human brain. Invasive BCI research plays a very important role

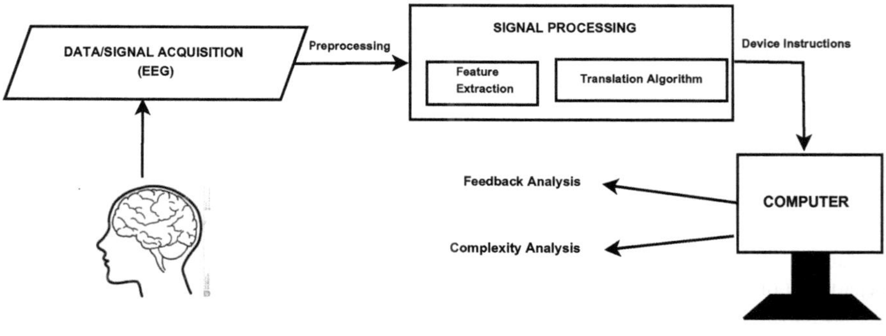

Fig. 1 A typical Brain-computer system architecture

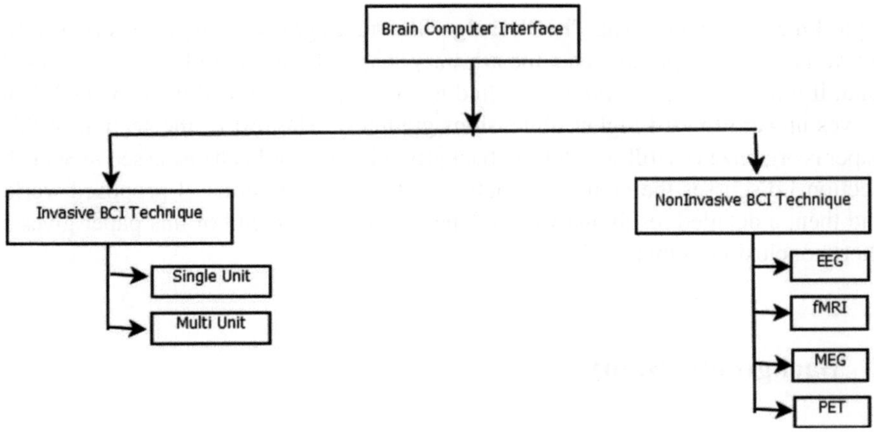

Fig. 2 Hierarchy of a typical BCI control system

to repair the injured sight of physically disabled people. But our proposed work is totally based on noninvasive neuroimaging technologies as interfaces. This is also called as an EEG-based BCIs. The hierarchy of a typical BCI system with different functional units is described in Fig. 2. Various neuroimaging method exists under invasive BCI: like(multi-unit(MU), single-unit (SU)). In the single-unit invasive recording, an only single electrode is implanted in the brain; when in case of multi-unit invasive technique, multiple electrode arrays are used to capture the brain signal. Electroencephalography (EEG), functional magnetic resonance imaging (fMRI), magnetoencephalography (MEG), positron emission tomography (PET) are used for noninvasive BCI techniques. EEG provides the recording of electrical activity of the brain from the surface of the scalp. MEG transfers electrical activity of the brain into the magnetic field and produces better spatial resolution than EEG. In fMRI technique, the brain activity is measured by detecting the changes in blood flow in the brain. PET is a nuclear imaging technique used in medicine to observe different processes, such as blood flow and metabolism. In a typical BCI system, a mechanical device is implanted inside the brain which controls that device as a natural part of its representation of the body. Many recent research works are focused on the potential of noninvasive BCI [9]. A multi-target consists of multiple image targets in a defined geometric arrangement, and using BCI, we will get data sets for identification of the same. In our paper, we work upon identification of multi-targets through cursor movement from BCI data.

However, we introduce here a special clustering method to achieve this work. Clustering is a task of grouping a set of similar objects called clusters, whereas the dissimilar objects are called outliers. There are various clustering algorithms available such as K-means, DBSCAN, hierarchical [10–15]. Traditional clustering algorithms like K-means, CLARANS, BIRCH, DBSCAN are unable to manage high-dimensional data [16]. However, this paper presents an important subspace clustering algorithm referred as CLIQUE, which can efficiently perform the clustering of high-dimensional data. The CLIQUE algorithm takes grid size and a density threshold value as user input to create the arbitrary shape of clusters in large-dimensional data. It initially starts the process of finding clusters at a single dimension and then moves upward toward higher dimensions gradually. The rest of the section of this paper is organized as follows. A brief background study has been discussed in Sect. 2. Section 3 discusses the main contribution of this work in terms of proposed work, and then, a detailed result analysis is done in Sect. 4. The end of this paper gives a short conclusion of this work.

2 Background Study

A number of techniques have been developed to establish a way of communication between the human brain and computer systems. There are several works which show how EEG signal from human brain is used for various actions into a computer system. In this section, we are dealing with such BCI-based communication techniques. A study at the Wadsworth Center [17] shows that a paralyzed people can monitor the cursor on a computer screen using μ and β rhythms EEG activity recorded from the scalp over sensorimotor cortex. They used specific ranges of μ and β rhythms for communication. In paper [18], a subject is trained to extract 8–12 Hz μ rhythm from scalp over the central sulcus of one hemisphere to move the cursor from central screen to the desired target location (either top or bottom). This μ rhythm is analyzed by high and low frequencies. Larger amplitudes (high frequencies) moved the cursor up, and smaller amplitudes (low frequencies) moved it down. The speed and accuracy of cursor control by this proposed system depend on the proper selection of voltage ranges and movements. The proper selection leads to proper up and down distributions of the cursor. The main objective of this work is to determine how disabled people can use his/her μ rhythm to control the movement of a cursor on a computer screen rapidly and reliably. This strategy is applicable for both one-dimensional and two-dimensional cursor movement control on a computer screen.

In paper [8], a new combination strategy is proposed which allows users to move a cursor to any position in two-dimensional space with motor imagery BCI tasks. In this strategy, users can combine multiple motor imagery BCI tasks simultaneously. This system is used as a tool to interact with the outside world. The main goal of this work depends on the combination of three motor imagery BCI tasks such as right-hand motor imagery, left-hand motor imagery, and feet motor imagery. By combining the above motor imagery tasks with high hit rates and short trajectories, users can perform the continuous cursor control of 2D movement. To perform this entire task, three well-trained participants are involved.

In paper [6], a suitable algorithm is introduced to move the cursor control toward the desired target on the computer screen efficiently and effectively. This proposed algorithm is referred as "Find-Target" algorithm. This proposed algorithm is based on noninvasive EEG-based BCI system. First, it receives three different user input, such as cursor location, desired target location, and five various amplitudes collected from EEG-based brain signals. Then, it stores those brain signal amplitudes into a ready queue in descending order. Second, it will evaluate the initial distance of cursor and target values. Then, it picks up the highest amplitude value from the ready queue and compares it with that initial distance. If the initial distance is larger, then the cursor location is upgraded with the help of the highest amplitude value. However, it determines a new cursor location that checks the difference again and resets the cursor location. In this way, the above process is going on recursively and it stops when the difference between new cursor location and the desired target location reaches zero. The main benefits of this approach over the previous one are that it is a multithreaded dynamic cursor movement which leads to an increase computational speedup. It follows a parallel execution approach rather than a sequential one. Unlike previous approaches, it does not depends on trial prediction or subjects training phase. That's why the cursor can reach the target very quickly in the first trial; then, immediately cursor will be ready for the next trial.

3 Multi-target Cursor Movement Using CLIQUE

The goal of this paper is to implement an effective technique of multi-target-based cursor movement using CLIQUE clustering. CLIQUE clustering is a combination of spatial and dense clustering where the input space is divided into $N * M$-dimensional cells based on configurable cell size [19]. The algorithm takes input position of the cursor and multiple targets (randomly placed in the application interface), the cell size of the grid in $N * M$-dimensional space, density threshold points for finding the dense cells. This clustering checks the targets within each cell using the DBSCAN algorithm and classifies the dense cell. Once the dense cells are identified, then we calculate the distance between the mean and the closest, furthest target in each cell, and based on the calculated distance, we merged two or more dense cells to form a new cluster of higher dimension. Here, we have used "Euclidean" distance measure to calculate the distance between mean and closest, furthest target. If the "Euclidean" distance between furthest and closest targets of two cells is less than the input cell size, then two dense cells of lower-dimensional space are merged into a single cluster of higher-dimensional space. After the formation of the cluster, we call "FindTarget" [6] algorithm of cursor movement for all the targets within the cluster. The "FindTarget" algorithm is an efficient algorithm for cursor movement where the ratio of cursor and target is 1:1. For inputs, the "FindTarget" algorithm takes the position of the cursor, target, and five mostly used amplitudes of the subject for a short time interval. After taking the inputs, the algorithm computes the distance between cursor and target and try to minimize them by resetting the cursor. For multi-target-based cursor movement, the "FindTarget" algorithm is called for several times for different clusters. Once the cursor reaches all the targets, then the cluster itself gets vanished. The same process is repeated for other clusters. The entire proposed algorithm of multi-target-based cursor movement using CLIQUE clustering has been discussed in algorithm 1.

Algorithm 1: Cursor Movement to Multi-target using CLIQUE Clustering Algorithm.

```
Input  : The initial position of the cursor and multiple targets in the
         N-dimensional cell, cell size(x), density threshold points(Th_c) within
         the cell.
Output: All the targets get vanished as the cursor reaches to all of them.
Setting the cursor as a center of the cell, apply DBSCAN algorithm on the
  input data set along with minimum targets per cell and store the result in a 2D
  array X.
while (length(X) ≠ NULL) do
    if (Number of targets (T) in the cell ≥ (Th_c)) then
        Make those cell as dense cells.
    else
        Consider as noise.
    end
end
foreach (dense cell D1[m][n] and D2[n][k]) do
    Compute the mean position (m1,m2) of targets in both the dense cells.
    foreach targets in dense cell D1 do
        p1= Maximum Distance(D1,m1)
    end
    foreach targets in dense cell D2 do
        p2= Minimum Distance(D2,m2)
    end
end
foreach (dense cell D1 [m][n] and D2[m][k]) do
    if (Euclidian distance(p1,p2)<= Cellsize(x)) then
        New dense cell of higher dimension D3[m][n][k]<- Merge(D1[m][n] ,
        D2[n][k] ).
        New cluster (c1) with higher dimensional space= D3[m][n][k].

end
Call FindTarget algorithm [6] for cursor movement to the nearest target within
  the cluster(c1).
When the cursor reaches the target, then that target gets vanished. Once all the
  targets within a cluster (c1) has vanished, then call the same process for
  another cluster.
```

4 Result and Analysis

The experimental analysis of the proposed algorithm has done on the following computing platform.

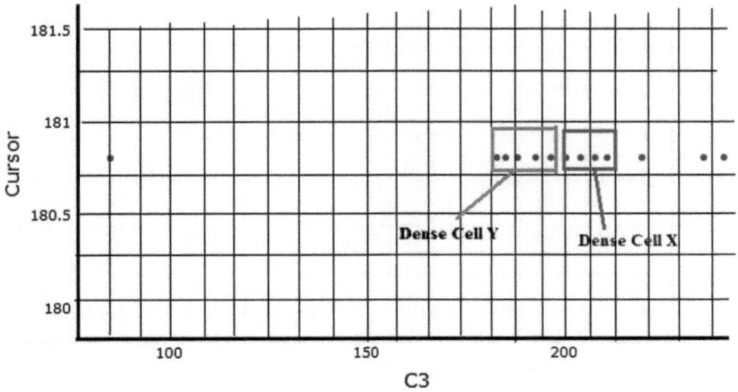

Fig. 3 Initial dense cells in the grid

Application Environment: Python 3.5.2

Hardware Environment: Operating system—Ubuntu16, Processor-3x Intel Core (TM) i5, RAM-2 GB, Clock-1.70 GHZ. CLIQUE clustering is a grid-based clustering technique. In this clustering, we have divided the entire input space into M * N-dimensional space and then apply CLIQUE clustering for cursor movement within each dense cell. Our proposed algorithm first identifies the dense cell using the DBSCAN algorithm. Then, find the optimum distance between each dense cell and try to merge the two or more dense cells to make a large cluster. After cluster formation, the cursor moves to all the targets using "FindTarget" algorithm [6] within the cluster. Once the cursor has reached all the targets within the cluster, then the cluster itself gets vanished. The detailed analysis has described in Figs. 3, 4, 5, and 6.

Step1: Identify the dense cells (cells having targets more than threshold points)in the grid. The dense cells are plotted in the Fig. 3.

Step 2: Merge two or more adjacent dense cells to form a large cluster. Here, we have merged two adjacent dense cell X and dense cell Y and create a new cluster with the higher dimension. Here, we choose the threshold points per cell = 2, so any cell having the number of targets more than threshold points is called the dense cell. The merging of two dense cells has been shown in the Fig. 4.

Step 3: Once the final cluster has been created, then we call "Find target" algorithm [6] to all the targets (in the dense region) within the final cluster. The cursor movement within the cluster has been described in the Fig. 5 and 6.

In Fig. 5, the cursor has reached the two targets in the leftmost grid cell, so they get vanished. In Fig. 6, the cursor has reached all the targets in the grid cell, so the entire cluster along with all the targets has vanished.

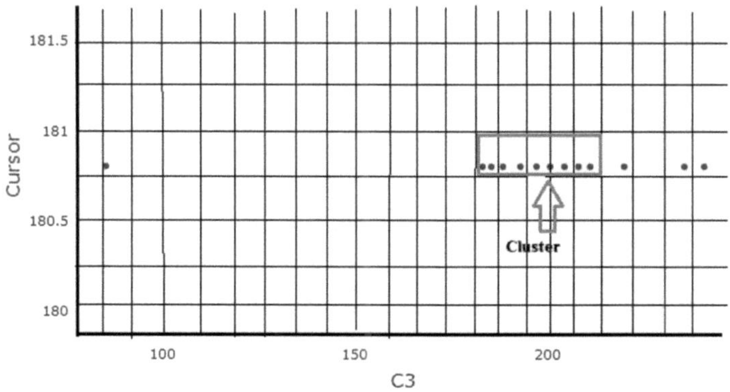

Fig. 4 Final cluster using CLIQUE

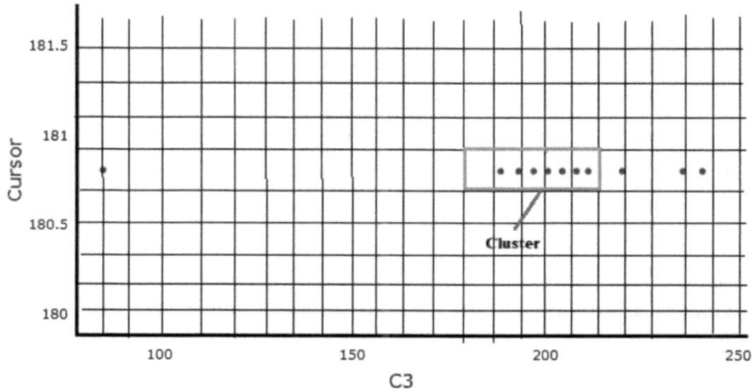

Fig. 5 Cursor movement within the cluster

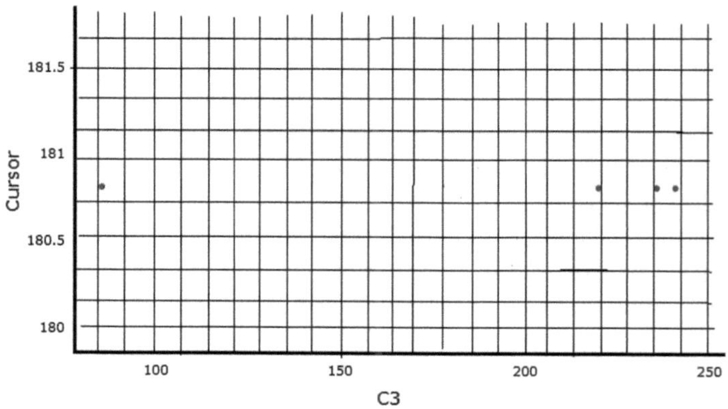

Fig. 6 The cursor has reached all the targets within the cluster, so the cluster itself has been vanished

5 Conclusion

In this paper, we combine cursor movement in BCI with clustering methods. The clustering-based multi-target BCI system is useful for paralyzed people as the cursor moves all the targets within the cluster in an efficient manner and gives the faster response to the user. We have implemented our proposed algorithm using CLIQUE clustering technique and analyzed the performance in terms of cursor movements among dense cell environment. If the cursor reaches the targets in the lower-dimensional space, then they can also be efficiently implemented for higher-dimensional space which can improve the bit transfer rate of cursor movement in the multi-target-based BCI.

References

1. Xia, B., Yang, J., Cheng, C., Xie, H.: A motor imagery based brain—computer interface speller. In: 2013 Advances in Computational Intelligence, pp. 413–421. Springer, Berlin (2013)
2. Donchin, E., Spencer, K.M., Wijesinghe, R.: The mental prosthesis: assessing the speed of a p300-based brain—computer interface. IEEE Trans. Rehabil. Eng. **8**, 174–179 (2000)
3. Huang, D., Qian, K., Fei, D.-Y., Jia, W., Chen, X., Bai, O.: Electroencephalography (EEG)-based brain-computer interface (BCI): a 2-D virtual wheelchair control based on event-related desynchronization/synchronization and state control. IEEE Trans Neural Syst. Rehab. Eng. **20**, 379–388 (2012)
4. Li, J., Liang, J., Zhao, Q., Li, J., Hong, K., Zhang, L.: Design of assistive wheelchair system directly steered by human thoughts. Int. J. Neural Syst. **23**, 1350013 (2013)
5. Li, J., Ji, H., Cao, L., Zang, D., Gu, R., Xia, B., Wu, Q.: Evaluation and application of a hybrid brain computer interface for real wheelchair parallel control with multi-degree of freedom. Int. J. Neural Syst. **24**, 1450014 (2014)
6. Chakladar, D.D., and Chakraborty, S.: Study and analysis of a fast moving cursor control in a multithreaded way in brain computer interface. In: International Conference on Computational Intelligence, Communications, and Business Analytics, pp. 44–56. Springer, Singapore, March 2017
7. Fabiani, G.E., McFarland, D.J., Wolpaw, J.R., Pfurtscheller, G.: Conversion of EEG activity into cursor movement by a brain-computer interface (BCI). IEEE Trans. Neural Syst. Rehabil. Eng. **12**(3), 331–338 (2004)
8. Xia, B., Maysam, O., Veser, S., Cao, L., Li, J., Jia, J., Xie, H., Birbaumer, N.: A combination strategy based brain—computer interface for two-dimensional movement control. J. Neural Eng. **12**(4), 046021 (2015)
9. Ortiz-Rosario, A., Adeli, H.: Brain-computer interface technologies: from signal to action. Rev. Neurosci. **24**(5), 537–552 (2013)
10. Lotte, F., Congedo, M., Lecuyer, A., Lamarche, F., Arnaldi, B.: A review of classification algorithms for EEG-based brain—computer interfaces. J. Neural Eng. **4**(2), R1 (2007)
11. Steinbach, M., Karypis, G., Kumar, V.: A comparison of document clustering techniques. KDD Workshop Text Min. **400**(1), 525–526 (2000)
12. Tajunisha, S., Saravanan, V.: Performance analysis of k-means with different initialization methods for high dimensional datasets. Int. J. Artif. Intell. Appl. (IJAIA) **1**(4), 44–52 (2010)
13. Schikuta, E.: Grid-clustering: an efficient hierarchical clustering method for very large data sets. In: Proceedings of the 13th International Conference on Pattern Recognition, 1996, vol. 2, pp. 101–105. IEEE (1996)

14. Andrade, G., Ramos, G., Madeira, D., Sachetto, R., Ferreira, R., Rocha, L.: G-DBSCAN: a GPU accelerated algorithm for density-based clustering. Procedia Comput. Sci. **18**, 369–378 (2013)
15. Aho, A.V., Hopcroft, J.E., Ullman, J.D.: The Design and Analysis of Computer Algorithms. Addison-Wesley, Reading, MA (1974)
16. Cao, F., Martin E., Weining Q., Aoying Z.: Density-based clustering over an evolving data stream with noise. In: Proceedings of the 2006 SIAM International Conference on Data Mining, pp. 328–339. Society for Industrial and Applied Mathematics (2006)
17. Wolpaw, J.R., McFarland, D.J., Vaughan, T.M.: Brain-computer interface research at the wadsworth center. IEEE Trans. Rehabil. Eng. **8**(2), 222–226 (2012)
18. Wolpaw, J.R., McFarland, D.J., Neat, G.W., Forneris, C.A.: An EEG-based brain-computer interface for cursor control. Electroencephalogr. Clin. Neurophysiol. **78**, 252–259 (1991)
19. Yadav, J., Kumar, D.: Sub space Clustering using CLIQUE: an exploratory study. Int. J. Adv. Res. Comput. Eng. Technol. (IJARCET) **3** (2014)

Part IX
Session 3C: Data Analysis

Data Mining in High-Performance Computing: A Survey of Related Algorithms

Pradip Kumar Majumder and Mohuya Chakraborty

Abstract Present days parallel, distributed or cloud computing technologies have been able to regulate large data sets efficiently. An important part of information technology is extensive data processing. This is because of availability and accelerated surge of data. Data mining is the process of examining large preexisting databases or raw data to generate new information for further use. Data mining algorithms must be efficient and effective in order to produce some meaningful output. The applications for these are limitless, from predicting a specific disease to very complex applications. In this research paper, we have compared most popular data mining algorithms with respect to conceptual architectures and advantages and disadvantages of them. Finally, we have proposed a new efficient data mining algorithm named as sifted K-means with independent K-value (SKIK) algorithm.

Keywords Cloud computing · Distributed/cluster technology
High-performance clusters (HPC) · Data mining algorithms
K-means algorithm

1 Introduction

Cloud computing is a new technology containing pool of resources with large number of computers. The computation task is distributed to this pool and also provides us with unlimited storage and computing power which will benefit us to mine large amount of data.

Tightly coupled systems including shared memory systems (SMS), distributed memory machines (DMM), or clusters of SMS workstations are connected with

P. K. Majumder (✉)
Department of IT, University of Engineering & Management, Kolkata, West Bengal, India
e-mail: majumder.pk@gmail.com

M. Chakraborty
Department of IT, Institute of Engineering & Management, Kolkata, West Bengal, India
e-mail: mohuyacb@iemcal.com

M. Chakraborty et al. (eds.), *Proceedings of International Ethical Hacking Conference 2018*, Advances in Intelligent Systems and Computing 811,
https://doi.org/10.1007/978-981-13-1544-2_35

a fast network in a parallel data mining environment. Loosely coupled processing nodes/computers are connected by the high-speed network in a distributed computing environment. Each node contributes to the execution or distribution/replication of data. It is generally called as a cluster of nodes. Usually, a cluster framework is used to set up a cluster.

The HPC clusters exploit parallel computing to exert more computation power for the resolution of a problem. HPC clusters have a large number of computers called "nodes," and mostly these nodes would be configured identically. Externally, the cluster looks like one system. Client programs that run on a node are called jobs, and they are constantly monitored through a queuing technique for proper use of every accessible resource. HPC jobs include replication of numerical models or study of information from logical instrumentation which allow scientists to construct new science at the use of high-performance computing [1].

Data mining is the process of examining large preexisting databases or raw data to generate new information for further use. Data mining algorithms must be efficient and effective in order to produce some meaningful output. Among the numerous available data mining algorithms, the popular ones are Apriori, DIC, GSP, SPADE, SPRINT, and K-means algorithms. In earlier days, due to limitations in computing power, data mining process was slower. Nowadays, the data mining process has speed up many folds due to the presence of high-performance parallel and distributed computing environments. However, the data available today are very large and growing in an exponential rate, which need more effective and accurate data mining algorithm to be deployed. K-means algorithm, despite being one of the most effective algorithms to be used in a parallel computing environment, has some major limitations, and we have worked on proposed SKIK algorithm to overcome one such limitation.

The organization of the paper is as follows: Sect. 1 holds the introduction, and Sect. 2 contains brief discussion of data mining concepts used in HPC environment along with their pros and cons. Section 3 describes the proposed sifted K-means with independent K-value (SKIK) algorithm created based on the advantages and disadvantage of existing algorithms, and Sect. 4 shows the complexity measurement of SKIK algorithm. Section 5 highlights conclusion to this paper with few focus points on imminent works.

2 Data Mining Concepts and Related Algorithms

Different data mining concepts including types of machine, parallelisms, load balance, database layouts, and candidates are discussed in detail in Sect. 2.1. Performance analyses of some of the most popular algorithms vis-a-vis concepts are provided in Table 1. In Sect. 2.2, various related and most popular algorithms are explained briefly. A comparative analysis of these algorithms highlighting their advantages and disadvantages are pointed out in Sect. 2.3.

Table 1 Comparisons among common concepts used with data mining algorithms [2]

Algorithm	Base algorithm	Type of machine	Parallelism type	Load balance type	DB layout	Concepts	DB type
CD	Apriori	DMM	Data	Static	Horizontal	Replicated	Partitioned
PDM	Apriori	DMM	Data	Static	Horizontal	Replicated	Partitioned
FDM	Apriori	DMM	Data	Static	Horizontal	Replicated	Partitioned
IDD	Apriori	DMM	Task	Static	Horizontal	Partitioned	Partitioned
HD	Apriori	DMM	Hybrid	Hybrid	Horizontal	Hybrid	Partitioned
CCPD	Apriori	SMS	Data	Static	Horizontal	Shared	Partitioned
PCCD	Apriori	SMS	Task	Task	Horizontal	Partitioned	Shared
HPA	Apriori	DMM	Task	Static	Horizontal	Partitioned	Partially replicated
APM	DIC	SMS	Task	Static	Horizontal	Shared	Partitioned
HPSPM	GSP	DMM	Task	Static	Horizontal	Partitioned	Partially replicated
SPADE	SPADE	SMS	Task	Dynamic	Vertical	Partitioned	Shared
D-MSDD	MSDD	DMM	Task	Static	Horizontal	Partitioned	Replicated
SPRINT	SPRINT	DMM	Data	Static	Vertical	Replicated	Partitioned
PDT	C4.5	DMM	Data	Static	Horizontal	Replicated	Partitioned
MWK	SPRINT	SMS	Data	Dynamic	Vertical	Shared	Shared
SUBTREE	SPRINT	SMS	Hybrid	Dynamic	Vertical	Partitioned	Partitioned
HTF	SPRINT	DMM	Hybrid	Dynamic	Vertical	Partitioned	Partitioned
P-Cluster	K-means	DMM	Data	Static	Horizontal	Replicated	Partitioned

2.1 Concepts

Type of machine used. The main two types of machines are distributed memory machines (DMM) and shared memory systems (SMS). In DMM, the effort is communication optimization and hence synchronization is implicit in message passing. For SMS, synchronization occurs via locks and barriers, and the aim is to minimize these points. Data decomposition is very important for DMM, but not for SMS. SMS typically use serial I/O, while DMM use parallel I/O [2].

Parallelism type. Task and data parallelism are two major parallelisms used. In data parallelism, the database is partitioned among P processors. Each processor performs evaluating candidate patterns/models on its local part of the database. In task parallelism, the processors perform different computations independently, but have/need access to the entire database. SMS can connect to whole data, but for DMM can do this through careful reproduction or specific connection to the local data. It is also possible for a hybrid parallelism having properties of both task and data parallelisms.

Load balance type. Two main load balancing types are static and dynamic load balancing. In static load balancing, work is partitioned among the processors using heuristic cost function, and there is no subsequent correction of load imbalances resulting from the dynamic nature of mining algorithms. Dynamic load balancing distributes work from heavily loaded processors to lightly loaded ones. Dynamic load balancing is important in multi-user environments and in heterogeneous platforms, which have different processor and network speeds.

Database layout type. Usually, the recommended database for data mining is a relational table having R rows, called records, and C columns, called attributes. Horizontal database design is used in numerous data mining algorithms. Here, they collect transaction id (tid) as a unit including attribute values for that transaction. Other procedures use a vertical database design. Here, they collect a list of all tids (called tidlist) containing the item with each attribute and the related attribute value.

Candidate concepts. Dissimilar mining procedures use either shared or replicated or partitioned candidate concept generation and evaluation. All processors check out a single copy of the candidate set in shared concept. The candidate concepts are copied on each system, and checked locally, before overall outcomes are achieved by fusing them in replicated concept. Each processor creates and examines a dislocated candidate set in the partitioned concept.

Database type. The database itself can be shared (in SMS or shared-disk architectures), partitioned (using round robin, hash, or range scheduling) among the available nodes (in DMM) or partially or totally replicated.

Table 1 shows a comparison among the common concepts used with most popular data mining algorithms.

2.2 *Most Popular Algorithms*

Apriori Algorithm. This algorithm is used for mining common itemsets in large data sets. The point of view is "bottom up." We call it candidate generation, where frequent subsets are available one item at a time, and groups of candidates are checked against the data. It is aimed to operate on transaction database.

Frequent Itemsets: All the sets holding the item with the least support (designated by D_i for ith itemset).

Apriori Property: All subgroups of frequent itemset have to be frequent.

Join Operation: A set of candidate k-itemsets are generated by joining D_{k-1} with itself to find D_k.

Prune Step: Any sparse (k − 1)-itemset cannot be a subset of a frequent k-itemset.

C_k: Candidate itemset of size k
D_k: frequent itemset of size k
D_1 = {frequent items};

STEP 1: Have the support S of each 1-itemset by examining the given database, correlate S with sup_{min}, and prepare a support of 1-itemsets, D_1

STEP 2: Use D_{k-1} and join D_{k-1} to create a set of candidate k-itemsets. And use Apriori property to prune the infrequent k-itemsets from this set.
STEP 3: Examine the given database to find the support S of each candidate k-itemset in the find set, correlate S with sup_{min}, and prepare a set of frequent k-itemsets D_k
STEP 4: Is the candidate set = Null, if YES go to STEP 5 else go to STEP 2
STEP 5: Produce all nonempty subsets of 1 for every common itemset 1,
STEP 6: If confidence C of the rule "s => (1 − s)" (=support of 1/support S of s)' min_conf, output the rule "s => (1 − s)", for every nonempty subset s of 1 [3].

Dynamic Itemset Counting Algorithm (DIC). It is an alternative to Apriori algorithm. As the transactions are read, the itemsets are dynamically inserted and removed. Assumptions are made that all subgroups of frequent itemset have to be frequent. After every T transactions, algorithm stops to add more itemsets. Itemsets are tagged in four different ways as they are counted:

Solid box: □ confirmed frequent itemset—an itemset we have completed counting and exceeds the support threshold sup_{min}

Solid circle: ○ confirmed infrequent itemset—we have completed counting and it is below sup_{min}.

Dashed box: ⬚ imagined frequent itemset—an itemset being counted that surpasses sup_{min}

Dashed circle: ◌ imagined uncommon itemset—an itemset being counted that is below sup_{min}

STEP 1: Tag the empty itemset with a solid square. Tag the 1-itemsets with dashed circles. Discard all other itemsets untagged.
STEP 2: While any dashed itemsets remain.

1. Read *M* transactions (if at the end of the transaction file, continue from the beginning). For each transaction, step up the corresponding counters for the itemsets that appear in the transaction and are tagged with dashes.
2. If a dashed circle's count surpasses sup_{min}, make it a dashed square. Insert a new counter for it and make it a dashed circle if any next superset of it has all subsets as solid or dashed squares.
3. If a dashed itemset has already been counted through all the transactions, make it solid and stop counting it [4].

Generalized Sequential Pattern Algorithm (GSP). A sequence database is formed of ordered elements or events. In GSP algorithm, horizontal data format is used and the candidates are generated and pruned from frequent sequences using Apriori algorithm.

STEP 1: Each item in database is a candidate of magnitude 1 at the beginning.
STEP 2: for each level (i.e., order of magnitude k) do

1. Examine database to gather support count for every candidate order.
2. Generate candidate magnitude (k + 1) orders from magnitude k frequent orders using Apriori.

STEP 3: Do it over till no common order or no candidate can be found [5].

Sequential PAttern Discovery using Equivalence classes (SPADE). It is an algorithm to frequent sequence mining using vertical ID list database format, where each sequence is related with a list of objects in which it appears. Then, frequent sequences can be found surely using intersections on ID lists. The procedure lowers the number of database scans and hence also lowers the execution time.

STEP 1: Sequences having singular item, in a single database scan, are measured as the number of 1-sequences.
STEP 2: Convert the vertical depiction into horizontal depiction in memory and measure the number of sequences for each pair of items using a two-dimensional matrix for 2-sequences calculation. Thus, this step can also be performed in only one scan.
STEP 3: Following n-sequences can then be formed by joining (n − 1)-sequences using their ID lists. The size of the ID lists is the count of sequences in which an item occurs. If this number is higher than minsup, the sequence is a frequent one.
STEP 4: If no frequent sequences available, the algorithm stops.

The algorithm can use a breadth-first or a depth-first search procedure to discover new sequences [6].

Scalable PaRallelizable INduction of decision Trees (SPRINT) Algorithm. This algorithm builds a model of the classifying characteristics based upon the other attributes. Classifications provided are called a training set of records having several attributes. Attributes are either continuous or categorical.

SPRINT algorithm frees all of memory restrictions in contrast with SLIQ algorithm. This is also fast and scalable and can be easily parallelized.

Original SPRINT algorithm has the following steps:

```
Division(Data D)
if (every point in D are of same class) then
   Return;
for every attribute A do
   calculate splits on attribute A;
Use best split found to divide D into D1 and D2;
Division(D1);
Division(D2);
```

Original Call: Division(Training Dataset)
Prune step is done using SLIQ algorithm [7].

K-means Clustering Algorithm. K-means is an unsupervised learning algorithm classifies a given data set through a certain number of clusters (assume k clusters) fixed a priori. Different cluster positions will cause different outcomes. So we must define k centers, one per cluster which should be placed in a clever manner. So, placing them as far as possible from each other seems a better choice. This algorithm tries to minimize “squared error function” given by:

$$J(X) = \sum_{i=1\rightarrow S} \sum_{j=1\rightarrow Si} \left(||w_i - x_j||\right)^2$$

where

"$||w_i - x_j||$" is the Euclidean distance between w_i and x_j.
"s_i" is the number of data points in ith cluster.
"s" is the number of cluster centers

Let $W = \{w_1, w_2, \ldots, w_n\}$ be the set of data points and $X = \{x_1, x_2, \ldots, x_s\}$ be the set of centers.

STEP 1: Randomly select "s" cluster centers.
STEP 2: Calculate the distance between each data point and cluster centers.
STEP 3: Assign the data point to the cluster center whose distance from the cluster center is minimum of all the cluster centers.
STEP 4: Recalculate the new cluster center using:

$$x_i = (1/s_i) \sum_{j=1\rightarrow Si} w_j$$

where "s_i" means the data points number in ith cluster.

STEP 5: Recalculate the distance between each data point and new obtained cluster centers.
STEP 6: If no data point was reassigned, then stop, otherwise repeat from STEP 3 [8, 9].

2.3 Advantages and Disadvantages of the Algorithms

Table 2 shows a comparative study about pros and cons of the above-mentioned algorithms.

3 Proposed SKIK Algorithm

K-means algorithm generates K clusters of the known data set, where every cluster can be expressed by a centroid which is a concise expression of all the objects present in a cluster. The main flaws of K-means algorithm are: (i) it is difficult to anticipate the number of clusters (value of K) and (ii) initial centroids have a big effect on the concluding outcome. Here, we are introducing a new algorithm sifted K-means with independent K-value (SKIK) algorithm to overcome these issues.

In data mining, we work on very large data set. We have proposed to sort these data based on any attribute as per user requirement. We would use parallel heap sort

Table 2 Advantages and disadvantages of popular algorithms

Algorithm	Advantages	Disadvantages
Apriori [10]	1. It is easy to apply and easy to figure out 2. It can be used on large itemsets	1. Sometimes, a large number of candidate rules are required which can be costly to compute 2. It goes through entire DB, hence calculating support is costly
DIC [11]	1. If the data are similar throughout the file and the interval is fairly small, it normally makes on the order of two passes 2. It can add and delete itemsets on the fly and thus extended to parallel and incremental versions	1. It is very delicate to dissimilar data 2. If the date are very associated, DIC counts most of the DB and itemset is realized to be actually large 3. It has performance issues as to how to increment the relevant counters for a specific transaction
GSP [12, 13]	1. We can enter bounds on the time separation between adjoining items in an arrangement 2. The items present in the pattern element can stretch a transaction set within a time frame specified by user 3. Detection of frequent patterns in various levels as needed by user is possible. Detecting generalized sequential patterns is also possible	1. Repeated DB scanning to compute the support of candidate patterns is costly for a large database 2. It may create patterns that do not exist in the DB as it generates candidates by linking smaller patterns without accessing the DB 3. All frequent sequences of length k is kept in memory to create patterns of length k+1 as it is BFS pattern mining algorithm. It consumes a good memory
SPADE [14, 15]	1. It uses vertical DB layout 2. The search space is expressed as lattice formation and breaks up original lattice into sub-lattices to process using either BFS or DFS 3. Efficient support counting method based on the idlist structure is used and is nearly twice faster than GSP algorithm. It shows linear scalability w.r.t. the number of sequences	1. A huge set of candidates generated, specially 2-item candidate sequence 2. Multiple scans of database in mining. magnitude of every candidate increases by one at every database scan 3. Inefficient for mining long sequential patterns. A long pattern causes an exponential number of short candidates
SPRINT [16, 17]	1. It removes memory constraints that limit existing decision tree algorithms 2. It consciously averts the need for any centralized, memory resident data structure 3. It allows analysis of nearly any sized data set, and it is fast 4. It is easily parallelized which needs few inclusions to serial algorithm	1. High workload in precisely locating the optimal splitting point 2. The main time expense is used to sort all records of the attribute table in the entire process
K-means [18]	1. Easy to deploy with a massive number of variables, and it is faster than hierarchical clustering (if K is small) 2. K-means can produce tighter clusters than hierarchical clustering 3. An instance can change cluster when the centroids are recalculated	1. Difficult to predict the number of clusters (K-value) 2. Initial seeds have a strong impact on the final results 3. Rescaling the data sets (normalization or standardization) will completely change results

[19] to sort as it uses a parallel approach across the cluster utilizing the available architecture.

Steps to find initial centroids:

1. From n objects, determine a point by arithmetic mean. This is the first initial centroid.
2. From n objects, decide next centroids so that the Euclidean distance of that object is highest from other decided original centroids. Keep a count of the centroids.
3. Repeat Step 2 until $n \leq 3$ [20].

We will get initial centroids from here and can use them in the proposed algorithm to calculate "optimal" centroids and K-value.

Determination of K:

The K-means algorithm creates compact clusters to minimize the sum of squared distances from all points to their cluster centers. We can thus use the distances of the points from their cluster center to measure if the clusters are compact. Thus, we adopt the inner-cluster distance, which is usually the distance between a point and its cluster center. We will take the median of all of these distances, described as

$$D_{wc} = (1/N) \sum_{i=1 \to k} \sum_{w \, \varepsilon \, Si} ||w - c_i||^2$$

where N is the count of components in the data set, k is the count of original clusters equal to the number of originally determined centroids, and c_i is the center of cluster s_i.

We can also measure the between-cluster distance. We take the minimum of the distance between cluster centers, defined as

$$D_{bc} = \min\left(||c_i - c_j||^2\right), \text{ where } i = 1, 2, \ldots, k - 1 \text{ and } j = i + 1, \ldots, k.$$

Now genuineness, $G = D_{wc}/D_{bc}$.

We need to decrease the inner-cluster distance, and this measure is in the numerator. So we need to decrease the genuineness measure. The between-cluster distance measure needs to be increased. Being in the denominator, we need to decrease the genuineness measure. Hence, clustering with a "lowest value" for the genuineness measure will provide us the "optimal value" of K for the K-means procedure [21].

We can also evaluate the "optimal" K using both inner-cluster and between-cluster scatter using the method proposed by Kim and Park [22].

Steps of SKIK:

1. Start.
2. Load the data set.
3. Sort the data using parallel heap sort.
4. Find initial centroids using previously mentioned procedure.
5. Determine K (number of clusters) from the centroids.
6. Calculate the distance between each data point and cluster centers.

7. Assign the data point to the cluster center whose distance from the cluster center is minimum of all the cluster centers.
8. Recalculate the new cluster center using:

$$x_i = (1/s_i) \sum_{j=1 \rightarrow Si} w_j$$

where “s_i” means the data points number in *i*th cluster.

9. Recalculate the distance between each data point and new obtained cluster centers.
10. If no data point was reassigned, then stop, otherwise repeat from Step 7.

4 Complexity Measurement of SKIK

Sorting may imply initial workload, but once done it will decrease computation time in many folds.

Time complexity of sorting at Step 3 with n data elements [19]

$$O(n\log n).$$

For the Step 4 to find initial centroids, time complexity for segregating the n data items into k parts and deciding the mean of each part is O(n). Thus, the total time complexity for discovering the initial centroids of a data set containing n elements and m attributes (where m is way less than n) is

$$O(n\log n).$$

Step 5 is again a partitioning procedure having complexity [22]

$$O(n\log n).$$

Steps 6–10 are same as the original K-means algorithm and hence take time

$$O(nKR).$$

where n is the number of data points, K is the number of clusters, and R is the number of iterations. The algorithm converges in very less number of iterations as the initial centroids are calculated in a clever method in harmony with the data dispersion.

So, the general complexity of SKIK is the maximum of

$$\{O(nlogn) + O(nlogn) + O(nlogn) + O(nKR)\}$$
$$\text{i.e., } O(nlogn + nKR)$$
$$\text{i.e., } O(n(logn + KR))$$

5 Conclusion

This paper has provided a detailed comparison among six most popular data mining algorithms which have significant contribution in high-performance cluster computation and artificial intelligence. The algorithms are Apriori, DIC, GSP, SPADE, SPRINT, and K-means. The paper presents short algorithmic steps about the main algorithms, explanation of their features, and respective advantages and disadvantages. Several variations of the algorithms exist, and they have been proved to be suitable based on certain scenarios. In the present days, research has been progressing with the most effective data mining algorithms, applicable with parallel and high-performance cloud computing, like SPRINT and K-means. We have proposed SKIK algorithm to improve K-means algorithm to be used with large data set and HPC architecture. We have shown the measurement of complexity of SKIK as well. In-depth research work needs to be conducted for extending the capabilities and complete performance analysis of the SKIK algorithm with respect to other available variations of K-means algorithm.

References

1. Arefin, A.: Introduction to High Performance Computing (HPC) Clusters. https://learn.scientificprogramming.io/introduction-to-high-performance-computing-hpc-clusters-9189e9daba5a
2. Zaki, M.: Parallel and Distributed Data Mining: An Introduction. LNCS, vol. 1759, p. 3 (2002)
3. INSOFE: Apriori Algorithm (2014). https://www.slideshare.net/INSOFE/apriori-algorithm-36054672
4. Brin, M., Ullman, T.: Dynamic itemset counting and implication rules for market basket data. SIGMOD Rec. **6**(2), 255–264 (1997)
5. Mining Sequential Patterns. https://www.slideshare.net/Krish_ver2/53-mining-sequential-patterns?next_slideshow=1, slide 7
6. Zaki, M.J.: Spade: an efficient algorithm for mining frequent sequences. In: Machine Learning, vol. 42, pp. 31–60 (2001)
7. Wang, Z., et al.: A searching method of candidate segmentation point in SPRINT classification. J. ECE **2016**, Article ID 2168478 (2016)
8. k-means clustering algorithm. https://sites.google.com/site/dataclusteringalgorithms/k-means-clustering-algorithm
9. Kanungo, T., Mount, D.M., Netanyahu, N.S., Piatko, C.D., Silverman, R., Wu, A.Y.: An efficient k-means clustering algorithm: analysis and implementation. IEEE Trans. Pattern Anal. Mach. Intell. **24**(7) (2002)
10. Jain, R.: A beginner's tutorial on the apriori algorithm in data mining with R implementation. http://blog.hackerearth.com/beginners-tutorial-apriori-algorithm-data-mining-r-implementation

11. Brin, S., Motwani, R., Ullman, J.D., Tsur, S.: Dynamic itemset counting and implication rules for market basket data. SIGMOD Rec. **6**(2), 255–264 (1997)
12. Grover, N.: Comparative study of various sequential pattern mining algorithms. Int. J. Comput. Appl. (0975–8887) **90**(17) (2014)
13. Fournier-Viger, P., et al.: A survey of sequential pattern mining. Ubiquitous Int. **1**(1), 6 (2017)
14. Slimani, T., Lazzez, A.: Sequential mining: patterns and algorithms analysis. IJCER **2**(5), 4 (2013)
15. Georgia Institute of Technology: Sequential Pattern Mining. https://www.cc.gatech.edu/~hic/CS7616/pdf/lecture13.pdf
16. Shafer, J., Agrawal, R., Mehta, M.: SPRINT: a scalable parallel classifier for data mining. In: Proceedings of the 22nd VLDB Conference, Mumbai, India, pp. 544–555 (1996)
17. Ding, Y., Zheng, Z., Ma, R.: Improved SPRINT algorithm and its application in the physical data analysis. TELKOMNIKA Indones. J. Electr. Eng. **12**(9), 6909–6920 (2014)
18. Santini, M.: Advantages & Disadvantages of k-Means and Hierarchical Clustering (Unsupervised Learning), ML4LT, p. 3. Department of Linguistics and Philology, Uppsala University (2016)
19. Zhenhua, W., Zhifeng, L., Guoliang, L.: Parallel optimization strategy of heap sort algorithm under multi-core environment. In: 7th ICMTMA, June 2015. https://doi.org/10.1109/icmtma.2015.190
20. Baswade, M., Nalwade, P.S.: Selection of initial centroids for k-means algorithm. IJCSMC **2**(7), 161–164 (2013)
21. Ray, S., Turi, R.H.: Determination of number of clusters in K-means clustering and application in colour image segmentation. In: The 4th International Conference on Advances in Pattern Recognition and Digital Techniques, pp. 137–143 (1999)
22. Kim, D.J., Park, Y.W.: A novel validity index for determination of the optimal number of clusters. IEICE Trans. Inf. **E84-D**(2), 281–285 (2001)
23. Ortega, J.P., Rocio, M.D., Rojas, B., Garcia, M.J.S.: Research issues on K-means algorithm: an experimental trial using Matlab. In: CEUR Workshop Proceedings, vol. 534 (2009)
24. Tan, S., Ghosh, K.: The k-Means Algorithm—Notes
25. Qin, J., Fu, W., Gao, H., Zheng, W.X.: Distributed k-means algorithm and fuzzy c-means algorithm for sensor networks based on multiagent consensus theory. IEEE Trans. Cybern. **47**(3), 772–783 (2017)
26. Xu, Y.H., et al: Implementation of the K-Means Algorithm on Heterogeneous Devices: A Use Case Based on an Industrial Dataset. Advances in Parallel Computing, vol. 32: Parallel Computing is Everywhere, pp. 642–651. IOS Press (2018)
27. Qi, H., Di, X., Li, J., Ma, H.: Improved K-means algorithm and its application to vehicle steering identification. In: Advanced Hybrid Information Processing, pp. 378–386. Springer International Publishing (2018)
28. Goel, L., Jain, N., Srivastava, S.: A novel PSO based algorithm to find initial seeds for the k-means clustering algorithm. In: Proceedings of the Communication and Computing Systems ICCCS 2016, Gurgaon, India, p. 159
29. Bai, L., Cheng, X., Liang, J., Shen, H., Guo, Y.: Fast density clustering strategies based on the k-means algorithm. Pattern Recognit. **71**, 375–386 (2017)

Personalized Product Recommendation Using Aspect-Based Opinion Mining of Reviews

Anand S. Tewari, Raunak Jain, Jyoti P. Singh and Asim G. Barman

Abstract Recently, recommender systems have been popularly used to handle massive data collected from applications such as movies, music, news, books, and research articles in a very efficient way. In practice, users generally prefer to take other people's opinions before buying or using any product. A rating is a numerical ranking of items based on a parallel estimation of their quality, standards, and performance. Ratings do not elaborate many things about the product. On the contrary, reviews are formal text evaluation of products where reviewers freely mention pros and cons. Reviews are more important as they provide insight and help in making informed decisions. Today the internet works as an exceptional originator of consumer reviews. The amount of opinionated data is increasing speedily, which is making it impractical for users to read all reviews to come to a conclusion. The proposed approach uses opinion mining which analyzes reviews and extracts different products features. Every user does not have the same preference for every feature. Some users prefer one feature, while some go for other features of the product. The proposed approach finds users' inclination toward different features of products and based on that analysis it recommends products to users.

Keywords Opinion mining · Product aspect · Aspect extraction · Recommender system

1 Introduction

In today's data rich era, new information is added at faster speed compared to the rate at which that data is being used by users to gain some valuable information. The difference between these speeds can be fixed by recommender systems (RS).

A. S. Tewari · R. Jain (✉) · J. P. Singh
Department of CSE, National Institute of Technology Patna, Patna, India
e-mail: rjraunakjain03@gmail.com

A. G. Barman
Department of ME, National Institute of Technology Patna, Patna, India

M. Chakraborty et al. (eds.), *Proceedings of International Ethical Hacking Conference 2018*, Advances in Intelligent Systems and Computing 811,
https://doi.org/10.1007/978-981-13-1544-2_36

RS mainly suggest items to an individual based on his past actions like, ratings given to different items, search history and similar users' activities [1].

Most common RS techniques are content-based filtering and collaborative filtering [2]. Content-based filtering, also known as cognitive filtering, works by comparing the content descriptions of items with the user profiles. User profiles have information about users and their taste. The technology behind *Netflix* and *Pandora* is content-based filtering [2]. On the other hand, the process of refining or evaluating products using the judgement of other people is known as collaborative filtering (CF). In recommender systems, collaborative filtering ages not more than a decade, but its basics are very old, have been used socially, i.e., sharing opinions. *Last.fm* recommendation engine uses collaborative filtering [3].

Before emergence of World Wide Web, we generally used to ask our neighbors, friends or specialists to be sure about any product before purchasing it [4]. Now in modern era, whenever a user wants to ask anything he asks cyber world. In today's world market, users have great choice among different products to consume, according to needs and interests. Reviews and ratings, available on Internet, are best way to come across other's opinion [5]. Using people's opinions about different items, RS helps target user to make mindset about different merchandises.

Ratings are always a numerical value defined in some certain fixed bounds. Usually lower ratings imply that product is not good enough but it doesn't mean that there is nothing good in that product, and it is not possible to mention this via ratings. Ratings supply a generalized summary about product. Reviews are texted opinions, having all positives and negatives about products. Reviews provide constructive criticism about products which helps consumers as well as manufacturers.

Generally, reviews are about different aspects or features of products [6]. Aspects are characteristics of the product. Few aspects are more promising than others and have great projection on decision making. It is perplexing to find aspects that are better in a particular product from reviews. The user opinionated data present on Internet are increasing massively. Users often have no choice, but to browse massive texts to find interesting information. Browsing them requires a lot of time and energy. However, investing that much time and efforts does not guarantee that one can get correct knowledge about products [7].

An automated system is required to address this problem so that knowledge can be withdrawn from that data. This paper presents an approach to manage this problem. The proposed approach recommends items based on features user is interested in, by combining collaborative filtering, opinion mining and SentiWordNet. The proposed method not only finds different features of products but also features in which user is interested, based on his past reviews. The proposed approach uses aspect-based opinion mining to find primary features of an item and then predicts score of those features from product reviews. Natural language processing is used to extract features using *Part-Of-Speech* (POS) tagging. This approach calculates score of these features using a lexical resource, i.e., SentiWordNet. These calculated scores along with information of similar users and features target user has liked are then used to recommend most probable items.

The remaining paper is structured as follows: Sect. 2 describes background knowledge and related work. Our approach, flow diagram and its details are described in Sect. 3. The practical implementation of proposed approach is defined in Sect. 4. Finally, Sect. 5 concludes the paper.

2 Background Knowledge and Related Work

2.1 Opinion Mining

Web 2.0 has given opportunity to people to write their experiences about various products, services and other things in the form of reviews at e-commerce sites, forums, blogs, etc. [8]. Now Web is full of opinionated text. However, this large amount of text has made it very challenging to extract useful data easily. Surveying these many reviews will confuse users about product as well as wastes their time. This requires a technique which can analyze these reviews and provide fruitful knowledge to users. The technique used is opinion mining.

The science of analyzing and extracting valuable information from text data is known as opinion mining. Opinion mining is a mixture of information retrieval and computational linguistics which is bothered not with the title of text, but with the opinions it conveys [9]. Generally, to perform opinion mining it is required to design automated systems using machine learning, artificial intelligence, data mining and natural language processing techniques [10]. These systems can collect and classify opinions about product available in opinionated data.

Sometimes opinion mining also referred as sentiment analysis, but sentiment analysis is different. Sentiment is defined as an attitude, thought or judgment prompted by feeling [11], whereas opinion is defined as a view, judgment formed in the mind about a particular matter [8]. The definitions indicate that an opinion is more of a person's view about something, whereas a sentiment is more of a feeling. However, in most cases opinions imply positive or negative sentiments. Sentiment analysis and opinion mining are almost same thing; however, there is minor difference between them that is opinion mining extracts and analyzes people's opinion about an entity, while sentiment analysis searches for the sentiment words/expression in a text and then analyzes it [12].

Opinion mining is useful in multiple ways for consumers as well as manufacturers [13]. Customers can make up their mind about a product that whether they should buy it or not. They get to know about pros and cons of product. Similarly, manufacturers evaluate their products using opinion mining of public reviews. Features which require modifications or improvements can easily be identified by makers using opinion mining. This helps in deciding features, products and services which are liked or disliked in a particular region. It helps businesses finding reasons behind low sales and possible solutions based on people's views [10]. Companies can anticipate market trends by tracking consumer views.

According to Pang and Lee [10], there are three major areas of opinion mining which are identification of sentiment polarity, detection of subjectivity and joint topic-sentiment analysis. Similarly, Liu [13] in his book narrates three different mining assignments. Later, he expands his grouping in his handbook [14] as sentiment and subjectivity classification, aspect-based opinion mining, sentiment analysis of comparative sentences, opinion search and retrieval and spam opinion discovery. Lastly, in his latest book [9] he specifies three generic categories of opinionated text mining as document-level, sentence-level, and phrase-level opinion mining.

In [5], authors have found semantic orientation of reviews to classify them using pointwise mutual information between bigrams and seed words. In [15], author's approach uses Naïve Bayes classification method to classify customer reviews. Limitation of this work is having incomplete training dataset and attribute independence.

2.2 Aspect-Based Opinion Mining

In some cases, document-level opinion mining or sentence-level opinion mining are fruitful, but when it is used in conclusive process, then these levels of information are not enough [8]. For example, a positive review of a product does not denote that reviewer is delighted with each and every aspect of the product. In the same way, a negative review does not say that reviewer has objection for every aspect. Typically, the reviewer mentions both positive and negative sides of item, still his broader viewpoint may be positive or negative. Indeed, document-level and sentence-level mining cannot produce detailed decisive data. So, it is required to grind opinions in-depth. In-depth excavating of reviews is aspect-based opinion mining [9]. Phrase-level mining aids in uncovering numerous features of various product from reviews.

Two main aims are there in the problem of aspect-based opinion mining: first is aspect extraction and other one is rating prediction. Bing Liu in his book [9] divided aspect extraction method into four categories: finding frequent nouns and noun phrases, using opinion and target relations, using supervised learning and using topic models. In [11] aspect-based opinion mining, approaches have been categorized into frequency-based, relation-based, and model-based types.

Most frequent aspects of product which have been discussed by many people in their reviews are recognized in frequency-based methods. Popescu and Etzioni [16] has designed an unsupervised knowledge extraction system called OPINE which mines reviews to select substantial product features. It counts the frequency of each noun and keeps only them which have value greater than the threshold. All these noun phrases now judged by calculating the pointwise mutual information between the phrase and associated discriminators. Hu and Liu [17] extracted frequent features from reviews using part-of-speech tagging. Scanffidi et al. [18] in their work compares the repetition of derived words with occurrence rate of these words.

Relation-based techniques find the correlation between the words and sentiments to identify aspects. In this type of methods, generally part-of-speech tagging is used to find aspects. Liu et al. [19] presented "opinion observer," for the visual comparison of customer reviews. It is used only for short comments. A supervised algorithm is used for feature extraction. Initially a training dataset is tagged using a POS tagger. Now actual aspects are manually identified in training dataset and replaced by a specific tag. Afterward, association rule is used to find POS patterns which are probable features. All of the derived patterns are not valuable, and some constraints are used to discard less important phrases.

In model-based methods, models are designed to pull out features. Hidden Markov model (HMM) and conditional random field (CRF) are most frequently used mathematical models based on supervised learning, and unsupervised topic modeling methodologies are probabilistic latent semantic indexing (PLSI) and latent Dirichlet allocation (LDA) [10]. Jin et al. [20] presented the model "Opinion Miner" based on HMM to find features, opinions and their polarities. The EnsembleTextHMM, a new supervised sentiment analysis using an ensemble of text-based hidden Markov models, method has been presented in [21].

An ontology-based sentiment classification method to detect features concerning financial news has been given in [22]. In [23], a text summarization technique has been proposed to determine the top-k most informative sentences about hotel from online reviews by calculating sentence importance. Authors in [24] have combined similarity and sentiment of reviews and proposed a recommendation ranking strategy to suggest items similar but superior to a query product. Techniques used for opinion mining and some simple assumptions are their limitations.

2.3 *SentiWordNet*

SentiWordNet is a publicly available lexical resource absolutely designed to support sentiment classification and opinion mining. It is the outcome of self-annotation of all the synsets of WordNet on the basis of "positivity," "negativity" and "neutrality." Three different mathematical values *Pos(s), Neg(s)* and *Obj(s),* correlated with every synsets, decide the positivity, negativity and objectivity of all the terms present in synset. Different meanings of same term may have different sentiment score. For each synset, sum of all three scores is always 1.0 and all three values lie in interval [0.0, 1.0]. It implies that a synset can have nonzero value for all three classes. This signifies that terms present in synset contain all three, opinion polarity up to a certain degree. SentiWordNet is generated in two steps: first one is semi-supervised learning step, and second one is random-walk step [25].

3 Proposed Approach

The proposed approach can be viewed as a two-phase process which is as follows:

3.1 Data Preprocessing and Opining Mining

In the first phase, customer reviews are mined and features are extracted. Major steps during this phase are as follows:

1. Perform preprocessing of reviews, which involves removal of meaningless, unwanted symbols and stop words.
2. Perform *part-of-speech* tagging on preprocessed reviews using POS tagger.
3. Based on these POS tags, perform feature extraction by finding bi-word phrases in which mostly features are represented by *noun* tags and corresponding *adjective* words are opinion about these features.
4. Store user's opinion about features, obtained in previous step, in the database. Schematic diagram of this phase is shown in Fig. 1.

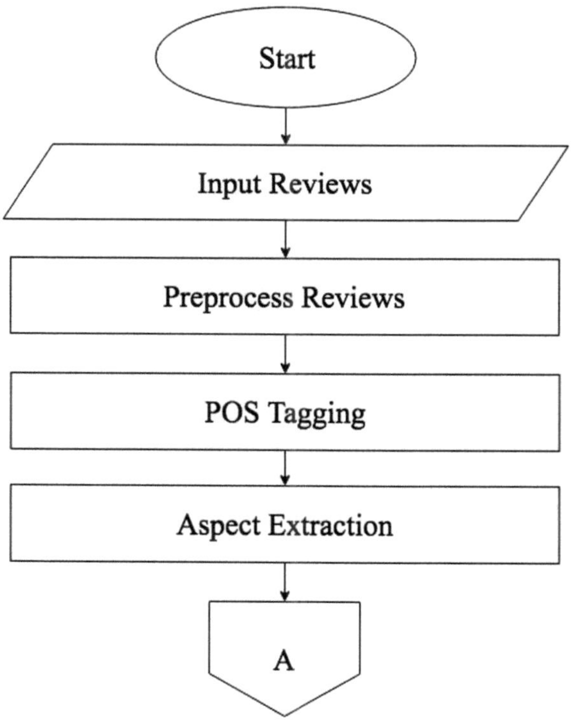

Fig. 1 Schematic diagram of the first phase

3.2 Score Calculation and Recommendation

The second phase generates personalized product recommendations for the target user, based on opinions of other similar users and features in which target user is interested. The major steps of this phase are as follows:

1. Now consider those extracted features which have frequency greater than some threshold value, specified by domain expert, across all items.
2. Calculate *average sentiment score* of each feature of every product using *Senti-WordNet.*
3. Similarly, calculate *average sentiment score* of each feature for every user and store it in database.
4. Calculate cosine-based similarity of every user with other users based on *average sentiment score* values for different features using Eq. (1).

$$\text{Similarity}\left(\overrightarrow{U_A}, \overrightarrow{U_B}\right) = \frac{\overrightarrow{U_A} \cdot \overrightarrow{U_B}}{\left|\overrightarrow{U_A}\right|\left|\overrightarrow{U_B}\right|} = \frac{\sum_{i=1}^{n} U_{A_i}.U_{B_i}}{\sqrt{\sum_{i=1}^{n} U_{A_i}^2} \cdot \sqrt{\sum_{i=1}^{n} U_{B_i}^2}} \quad (1)$$

where U_A and U_B are feature vectors of user *A* and *B*. Similarly, *i* is the number of commonly scored features between user *A* and *B*.

5. Select those products that have not been reviewed by the target user but have been reviewed by its similar users.
6. Find the occurrence frequency of each feature, discussed by the target user in his past reviews and arrange features in decreasing order of their frequency.
7. Now recommend products obtained from *step-5* in the order of scores for features in the same sequence as obtained in *step-6*. Schematic diagram of this phase is shown in Fig. 2.

4 Practical Implementation

Every review goes through preprocessing and POS tagging activity. In this work, *Stanford Core NLP* tagger has been used. For example, the output of the review *“Shoes are nice and good looking.”*, after performing preprocessing and POS tagging activity is shown in Fig. 3.

The Stanford tagger uses some standard tags to define part-of-speech of each word, such as NN and JJ. Some useful tags have been shown in (Table 1).

Based on these POS tags, features are extracted from all reviews. In this work, fifteen features which are most frequent in all reviews have been considered. After performing all steps described in proposed approach, the system recommends products, for example, let us say there are total 10 users and 10 products. After performing feature extraction, 15 features are found, but only those features are considered whose frequencies are higher and these are f_1, f_2, f_3, f_4, f_5 and f_6. Suppose the target user

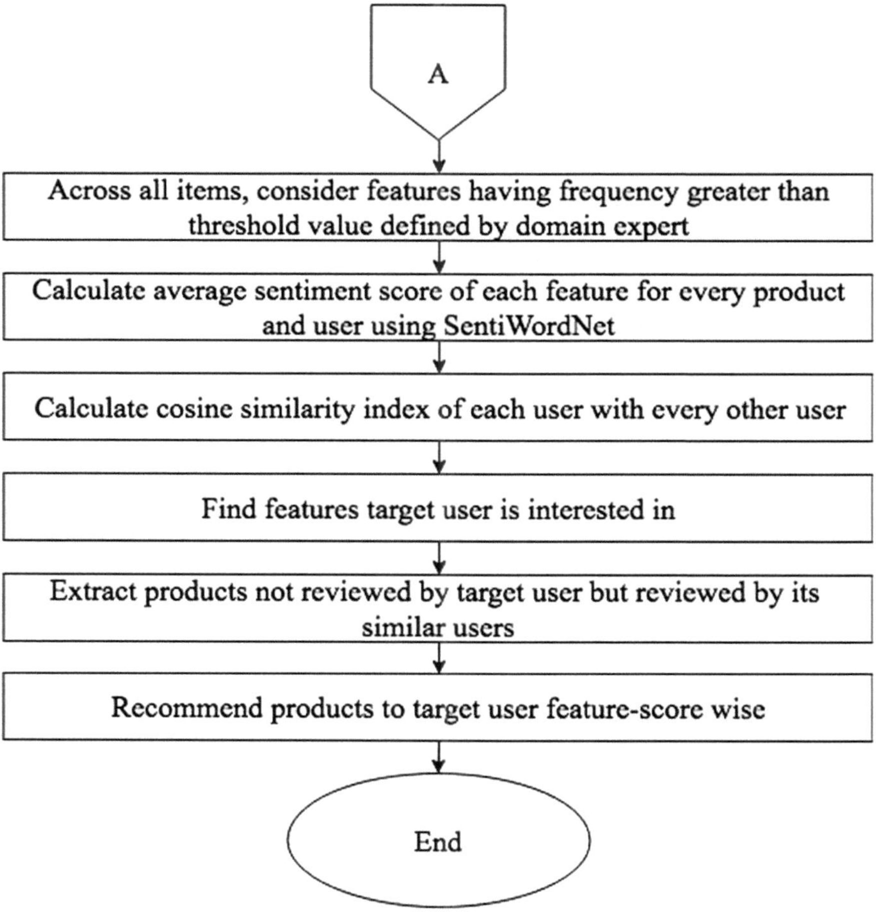

Fig. 2 Schematic diagram of the second phase

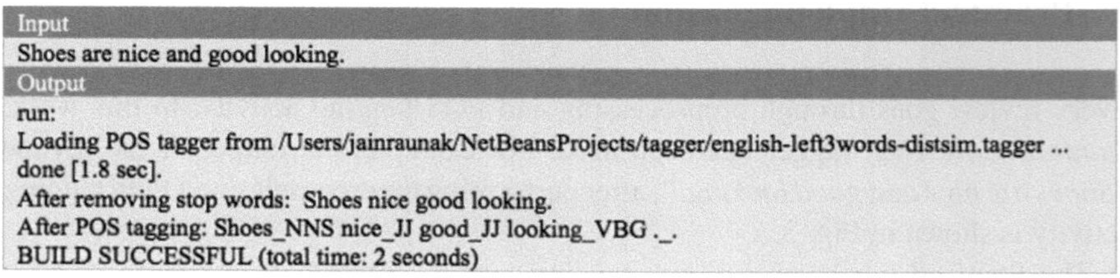

Fig. 3 Output after preprocessing and POS tagging

is u_1 and his similar users are u_3, u_5, u_8, u_9, and u_7, products obtained from *step-5* of second phase of recommendation for target user are i_8, i_1, i_2, i_6, and i_{10}. Scores of different features of these products are shown in Table 2.

Table 1 Commonly used POS tags

Tag	Description	Tag	Description
NN, NNS, NNP	Noun	JJ, JJR, JJS	Adjective
VB, VBD, VBG, VBN, VBP, VBZ	Verb	RB, RBR, RBS	Adverb

Table 2 Product-features average score

Product_Id	f_1	f_2	f_3	f_4	f_5	f_6
i1	−0.569	0.601	0.796	0.387	0.427	0.489
i2	0.544	−0.230	−0.379	−0.420	– 0.101	−0.506
i6	0.340	0.870	0.509	−0.418	−0.689	0.356
i8	−0.297	0.367	−0.489	0.378	0.768	0.601
i10	0.507	−0.601	−0.126	0.769	0.456	0.732

Table 3 Frequency count of features for target user

f_1	f_2	f_3	f_4	f_5	f_6
3	10	8	6	5	*0*

The frequency of features obtained from *step-6* of second phase of recommendation for the target user u_1 is shown in Table 3.

So using *step-6* of phase-two, the sequence obtained is $f_2, f_3, f_4, f_5, f_1, f_6$. Now *step-7* checks the scores of features in items obtained from *step-5*. So, accordingly first of all it is checked that which product has highest score for feature f_2, that product is recommended first, next checked for feature f_3 and so on. Therefore, final recommendation order of products is i_6, i_1, i_{10}, i_8, i_2.

5 Conclusion

Now almost every user is writing their experiences about different products in the form of reviews. Opinion mining handles this huge user generated opinionated data. A comparatively recent sub-area of opinion mining is aspect-based opinion mining, and it extracts products features and users' opinion about those features from reviews. In this paper, a novel approach has been proposed to recommend products to users on the basis of features on which they have shown interest in their past reviews by arranging products in the order of score of features. The proposed approach extracts features and opinions from reviews using part-of-speech tagging and then calculates their score using a publicly available lexical resource SentiWordNet. Afterward, it recommends product to the target user with the collaboration of other similar users. In future, this work can be expanded by refining feature extraction and selection method so that only those features will be considered which affect the decision making most.

As in review "The hotel is expensive," the reviewer is talking about the price of hotel but the word price is mentioned nowhere in text. As a future enhancement to this work, all such problems in aspect extraction have to be addressed.

References

1. Aggarwal, C.C.: Recommender Systems: The Textbook. 1st edn. Springer International Publishing, USA (2016)
2. Ricci, F., Rokach, L., Shapira, B.: Recommender Systems Handbook, 1st edn. Springer, USA (2011)
3. Schafer, J.B., Frankowski, D., Herlocker, J., Sen, S.: The Adaptive Web: Methods and Strategies of Web Personalization, 1st edn. Springer, Heidelberg (2007)
4. Hasan, K.A., Sabuj, M.S., Afrin, Z.: Opinion mining using Naive Bayes. In: IEEE International WIE Conference on Electrical and Computer Engineering (WIECON-ECE) 2015, pp. 511–514. IEEE, Bangladesh (2015)
5. Nakade, S.M., Deshmukh, S.N.: Observing performance measurements of unsupervised PMI algorithm. Int. J. Eng. Sci. 7563–7568 (2016)
6. Zhang, Y.: Incorporating phrase-level sentiment analysis on textual reviews for personalized recommendation. In: Proceedings of the Eighth ACM International Conference on Web Search and Data Mining, pp. 435–440. ACM, Shanghai (2015)
7. Ganu, G., Elhadad, N., Marian, A.: Beyond the stars: improving rating predictions using review text content. In: 12th International Workshop on the Web and Databases, pp. 1–6. USA (2009)
8. Tewari, A.S., Barman, A.G.: Collaborative recommendation system using dynamic content based filtering, association rule mining and opinion mining. Int. J. Intell. Eng. Syst. **10**(5), 57–66 (2017)
9. Liu, B.: Sentiment analysis and opinion mining. Synth. Lect. Human Lang. Technol. **5**(1), 1–167 (2012)
10. Pang, B., Lee, L.: Opinion mining and sentiment analysis. Foundat. Trends Informat. Retrieval **2**(1–2), 1–135 (2008)
11. Kreuz, R.J., Glucksberg, S.: How to be sarcastic: the echoic reminder theory of verbal irony. J. Exp. Psychol. Gen. **118**(4), 374–386 (1989)
12. Pang, B., Lee, L.: A sentimental education: Sentiment analysis using subjectivity summarization based on minimum cuts. In: Proceedings of the 42nd Annual Meeting on Association for Computational Linguistics 2004, pp. 271–278. Association for Computational Linguistics, Barcelona (2004)
13. Liu, B.: Web Data Mining: Exploring Hyperlinks, Contents, and Usage Data, 1st edn. Springer Science & Business Media, Heidelberg (2007)
14. Liu, B.: Handbook of Natural Language Processing, 2nd edn. Taylor & Francis, Boca Raton (2010)
15. Pisote, A., Bhuyar, V.: Review article on opinion mining using Naïve Bayes classifier. Advanc. Comput. Res. **7**(1), 259–261 (2015)
16. Popescu, A.M., Etzioni, O.: Extracting product features and opinions from reviews. In: Kao, A., Poteet, S.R. (eds.) Natural Language Processing and Text Mining 2007, pp. 9–28. Springer, London (2007)
17. Hu, M., Liu, B.: Mining and summarizing customer reviews. In: Proceedings of the Tenth ACM SIGKDD International Conference on Knowledge Discovery and Data Mining 2004, pp. 168–177. ACM, Seattle (2004)
18. Scaffidi, C., Bierhoff, K., Chang, E., Felker, M., Ng, H., Jin, C.: Red opal: product-feature scoring from reviews. In: Proceedings of the 8th ACM Conference on Electronic Commerce 2007, pp. 182–191. ACM, San Diego (2007)

19. Liu, B., Hu, M., Cheng. J.: Opinion observer: analyzing and comparing opinions on the web. In: Proceedings of the 14th International Conference on World Wide Web 2005, pp. 342–351. ACM, Chiba (2005)
20. Jin, W., Ho, H.H., Srihari, R.K.: OpinionMiner: a novel machine learning system for web opinion mining and extraction. In: Proceedings of the 15th ACM SIGKDD International Conference on Knowledge Discovery and Data Mining 2009, pp. 1195–1204. ACM, Paris (2009)
21. Kang, M., Ahn, J., Lee, K.: Opinion mining using ensemble text hidden Markov models for text classification. Exp. Syst. Appl. **94**(2018), 218–227 (2018)
22. Salas-Zárate, M.D.P., Valencia-García, R., Ruiz-Martínez, A., Colomo-Palacios, R.: Feature-based opinion mining in financial news: an ontology-driven approach. J. Informat. Sci. **43**(4), 458–479 (2017)
23. Hu, Y.H., Chen, Y.L., Chou, H.L.: Opinion mining from online hotel reviews—a text summarization approach. Inf. Process. Manag. **53**(2), 436–449 (2017)
24. Dong, R., O'Mahony, M.P., Schaal, M., McCarthy, K., Smyth, B.: Combining similarity and sentiment in opinion mining for product recommendation. J. Intell. Informat. Syst. **46**(2), 285–312 (2016)
25. Baccianella, S., Esuli, A., Sebastiani, F.: Sentiwordnet 3.0: an enhanced lexical resource for sentiment analysis and opinion mining. In: Calzolari, N., Choukri, K. (eds.) Proceedings of the Seventh conference on International Language Resources and Evaluation, 2010, LREC, vol. 10, pp. 2200–2204. ELRA Malta (2010)

Quantitative Rainfall Prediction: Deep Neural Network-Based Approach

Debraj Dhar, Sougato Bagchi, Chayan Kumar Kayal, Soham Mukherjee and Sankhadeep Chatterjee

Abstract Forecasting the weather has always been a challenge using conventional methods of climatology, analogue and numerical weather prediction. To improvise the prediction of weather much further, the proposed method can be used. In this work, authors proposed a method which uses the advantages of deep neural network to achieve high degree of performance and accuracy compared to the old conventional ways of forecasting the weather. It is done by feeding the perceptrons of the DNN some specific features like temperature, relative humidity, vapor and pressure. The output generated is a highly accurate amount of the rainfall based on the given input data.

Keywords Forecast · Rainfall prediction · Deep neural network · Perceptrons

1 Introduction

A meteorologist's biggest job is to predict how the weather will change depending upon the climate changing parameters and it has always been a challenging job to predict this on a higher scale of accuracy [1]. Using DNN (deep neural network) [2], the task of predicting the weather [3] can be achieved with much greater accuracy.

D. Dhar · S. Bagchi · C. K. Kayal · S. Mukherjee · S. Chatterjee (✉)
Department of Computer Science & Engineering, University of Engineering & Management, Kolkata, India
e-mail: chatterjeesankhadeep.cu@gmail.com

D. Dhar
e-mail: debrajdhar100@gmail.com

S. Bagchi
e-mail: sougato97@gmail.com

C. K. Kayal
e-mail: chayankayal32@gmail.com

S. Mukherjee
e-mail: mukherjee.soham56@gmail.com

M. Chakraborty et al. (eds.), *Proceedings of International Ethical Hacking Conference 2018*, Advances in Intelligent Systems and Computing 811,
https://doi.org/10.1007/978-981-13-1544-2_37

The rapid growth in technological development has made it possible to use DNN model to forecast weather at any given point of time.

An artificial neural network (ANN) [4–8] mainly consists of an input layer, an output layer and a hidden layer. A hidden layer contains many neurons also known as perceptron. Features or attributes act as the input for the input layers, and output layer generates the result, while hidden layer performs various operations and finds the co-relation among the attributes to produce the output [9–18]. In the proposed work, authors have compared this DNN architecture with other conventional approaches like multiple linear regression model (MLR) and ANN to establish the effectiveness of the DNN model.

Multiple linear regression is the most common form of linear regression analysis. As a predictive analysis, the multiple linear regression is used to explain the relationship between one continuous dependent variable and two or more independent variables. In traditional ways of predicting the rainfall, MLR was often used.

In the current study, a deep neural network model-based rainfall prediction model has been proposed. The problem has been framed as a regression problem. The proposed model is then compared with well-known ANN and MLR models. Experimental results indicated that DNN could be a suitable choice over the traditional models.

2 Proposed Method

In this work, we have made a DNN regressor model [2], consisting of six hidden layers and each layer having 100 neurons. The number of hidden layers and neurons taken are completely arbitrary. At first, all the features, i.e., the parameters responsible for weather prediction are given to the Input layer to train the regressor model. The following steps are taken as a basic flow of experiments:

1. Preprocessing: The following preprocessing is done in the dataset before the regression.
   - (i) Feature Extraction—This step involves extraction of significant features which are most important in regression using the statistical correlation method.
   - (ii) Data Cleaning—The dataset might contain missing or inconsistent values. To deal with such issues, statistical methods are used in this step of preprocessing.
   - iii) Data Normalization—The normalization of the dataset is carried out to reduce the distance between attribute values. It is generally achieved by keeping the value range in between -1 and $+1$.
2. After the preprocessing of the database, the database was then split into training and testing set in a ratio of 70:30. Both the training and testing set data had been shuffled to get the optimal result at the end of the experiment.

3. In the training phase, the training dataset was supplied to the DNN regressor model authors had previously built to train the model. The optimizer used to reduce the generated error was "Adagrad" optimizer, and the activation function used in the work was "RELU" with a learning rate of 0.03.
4. In the testing phase, the test dataset was supplied to the DNN model to evaluate the performance of the whole process.

To measure the performance, some statistical approaches were used to get the error loss and accuracy of the model. Figure 1 shows a flowchart of the working method used by the authors.

3 Dataset Description

The dataset of our work has a total of 2485 rows and 14 columns. It contains information regarding the rainfall over a period of 7 years from 1989 to 1995. There are a total of 14 attributes in the dataset. Out of those 14 attributes, we have used eight attributes like temperature, vapor, relative humidity and pressure, which are some significant features in forecasting the rainfall. Figures 2, 3, 4 and 5 depict the relation between the different attributes and the rainfall based on the actual dataset.

4 Results and Discussions

In the current study, rainfall prediction has been done using deep neural network. The problem has been framed as a regression problem. For experimental purpose, well-known Tensorflow [16] library has been used. After training a total 1738 rows in a shuffled order for 5000 steps, the DNN model produced a final loss of 0.00589925. During the evaluation of the test dataset, the model evaluated as following:

Average loss	0.0041428246
Global steps	5000
Loss	0.03287816

Figure 6 depicts the plot between all the features (each color representing one feature) and true values of rainfall, and Fig. 7 depicts the plot between all the features and the predicted rainfall values. In Fig. 8, original values of the rainfall and predicted values of the rainfall have been plotted.

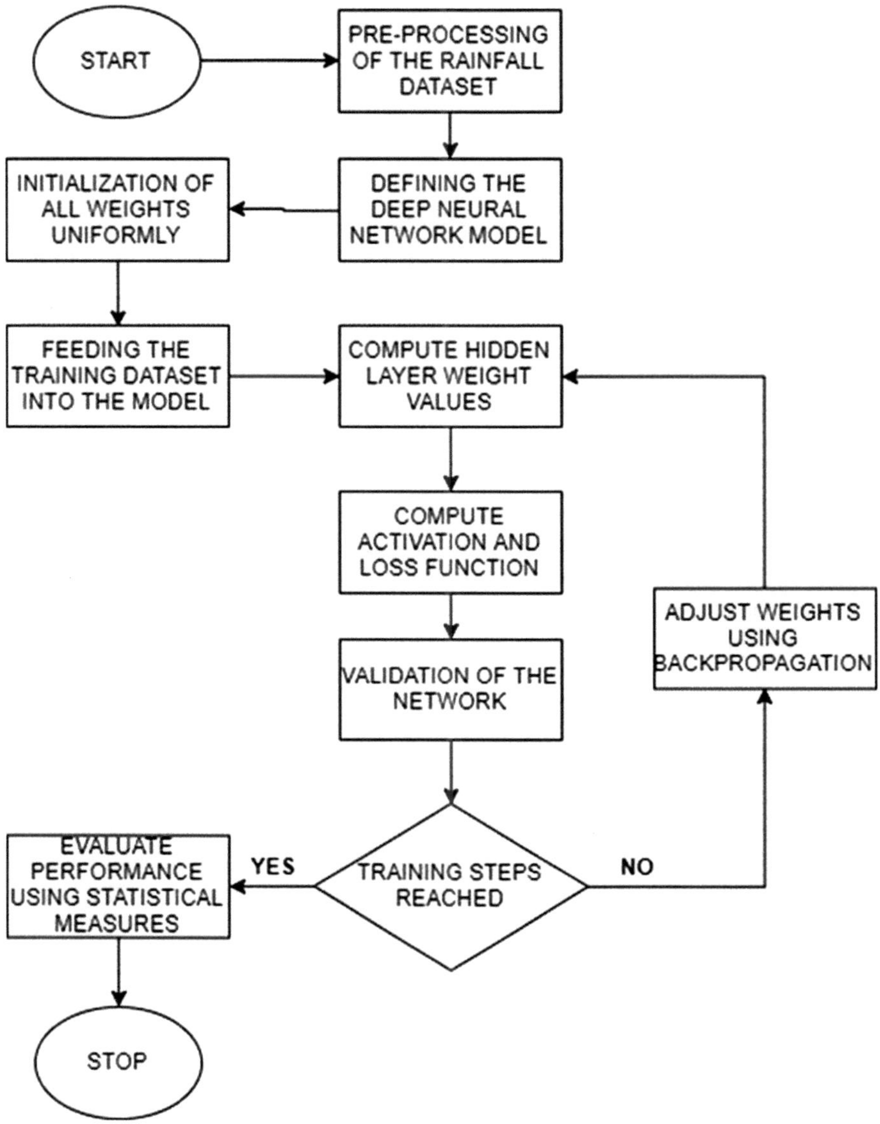

Fig. 1 Flowchart of the methodology of the DNN model

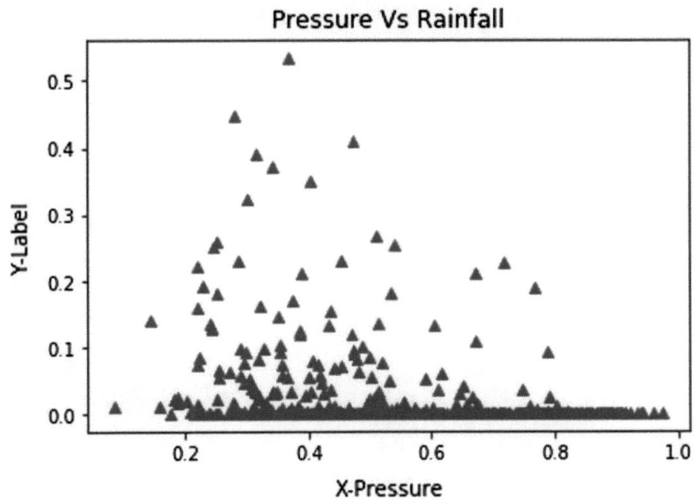

Fig. 2 Plot of pressure versus rainfall

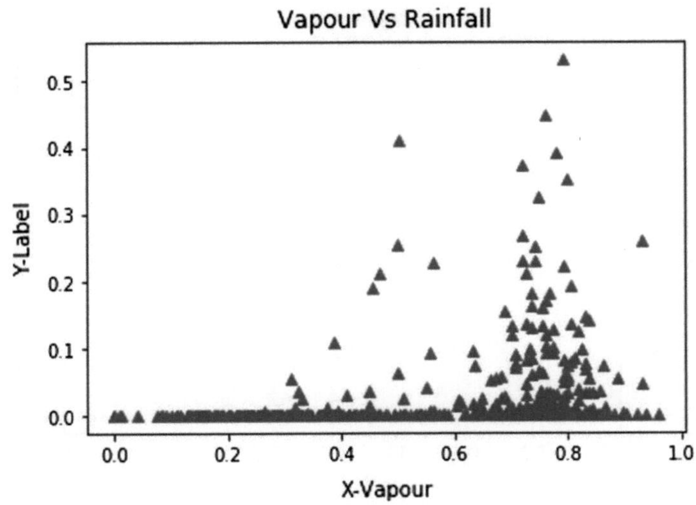

Fig. 3 Plot of vapor versus rainfall

The min, max, mean and standard deviation of estimated and observed rainfall are shown in Table 1. These statistics are more or less similar. After comparing the results of DNN model to an ANN Model and a traditional MLR model in terms of MSE, RMSE and R^2 Parameters, the results have been reported in Table 2 which reveals that the loss on the DNN model is lower compared to both the ANN and the MLR model.

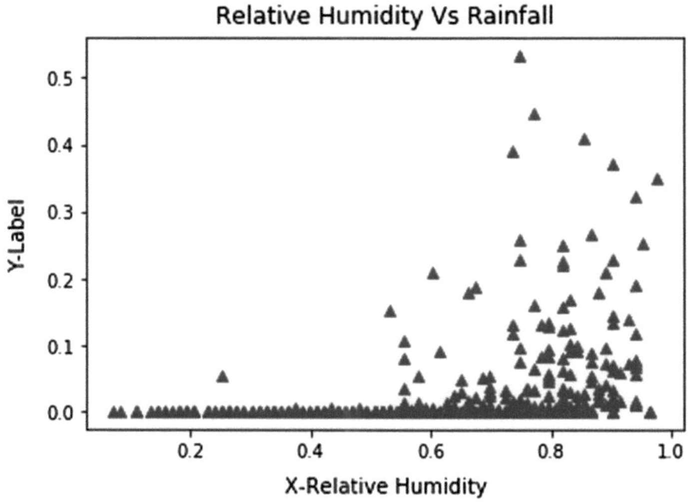

Fig. 4 Plot of relative humidity versus rainfall

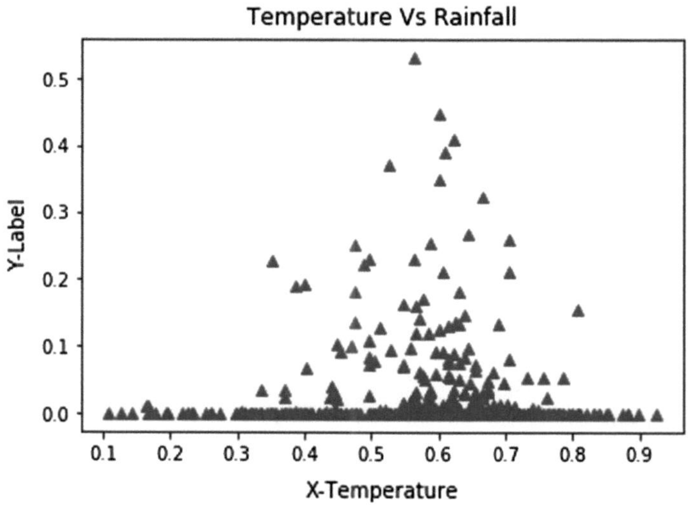

Fig. 5 Plot of temperature versus rainfall

5 Conclusion

From all the experiments conducted over this dataset with the DNN regressor model, the results have been proven to be of satisfactory amount of improvement over the existing forecasting methods of numerical and statistical predicting methods. But even though the loss is much less than conventional models, the implementation of this model for large datasets will require a lot of processing power to do so.

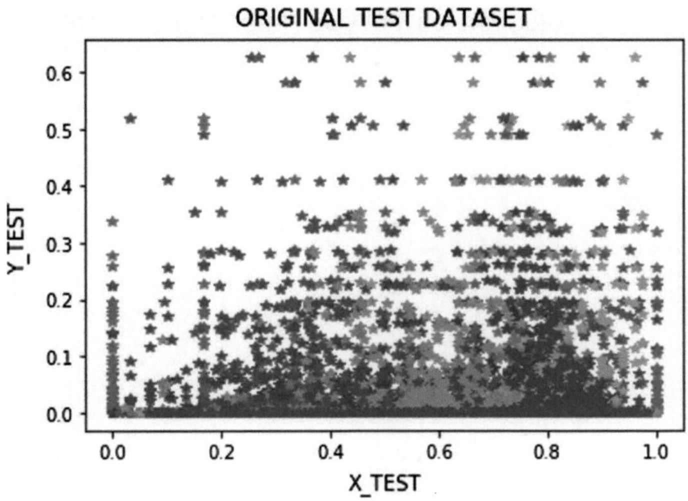

Fig. 6 Plot of the original test dataset

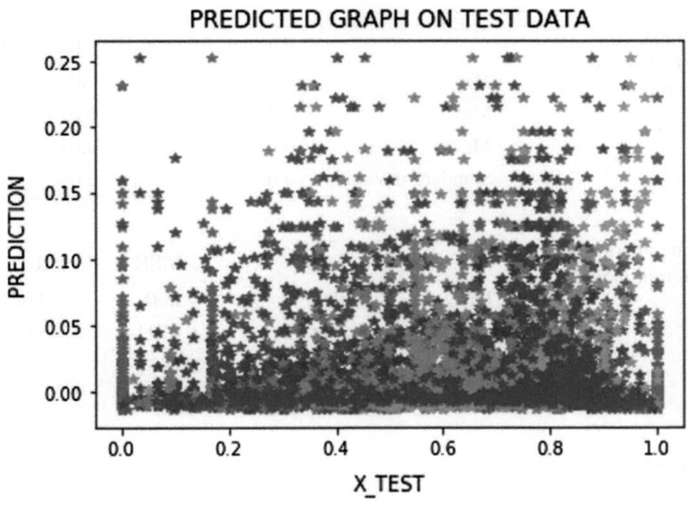

Fig. 7 Plot of the predicted test dataset

The present work has analyzed the performances of the model in terms of several performance measuring parameters like mse, rmse, r-squared to provide a vivid picture of the exact performance of the DNN model. In future, using DNN model, other weather parameters like radiation can also be calculated.

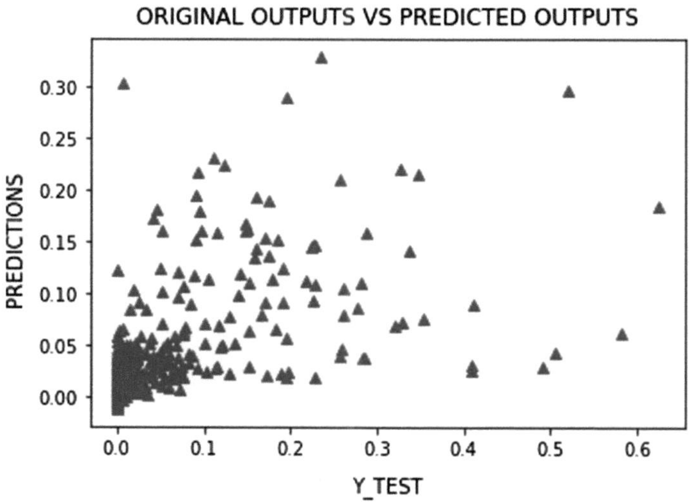

Fig. 8 Plot of the original versus test dataset

Table 1 Comparison between estimated and observed rainfall

	Estimated rainfall	Observed rainfall
Min	0.002	0.0
Max	0.52	0.62
Mean	0.03	0.03
Standard deviation	0.05	0.07

Table 2 Comparison between different models

	MSE	RMSE	R^2
DNN	0.0037	0.0609	37.51
ANN	0.0052	0.0721	39.73
MLR	0.0098	0.0989	35.24

References

1. Gong, P., Howarth, P.J.: Frequency-based contextual classification and gray-level vector reduction for land-use identification. Photogrammetric Engineering and Remote Sensing, 58, pp. 423–437 (1992). https://ci.nii.ac.jp/naid/80006391998/
2. Xu, Y., Du, J., Dai, L. R., Lee, C.H.: A regression approach to speech enhancement based on deep neural networks. IEEE/ACM Trans. Audio, Speech Lang. Process. (TASLP), **23**(1), 7–19 (2015)
3. Dalto, M., Matuško, J., Vašak, M.: Deep neural networks for ultra-short-term wind forecasting. In; 2015 IEEE International Conference on Industrial Technology (ICIT), pp. 1657–1663. IEEE (2015)
4. Chatterjee, S., Sarkar, S., Hore, S., Dey, N., Ashour, A.S., Shi, F., Le, D.N.: Structural failure classification for reinforced concrete buildings using trained neural network based multi-objective genetic algorithm. Struct. Eng. Mech. **63**(4), 429–438 (2017)

5. Chatterjee, S., Dey, N., Shi, F., Ashour, A.S., Fong, S.J., Sen, S.: Clinical application of modified bag-of-features coupled with hybrid neural-based classifier in dengue fever classification using gene expression data. Med. Biol. Eng. Comput. 1–12 (2017)
6. Chatterjee, S., Sarkar, S., Dey, N., Ashour, A.S., Sen, S., Hassanien, A.E.: Application of cuckoo search in water quality prediction using artificial neural network. Int. J. Comput. Intell. Stud. **6**(2–3), 229–244 (2017)
7. Chatterjee, S., Banerjee, S., Mazumdar, K.G., Bose, S., Sen, S.: Non-dominated sorting genetic algorithm—II supported neural network in classifying forest types. In: 2017 1st International Conference on Electronics, Materials Engineering and Nano-Technology (IEMENTech), pp. 1–6. IEEE (2017)
8. Chatterjee, S., Banerjee, S., Basu, P., Debnath, M., Sen, S.: Cuckoo search coupled artificial neural network in detection of chronic kidney disease. In: 2017 1st International Conference on Electronics, Materials Engineering and Nano-Technology (IEMENTech), pp. 1–4. IEEE (2017)
9. Chatterjee, S., Dey, N., Ashour, A.S., Drugarin, C.V.A.: Electrical energy output prediction using cuckoo search based artificial neural network. In: Smart Trends in Systems, Security and Sustainability, pp. 277–285. Springer, Singapore (2017)
10. Chakraborty, S., Dey, N., Chatterjee, S., Ashour, A.S.: Gradient Approximation in Retinal Blood Vessel Segmentation
11. Chatterjee, S., Sarkar, S., Dey, N., Ashour, A.S., Sen, S.: Hybrid Non-dominated sorting genetic algorithm: II-neural network approach. Advanc. Appl. Metaheuristic Comput. 264 (2017)
12. Chatterjee, S., Sarkar, S., Hore, S., Dey, N., Ashour, A.S., Balas, V.E.: Particle swarm optimization trained neural network for structural failure prediction of multistoried RC buildings. Neural Comput. Appl. **28**(8), 2005–2016 (2017)
13. Chatterjee, S., Ghosh, S., Dawn, S., Hore, S., Dey, N.: Forest type classification: a hybrid NN-GA model based approach. In: Information Systems Design and intelligent applications, pp. 227–236. Springer, New Delhi
14. Chatterjee, S., Hore, S., Dey, N., Chakraborty, S., Ashour, A.S.: Dengue fever classification using gene expression data: a PSO based artificial neural network approach. In: Proceedings of the 5th International Conference on Frontiers in Intelligent Computing: Theory and Applications, pp. 331–341. Springer, Singapore (2017)
15. Chatterjee, S., Datta, B., Sen, S., Dey, N., Debnath, N.C.: Rainfall prediction using hybrid neural network approach. In; 2nd International Conference on Recent Advances in Signal Processing, Telecommunications & Computing (SigTelCom)—2018, Vietnam (In press)
16. Abadi, M., Barham, P., Chen, J., Chen, Z., Davis, A., Dean, J., Kudlur, M.: TensorFlow: a system for large-scale machine learning. In: OSDI, vol. 16, pp. 265–283
17. Mathie, M., Coster, A., Lovell, N., Celler, B.: Accelerometry: Providing an integrated, practical method for long-term, ambulatory monitoring of human movement. Physiol. Meas. **25**, 1–20 (2004)
18. Foerster, F., Fahrenberg, J.: Motion pattern and posture: correctly assessed by calibrated accelerometers. Behav. Res. Methods Instrum. Comput. **32**(3), 450–457 (2000)

Prediction of Benzene Concentration of Air in Urban Area Using Deep Neural Network

Radhika Ray, Siddhartha Haldar, Subhadeep Biswas, Ruptirtha Mukherjee, Shayan Banerjee and Sankhadeep Chatterjee

Abstract Recent studies have revealed the adverse effect of benzene as an air pollutant. Benzene has been proved to be causing several health hazards in unbar areas. Researchers have employed machine learning methods to predict the available benzene concentration in a particular area. Motivated by the recent advancements in the field of machine learning, the authors have proposed a deep learning-based model to predict benzene quantity in order to determine the quality of air as well. Benzene quantity prediction in the atmosphere has been accomplished with respect to certain specified elements (like carbon monoxide, PT08.S1, PT08.S2) that coexist along with benzene (C_6H_6). A feature selection stage has been employed using correlation analysis to find the most suitable set of features. Six features have been selected for the experimental purpose. Further, the proposed model has been compared with well-known machine learning models such as linear regression, polynomial regression, K-nearest neighbor, multilayer perceptron feedforward network (MLP-FFN) in terms of RMSE. Experimental results have suggested that the proposed deep learning-based model is superior to the other models under current study.

Keywords Deep learning · Benzene prediction · Air quality

R. Ray · S. Haldar · S. Biswas · R. Mukherjee · S. Banerjee (✉) · S. Chatterjee (✉)
Department of Computer Science & Engineering, University of Engineering & Management, Kolkata, India
e-mail: shayanbanerjee96@gmail.com

S. Chatterjee
e-mail: chatterjeesankhadeep.cu@gmail.com

R. Ray
e-mail: rray6797@gmail.com

S. Haldar
e-mail: sidhaldar98@gmail.com

S. Biswas
e-mail: subhadeepbiswas250@gmail.com

R. Mukherjee
e-mail: rupje65@gmail.com

M. Chakraborty et al. (eds.), *Proceedings of International Ethical Hacking Conference 2018*, Advances in Intelligent Systems and Computing 811,
https://doi.org/10.1007/978-981-13-1544-2_38

1 Introduction

Deep learning is an increasingly growing topic of discussion in both workplace and outside of it. It has the potential to change or affect not only the way we live but also how we work. Learning can be supervised, semi-supervised, or unsupervised. Machine learning [1–6] has been found to be suitable for different real-life problem-solving. There are many existing models of machine learning, out of which linear regression, KNN regression, polynomial regression, and multilayered perceptron solve the purpose of this paper. Due to the improved data processing models, deep learning generates actionable results when solving data science tasks. The capacity to determine the most important features allows deep learning to efficiently provide data scientists with concise and reliable analyzed results.

Nikolaos G. Paterakis et al. have drawn a comparison between traditional machine learning approaches and the more advanced deep learning methods. The use of multilayer perceptrons enhanced with deep learning potential has been proposed as well. The results thus obtained from the comparisons show that multilayer perceptrons outperform the other traditional methods like support vector machine, linear regression. Krishnan et al. have examined the relationship between total dissolved solids (TDS), pH, and conductive water quality parameters (WQPs) which are involved in detecting the hexavalent chromium contamination in the drinking water distribution system. Multiple linear regression model was used to determine the correlation between the actual and estimated WQPs. It was found that errors between the estimated and the actual WQPs lie between 0.33 and 19.18%. Wint Thida Zaw et al. have proposed the use of multivariable polynomial regression (MPR) in implementing the precipitation model over Myanmar. The results obtained from the proposed model were compared with the results obtained from a multiple linear regression model (MLR), and it was found that MPR achieved a closer agreement between the actual and the estimated rainfall than MLR. Liming Zhong et al. have presented a k-nearest neighbor (KNN) regression method for predicting CT image from MRI data, where the nearest neighbors of the individual MR image patch were searched in the constraint spatial range. To further improve the accuracy, the use of supervised descriptor learning method has also been proposed. The results thus obtained show that the proposed method is more effective at predicting CT images from MRI data than two state-of-the-art methods for CT prediction. Mladen Dalto et al. have presented the use of deep neural networks and input variable selection algorithm for ultra-short-term wind forecasting. For a set of different locations, shallow and deep neural networks along with input variable selection algorithm were compared on predicting the ultra-short-term wind. Results showed that carefully selected deep neural networks outperformed the shallow ones.

2 Proposed Methodology

2.1 Data Analysis and Data Collection

To evaluate the performance of various prediction methods, a large dataset provided by ENEA—National Agency for New Technologies, Energy and Sustainable Economic Development was employed. The dataset contains the responses of a gas multisensory device deployed on the field in an Italian city. It contains 9358 instances of hourly averaged responses from an array of five metal oxide chemical sensors embedded in an air quality chemical multisensory device. The device was located on the field in a significantly polluted area, at road level, within an Italian city. Data were recorded from March 2004 to February 2005 (one year), representing the longest freely available recordings of on-field deployed air quality chemical sensor devices' responses. The attributes that were present in the dataset are as follows:

i. Date (DD/MM/YYYY)
ii. Time (HH.MM.SS)
iii. True hourly averaged concentration Co in mg/mg^3
iv. PT08.S1 (tin oxide) hourly averaged sensor response
v. True hourly averaged overall non-methane hydrocarbons concentration in mg/mg^3
vi. True hourly averaged benzene concentration in mg/mg^3
vii. PT08.S2 (titania) hourly averaged sensor response
viii. True hourly averaged NOx concentration in ppb
ix. PT08.S3 (tungsten oxide) hourly averaged sensor response
x. True hourly averaged NO_2 concentration in mg/mg^3
xi. PT08.S4 (tungsten oxide) hourly averaged sensor response
xii. PT08.S5 (indium oxide) hourly averaged sensor response
xiii. Temperature in Â°C
xiv. Relative Humidity (%)
xv. Absolute Humidity (AH)

The date, time was not needed for the evaluation process and therefore was dropped from the dataset. Moreover, the non-methane hydrocarbon records were mostly missing and −200 (undesirable) values and were also dropped from the dataset. Last 114 records of all the attributes of the dataset were NaN values and were also dropped. In Fig. 1, the heat map shows the correlation of different attributes among themselves. The correlation map helps in selecting the attributes that will make the most difference while evaluation. The attributes PT08.S3 (NOx), NO_2(GT), Temperature (T), Relative Humidity (RH), and Absolute Humidity (AH) show least relation with benzene (C_6H_6), i.e., less than 0.6, and were finally dropped from the dataset to produce the desired dataset. Figure 2 shows the "heat map" which describes the correlation among the selected attributes. The resultant six attributes except benzene will be fed into the day-ahead prediction model as inputs, whereas benzene (C_6H_6) being the output.

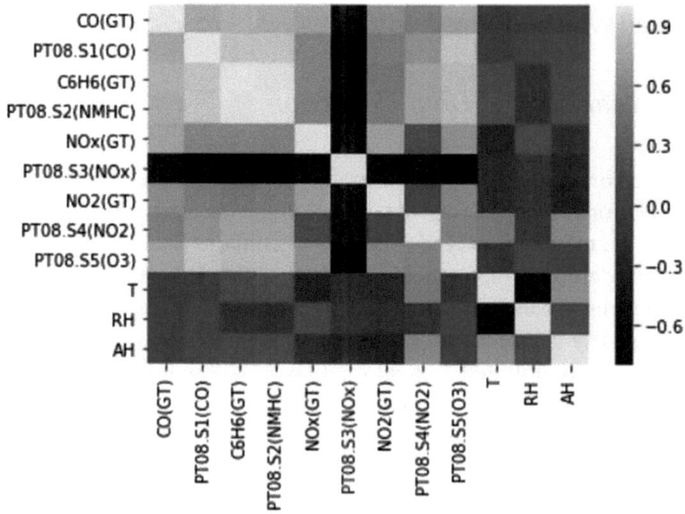

Fig. 1 Correlation result of ENEA dataset

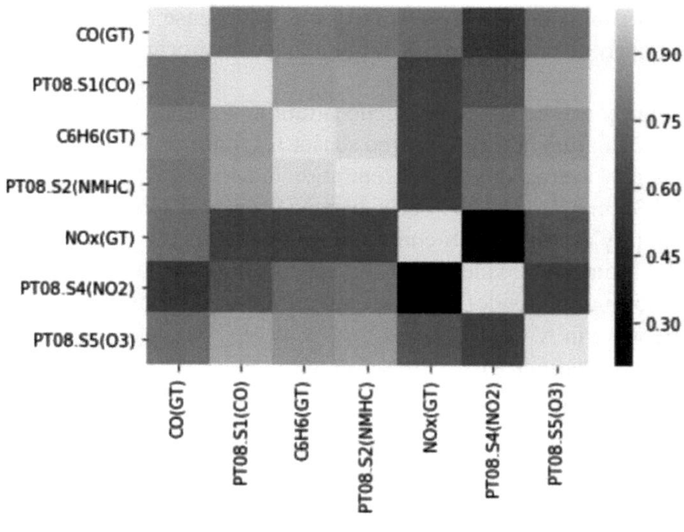

Fig. 2 Correlation result of the cleaned ENEA dataset

2.2 *Prediction Models*

2.2.1 Linear Regression Model

Linear regression [7, 8] is one of the basic and the most commonly used algorithms in machine learning. The linear regression model estimates the linear relationship between the independent variables (features) and the dependent or target variable (output). The model follows a linear equation which assigns a scale factor to each of the input variables and an additional coefficient called the bias to give an additional degree of freedom to the line. For a single input and a single output, the equation is as follows:

$$y = \beta 0 + \beta 1 * x$$

Once the model learns the coefficients, they can be used to predict the output values for specific input values. The best way to predict these coefficients is to minimize the cost function. This is done by using an efficient optimization algorithm called the gradient descent, which helps the model to make changes to the coefficients in the equation in such a way so that the error (difference between actual output and predicted output) is minimized.

2.2.2 Polynomial Regression Model

Polynomial regression [9] is a form of regression analysis in which the relationship between the independent variable x and the dependent variable y is modeled as an nth degree polynomial in x. For example:

$$y = \beta 0 + \beta 1x + \beta 1x2 + \varepsilon$$

The main difference between polynomial regression and multiple linear regression is the number of independent variables: the former has one and varied powers while the later has many different dependent variables with same power. Polynomial regression is always better than linear regression as it finds better best-fit lines for curved functional line, by the process of curve fitting. In the following project, we have used polynomial linear regression model to train and fit the dataset so that it can predict the best-fit line.

2.2.3 KNN Regression

K-nearest neighbor [10] is one of the basic and popular learning models. The subset, called the validation set, can be used to select the distance-weighted average of the k-nearest neighbors. Since an array of continuous variables is being used, the distance between the nearest neighbor variables is calculated using Manhattan Distance. The test error rate has been estimated by holding out a subset of appropriate level of flexibility of our algorithm. Here, the most popular validation approach called k-cross-fold validation has been used. K-fold divides all the samples into k groups of samples, called folds, of equal sizes. The number "k" decides how many neighbors influence the prediction. If k = 1, then the algorithm is simply called the nearest neighbor algorithm.

2.2.4 Multilayer Perceptron (MLP)

Multilayer perceptron (MLP) [11, 12] is a feedforward artificial neural network [13–19] with one or more layers between input and output layer. The perceptron computes a single output from multiple real-valued inputs by forming a linear combination according to its input weights and then possibly putting the output through some linear activation function. An MLP network is typically trained using supervised learning which consists of two steps: the feedforward and the backpropagation. In the feedforward, the data flows forward layer by layer through the activation function and reaches the last layer, while the backpropagation deals with going backward in the model and manipulating the weights with an intention to reduce the loss calculated using cost function. This one cycle of feedforward and backpropagation is known as one epoch. Several epochs result in the total training of the model which makes the perceptron powerful enough to predict the desirable results. The MLP we have used relies on a network having an input layer, an output layer, and two hidden layers with 125, 105 nodes, respectively. The optimum number of nodes was determined using an exhaustive search, and resultant configuration was trained further using the above-mentioned learning process. The network is further powered by the ReLU activation function and Adam optimizer which help in feedforward and backpropagation process.

2.2.5 Deep Neural Network

Deep neural network (DNN) is an artificial neural network which is inspired by the biological neural networks that resemble the human brain. DNNs can model complex nonlinear relationships. The DNN architecture generates compositional models which are expressed as a layered composition of primitive data types. DNN models generally have more than two layers of neurons. In this particular model, we have used a three-layered DNN. DNNs are typically feedforward networks in which data flows from the input layer to the output layer without looping back; however, Tensorflow by default chooses backpropagation algorithm for learning. The learning rate is how quickly a network abandons old beliefs for new ones. Neural networks are often trained by gradient descent on the weights. This means at each iteration we use backpropagation to calculate the derivative of the loss function with respect to each weight and subtract it from that weight. For our model, we had learning rate of 0.01.

Deep learning [20–22] architectures resort to a layer-by-layer approach in a lion's share of the cases. Deep learning helps to resolve abstractions and chooses the most optimum features to maximize performance. The layers comprise several nonlinear processing units which produce results that have a direct impact on the outputs of not only the subsequent layers but also the last and final layer. A DNN tunes itself in response to the error generated at each epoch and hence keeps getting better with every training session. The model we have used relies on a three-layered network having 125, 105, and 100 nodes at the respective layers. The optimum number of nodes has been determined using an exhaustive search, and the network has been trained further on the chosen configuration. The network is powered by the ReLU activation function and also harnesses the AdaGrad optimizer to learn and correct itself over time.

3 Results

In the current study, the proposed deep neural network-based model has been compared with other well-known models as discussed in Sect. 2. The comparison has been done in terms of root-mean-square error (RMSE) of the models and accordingly describes which among them is the most desirable. The root-mean-square deviation (RMSD) or root-mean-square error (RMSE) (or sometimes root-mean-squared error) is a frequently used measure of the differences between values (sample and population values) predicted by a model or an estimator and the values actually observed. Figures 1, 2, 3, 4, and 5 depict the regression plots for the models under current study.

Fig. 3 Regression plot for linear regression

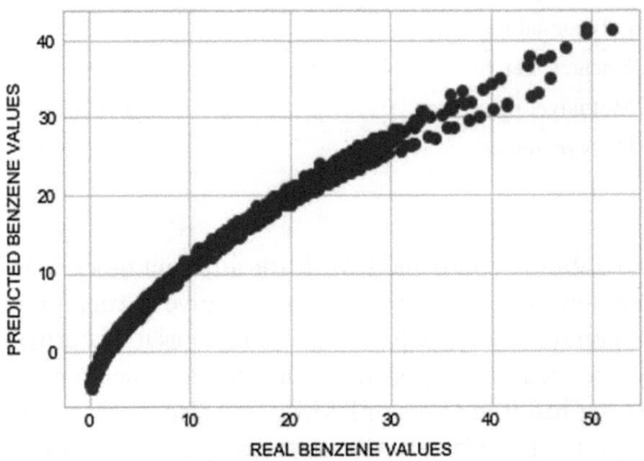

Fig. 4 Regression plot for polynomial regression

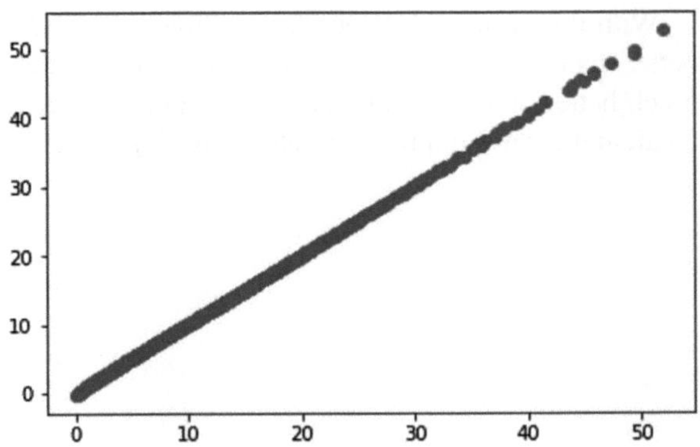

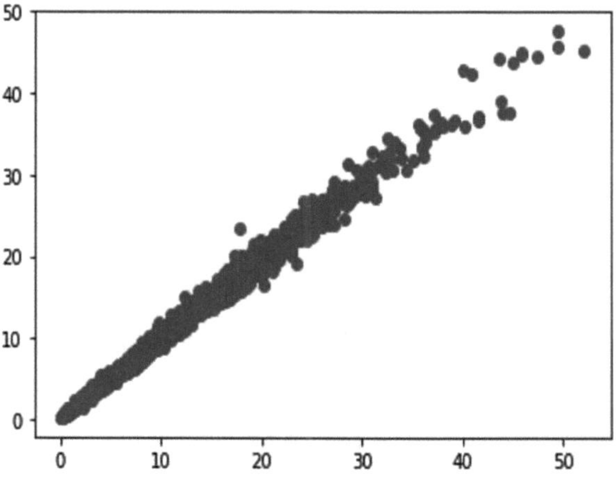

Fig. 5 Regression plot for k-nearest neighbors

Table 1 Comparison of different predictive model on basis of RMSE values

Model	RMSE value
Linear regression	1.40953404943
Polynomial regression	0.0731271602827
k-nearest neighbor (*k*NN) algorithm	0.986
Multilayer perceptron	0.492628078723
DNN regressor	0.405181022925

Table 1 reports the RMSE for different models. It has revealed that the different models which are present already give a very higher RMSE compared to DNN regressor (except polynomial regressor as it is overfit). The DNN regressor is actually a deep learning algorithm or technique which among the other models gives much better result in terms of RMSE, and also, a remarkable capacity of this algorithm is that it can perform better on retraining as it retains the previous result and ensures that the system does not give the worse result.

With this result, we hence try to show that DNN or deep learning algorithm is better than those existing models in terms of prediction with less errors and hence much better precision, and the most important fact is that it can remember its past result and always improve on retraining (Figs. 6 and 7).

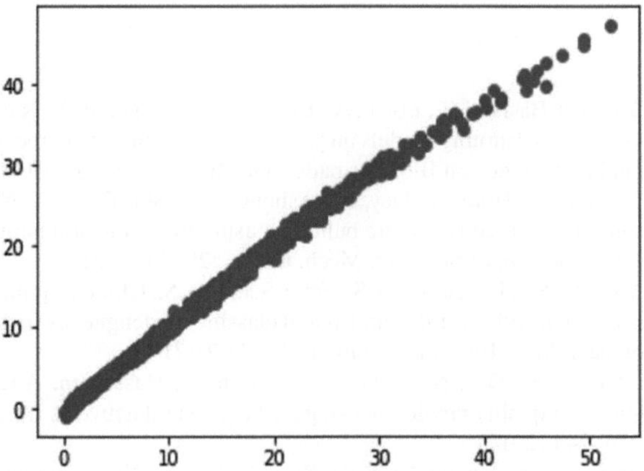

Fig. 6 Regression plot for MLP-FFN

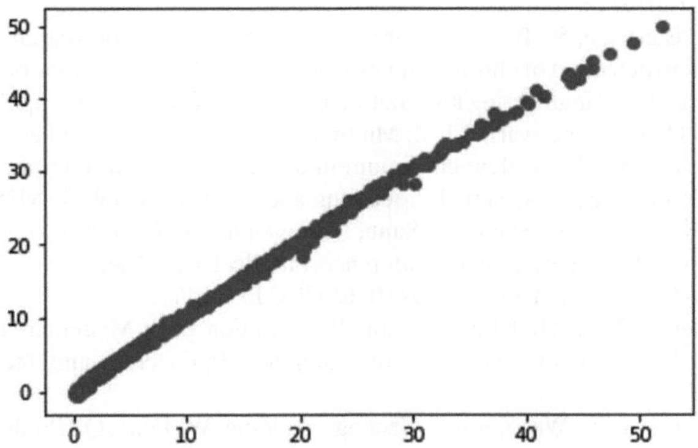

Fig. 7 Regression plot for DNN regressor

4 Conclusion

The current article has proposed a deep learning-based prediction model to predict benzene concentration in the urban area. An efficient feature selection method based on correlation analysis has been employed as well. The proposed method has been compared in terms of RMSE, with several well-known prediction models to prove its ingenuity. For experimental purpose, a publicly available dataset has been used consisting of data from an Italian city. Simulated results have suggested that the proposed model is superior to other models with an RMSE of 0.41. Nevertheless, a future study can be conducted to build a more trustworthy model.

References

1. Davoudi, A., Ozrazgat-Baslanti, T., Ebadi, A., Bursian, A.C., Bihorac, A., Rashidi, P.: Delirium prediction using machine learning models on predictive electronic health records data. In: IEEE 17th International Conference on Bioinformatics and Bioengineering (BIBE) (2017)
2. Chatterjee, S., Sarkar, S., Hore, S., Dey, N., Ashour, A.S., Shi, F., Le, D.N.: Structural failure classification for reinforced concrete buildings using trained neural network based multi-objective genetic algorithm. Struct. Eng. Mech. **63**(4), 429–438 (2017)
3. Chatterjee, S., Dey, N., Shi, F., Ashour, A.S., Fong, S.J., Sen, S.: Clinical application of modified bag-of-features coupled with hybrid neural-based classifier in dengue fever classification using gene expression data. Med. Biol. Eng. Comput. 1–12 (2017)
4. Chatterjee, S., Sarkar, S., Dey, N., Ashour, A.S., Sen, S., Hassanien, A.E.: Application of cuckoo search in water quality prediction using artificial neural network. Int. J. Comput. Intell. Stud. **6**(2–3), 229–244 (2017)
5. Chatterjee, S., Banerjee, S., Mazumdar, K.G., Bose, S., Sen, S.: Non-dominated sorting genetic algorithm—II supported neural network in classifying forest types. In: 2017 1st International Conference on Electronics, Materials Engineering and Nano-technology (IEMENTech), pp. 1–6. IEEE (2017)
6. Chatterjee, S., Banerjee, S., Basu, P., Debnath, M., Sen, S.: Cuckoo search coupled artificial neural network in detection of chronic kidney disease. In: 2017 1st International Conference on Electronics, Materials Engineering and Nano-technology (IEMENTech), pp. 1–4. IEEE (2017)
7. Krishnan, K.S.D., Bhuvaneswari, P.T.V.: Multiple linear regression based water quality parameter modeling to detect hexavalent chromium in drinking water. In: International Conference on Wireless Communications, Signal Processing and Networking (WiSPNET) (2017)
8. Menon, S.P., Bharadwaj, R., Shetty, P., Sanu, P., Nagendra, S.: Prediction of temperature using linear regression. In: International Conference on Electrical, Electronics, Communication, Computer, and Optimization Techniques (ICEECCOT) (2017)
9. Zaw, W.T., Naing, T.T.: Modeling of rainfall prediction over Myanmar using polynomial regression. In: International Conference on Computer Engineering and Technology. ICCET ‘09 (2009)
10. Zhong, L., Lin, L., Lu, Z., Wu, Y., Lu, Z., Huang, M., Yang, W., Feng, Q.: Predict CT image from MRI data using KNN-regression with learned local descriptors. In: IEEE 13th International Symposium on Biomedical Imaging (ISBI) (2016)
11. Bounds, D.G., Lloyd, P.J., Mathew, B., Waddell, G.: A multilayer perceptron network for the diagnosis of low back pain. In: IEEE International Conference on Neural Networks (1988)
12. Mukhlishin, M.F., Saputra, R., Wibowo, A.: Predicting house sale price using fuzzy logic, artificial neural network and K-Nearest neighbor. In: 1st International Conference on Informatics and Computational Sciences (ICICoS) (2017)
13. Chatterjee, S., Dey, N., Ashour, A.S., Drugarin, C.V.A.: Electrical energy output prediction using cuckoo search based artificial neural network. In: Smart Trends in Systems, Security and Sustainability, pp. 277–285. Springer, Singapore (2017)
14. Chakraborty, S., Dey, N., Chatterjee, S., Ashour, A.S.: Gradient Approximation in Retinal Blood Vessel Segmentation
15. Chatterjee, S., Sarkar, S., Dey, N., Ashour, A.S., Sen, S.: Hybrid Non-dominated sorting genetic algorithm: II-neural network approach. Advanc. Appl. Metaheur. Comput. 264 (2017)
16. Chatterjee, S., Sarkar, S., Hore, S., Dey, N., Ashour, A.S., Balas, V.E.: Particle swarm optimization trained neural network for structural failure prediction of multistoried RC buildings. Neural Comput. Appl. **28**(8), 2005–2016 (2017)
17. Chatterjee, S., Ghosh, S., Dawn, S., Hore, S., Dey, N.: Forest type classification: a hybrid NN-GA model based approach. In: Information systems design and intelligent applications, pp. 227–236. Springer, New Delhi (2016)

18. Chatterjee, S., Hore, S., Dey, N., Chakraborty, S., Ashour, A.S.: Dengue fever classification using gene expression data: a PSO based artificial neural network approach. In: Proceedings of the 5th International Conference on Frontiers in Intelligent Computing: Theory and Applications, pp. 331–341. Springer, Singapore (2017)
19. Chatterjee, S., Datta, B., Sen, S., Dey, N., Debnath, N.C.: Rainfall prediction using hybrid neural network approach. In: 2nd International Conference on Recent Advances in Signal Processing, Telecommunications & Computing (SigTelCom)—2018, Vietnam (In press)
20. Dalto, M., Matuško, J., Vašak, M.: Deep neural networks for ultra-short-term wind forecasting. In: IEEE International Conference on Industrial Technology (ICIT) (2015)
21. Paterakis, N.G., Mocanu, E., Gibescu, M., Stappers, B., van Alst, W.: Deep learning versus traditional machine learning methods for aggregated energy demand prediction. In: IEEE PES on Innovative Smart Grid Technologies Conference Europe (ISGT-Europe) (2017)
22. Abel, J., Fingscheidt, T.: A DNN regression approach to speech enhancement by artificial bandwidth extension. In: IEEE Workshop on Applications of Signal Processing to Audio and Acoustics (WASPAA) (2017)

Rough Set-Based Feature Subset Selection Technique Using Jaccard's Similarity Index

Bhawna Tibrewal, Gargi Sur Chaudhury, Sanjay Chakraborty and Animesh Kairi

Abstract Feature selection is the tool required to study data with high dimensions in an easy way. It involves extracting attributes from a dataset having a large number of attributes in such a way so as the reduced attribute set can describe the dataset in a manner similar to that of the entire attribute set. Reducing the features of the data and selecting only the more relevant features reduce the computational and storage requirements which are needed to process the entire dataset. Rough set is the approach of approximating a conventional set. It is used in data mining for reduction of datasets and to find hidden pattern in datasets. This paper aims to devise an algorithm which performs feature selection on a given dataset using the concepts of rough set.

Keywords Feature subset selection · Machine learning · Jaccard coefficient Rough set · Adjacency matrix · Indiscernibility relation

1 Introduction

Feature selection is the process of extracting a subset from a given set of attributes or features so that the study of the behavior of objects can be done on the basis of the reduced set of attributes [1, 2]. Working with a smaller set of attributes simplifies the

B. Tibrewal (✉) · G. S. Chaudhury
Computer Science and Engineering Department,
Institute of Engineering and Management, Kolkata, India
e-mail: bhawnatibrewal1@gmail.com

G. S. Chaudhury
e-mail: gargisurchaudhury@gmail.com

S. Chakraborty
Department of Information Technology, TechnoIndia, Salt Lake, Kolkata, India
e-mail: schakraborty770@gmail.com

A. Kairi
Department of Information Technology, Institute of Engineering and Management,
Salt Lake, Kolkata, West Bengal, India
e-mail: animesh.kairi@iemcal.com

M. Chakraborty et al. (eds.), *Proceedings of International Ethical Hacking Conference 2018*, Advances in Intelligent Systems and Computing 811,
https://doi.org/10.1007/978-981-13-1544-2_39

data mining process. The basic principle of feature selection is that data can contain redundant attributes which can be removed without affecting the characteristics of the dataset [3]. Predictive modeling is the process of predicting outcomes using concepts of data mining and probability [4]. Each model consists of variables which can affect results. The variables are called predictors. A statistical model is formulated after collecting data for the predictors required. Some attributes of the data do not prove to be useful for designing the predictive model. In some cases, such attributes may also reduce the accuracy of the model. Hence, feature selection methods can be used to identify which attributes are undesirable and remove them. Also if a fewer number of attributes are used to design the predictive model, then the complexity of the model is reduced and also a simpler model can be built which will be easier to be understood and analyzed. Feature selection and dimensionality reduction are different from each other [5]. The similarity between them is both aim at reducing the number of attributes of the dataset. But in dimensionality reduction, this is done by creating new combinations of attributes, while feature selection methods include and exclude attributes of the dataset without modifying them. Rough set is a mathematical tool that is used to find the dependencies among data and to eliminate the redundant attributes with minimal information loss [6].

Rough set theory was proposed by Pawlak [7]. Rough sets are approximations of crisp or conventional sets. Whether an element belongs in the rough set or not is decided based on approximations. Two such approximations of the set which are considered are the upper and lower approximations. Rough set theory can be applied to discrete valued attributes of datasets [8]. Hence, if the dataset contains continuous-valued attributes, they must be discretized before applying the rough set theory. It is used for classification to discover structural relationships within noisy data. Rough set theory involves breaking the given training dataset into equivalence classes. In real-world data, many time it happens that some classes are identical to one another on the basis of the available attributes. In such cases, rough sets are used to approximate the data. For a given class, C, the rough set theory considers two sets. Firstly, a lower approximation of C and secondly upper approximation of C. The lower approximation of C contains the data tuples that, belong to C without ambiguity based on the attribute information. The upper approximation of C contains the data tuples which cannot be described as not belonging to C, based on the attribute information. Rough sets can also be used for attribute subset selection or feature reduction, where attributes that are not important for the classification of the given training data and hence can be removed. The problem of finding the minimal subsets (reducts) of attributes that can fully describe the dataset without loss in useful information is NP-hard [9]. However, algorithms have been formulated which reduce the computation intensity of this problem. In one such method, a discernibility matrix that is used to store the differences between attribute values for each pair of data tuples. Instead of searching the entire training set, redundant attributes are identified using the matrix [10, 11]. In rough set theory, an information system is defined as $T = (U, A)$ where U and A are finite non-empty sets, U is the set of objects and A is the set of the attributes of the objects. Each feature $a \in A$ is associated with a set V_a, called the domain of a. The P-indiscernibility relation $IND(P)$ where $P \subset A$ is calculated using the formula,

$$IND(P) = \{(x, y) \in U^2 : \forall a \in P, a(x) = a(y)\}$$

$x[P]$ or $U|D$ denotes the equivalence classes formed due to partitioning of U by IND(P). Using the indiscernibility relation, redundant attributes can be defined. The attributes which when removed do not disturb the indiscernibility relation are redundant attributes. The removal of attributes does not affect the classification of data. Earlier the power of rough set theory was not realized fully.

At present times, rough set theory has shown its strong presence in the field of data analysis both as an independent entity and also as an overlap with other concepts like fuzzy sets, cluster analysis, and statistics. In the areas of artificial intelligence, rough set theory is essential for machine learning, expert systems, pattern recognition and data mining. It is also useful for decision analysis and knowledge discovery from databases. In real-world scenarios of banking, market study, finance, and medicine, rough set theory has found many applications. The fundamental application of the rough set theory is to study and categorize information which is incomplete and having redundancies. The two core concepts of this theory are reduct and core. For a small dataset, using rough set theory, the minimal reduct can be easily calculated. However, the general solution for finding the minimal reduct is NP-hard; i.e., it cannot be solved in polynomial time. When datasets grow in both volume and dimensionality, applying the rough set theory to determine the minimal reduct becomes computationally infeasible. Rough set theory is most popularly used for achieving the following functionalities: (i) establish relationships that cannot be accomplished using standard statistical approaches; (ii) generate decision rules from data; (iii) devise algorithms which detect hidden patterns in datasets; (iv) reduction of datasets. The models developed using rough set theory are straightforward and easily understandable.

The Jaccard coefficient or the Jaccard similarity index compares the elements of two sets to check which elements are common to two sets and which are distinct [12]. It measures the similarity which exists between two sets, with a range of minimum similarity (0%) to maximum similarity (100%). A Jaccard coefficient of 0 indicates that the two sets have no common elements and are overlapped and the Jaccard coefficient of 1 indicated that the two sets are composed of the same elements and are identical [13]. Jaccard Index = (elements present in both sets)/(elements present in either set) * 100 Or $J(X, Y) = |X \cap Y|/|X \cup Y|$. The Jaccard distance measures dissimilarity between sample sets. It is complementary to the Jaccard coefficient and is obtained by subtracting the Jaccard coefficient from 1. The Jaccard similarity measures how closely related two sets are and its value is higher when the sets are more alike to each other. The Jaccard distance is useful to evaluate the distance between two data objects when the feature set of those two objects is sparse. This is because the Jaccard approach considers only those features which are present in at least one of the two data objects instead of considering the entire set of features available.

Similarity measures to what extent two data objects are alike. In the field of data mining, similarity can be described as a distance such that it has dimensions represented by the features of the data objects. A small distance indicates a high degree of similarity and a large distance indicates a low degree of similarity. Similarity measure or similarity function is a real-valued function that is used to quantitatively describe the similarity which exists between two objects. Such functions are often, in some way, determined by the inverse of distance metrics. Hence for similar objects which have smaller distances between them, these functions take up large values and for dissimilar objects which lie at greater distances from one another, these functions take up small values.

The rest of the paper is organized as follows. In Sect. 2, we have described the proposed work and the algorithm of rough set-based feature subset selection. In Sect. 3, we illustrate the application of our proposed method on two demo datasets in terms of feature subset selection. Section 4 gives the discussion of the entire approach. Finally, Sect. 5 gives the conclusion of this paper.

2 Proposed Work

In this paper, we aim to devise an algorithm which uses rough sets to perform feature selection. We are provided with the decision table as an input which contains the values of objects corresponding to a certain set of attributes. Firstly, the set of objects whose attributes are being studied and divided into classes using the indiscernibility relation in which two data tuples belong to a particular class if they have the same values for the corresponding attribute subset of the total set of attributes. Then the redundant attributes of the dataset are determined using the lower and upper approximation of the equivalence classed thus formed. An attribute is considered as redundant if the positive region of the decision attribute D is unaffected on removing that particular attribute. After identifying the redundant attributes, they are removed from the attribute set. Binary dummy variables are created for each value in the domain of the reduced attribute set [14]. The contingency table is constructed for every pair of distinct objects based on certain rules (described in our proposed algorithm). We calculate the Jaccard's distance and Jaccard's coefficient for each pair of distinct objects and then compute mean Jaccard's coefficient. Corresponding adjacency relation between pair of objects is determined using the mean Jaccard's coefficient. Based on the above adjacency matrix, a similarity graph is generated which contains the objects as the vertices and edges are present where the inter-object similarity is quite high.

3 Mathematical Illustration on Demo Datasets

We analyze the above-proposed algorithm by applying it to two demo datasets.

DEMO SET 1

Let B = {$e1$, $e2$, $e3$, $e4$, $e5$, $e6$} be the set of 6 objects. The set of Condition attributes of Information System C = {'Headache', 'Muslce_pain' and 'Temperature'} The set of Decision attribute of information system D = {Flu}

Table 1 shows the information matrix where the values of objects (representing patients) e1–e6 are shown for attributes 'Headache', 'Muslce_pain', and 'Temperature'.

Table 2 represents the decision matrix which is obtained from Table 1 by including the values of objects e1–e6 for the decision attribute 'Flu'. Calculating IND(X) gives:

$$IND(Headache) = e1, e2, e3, e4, e5, e6$$
$$IND(Muscle_pain) = e1, e2, e3, e4, e6, e5$$
$$IND(Temperature) = e1, e4, e2, e5, e3, e6$$
$$IND(Headache, Muscle_pain) = e1, e2, e3, e4, e6, e5$$
$$IND(Headache, Temperature) = e1, e2, e3, e4, e5, e6$$
$$IND(Muscle_pain, Temperature) = e1, e4, e2, e3, e6, e5$$
$$IND(Headache, Muscle_pain, Temperature) = e1, e2, e3, e4, e5, e6$$

Table 1 Information system for demo set 1

Objects	Attributes		
	Headache	Muscle_pain	Temperature
e1	Yes	Yes	Normal
e2	Yes	Yes	High
e3	Yes	Yes	Very_high
e4	No	Yes	Normal
e5	No	No	High
e6	No	Yes	Very_high

Table 2 Decision matrix

Objects	Attributes			Decision flu
	Headache	Muscle_pain	Temperature	
e1	Yes	Yes	Normal	no
e2	Yes	Yes	High	yes
e3	Yes	Yes	Very_high	yes
e4	no	Yes	Normal	no
e5	no	No	High	no
e6	no	Yes	Very_high	yes

IND(Headache,Temperature) = IND(Headache, Muslce_pain, Temperature)
Thus the attribute 'Muslce_pain' is found to be redundant and is hence removed.

Table 3 represents the knowledge matrix which is constructed for objects e1–e6 against all values taken up by the attributes of the reduced attribute set. For example, e1 has headache and normal temperature and hence has HYes and TNormal as 1 while HNo, THigh and TVHigh are 0 for e1.

Table 4 represents the contingency matrices for every pair of data objects.

Tables 5 and 6 represent the distance matrix and similarity matrix calculated for every pair of data objects using the contingency matrices.

Computed Mean_coefficient = 0.198
Table 7 represents the adjacency matrix for the data sets of table 1.

Table 3 Knowledge matrix

Objects	HYes	HNo	TNormal	TVHigh	THigh
e1	1	0	1	0	0
e2	1	0	0	0	1
e3	1	0	0	1	0
e4	0	1	1	0	0
e5	0	1	0	0	1
e6	0	1	0	1	0

Table 4 Contingency matrices

C(e1, e2)	0	1	C(e1, e3)	0	1	C(e1, e4)	0	1	C(e1, e5)	0	1
0	2	1	0	2	1	0	1	1	0	1	2
1	1	1	1	1	1	1	1	1	1	2	0
C(e1, e6)	0	1	C(e2, e3)	0	1	C(e2, e4)	0	1	C(e2, e5)	0	1
0	1	2	0	2	1	0	1	2	0	2	1
1	2	0	1	1	1	1	2	0	1	1	1
C(e2, e6)	0	1	C(e3, e4)	0	1	C(e3, e5)	0	1	C(e3, e6)	0	1
0	1	2	0	1	2	0	1	2	0	2	1
1	2	0	1	2	0	1	2	0	1	1	1
C(e4, e5)	0	1	C(e4, e6)	0	1	C(e5, e6)	0	1			
0	2	1	0	2	1	0	2	1			
1	1	1	1	1	1	1	1	1			

Algorithm 1: Computing Inter-object Similarity using rough set-based feature selection and Jaccard's index

Input : Set of objects U[m] and set of attributes A[n]
Output: Adjacency matrix for the set of objects U[m]

1 Construct information matrix I[m] [n] where I[i] [j] = value of U[i] under A[j].
2 Construct decision matrix D[m] [$n+1$] as

$$D[i][j] = \begin{cases} I[i][j], & j \neq (n+1) \\ d[i], & j = (n+1) \end{cases}$$

where d[i] is the value of U[i] under the decision variable.
3 Consider all non-empty subsets X of set A[n].
Calculate the X-Indiscernibility, IND(X) where IND(X) = set of all (U[a],U[b]) such that I[a] [k]=I[b] [k] where A[k] $\in X$.
4 Search for Y where $IND(Y) = IND(A)$ such that Y is of lowest possible size.
5 Remove attributes A[i] where A[i] $\notin Y$.
6 Remove the duplicate rows from the information system, because duplicate rows or cases do not add to any information.
7 Consider the reduced information matrix I[m] [$n-1$], assign for each unique value of attributes $\in$ domain(A [i]) a binary dummy attribute, and construct a knowledge matrix M[m] [v] where v is the total number of dummy attributes created.
8 Construct Contingency table $C_{ij}[2][2]$ for each pair of distinct object U[i] and U[j] where
– $C_{ij}[0][0] = t =$ number of attributes k, such that $M[i][k] = M[j][k] = 0$
– $C_{ij}[0][1] = s =$ number of attributes k, such that $M[i][k] = 1$ and $M[j][k] = 0$
– $C_{ij}[1][0] = r =$number of attributes k, such that $M[i][k] = 0$ and $M[j][k] = 1$
– $C_{ij}[1][1] = q =$number of attributes k, such that $M[i][k] = M[j][k] = 1$
9 Calculate the Jaccards distance for each pair of objects (i, j) using C_{ij} where $d(i, j) = (r+s)/(q+r+s)$ for $i \neq j$ and store the value in matrix Dist[m] [m] where

$$Dist[i][j] = \begin{cases} d(i, j), & i \neq j \\ 0, & otherwise \end{cases}$$

10 Calculate Jaccards coefficient $s(i, j) = q/(q+r+s)$ and store in $Similarity$ matrix S [m] [m] where

$$S[i][j] = \begin{cases} s(i, j), & i \neq j \\ 0, & otherwise \end{cases}$$

11 Calculate the $Mean\_coefficient = \overline{S[i][j]}$
12 Construct adjacency matrix Adj[m] [m] where

$$Adj[i][j] = \begin{cases} 1, & S[i][j] > Mean\_coefficient \\ 0, & otherwise \end{cases}$$

13 Based on the above adjacency matrix, a similarity graph is generated which contains the objects as the vertices and edges are present where the inter-object similarity is quite high.

For demonstration, we consider a subset of the objects given in the adjacency matrix and draw a similarity graph as shown in Fig. 1.

Table 5 Distance matrix

Objects	e1	e2	e3	e4	e5	e6
e1	0	0.66	0.66	0.66	1	1
e2	0.66	0	0.66	1	0.66	1
e3	0.66	0.66	0	1	1	0.66
e4	0.66	1	1	0	0.66	0.66
e5	1	0.66	1	0.66	0	0.66
e6	1	1	0.66	0.66	0.66	0

Table 6 Similarity matrix

Objects	e1	e2	e3	e4	e5	e6
e1	0	0.33	0.33	0.33	0	0
e2	0.33	0	0.33	0	0.33	0
e3	0.33	0.33	0	0	0	0.33
e4	0.33	0	0	0	0.33	0.33
e5	0	0.33	0	0.33	0	0.33
e6	0	0	0.33	0.33	0.33	0

Table 7 Adjacency matrix

Objects	e1	e2	e3	e4	e5	e6
e1	0	1	1	1	0	0
e2	1	0	1	0	1	0
e3	1	1	0	0	0	1
e4	1	0	0	0	1	1
e5	0	1	0	1	0	1
e6	0	0	1	1	1	0

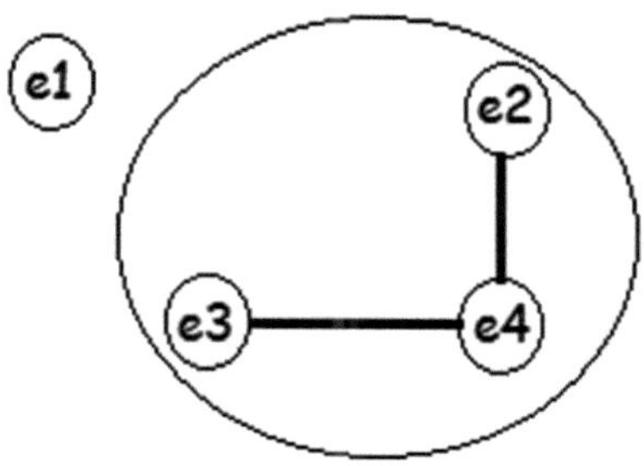

Fig. 1 Similarity graph for demo set 1

Figure 1 is the similarity graph generated based on the adjacency matrix given in Table 7. We can observe that the similarity between objects {e2, e3, e4} is high and e1 does not have a high correlation with any other object.

DEMO SET 2

Let B = {$P1, P2, P3, P4, P5, P6, P7, P8, P9, P10, P11, P12, P13, P14, P15$} be the set of 15 patients.

The set of Condition attributes of Information System C = {Heart Palpitation, Blood Pressure, Chest Pain, Cholesterol, Fatigue, Shortness of Breath, Rapid Weight Gain} The set of Decision attribute of information system Patient D = {Heart Problem}.

Table 8 shows the information matrix where the values of objects (representing patients) P1–P15 are shown for attributes 'H.P', 'B.P', 'C.P', 'Cholesterol', 'Fatigue', 'S.O.B', and 'R.W.G'.

IND(H.P, B.P, C.P, Cholesterol, Fatigue, S.O.B, R.W.G) = {P1, P2, P3, (P4, P10), (P5, P7), P6, (P8, P9), P11, P12, P13, P14, P15}

Applying our algorithm to the dataset given in Table 8 we can have three reduct sets for patient information system:

1. {Blood Pressure, Chest Pain, Shortness of Breath, Cholesterol}
2. {Heart Palpitation, Chest Pain, Cholesterol}
3. {Blood Pressure, Chest Pain, Fatigue, Cholesterol}

We consider the reduct {Heart Palpitation, Chest Pain, Cholesterol} since it has the least number of attributes.

Table 8 Patient information system

Objects	Conditional							Decision
Patient	H.P	B.P	C.P	Cholesterol	Fatigue	S.O.B	R.W.G	Heart problem
P1	High	High	High	Normal	Low	Low	High	Yes
P2	Low	Low	High	Normal	Low	Normal	Low	No
P3	Low	High	Low	Low	Normal	Low	Low	No
P4	V_high	Low	V_high	Low	Low	Low	High	Yes
P5	High	High	V_high	Low	Low	Low	High	Yes
P6	Low	High	Low	High	Low	Normal	Low	No
P7	High	High	V_high	Low	Low	Low	High	Yes
P8	High	High	High	Low	Low	Low	High	Yes
P9	High	High	High	Low	Low	Low	High	Yes
P10	High	Low	V_high	Low	Low	Normal	High	Yes
P11	V_high	High	High	Low	Normal	Normal	High	Yes
P12	V_high	Low	High	High	Low	Low	High	Yes
P13	High	Low	Low	Low	High	Low	Low	No
P14	Low	Low	High	High	Low	High	Low	No
P15	High	V_high	V_high	Normal	Low	Normal	High	Yes

Table 9 Adjaceny matrix for patients data set

Patient	1	2	3	4	5	6	8	11	12	13	14	15
1	0	1	0	0	1	0	1	1	1	1	1	1
2	1	0	1	0	0	1	1	1	1	0	1	1
3	0	1	0	1	1	1	1	1	0	1	1	0
4	0	0	1	0	1	0	1	1	1	1	0	1
5	1	0	1	1	0	0	1	1	0	1	0	1
6	0	1	1	0	0	0	0	0	1	1	1	0
8	1	1	1	1	1	0	0	1	1	1	1	0
11	1	1	1	1	1	0	1	0	1	1	0	1
12	1	1	0	1	0	1	1	1	0	0	1	0
13	1	0	1	1	1	1	1	1	0	0	0	1
14	1	1	1	0	0	1	1	0	1	0	0	0
15	1	1	0	1	1	0	0	1	0	1	0	0

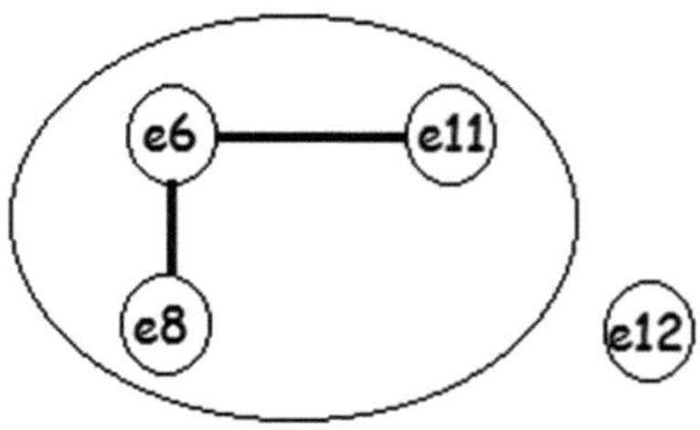

Fig. 2 Similarity graph for demo set 2

We remove patient objects {P7, P9, P10} from the given set as {P5, P7},{P8, P9}, and {P4, P10} are duplicate rows and hence add no information.

Table 9 represents the adjacency matrix for the patient data set. For demonstration, we consider a subset of the objects given in the adjacency matrix and draw a similarity graph as shown in Fig. 2. Figure 2 is the similarity graph generated based on the adjacency matrix given in Table 9. We can observe that the similarity between objects {e6, e11, e8} is high and e12 does not have a high correlation with any other object.

4 Discussion

The proposed algorithm in this paper requires to obtain all possible subsets of the entire set if attributes and compare the indiscernibility classes obtained for all those subsets. For a very large dataset, this may lead to high computational costs. Other algorithms have been proposed to perform feature selection more optimally [15] like wrapper methods [16]. A drawback in this approach still remains due to the fact that the best subset cannot be solved in polynomial time. As a future work, the possibility of making the approach more robust can be explored.

5 Conclusion

Based on the proposed algorithm, two demo datasets have been presented and the algorithm has been implemented on those datasets. It is found that using the indiscernibility relation with the concept of rough sets, we can determine the presence of redundant attributes in the original attribute and hence removed. After obtaining the reduced attribute set, the contingency matrices are constructed for every pair of data objects which helps to obtain the desired distance and similarity matrices. The adjacency matrix developed in the final stages of the algorithm helps us to draw a vertex-edge graph of the data objects. This graph can be used to visualize how closely related two data objects are.

References

1. Singh, B., Kushwaha, N., Vyas, O.P.: A feature subset selection technique for high dimensional data using symmetric uncertainty. J. Data Anal. Informat. Process. **2**(4), 95 (2014)
2. Jovic, A., Brkic, K., Bogunovic, N.: A review of feature selection methods with applications. In: 38th International Convention on Information and Communication Technology, Electronics and Microelectronics (MIPRO), pp. 1200–1205. IEEE (2015)
3. Rudnicki, W.R., Wrzesień, M., Paja, W.: All relevant feature selection methods and applications. In: Feature Selection for Data and Pattern Recognition, pp. 11–28. Springer (2015)
4. Ramaswami, M., Bhaskaran, R.: A Study On Feature Selection Techniques in Educational Data Mining. arXiv:0912.3924 (2009)
5. Mladenić, D.: Feature selection for dimensionality reduction. In: Subspace, Latent Structure and Feature Selection, pp. 84–102, Springer (2006)
6. Caballero, Y., Alvarez, D., Bello, R., Garcia, M.M.: Feature selection algorithms using rough set theory. In: ISDA Seventh International Conference on Intelligent Systems Design and Applications, pp. 407–411. IEEE
7. Pawlak, Z.: Rough sets. Int. J. Comput. Informat. Sci. **11**(5), 341–356 (1982)
8. Pawlak, Z.: Rough Sets: Theoretical Aspects of Reasoning About Data. Springer Science and Business Media, vol. 9 (2012)
9. Al-Radaideh, Q.A., Sulaiman, M.N., Selamat, M.H., Ibrahim, H.: Approximate reduct computation by rough sets based attribute weighting. In: IEEE International Conference on Granular Computing, vol. 2, pp. 383–386. IEEE (2005)
10. Zhang, M., Yao, J.: A rough sets based approach to feature selection. In: IEEE Annual Meeting of the Fuzzy Information, Processing NAFIPS'04, vol. 1, pp. 434–439 (2004)
11. Vijayabalaji, S., Balaji, P.: Rough matrix theory and its decision making. Int. J. Pure Appl. Math., **87**(6), 845–853
12. Maheswari, D.U., Gunasundari, R.: User interesting navigation pattern discovery using fuzzy correlation based rule mining. Int. J. Appl. Eng. Res. **12**(22), 11818–11823 (2017)
13. Chelvan, M.P., Perumal, K.: On feature selection algorithms and feature selection stability measures. Comp. Anal. **9**(06), 159–168
14. Skrivanek, S.: The Use of Dummy Variables in Regression Analysis. More Steam, LLC (2009)
15. Koller, D., Sahami, M.: Toward Optimal Feature Selection. Technical Reports, Stanford InfoLab (1996)
16. Kohavi, R., John, G.H.: Wrappers for feature subset selection. Artif. Intell. **97**(1–2), 273–324 (1997)

An Approach Towards Development of a Predictive Model for Female Kidnapping in India Using R Programming

Sumit Chatterjee, Surojit Das, Sourav Banerjee and Utpal Biswas

Abstract The major concern of the present world is the increase in criminal activities that are taking place throughout the world. The criminal activities in today's world include murder, theft, rape, women exploitation, human trafficking, possession of illegal properties, kidnapping. This paper summarizes the criminal activities related to female kidnapping in India. This paper highlights the statistical analysis of female kidnapping in India and thereby develops a predictive model to envisage the purpose of kidnapping of an individual female based on certain parameters. The model is developed using the decision tree technique by applying Iterative Dichotomizer (ID3) algorithm. The ID3 algorithm uses the entropy measure as a criterion for selecting classifiers for branching. The raw data set is converted to an appropriate one by converting the categorical values to numerical values using label and one hot encoding. This model is then trained with the appropriate training data set, and then, its performance is evaluated with a testing data set. The efficiency of the model is detected with the measures of accuracy, precision, recall, F1, AUC scores.

Keywords Human trafficking · Kidnapping · Statistical analysis · Decision tree ID3 algorithm · Entropy · Classifier · Label and one hot encoding · Accuracy Precision · Recall · F1 · AUC

S. Chatterjee (✉) · S. Banerjee
Kalyani Government Engineering College, Kalyani, Nadia 741235, India
e-mail: sumitc219@rediffmail.com

S. Banerjee
e-mail: mr.sourav.banerjee@ieee.org

S. Das
Brainware University, Kolkata 700124, West Bengal, India
e-mail: surojitdas90@gmail.com

U. Biswas
University of Kalyani, Kalyani, Nadia 741235, India
e-mail: utpal01in@yahoo.com

M. Chakraborty et al. (eds.), *Proceedings of International Ethical Hacking Conference 2018*, Advances in Intelligent Systems and Computing 811,
https://doi.org/10.1007/978-981-13-1544-2_40

1 Introduction

The crime is not a problem of an individual, but it is a problem of the entire society [1]. It influences the socio, political and economic aspects of a country. The various forms of crime in India are murder, theft, women exploitation, human trafficking, prostitution, etc., [1]. The social problems can occur due to unpredictable social changes and unwillingness to adapt in new social environment [1]. The rapid changes in social environment often trigger crime. Though different societies have different changes, they may stimulate the same kind of crime rates [1]. The women in India are victims of crime, exploitation and violence. According to the statistical report of National Crime Record Bureau in 2015, in every eighth minute a woman is being kidnapped in India [2]. The criminal activities are also spread using illicit Web domain (dark nets) and social networks [3]. The websites of dark nets provide advertisements which are actually traps for women trafficking [3]. The women are kidnapped for various purposes like adoption, begging, marriage, illicit intercourse, camel racing, prostitution, ransom. This paper has focussed on the evaluation of the statistics relating to the mentioned purposes of kidnapping. The women trafficking are ranked third in organized crime list. According to the Human Rights Commission of India, every year 40,000 children are kidnapped and 11,000 remain untraced [4, 5].

2 Related Work

The crime analysis using statistical tools and machine learning algorithms is an interesting area of research nowadays [6]. But not too many research works have been done in this field. There have been works done to identify behavioural patterns related to human trafficking by extracting information from newspaper, social networks and illicit Web domains and generating corpus vector containing details about human trafficking [7, 8]. The techniques which were used were data mining for finding the hidden information, data pre-processing to gather all the information, parsing for obtaining a semantic meaning for the collected information, using techniques like deep learning [9] and NLP [9] to classify information and find the words related to trafficking. In another project [10], a website called 'Backpage' which is used for classified advertisement has been used as a source of information for many of the research works [11, 12]. The machine learning algorithms (semi-supervised) [10] are used to generate predictive model which is trained with labelled and unlabelled data and then tested on unseen data to evaluate their efficiency. 'Backpage' has also been used by a research team of undergraduate students for examining sex traffic in Pennsylvania [11–13]. The prostitution advertisements of different countries were analysed, and the major countries involved in trafficking were identified. Females involved were identified and formed three subnetworks, namely disconnected subnetworks, high-density interconnected subnetworks and sparsely interconnected subnetworks [13].

Open-source data available [14] from the Internet have been also used to design a framework for indicating human trafficking. A three-level model has been proposed by Glynn Rankin [14] to understand and report all the illegal activities related to trafficking across the network. The level 1 deals with the credible and accepted indicators, the level 2 describes the materials produced as a result of anti-trafficking activities, and the level 3 deals with social media and citizen-created content. This model was developed in challenging Boisot's knowledge management model [15] or iSpace [15] which is a three-dimensional cube-like structure.

In a project of domestic human trafficking, movements along the circuits (systematic movement of provider to various cities for promoting sex activity) have been given stress [16, 17]. The network analysis methods [16] were used to find the circuits from GIS data. Various escort advertisements were collected with a count of about 90 advertisements per day, duplicate advertisements were eliminated, and then, the location of each advertisement was analysed to identify intra-state circuit. In another paper [18] related to crime analysis in Haryana, a spatio-temporal analysis of abduction of women in the city of Chandigarh has been carried out to find out the areal variation and distribution. The factors like socio-culture, economic condition and population explosion have been identified as one of the major causes for such crimes. This project also proposes [18] a formula:

$$Crime\ Rate = Cri = (CXi/TFPi) * 100{,}000 \tag{1}$$

where

CRi rate of crime 'X' in wards I,
CXi crime 'X' in wards I,
TFPi total female population in wards I

Another research paper [19] says that the sex traffickers target children more than the adults. Thus, an initiative can be taken to prevent child trafficking by employing a child safety system with the use of smart identity card. The smart identity card is enabled with radio frequency identification device (RFID) which can help in tagging the school children. The objective of this paper was to ensure security by reducing the gap between parents, children and teachers. A mobile safety monitoring system [20] was proposed in this paper that informs the guardians about the information relating to the safety of their children. A school security system (SSS) [21] prototype model was also proposed based on Web-based development using PHP, VB.net, Apache Web server and MySQL. The out time and in time of each student will be recorded successfully in the system, and the GPS technology will enable the parents to receive continuous information about their children.

Sometimes, analysing the advertisements of the sex workers can too work. The rigorous content analysis of online advertisements of sex workers can provide certain behavioural patterns that can identify the victims of human trafficking [22]. Statistics suggest that about 75% of the advertisements possess more than one primary indicator. The virtual indicators that are discovered are: movement, shared management, controlled/restricted movement, advertisements posted by third party, advertisement ethnicity/nationality. To provide answers to entry-centric questions on human traffic data to help the investigators is a matter of challenge. An entity-centric knowledge graph can be developed to build a semantic search engine to help the investigators [23]. The idea is to take a large number of advertisements from the Web and create a knowledge graph. This allows the investigators to search their queries with a well-built semantic search engine.

3 Proposed Work

3.1 Statistical Analysis of the Female Kidnapping Records in India

The paper uses a data set which contains the records of kidnapping of women of different ages for various purposes. Based on these records, a statistical analysis has been performed to find out the number of females kidnapped of different age categories and for different purposes like begging, prostitution.

3.2 Design of a Predictive Model for Female Kidnapping

Certain parameters have been identified which can act as classifiers for designing a predictive model for female kidnapping. These parameters are age, financial status, glamour and marital status. A predictive model has been designed using these parameters as classifiers by constructing a decision tree [24]. A decision tree is a supervised machine learning technique used for solving classification problems. It can work with both categorical and continuous data and is widely used for multilevel classification problems. In a decision tree, the population is split into two subpopulations based on the significant classifier (parameter). The algorithm which is used to construct the decision tree is Iterative Dichotomizer (ID3) [25] algorithm. It uses a top-down approach to construct the tree. It is based on maximizing the information gain and minimizing the entropy [26]. At each node, every possible property is tried and the property which maximizes the information gain and minimizes the entropy is selected to split the node. This process is recursively repeated until all the leaf nodes are homogeneous (contains objects of same classes). It is a greedy algorithm, always uses entropy as a criterion, and never looks for alternative choices like Gini index, classification error. The predictive model can predict for what purpose (begging, prostitution, etc.) an individual has been kidnapped based on these parameters.

Figure 1 shows the decision tree of classification of iris data set. Iris data set contains the features of three classes (species) of flowers, namely setosa, versicolor and virginica. Four features have been specified in the data set, namely sepal length, sepal width, petal length and petal width. The data set has 150 samples with 50

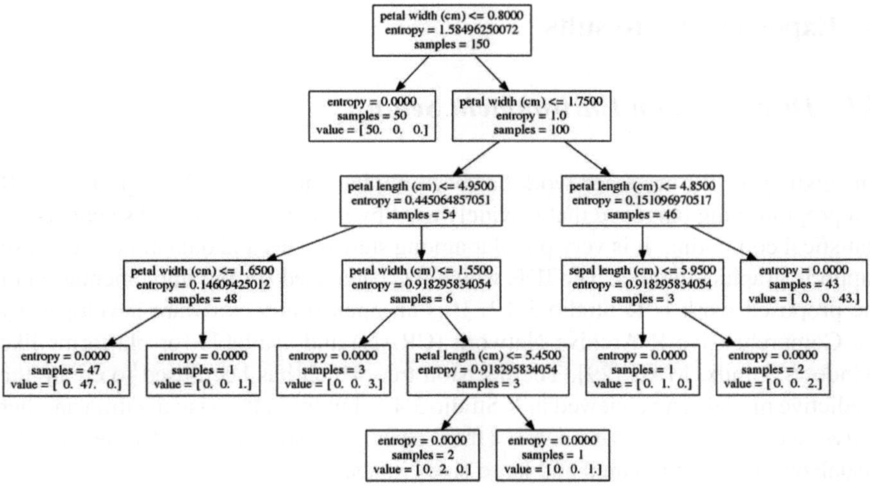

Fig. 1 An Illustration of decision tree for classification of iris data set [27]

samples belonging to each of the classes. The root node selects petal width as one of the classifiers based on the value of entropy. At each step, the decision tree classifies the samples into different classes by selecting different classifiers with a motive to minimize the entropy. The leaf nodes denote the final decision regarding the classification. This kind of approach has also been used in the decision of the proposed model discussed later in the Sect. 4.3.

All the related works which have been mentioned in Sect. 2 have been done on the data available from the Internet. This paper uses the manual data obtained from the records of the Government of India [28] and hence covers the cases which include all kinds of kidnapping and trafficking of women including those via the Internet. Thus, the scope of applicability of this paper is comparatively more because it deals with larger data set. Moreover, the proposed model also predicts the purposes of kidnapping which can help the police to investigate and trace the criminals who deal with that particular purpose. Thus, this paper has a unique importance in comparison with the papers mentioned in the related works. The predictive model that has been designed is also beneficial to the society as police can easily search among the criminals involved in a specific purpose for kidnapping thereby enhancing the investigation process and reducing the time.

4 Experimental Results

4.1 Discussion on Environment Setup

The coding for the proposed work has been implemented using R programming. R is a programming language that is widely used by mathematicians and scientists for statistical computing. It is very popular among statisticians and data miners and also supports graphical works. The IDE which has been used for the implementation of the proposed work is R Studio 3.4.1. It is an open-source software developed by the Comprehensive R Archive Network (CRAN) and available for platforms like Windows, Linux, MAC [29]. The decision tree which has been used to create the predictive model can be viewed in R Studio 3.4.1, but for better visualization another software Graphviz 2.83 is used. It is a third-party software and is used widely for the visualization of complicated and large-sized images.

4.2 Data Set Used

The data set provides the information that among the major purposes behind female kidnapping, the most predominant purposes are: begging, marriage, illicit intercourse, prostitution, ransom.

Volume of Data Set. Number of samples in training data = 150, number of samples in testing data = 50.

Clustering. Clustering is done based on the purposes of kidnapping. The clusters and their size are specified in Table 1.

The parameters and their ranges which have been selected to predict the purpose behind kidnapping are mentioned in Table 2.

The categorical data like financial status, glamour, marital status are converted to numerical values using label encoding [30]. The label encoding is used to transform non-numeric values into numeric values. The numeric values after encoding range from 0 to number of classes—1. The conventions used are mentioned in Table 3.

Table 1 Clustering of female kidnapping data

Cluster (class)	Number of samples in each cluster (class)
Begging	30
Marriage	30
Illicit intercourse	30
Prostitution	30
Ransom	30

Next one hot encoding or dummy encoding is used to generate new columns [30]. The name 'dummy' suggests the use of a duplicate variable to represent one level of categorical variable. If a level is present, then it is indicated by '1', and if a level is absent, then it is indicated by '0'. One dummy variable is created for each and every level. The application of one hot encoding on the data generates columns having binary values 0 and 1 for each of the kidnapping types/causes. Suppose begging is a cause for kidnapping. So a new column called begging is generated which has values of 1 for all the records (rows) which correspond to begging and 0 for the others. The similar work is done for other causes like marriage, illicit intercourse, prostitution, ransom.

4.3 Construction and Analysis of Decision Tree

A predictive model is created by analysing the pre-processed data set using a decision tree.

The predictive model designed using decision tree shown in Fig. 2 is trained using 150 sample records. The decision tree starts initially with 150 samples. Then, it has to choose a parameter as classifier for classifying the samples into specific categories. The parameter is chosen on the basis of the entropy value [31]. Entropy is the measure of impurity or uncertainty. So higher values of entropy mean more uncertainty, while lower entropy values mean less uncertainty. Thus, the target of the decision tree is to

Table 2 Data set of female kidnapping

Age	Below 10	10–15	15–18	18–30	30–50	Above 50
Financial status	Below 1 lakh		1–5 lakh		Above 5 lakh	
Glamour	Low		Medium		High	
Marital status	Unmarried		Married			

Table 3 Label encoding of categorical data

Parameters	Categorical values	Numerical values
Financial Status	Below 1 lakh	1
	1–5 lakh	2
	Above 5 lakh	3
Glamour	Low	1
	Medium	2
	High	3
Marital status	Unmarried	0
	Married	1

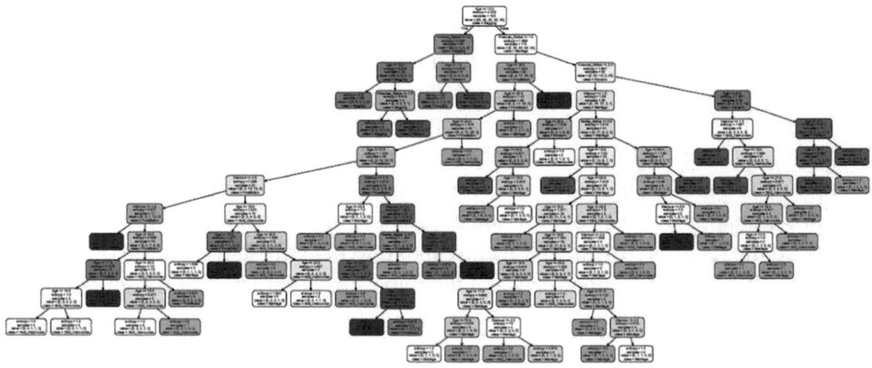

Fig. 2 Decision tree classification model of female kidnapping

minimize the value of entropy as far as possible. The minimum value of entropy is 0 which means that this condition has no uncertainty; i.e., it is certain.

The root node contains 150 sample records with 30 records for each of the kidnapping purposes (begging, marriage, illicit intercourse, prostitution, ransom). The class begging is chosen as label of the node arbitrarily as all of the classes (kidnapping purposes) has equal number of sample records; i.e., each has 30 sample records. The parameter age is chosen as the classifier with an entropy value of 2.322. The threshold value of age is selected as 13.5. At the next level of the tree, two child nodes are formed from the root node. One of the child nodes contains sample records with age values less than or equal to 13.5, and the other child node contains sample records with age values greater than 13.5. The class begging is chosen as label of the first child node as the majority number of records correspond to begging. The first child node has an entropy value of 0.592, and financial status is chosen as the classifier for the next level with a threshold value of 2.5. The class marriage is chosen as label of the second child node as the majority number of records correspond to marriage. The second child node has an entropy value of 1.996, and financial status is chosen as the classifier for the next level with a threshold value of 1.5. This process of generation of child nodes continues until the leaf nodes containing sample records belonging to a particular category (kidnapping purpose) are generated. The model is subjected to a testing set of data with 50 samples, and its efficiency of classification is measured based on the following scores:

Accuracy. It is defined as the number of correct predictions out of the total number of predictions [32].

$$Accuracy = (TP + TN)/(TP + TN + FP + FN) \tag{2}$$

Precision. It is defined as the number of actual true cases that has been detected out of the total number of predicted true cases [32].

$$Precision = TP/(TP + FP) \tag{3}$$

Fig. 3 ROC curve [33]

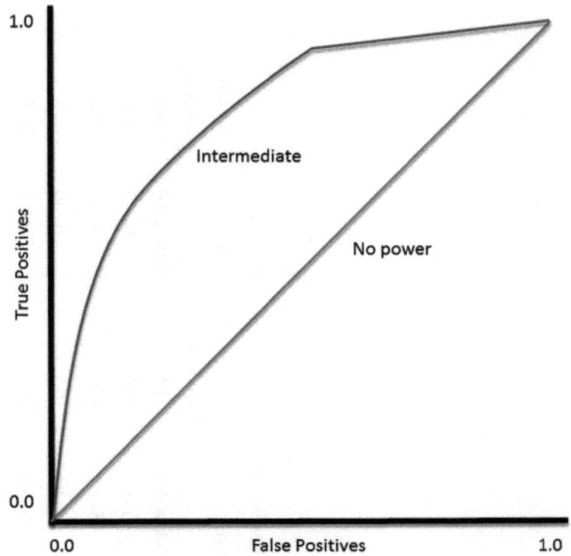

Recall. It is defined as the number of true cases which has been predicted correctly out of the total number of true cases [32]

$$Recall = TP/(TP + FN) \tag{4}$$

F1 Score. It is the harmonic average for precision and recall. It has a value of 1 for ideal values of precision and recall [32].

$$F1\ Score = 2 * (Precision * Recall)/(Precision + Recall) \tag{5}$$

AUC Score. It stands for area under receiver operator characteristics (ROC) curve. ROC curve is a graph plotting true positive versus false positive [32] (Fig. 3).

4.4 Result Set Statistical Analysis of the Female Kidnapping Records in India

The different purposes for kidnapping and their corresponding age categories and number of female kidnapping for each age category are presented in the following Table 4.

Tabular Analysis:
Table 4 clearly indicates that most of the females below the age of 10 are kidnapped for begging purposes and no women above age of 50 is ever kidnapped for beg-

Table 4 Statistical table for female kidnapping records

Begging		Marriage		Illicit Intercourse		Prostitution		Ransom	
Age	Number of females	Age	Number of females	Age	Number of Females	Age	Number of Females	Age	Number of females
Below 10	256	Below 10	149	Below 10	124	Below 10	13	Below 10	86
10–15	28	10–15	5696	10–15	179	10–15	228	10–15	60
15–18	6	15–18	23,418	15–18	44,046	15–18	826	15–18	64
18–30	36	18–30	75,877	18–30	13,848	18–30	7182	18–30	817
30–50	8	30–50	11,465	30–50	2994	30–50	300	30–50	207
Above 50	0	Above 50	127	Above 50	26	Above 50	51	Above 50	15

Table 5 Measurement of scores for model evaluation

Classes	Accuracy	Precision	Recall	F1 Score	AUC score
Begging	1.0	1.0	1.0	1.0	1.0
Marriage	0.86	1.0	0.33	0.50	0.66
Illicit intercourse	0.93	1.0	0.66	0.80	0.83
Prostitution	0.93	1.0	0.66	0.80	0.83
Ransom	1.0	1.0	1.0	1.0	1.0

ging. Women within age 18–30 are targeted majorly for marriage, and 15–18 are targeted for illicit intercourse. Women with 18–30 are mostly trafficked for prostitution. Statistics for ransom should not possess any relation with age, but still the girls of age 18–30 are mostly targeted. This may be due to the reason that if ransom is not paid then they can traffic those females for prostitution.

Graphical Analysis:
Figures 4a–e represent the graphical charts of Table 4. The charts clearly show that females of age below 10 have more chances for kidnapped for begging, females of age 18–30 are prone for being kidnapped for marriage and prostitution, and females of age 15–18 are targeted for illicit intercourse and ransom. Figure 4f shows the total number of women kidnapped for each of the purposes (begging, marriage, illicit intercourse, prostitution and ransom).

Design of a Predictive Model for Female Kidnapping. The predictive model designed by the application of decision tree using ID3 algorithm is trained properly and subjected to the testing data set. The performance of the evaluation model has to be measured by the use of the testing data set by counting the number of correct classifications and misclassifications. There are certain measures to evaluate a classification model. These measures actually deal with the number of correct classifications and misclassifications and are calculated as ratios. The efficiency of the predictive model has been presented with the following measures:

Tabular Analysis:
Table 5 clearly shows that the kidnappings for begging are accurately predicted, with no false positives and false negatives, with an ideal F1 score and an ideal AUC score. Kidnappings for marriage are predicted with 86% accuracy; prediction has no false positives but large false negatives giving an average F1 score and AUC score. Kidnappings for illicit intercourse are predicted with 93% accuracy, no false positives, moderate false negatives giving a satisfactory F1 score and AUC score. Kidnappings for prostitution and ransom have similar statistics with illicit intercourse and begging, respectively.

Final Average Scores:

- Average accuracy score = 0.94
- Average precision score = 1.0

Graphical Analysis:

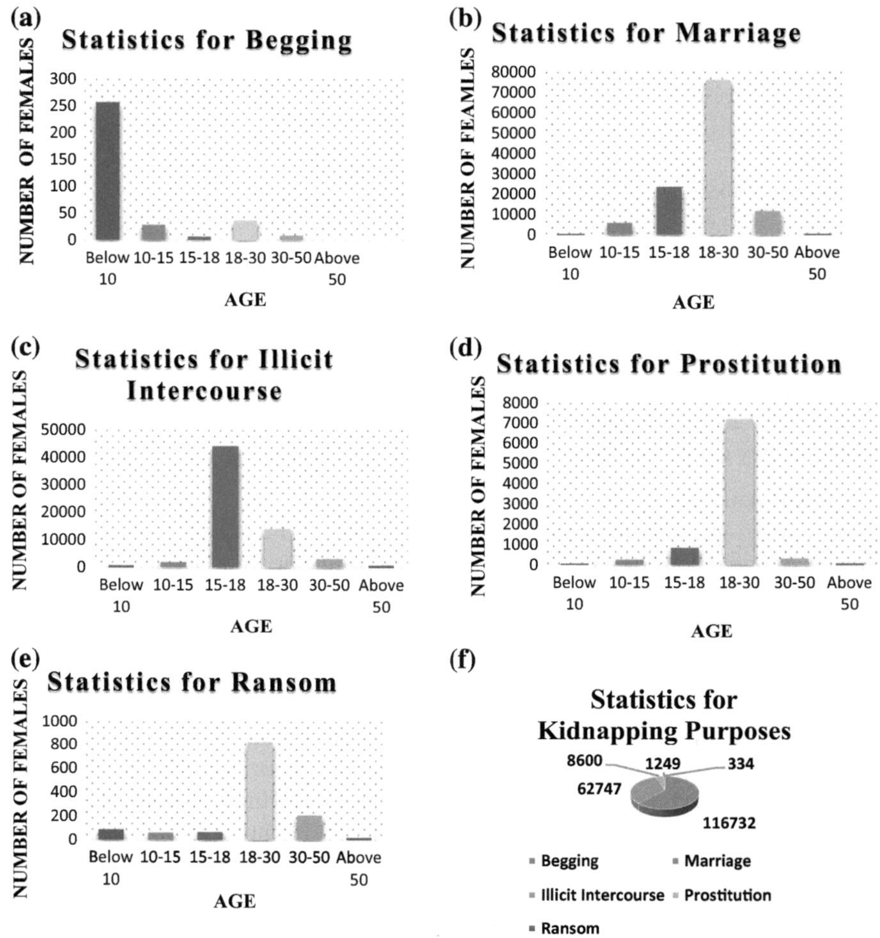

Fig. 4 **a** Begging. **b** Marriage. **c** Illicit intercourse. **d** Prostitution. **e** Ransom. **f** Total female kidnapping

- Average recall score = 0.73
- Average F1 score = 0.82
- Average AUC score = 0.86

Graphical Analysis:

The charts mentioned in Fig. 5a–e clearly state that kidnappings for begging and ransom are predicted with 100% accuracy, without any false positives and false negatives resulting in an ideal F1 score and AUC score. Kidnappings for marriage and illicit intercourse are predicted with 86% accuracy and 93% accuracy, respectively. There are large false negative predictions in cases of marriage and moderate false

Graphical Analysis:

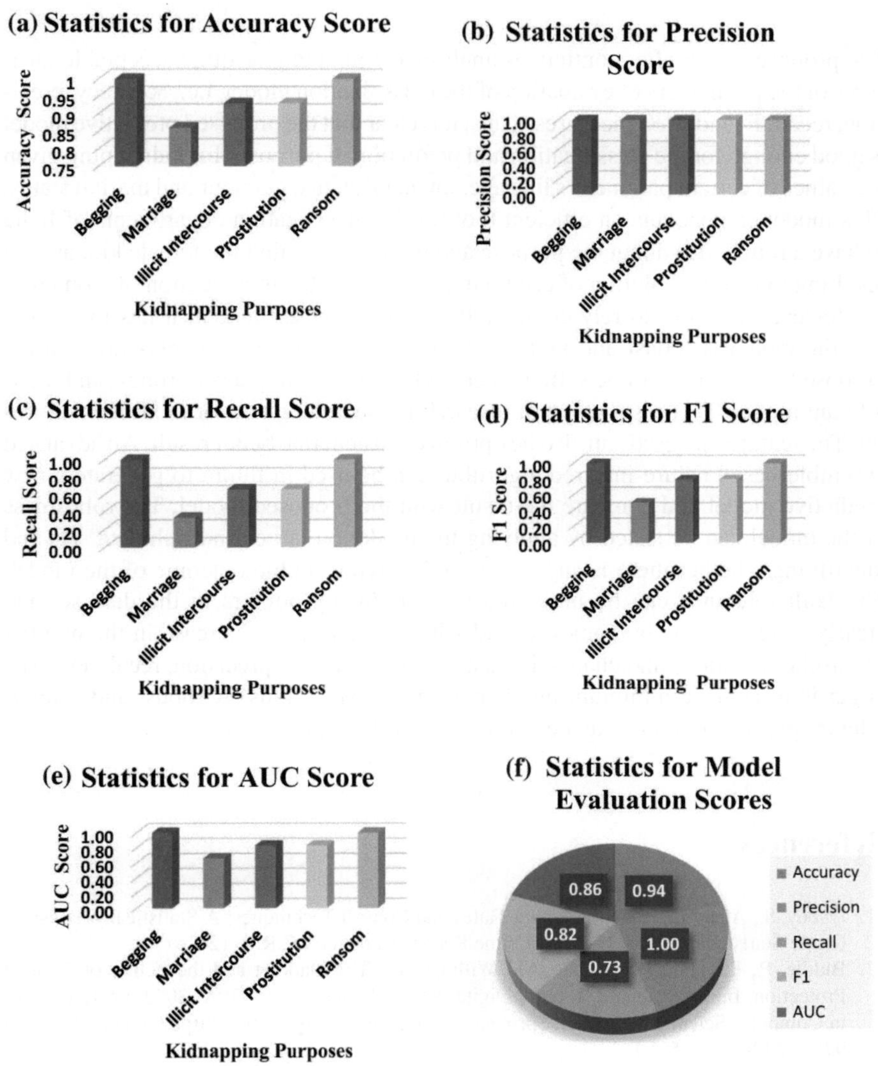

Fig. 5 **a** Accuracy score **b** Precision score **c** Recall score **d** F1 score **e** AUC score **f** Average of all scores

negative predictions for illicit intercourse. F1 and AUC scores for marriage cases are average, while they are satisfactory in cases of illicit intercourse. Statistics for prostitution and ransom are exactly the same as that of illicit intercourse and begging, respectively. Figure 5f gives a description of the average measures of accuracy, precision, recall, F1 and AUC scores after combining all the cases.

5 Conclusion and Future Work

The proposed work after continuous analysis is found to give quite acceptable measures of the parameters of evaluation of the classification model, i.e., accuracy, precision, recall, F1and ROC measures. Thus, it is clear that the proposed predictive model is good enough for the classification and prediction of purposes for kidnapping given the values of certain parameters like age, financial status, glamour and marital status. This model can become an efficient tool for the investigation departments of India to have a rough idea about the purpose and intentions behind the female kidnapping resulting in a better solution of criminal activities related to abduction of women.

However, in order to rely completely on the proposed model, it has to be sure that the model is robust and fault tolerant. The features used as classifiers in the proposed model may not be sufficient enough for generating a very robust and fault-tolerant model. There is a scope of research for searching new or modified features set. The features set used can also be optimized to generate better result. An advanced ensemble-based nature-inspired algorithm can be used in future to generate a new predictive model and compare the result with the proposed model. The robustness of the model can be tested by applying the model on larger and rich data sets and identifying whether there is any abnormal behaviour in the outcome of the model. The fault tolerance can be measured by introducing outliers in the data set and clearly observing the outcomes to find whether there is an increase in the number of misclassifications and change in measures of accuracy, precision, recall, etc. The target is to achieve minimum number of misclassifications. A robust and a fault-tolerant predictive model can be applied in real-life applications.

References

1. Dubey, S., Agarwal, P: Crime, Crime Rates and Control Techniques: A Statistical Analysis
2. Crime Statistic Report of National Crime Record Bureau (N.C.R.B) (2015)
3. Biddle, P., England, P., Peinado, M., Willman, B.: The Darknet and the Future of Content Protection. In: Feigenbaum, J. (eds) Digital Rights Management. DRM 2002. Lecture Notes in Computer Science, vol. 2696. Springer, Berlin, Heidelberg (2003). https://doi.org/10.1007/978-3-540-44993-5_10
4. Status of children in 14–18 years: An Report from National Commission for Protection of Child Rights—Government of India (Published Year 2014 and 2015)
5. NCPR. http://ncpcr.gov.in/showfile.php?lang=1&level=2&&sublinkid=300&lid=739. Accessed 27 Feb 2018
6. McClendon, L., Meghanathan, N.: Using Machine Learning Algorithms to Analyze Crime Data. Mach. Learn. Appl. Int. J. **2**, 1–12. https://doi.org/10.5121/mlaij.2015.2101
7. Burbano, D., Hernandez-Alvarez, M.: Identifying human trafficking patterns online, 2017 IEEE Second Ecuador Technical Chapters Meeting (ETCM), Salinas, 2017, pp. 1–6. https://doi.org/10.1109/etcm.2017.8247461
8. Feinerer, I.: Introduction to the TM Package Text Mining in R
9. Lopez, M., Kalita, J.: Deep Learning Applied to NLP. https://arxiv.org/abs/1703.03091
10. Alvari, H., Shakarian, P., Snyder, J.K.: A Non-parametric Learning Approach to Identify Online Human Trafficking, pp. 33–138 (2016). https://doi.org/10.1109/isi.2016.7745456

11. Dank, M.: Estimating the Size And Structure of the Underground Commercial Sex Economy in Eight Major US Cities (2014)
12. Backpage Homepage (2018). https://www.backpage.com. Accessed 27 Feb 2018
13. Giacobe, N.A., et al.: Characterizing sex trafficking in Pennsylvania for law enforcement. In: 2016 IEEE Symposium on Technologies for Homeland Security (HST), Waltham, MA, pp. 1–5 (2016). https://doi.org/10.1109/ths.2016.7568914
14. Brewster, B., Ingle, T., Rankin, G.: Crawling open-source data for indicators of human trafficking. In: 2014 IEEE/ACM 7th International Conference on Utility and Cloud Computing, London, pp. 714–719. https://doi.org/10.1109/ucc.2014.116, 2014
15. Boisot, M.: Knowledge Assets: Securing Competitive Advantage in the Information Economy (1998), ISBN-13: 978-0198296072, ISBN-10: 019829607X
16. Ibanez, M., Suthers, D.: Detection of domestic human trafficking indicators and movement trends using content available on open internet sources. In: 2014 47th Hawaii International Conference on System Sciences, Waikoloa, HI, pp. 1556–1565 (2014). https://doi.org/10.1109/hicss.2014.200
17. Kreyling, S., West, C., Olson, J.: Technology and Research Requirements for Combating Human, Pacific Northwest National Laboratory
18. Chhachhiya, V.: Spatio-temporal analysis of kidnapping and abduction of women in Chandigarh. Int. J. Sci. Res. Publ. **7**(7), 648, ISSN 2250-3153 (2017)
19. Rengaraj, V., Bijlani, K.: A study and implementation of smart ID card with M-learning and child security. In: 2016 2nd International Conference on Applied and Theoretical Computing and Communication Technology (iCATccT), Bangalore, pp. 305–311 (2016). https://doi.org/10.1109/ICATCCT.2016.7912013
20. Huang, Z., Gao, Z., Lu H., Zhang, J., Feng, Z., Xia, H.: An mobile safety monitoring system for children. In: 2014 10th International Conference on Mobile Ad-hoc and Sensor Networks (MSN), pp. 323–328 (2014). https://ieeecomputersociety.org/10.1109/MSN.2014.55
21. Deenadayalan, C., Murali, M., Baanupriya, L.R.: Implementing prototype model for school security system (SSS) using RFID. In: 2012 Third International Conference on Computing Communication & Networking Technologies (ICCCNT), pp. 1–6 (2012). https://doi.org/10.1109/icccnt.2012.6396090
22. Ibanez, M., Gazan, R.:Virtual indicators of sex trafficking to identify potential victims in online advertisements. In: 2016 IEEE/ACM International Conference on Advances in Social Networks Analysis and Mining (ASONAM), San Francisco, CA, pp. 818–824 (2016). https://doi.org/10.1109/asonam.2016.7752332
23. Kejriwal, M., Szekely, P.: Knowledge graphs for social good: an entity-centric search engine for the human trafficking domain. IEEE Trans. Big Data. https://doi.org/10.1109/tbdata.2017.2763164
24. Cristina, P., Napoca, C.: Decision Tree (2010)
25. CISE, https://www.cise.ufl.edu/~ddd/cap6635/Fall-97/Short-papers/2.htm. Accessed 27 Feb 2018
26. Packtub. https://www.packtpub.com/mapt/book/big_data_and_business_intelligence/9781784395803/6/ch06lvl1sec43/entropy-and-information-gain. Accessed 27 Feb 2018
27. The grimmscientist. https://www.thegrimmscientist.com/tutorial-decision-trees/. Accessed 27 Feb 2018
28. Open Government Data: https://data.gov.in/catalog/age-group-wise-victims-kidnapping-and-abduction. Accessed 27 Feb 2018
29. Cran Homepage. https://cran.r-project.org/. Accessed 27 Feb 2018
30. Machine Learning Mastery. https://machinelearningmastery.com/why-one-hot-encode-data-in-machine-learning/. Accessed 27 Feb 2018
31. Saedsayad. http://www.saedsayad.com/decision_tree.htm. Accessed 27 Feb 2018
32. Analytics Vidhya. https://www.analyticsvidhya.com/blog/2016/02/7-important-model-evaluation-error-metrics/. Accessed 27 Feb 2018
33. Codalab. https://codalab.lri.fr/competitions/132#learn_the_details-evaluation. Accessed 27 Feb 2018

Author Index

M. Chakraborty et al. (eds.), *Proceedings of International Ethical Hacking Conference 2018*, Advances in Intelligent Systems and Computing 811,
https://doi.org/10.1007/978-981-13-1544-2

MIX
Papier aus verantwortungsvollen Quellen
Paper from responsible sources
FSC® C105338

FSC
www.fsc.org

If you have any concerns about our products,
you can contact us on
ProductSafety@springernature.com

In case Publisher is established outside the EU,
the EU authorized representative is:
Springer Nature Customer Service Center GmbH
Europaplatz 3, 69115 Heidelberg, Germany

Printed by Libri Plureos GmbH
in Hamburg, Germany